Creators of Mathematical and Computational Sciences

Ravi P. Agarwal • Syamal K. Sen

Creators of Mathematical and Computational Sciences

Springer

Ravi P. Agarwal
Department of Mathematics
Texas A&M University-Kingsville
Kingsville, TX, USA

Syamal K. Sen
GVP-Prof. V. Lakshmikantham Institute
for Advanced Studies
GVP College of Engineering Campus
Visakhapatnam, India

ISBN 978-3-319-10869-8 ISBN 978-3-319-10870-4 (eBook)
DOI 10.1007/978-3-319-10870-4
Springer Cham Heidelberg New York Dordrecht London

Library of Congress Control Number: 2014949778

Mathematics Subject Classification (2010): 01A05, 01A32, 01A61, 01A85

Printed on acid-free paper

Springer is part of Springer Science+Business Media (www.springer.com)

Truth shall prevail despite all attempts at suppression

No subject loses more than mathematics
by any attempt to dissociate it from its history
James Whitbread Lee Glaisher (FRS, FRAS) (1848–1928)

Preface

Eric Temple Bell (1883–1960) remarked, "It would be an injustice to pioneers in mathematics to stress modern mathematical ideas with little reference to those who initiated the first and possibly the most difficult steps. Nearly everything useful that was discovered in mathematics before the seventeenth century has either been so greatly simplified that it is now part of every regular school course, or it has long since been absorbed as a detail in some work of greater generality." Devotees of mathematics scrutinize, memorize, and derive formulas and theorems every day of their lives, but not many of them realize that the current level of mathematical knowledge has resulted from the strenuous labors of countless generations. One cannot underestimate the influence of every culture, personality, philosophy, region, religion, society, and social status on mathematical development throughout the centuries. Hermann Hankel (1839–1873) observed, "In most sciences one generation tears down what another has built, and what one has established, another undoes. In mathematics alone each generation adds a new story to the old structure." Analogously, anthropologist Ralph Linton (1893–1953) stated hypothetically that "…if Einstein had been born into a primitive tribe which was unable to count beyond three, life-long application to mathematics probably would not have carried him beyond the development of a decimal system based on fingers and toes." Similar views were expressed later by Isaac Asimov (1920–1992): "Mathematics is a unique aspect of human thought, and its history differs in essence from all other histories. Only in mathematics there is no significant correction—only extensions. Each great mathematician adds to what came previously, but nothing needs to be uprooted."

We aim to bring the details of this struggle to light. This collection records the essential discoveries of mathematics in chronological order, following the birth of ideas on the basis of prior ideas ad infinitum. We examine contemporary events occurring side by side in different countries or cultures, reflecting some of the noblest thoughts of generations. We document the winding path of mathematical scholarship throughout history, and most importantly, the thought process of all individuals that resulted in the mastery of their subject. We have derived information from records from several locations, some of which have been unearthed only in recent years, throwing light on the very humanity of mathematicians who have very often been reduced to a theorem in a school textbook. Our book implicitly addresses the nature and character of all mathematicians as we try to understand their visible actions. We hope that this will enable the readers to understand their mode of thinking and perhaps even to emulate their virtues in life. In cases of controversy we have taken the most appealing and logical approach. Wherever possible, we have excluded sophisticated mathematical terms so that our book's content may be easily accessible to nonmathematicians. It must be noted that throughout this collection, amusing anecdotes and after-dinner jokes have been taken from the published literature, but it is difficult to vouch for the authenticity of most of them. Still, they are interesting because they reveal the human nature of mathematicians, who are very often believed to be eccentric individuals.

Certainly a book of this type cannot be written without deriving many valuable ideas from several sources. We express our indebtedness to all authors, too numerous to acknowledge individually, from whose specialized knowledge we have been benefitted. We have also immensely benefitted from several Web sites, especially en.wikipedia.org and www-history.mcs.st-andrews.ac.uk. We record our appreciation to our friends and colleagues, especially Bashir Ahmad (Saudi Arabia), Józef Banaś (Poland), Jaromir Bastinec (Czech Republic), Leonid Berezansky (Israel), Gabriele Bonanno (Italy), Alberto Cabada (Spain), Wing–Sum Cheung (Hong Kong), Manuel De la Sen (Spain), Josef Diblik (Czech Republic), Shusen Ding (USA), Alexander Domoshnitsky (Israel), Paul Eloe (USA), Vijay Gupta (India), Johnny Henderson (USA), Nan–jing Huang (China), Gennaro Infante (Italy), Billur Kaymakcalan (Turkey), George L. Karakostas (Greece), Ivan Kiguradze (Georgia Republic), Peter Kloeden (Germany), Yongwimon Lenbury (Thailand), Dumitru Motreanu (France), Juan J. Nieto (Spain), Donal O'Regan (Ireland), Victoria Otero-Espinar (Spain), Adrian Petrusel (Romania), Sandra Pinelas (Portugal), Irena Rachunkova (Czech Republic), Maria A. Ragusa (Italy), Simeon Reich (Israel), Cheon S. Ryoo (Korea), Masilamani Sambandham (USA), Alexander L. Skubachevskii (Russia), Svatoslav Stanek (Czech Republic), Ioannis P. Stavroulakis (Greece), Tomonari Suzuki (Japan), E. Thandapani (India), George Yin (USA), Patricia J.Y. Wong (Singapore), and Agacik Zafer (Turkey), for their encouragement, support, and constructive comments. Our sincere thanks to Lane Christiansen who read the entire manuscript several times and suggested improvements on almost every page. Special thanks to our wives Sadhna Agarwal and Ella Sen, whose continued encouragement and sacrifice deserve special mention. Last but not the least thanks to Mr. Marc Strauss (Springer, New York) for his interest in this project from the beginning till its publication.

Kingsville, TX, USA — Ravi P. Agarwal
Visakhapatnam, AP, India — Syamal K. Sen

Contents

Introduction

The following questions need to be addressed before one can appreciate the creations of the mathematicians and computational scientists:

- What is Mathematics?
- What is Mathematical Science?
- What is a Mathematical Proof?
- What is Computational Science/Mathematics?
- Is a proof in Computational Mathematics always possible?

The word 'mathematics' was coined by Pythagoras, who flourished around 500 BC. It meant 'a subject of instruction,' and its first part, 'math,' comes from an old Indo-European root that is related to the English word 'mind.' The Pythagoreans grouped arithmetic, astronomy, geometry, and music together and for several centuries mathematics referred to only these four subjects. However, as we proceed it will become clear that the study of arithmetic, astronomy, and geometry began long before Pythagoras. In fact, a fair majority of the biggest breakthroughs in mathematics were made possible through the work of people other than those who have been credited in the history books. As Sir Francis Darwin (1848–1925) observed, "But in science the credit goes to the man who convinces the world, not to the man to whom the idea first occurs." In many instances, the fault lies with the historians themselves, who inject their own opinion into their texts rather than reporting unbiased fact. Following the footsteps of these historians, we also believe that the origin of mathematics is the number sense, which has played a vital role in the development of these subjects.

Number Sense

Number sense can be defined as an intuitive understanding of the integers, their magnitude, their patterns and relationships, and how they are affected by the basic operations: addition, subtraction, multiplication, and division (which has always

R.P. Agarwal, S.K. Sen, *Creators of Mathematical and Computational Sciences*,
DOI 10.1007/978-3-319-10870-4_1

been the most difficult one). A child between 12 and 18 months is able to recognize if an object is removed from or added to a small collection. This simple number sense, the ability to distinguish 'plenty' and 'few' without counting, is a useful tool for a conscious being and a fundamental ability of humans. For primitive man and children, mathematics is simply a comparison of small collections. So far as nonhuman living beings are concerned, there are recorded incidences of birds, animals, insects, and aquatic creatures who show through their behavior a certain number sense, namely, comparing/sorting. These creatures possibly possess mathematical abilities that exceed our observations or beliefs about them. Were we to understand their minds, we would be probably be astonished at our ignorance of their mental and mathematical faculties.

Due to the needs of daily life in antiquity, man began comparing numbers by tallying ('tally' or 'tally stick' often refers to a notched stick, but here we mean to refer to any simple marks or scores). For this, he employed the method of one-to-one correspondence and used many different objects, for example, knucklebones, pebbles (in every corner of the globe), pellets, sticks, shells, scratches on stones, notches cut on wooden sticks or pieces of bone, and body parts, especially fingers and toes. The oldest known (possibly) mathematical object is the Lebombo bone, discovered in the Lebombo mountains of Swaziland, a landlocked country between South Africa and Mozambique, which has been dated to approximately 35,000 BC. It consists of 29 distinct notches cut into a baboon's fibula. In 1937, about 7 inches of bone from a young wolf was found in the Czech Republic that is estimated to be as old as 30,000 BC. This bone is engraved with 55 deeply cut notches, almost equal in length, arranged in groups of five (as tally marks traditionally are). The Ishango bone, unearthed in 1960 near the headwaters of the Nile River in central Africa's Congo, is estimated to be 22,000 years old. It is one of the oldest known objects that shows evidence of an understanding of numbers beyond counting. It contains three columns of marks: the middle column contains the sequence of numbers 3, 6, 4, 8, 10, 5, 5, 7, which add up to 48; the numbers in the left and right columns are all odd numbers, the left column lists the prime numbers between 10 and 20, i.e., 11, 13, 17, 19, and the right column consists of the numbers 11, 21, 19, 9; both side columns add up to 60. These numbers cannot be random and suggest an understanding of multiplication and division by two (and perhaps familiarity with prime numbers and the lunar cycle). The notched sticks used by our ancestors have remained much the same through many centuries of evolutionary, historical, and cultural change, right down to the present day. It is interesting to note that until the early nineteenth century, tally sticks were used in various roles in Britain, France, Germany, Switzerland, and throughout Scandinavia.

Around 4000 BC, special symbols (a symbol is a thing, perceivable by the senses, which stands for something else) for numbers appeared in the river valleys of the Indus (which was home to more than five million people), the Nile, the Tigris and Euphrates, and the Yangtze. These symbols were written on palm leaves, papyrus, clay tablets, bark or bamboo, and (oracle) bones, with some materials more perishable than others. These show that man began using symbols like strokes, circles, or dots to indicate numbers. A child of 3 or 4 years old naturally starts denoting numbers by strokes, circles, or dots. Gradually these became specialized, genuine numerical symbols, which were different for inhabitants of different locations. As these symbols evolved, their utility slowly improved and they began to assist man in assimilating, combining, distinguishing, remembering, and expressing large numbers. The next step forward was the creation of names for the numbers and the basic operations. Today every young child is able to count small numbers on his fingers, say up to 10, and learns the abstract concepts of addition and subtraction. In antiquity, counting was considered a talent as mystical and arcane as casting spells and calling the gods by name. In the Egyptian *Book of the Dead*, when a dead soul is challenged by Aqen, the ferryman who conveys departed spirits across a river in the netherworld, Aqen refuses to allow anyone aboard "who does not know the number of his fingers." The soul must then recite a counting rhyme to tally his fingers, satisfying the ferryman. However, as early as 3500 BC, the Egyptians had a fully developed number system that would allow counting to continue indefinitely with the introduction from time to time of new symbols. In the world, although there are about 4,000 languages of which several hundred are widespread, there are only several dozen alphabets and writing systems that represent them; however, we can safely say that there is but one single place-value system that uses zero and nine numerals, symbolically $0, 1, 2, 3, 4, 5, 6, 7, 8, 9$, by which any number can be expressed and understood rather easily. This gift of India has truly united the universe in the language of numbers. It is worth noting that there is hardly any aspect of our life in which numbers do not play a significant—though generally hidden—part.

Arithmetic

Arithmetic is the most useful and the simplest branch of all the sciences, and there is no other branch of human knowledge that has been more widely spread among the masses. It can be defined as the concrete knowledge of integers, rational numbers, irrational (illogical or unreasonable) numbers, real numbers, and mastery of their manipulation through the abstract operations. Thus arithmetic forms a very sophisticated realization of number sense. Presumably, it began with the need of primitive man to count his few possessions—his cattle or perhaps his wives. In elementary schools the study of arithmetic begins with the recitation of the integers and an intuitive recognition of serial order, and addition is the very first operation that one learns. By the time a child is 6 years old, he masters numbers as an aid to counting and learns basic operations. The ancient people of the Indus River Valley had a highly developed knowledge of arithmetic and a sophisticated system of weights and measures. There is some evidence of basic arithmetic in the petroglyphs at the Knowth and Newgrange burial mounds in Ireland, dating from about 3500 and 3200 BC, respectively. These utilize a repeated zigzag glyph for counting, a system that continued to be used in Britain and Ireland into the first millennium BC. The Egyptians and Babylonians used all the elementary arithmetic operations as early as 2000 BC. Cuneiform tablets from Mesopotamia contain multiplication tables and division problems. *Higher arithmetic*, which is a synonym for *number theory*, is the branch of arithmetic that is concerned with the properties and relationships of whole numbers for their own sake and hence has been classified as pure mathematics. Although the term 'number theory' (la theorie des nombres) was coined by Fermat during the seventeenth century, its history is at least 5,000 years old. However, Pythagoras and his disciples are generally given credit for initiating the actual theory of numbers. Pythagoras believed that all universal laws can be explained by numbers and hence he declared that "Everything is Number." The first systematic treatment of number theory was given by Euclid in books 7, 8, and 9 of his Elements. The ancient Greeks were fascinated by number theory; in fact, Plato wrote that "We must endeavor to persuade those who are to be the principal men of the state to go and learn arithmetic." Since antiquity, number theory has captivated the best minds of every era. During 1600–1850, Chebyshev, Dirichlet, Euler, Fermat, Gauss, Lagrange, and Legendre greatly enriched number theory. In 1801, Gauss asserted that number is purely a psychic reality, a free creation of mind, and he crowned number theory as "The Queen of Mathematics." According to Bell, "number theory is the last great uncivilized continent of mathematics". In recent years, number theory has been divided into (1) elementary or classical number theory, (2) algebraic number theory, (3) analytical number theory, (4) probabilistic number theory, and (5) computational number theory, which finds diverse applications. An important aspect of number theory is that challenging problems can be formulated in very simple terms; however, hidden within their simplicity is complexity. Some of these problems have been instrumental in the development of large parts of mathematics. Amateurs and professionals are on an almost equal footing in this field.

Astronomy

Astronomy is one of the most ancient (about 3.4 million years before 4500 BC) continuously pursued branches of natural science. Many historians believe it to be older than mathematical research because of the intense exploration of mathematics that it necessitates. It can be broadly defined as the study of all objects beyond our world, especially celestial bodies such as Comets, Galaxies, Moons, Nebulae, Planets, Stars, and the Sun. Astronomy also helps us understand the physical laws and origins of our universe. The moment a child becomes aware of the world around him, he begins to question the nature of his surroundings to whoever might be in his vicinity; he wants to know everything about the Moons, Planets, and the Sun. Primitive man was also curious about astronomical events, like the beauty and wonder of the night sky, the majesty of the daily visit of the Sun, and the strange waxing and waning of the Moon and its influence on various human activities. Of course, early astronomy was limited to the observations, patterns, and predictions of the motions of objects visible to the naked eye. While the movements of the Planets have been tracked around the world, astronomy has deep roots in China,

Egypt, India, Mesopotamia, Central America, and later in Greece. The primitive man used number sense (and eventually arithmetic) to prove that celestial bodies exhibit regularity of behavior and move with identifiable patterns. He observed that the Moon passes from one phase to another (from new Moon to half Moon, from half Moon to full Moon) in 7 days, that the Moon goes around the Earth every 28 days, and that the Sun completes its orbit about the Earth in 365 days. These observations helped him make primitive calendars so that he could keep track of the beginning and end of the rainy monsoon season, the annual snow melt and the consequent rise of the rivers, auspicious days for performing sacrifices, and the pattern of solar eclipses. In India more than 5000 years ago it was known that the distances of the Moon and Sun from the Earth are 108 (which is a sacred number in several eastern religions and cults) times the diameters of these heavenly bodies, respectively, (the observed values are 110.6 for the Moon and 107.6 for the Sun). By the middle of the second millennium BC, the Chinese had begun recording astronomical events on bone fragments, some of which are extant. Today, astronomy has become a detailed and vast scientific field with the most elaborate instrumentation. **Astrology**, which has been wrongly claimed to be older than astronomy, deals with the belief that there is a relationship between astronomical phenomena (especially the cycles of planets and the stars) and events in human life. Right from the beginning many cultures attached importance to astronomical events, particularly the Indians, the Chinese, and the Mayans, who all developed sophisticated systems for predicting terrestrial events based on celestial observations. Now almost everyone believes in astrology to some extent (except those individuals who do not want to understand how celestial objects affect life on Earth). Astrology was originally studied along with astronomy, but the two have been separated.

The word **Geometry** comes from two Greek words meaning 'earth' and 'measure,' which was known 5000 years ago as *sulba/sulva*, or *rule of chords*. Considering an object, geometry examines its shape, measures its size, especially its length, area, and volume, and makes sense of its spatial relationship with other objects and the properties of its surrounding space. Geometry is the basis of the survey of the Earth and the heavens, of all astronomy, of most laws in physics, art, architecture, and painting, and of all engineering and technology; its utility is unquestioned. In the beginning, arithmetic and geometry were closely interwoven. We notice that like primitive man a child at a very early age starts recognizing the difference between circles, squares, and triangles. The study of geometry originated in India for the purpose of constructing precisely measured altars for ritual sacrifices. Geometry was systematically studied by Egyptian priests because the periodical inundations of the Nile River obliterated property lines. Most of the historians of mathematics have falsely documented that "before 600 BC geometry lacked deductive structure, there were no theoretical results, nor any general rules of procedure. It supplied only calculations, and these sometimes approximate, for problems that had a practical bearing in construction and surveying." However, the 5000-year-old Sulvasutras list geometric constructions for squares, rectangles, parallelograms, and trapezia. It also shows how one may construct a square n times in area to a given square; a square of area equal to the sum of two squares; a square whose whole area is equal to the difference of two squares; a square equal to a rectangle; a triangle equal to a rectangle; a triangle equal to a rhombus; and a square equal to the sum of two triangles or two pentagons. Several theorems are also explicitly mentioned: the diagonal of a rectangle divides it into equal parts; the diagonals of a rectangle bisecting each other and opposite areas are equal; the perpendicular through the vertex of an isosceles triangle on the base divides the triangle into equal halves; a rectangle and a parallelogram on the same base and between the parallels are equal in area; and the diagonals of a rhombus bisect each other at right angles. Baudhayana contains one of the earliest references to what is known today as the Pythagorean theorem (with a convincing, valid proof): a rope that is stretched across the diagonal of a square produces an area double the size of the original square. This is a special case of the Pythagorean theorem for a 45° right triangle. The Pythagorean triples $(3,4,5)$, $(5,12,13)$, $(8,15,17)$, and $(12,35,37)$ are part of Apastamba's rules for altar construction. Certain kinds of Pythagorean triples are listed on Plimpton 322 (a Babylonian clay tablet), which is believed to be almost 4000 years old. The Katyayana, written later, gives a more general version of the Pythagorean theorem: a rope that is stretched along the length of the diagonal of a rectangle produces an area which the vertical and horizontal sides make together. In other words, the square of the hypotenuse equals the sum of the squares of the sides. The Pythagorean theorem was also known to the Chinese long before Pythagoras as the Gou-gu theorem. The geometrical books of Euclid's Elements are a systematic and logical compilation of the works of earlier writers. For several centuries this work remained the world's only acknowledged textbook of plane and solid geometry (Euclidean Geometry). In appreciation, Mohandas Karamchand Gandhi (Mahatma Gandhi, 1869–1948) in his *The Story of My Experiments with Truth* wrote "When however, with much effort I reached the thirteenth proposition of Euclid the utter simplicity of the subject was suddenly revealed to me. A subject which only required a pure and simple use of one's reasoning powers could not be difficult. Ever since that time geometry has been both easy and interesting to me." The geometrical books of the Elements coincide almost exactly with the usual secondary school course of geometry. Since the sixteenth century several different branches of geometry have emerged: affine geometry, algebraic geometry, analytic geometry/coordinate geometry, combinatorial geometry, computational geometry, descriptive geometry, differential geometry, discrete geometry, metric geometry, projective geometry, non-Euclidean geometry, and Riemannian geometry. Research in these geometries continues and from time to time revolutionary results are announced.

The word **Algebra** is derived from a Latin translation of the treatise *Hisab al-jabr w'al-muqabala* by al-Khwarizmi. Although it has now become a branch of mathematics that deals with structure, relation, and quantity, 5000 years ago it was confined to solving linear $ax = b$, quadratic $ax^2 + bx = c$, $ax^2 = bx+c$, $ax^2+c = bx$, and indeterminate equations $ax+c = by$ and $x^2+y^2 = z^2$, where a,b, and c are given positive integers. For linear and quadratic equations, the Indians in their Sulvasutras and Babylonians on their Cuneiform Tablets found exact (both determinate and indeterminate) positive solutions by the same method that we use today. The Egyptian Rhind Papyrus provides a method for the solution of linear equations that is known as the method of false position (regula falsi). To solve quadratic equations, the Vedic Indian mathematicians and the Greeks (from the time of Plato) used *Geometric Algebra* in which terms were represented by the sides of geometric objects. In the Sulvasutras, first degree indeterminate equations of the form $ax + c = by$ are examined for integer solutions of x and y. Later, Aryabhatta used the "Kuttaka" method to solve these equations. Second order indeterminate equations of the form $Nx^2 + 1 = y^2$, where N is an integer, were discussed by Brahmagupta. For their solution, he employed his "Bhavana" method and showed that they have infinitely many solutions. In current literature this is known as Pell's equation, and it has been recorded that Fermat was the first to assert that it has infinitely many solutions. Bhaskara II improved Brahmagupta's method with his "chakravala" (cyclic) method. In the early ninth century this ancient knowledge found a home in the Islamic world, where it was properly organized and then advanced to some extent by a number of Arabic mathematicians, especially al-Khwarizmi, who used geometric arguments. While some particular higher order (cubic and biquadratic) equations appeared as early as 2000 BC (e.g., on the Cuneiform Tablets), Omar Khayyám's most original work is his classification of nineteen types of cubic equations, which, following Menaechmus, Archimedes, Dionysodorus of Caunus (around 250–190 BC), Thabit ibn Qurra, and al-Haitham, he found positive solutions by means of intersecting conic sections (circles, hyperbolas, and parabolas). Omar believed that for cubic equations arithmetic solutions (explicit formulas in terms of the coefficients of the equations) were impossible. The search for such solutions continued until the beginning of the sixteenth century, when the Italian mathematicians Scipione del Ferro, Niccoló Fontana (Tartaglia), and Girolamo Cardano completely solved the general cubic equation, obtaining formulas involving cube roots and square roots. This was immediately followed by the explicit solutions of the biquadratic equations by Cardano's pupil, Ludovico Ferrari. This discovery has been considered to be the first major breakthrough in mathematics after Euclid's Elements; however, these complicated formulas have few applications and hence do not form a topic of any course. After the successful solution of the cubic and biquadratic equations, almost every mathematician tried to find solutions of quintic and higher degree equations in an analogous way. However, no progress was made for almost 300 years, other than several incorrect claims. Then in 1797, Gauss proved the qualitative result, namely, "the fundamental theorem of algebra," which states that every algebraic (polynomial) equation in one unknown has a real or complex root. In the year 1799, Paolo Ruffini, an Italian physician who also taught mathematics, provided a result which showed that the general quintic equation cannot be solved. His proof was the first step in the right direction, but had some minor faults. Finally the Norwegian genius Niels Abel, at the age of just 19, proved the unsolvability of the quintic equations with more precise arguments than Ruffini's. He published this discovery at his own expense in the year 1824, and it went unnoticed by contemporary mathematicians. In 1826, Abel provided the details of his work in the founding volume of the *Journal für die reine und angewandte mathematik* (Journal of Pure and Applied Mathematics). This result is known as the Abel–Ruffini theorem and states that it is impossible to find a general formula for the roots of a polynomial equation of degree five or higher in terms of the coefficients of the equation if only the basic operations of arithmetic and extraction of roots is allowed. However, this is not the end of the story; there are arbitrarily many special forms of equations of any degree that *are* solvable in this way. This was completely characterized by the French mathematician Evariste Galöis. His work remained in manuscript form and was published posthumously in 1846.

The Chinese mathematician Liu Hui, in his commentary on the *Jiuzhang Suanshu* (Nine Chapters on the Mathematical Art, which is believed to have been written around 1000 BC), reports in Chapter 7 a clever method (the rule of false position) for solving systems of two simultaneous linear equations with two unknowns, and in Chapter 8 he states a method (now known as Gaussian elimination) for finding the solutions of determinate and indeterminate systems of linear equations of higher order with both positive and negative numbers. Certain types of simultaneous linear equations with n unknowns were also solved by Thymaridas of Paros (around 400–350 BC). For this he introduced and used the method known as the "bloom of Thymaridas." Although determinants and matrices were used to solve systems of linear equations in Jiuzhang Suanshu, these concepts were, respectively, coined by Seki Kowa in 1683 (10 years before Leibniz) and Sylvester in 1850 (8 years before Cayley). Cayley and Sylvester continued their work on determinants and matrices, which later grew into its own subject through work done by several mathematicians and physicists: Binet, Bôcher, Cauchy, Cholesky, Dickson, Dodgson, Fredholm, Frobenius, Hadamard, Heisenberg, Hermite, Hurewicz, Jacobi, Jordan, Kowalewski, Perron, Olga Todd, Vandermonde, Wronski, and others. The work of the Indians, the Babylonians, and the Chinese on linear and quadratic equations and systems of two linear equations with two unknowns constitutes the present school course of *elementary algebra*, while matrices, determinants, and systems of $n(\geq 3)$ are taught in upper-level mathematics courses, ranging from middle school to college-level *linear algebra.*

Until the sixteenth century algebra was mostly *rhetorical*, that is, each equation was expressed in ordinary language. The next stage of algebra, which was initiated by Brahmagupta and Diophantus and continued by the Arabian mathematicians, has been called *syncopated* and in it symbols are partially used. Finally, *symbolic* algebra, which we use today, became established through the works of Vieta (he used the term 'analysis' to refer to algebra), Descartes, and Wallis. In the nineteenth century, symbols were used to write most mathematics as formulas. These formulas played a key role in advancing mathematics, particularly algebra, to great heights. In appreciation of these formulas, Heinrich Hertz (1857–1892), the discoverer of electromagnetic waves, said "One cannot escape the feeling that these mathematical formulas have an independent existence and an intelligence of their own, that they are wiser than we are, wiser even than their discoverers, that we get more out of them than was originally put into them." In recent years various different types of algebra have emerged: abstract algebra (with topics such as groups, fields, and rings), associative algebra, Banach algebra, Boolean algebra, Borel algebra, C^*-algebra, Clifford algebra, commutative algebra, computer algebra, Heyting algebra, homological algebra, Hopf algebra, Lie algebra (the theory of continuous groups), relational algebra, sigma-algebra, symmetric algebra, tensor algebra, universal algebra, vector algebra, von Neumann algebra, and the list continues. Algebraists keep themselves busy by searching for new qualitative and quantitative results and their applications in diverse fields.

The word **Trigonometry** comes from two Greek words meaning 'triangle' and 'measure.' It deals with triangles (in particular, right-angled triangles) and the relationship between the lengths of their sides and the angles between those sides, that is, the trigonometric functions sine (sin), cosine (cos), tangent (tan), cotangent (cot), secant (sec), and cosecant (cosec), and calculations based on these functions. **Spherical Trigonometry** studies triangles on spherical polygons defined by a number of intersecting great circles on the sphere. In antiquity, trigonometry was studied as a branch of astronomy, and for this purpose several astronomical rules for spherical triangles were discovered that are scattered all over ancient astronomical texts such as Surya Siddhanta and its commentaries:

$$\cos a = \cos b \cos c + \sin b \sin c \cos A$$
$$\cos A \sin c = \cos a \cos b - \sin a \cos b \cos C$$
$$\frac{\sin a}{\sin A} = \frac{\sin b}{\sin B} = \frac{\sin c}{\sin C},$$

where A, B, and C are the angles of a spherical triangle, of which the opposite sides are a, b, and c, respectively. In addition to expanding the mathematical work in the Siddhantas, Aryabhata constructed the oldest surviving tables of sine and versine $(1 - \cos)$ values, in $3.75°$ intervals from $0°$ to $90°$, to an accuracy of four decimal places. Some evidence of trigonometry can also be found in early Egyptian and Babylonian mathematics, for example, results on the ratios of the sides of similar triangles. The Greeks preferred to present trigonometric results in a geometric way. An early application of trigonometry independent of astronomy is due to Thales, who used proportionate right-angled triangles to calculate the height of a pyramid. The sharp trigonometric inequality $\sin\alpha/\sin\beta < \alpha/\beta < \tan\alpha/\tan\beta$, where $0° < \beta < \alpha < 90°$, was established by Aristarchus almost 2300 years ago. Hipparchus, who has been named "the father of trigonometry", compiled trigonometric tables and systematically used the $360°$ circle. Bhaskara I offered a formula for calculating the sine of an acute angle without the use of a table. He also gave the following point-wise approximation formula for $\sin x$, which has a relative error less than $1.9\,\%$

$$\sin x = \frac{16x(\pi - x)}{[5\pi^2 - 4x(\pi - x)]}.$$

This was followed by Varahamihira, who gave trigonometric formulae that correspond to $\sin x = \cos(\pi/2 - x)$, $\sin^2 x + \cos^2 x = 1$, and $(1 - \cos 2x)/2 = \sin^2 x$. Varahamihira also developed new interpolation methods for the production of sine tables that were even more accurate than Aryabhata's. A few years later,

Brahmagupta redeveloped the formula $1 - \sin^2 x = \cos^2 x = \sin^2(\pi/2 - x)$ and gave an interpolation formula for computing sine values. In the first quarter of the second century, Menelaus of Alexandria proved an important theorem for spherical triangles which still bears his name. Specifically, he established that the sum of the angles of a spherical triangle is greater than 180°. Ptolemaeus (Ptolemy) then derived the half-angle formula

$$\sin^2\left(\frac{x}{2}\right) = \frac{1 - \cos x}{2}$$

and used it to construct trigonometric tables, extending Hipparchus' work. Unfortunately, the tables made by Hipparchus and Ptolemy have not survived. Ptolemy's thirteen books of the *Almagest* are the most influential and significant trigonometric work of all antiquity. Almost three centuries after the Almagest, Bhaskara II developed spherical trigonometry systematically and discovered many trigonometric results, such as

$$\sin(a \pm b) = \sin a \cos b \pm \cos a \sin b.$$

Toward the end of the eighth century the knowledge of trigonometric functions and methods reached Persian and Arabic astronomers, who advanced it to its place as a major mathematical discipline independent of astronomy. Specifically, al-Khwarizmi produced accurate sine and cosine tables, the first table of tangents, and very general results in spherical trigonometry; al-Battani discovered the reciprocal functions of secant and cosecant and produced the first table of cosecants for each degree from 1° to 90°; Abu al-Wafa (940–998) computed sine tables in 0.25° increments to eight decimal places of accuracy, produced accurate tables of tangent values, and developed the trigonometric formula $\sin 2x = 2 \sin x \cos x$; al-Tusi listed the six distinct cases of a right-angled triangle in spherical trigonometry, stated the law of sines for plane and spherical triangles, and discovered the law of tangents for spherical triangles; and al-Kashi gave the formula $\sin 3x = 3 \sin x - 4 \sin^3 x$, which has been attributed to Viéte. Around 1400, Madhava used series expansions to produce sine and cosine tables to 12 and 9 decimal places of accuracy, respectively. A few years later, Jyesthadevan gave rules for finding the sines and cosines of the sum and difference of two angles. In the fifteenth century, trigonometry reached Europe via Latin translations of the writings of these Persian and Arabic astronomers. Regiomontanus was perhaps the first mathematician in Europe to treat trigonometry as a distinct mathematical discipline. In the seventeenth century, Newton and Stirling gave the general Newton–Stirling interpolation formula for trigonometric functions and Gregory developed trigonometric series. In the eighteenth century, Euler provided an analytic treatment of trigonometric functions and Maclaurin gave further expansions of trigonometric functions.

Besides astronomy, all of science, and engineering, trigonometry finds applications in a wide variety of disciplines, such as acoustics, audio synthesis, cartography, computer graphics, crystallography, economics, electronics, the financial market,

Fourier analysis, geodesy, land surveying, medical imaging (CAT scans, ultrasound, etc.), meteorology, music theory, navigation, number theory (and therefore cryptology), oceanography, optics, pharmacy, probability theory, seismology, statistics, the wave equation, and the list goes on. Trigonometry is usually taught in secondary/high schools, often as a precalculus course.

In antiquity, the general problem of finding area/volume of a geometric figure came to be known as the **quadrature problem**. The word quadrature implied construction of a square having the same area. In the Sulvasutras and the Cuneiform tablets we find formulas for the areas of squares, rectangles, right triangles, certain trapezoids, and the volumes of prisms, right circular cylinders, and the frustums of cones or square pyramids. Around 1850 BC, the Egyptians knew the correct formula for obtaining the volume of a frustum, that is,

$$V = \frac{1}{3}h(a^2 + ab + b^2);$$

here, h is the altitude, and a and b are the lengths of the sides of the square top and square base, respectively (when $a = 0$ this reduces to the well-known formula for the volume of the pyramid, $V = (1/3)hb^2$). This formula is a subject of debate even today. Another important exact formula for finding the area A of a triangle having sides a, b, and c, known in the literature as Heron's formula, was stated earlier by Bhaskara I:

$$A = \sqrt{s(s-a)(s-b)(s-c)},$$

where $2s = a + b + c$. This formula was later generalized by Brahmagupta for finding the area A of any cyclic quadrilateral given its four sides a, b, c, d as

$$A = \sqrt{(s-a)(s-b)(s-c)(s-d)},$$

where $2s = a + b + c + d$, which was later rediscovered by Snell. Heron is also credited for giving the formula for the volume of the frustum of a cone as

$$V = \frac{1}{4}\pi h(r + R)^2,$$

where h is the height and r and R are the radii of the bases. The exact area of the first curved shape figure, the lune (bounded by pair of circular arcs), was discovered by Hippocrates. He also discovered the formula πr^2 for the area of a circle in terms of its radius r. Democritus was the first to state that the volume of a cone is one-third that of a cylinder having the same base and equal height, and that the volume of a pyramid is one-third that of a prism having the same base and equal height. Archimedes showed that the volume of a sphere is two-thirds the volume of the smallest cylinder that can contain it. Liu Hui gave the formula for the volume of the truncated triangular prism, which has now been credited to Legendre. Sridhara gave the first correct formulas for the volumes of spheres and truncated cones. Tartaglia gave a formula for the volume of a tetrahedron in terms of the distance values measured pairwise between its four corners. In the Sulvasutras and the work of Aryabhatta, we find curious approximations of π, that is, the area of a circle of radius 1. Antiphon and Bryson tried to find the area of a circle by inscribing it by a regular polygon with an infinite number of sides. Around 240 BC, Archimedes used Aristotle's idea of the potential infinity to generalize *the method of exhaustion* of Eudoxus, which formalized the ideas of Antiphon and Bryson. He used this method to find areas bounded by parabolas and spirals, the volumes of cylinders, areas of segments of spheres, and especially to approximate π by bounding its value between $22/7$ and $223/71$. Although Archimedes' geometric method (the circumference of a circle lies between the perimeters of the inscribed and circumscribed regular polygons of n sides, and as n increases, the deviation of the circumference from the two perimeters becomes smaller) to approximate the value of π is heuristic (not regarded as final and strict but merely as provisional and plausible), it foreshadowed the concept of the limit. Archimedes' method was later refined for the purpose of making better approximations to π by Liu Hui, Zu Chongzhi, van Ceulen, and others. Zu Chongzhi also established a geometric method for finding the volume of a sphere that was similar to Archimedes' and would later be called Cavalieri's principle. Geometric and analytic methods for several other quadrature problems were employed by Pappus of Alexandria, Mahavira, Galileo, Kepler, Descartes, Cavalieri, Fermat, Gilles Personnier de Roberval (1602–1675), Torricelli, Wallis, Brouncker, Pascal, Huygens, Barrow, Gregory, and others. The quadrature problem eventually gave rise to **integral calculus**, which is the assimilation of the geometric and analytic methods and the understanding of *the calculus of infinitesimals* or *infinitesimal calculus*. This final step was taken independently by both Newton and Leibniz. The term 'integral' appeared for the first time in a paper produced by Jacob Bernoulli in 1690, and the term 'integral calculus' was introduced by Leibniz and Johann Bernoulli in 1698. Since this pioneering work by Riemann and Lebesgue, integral calculus has become a universal method for calculating area, volume, arc length, center of mass, work, and pressure. In mathematics, the word quadrature has survived and it refers to only numerical integration.

The word **tangent** comes from the Latin 'tangere,' to touch. The term **tangent line** is due to Leibniz, 1692, 1694, who defined it as the line through a pair of infinitely close points on the curve. In 1629, Fermat invented a general geometric procedure for drawing tangent lines, or simply, tangents on curves whose analytic forms were known. He made this great discovery while trying to find the maxima and minima of these curves. During 1664–1666, Barrow developed a new geometric method for determining tangents on curves. In fact, in the seventeenth century, the tangent line problem became one of the central questions in geometry. Descartes remarked "And I dare say that this is not only the most useful and most general problem in geometry that I know, but even that I have ever desired to know". In Europe, the tangent problem prefigured **differential calculus**, another term that was coined by Leibniz in 1684, which has been credited solely to Newton and Leibniz. For Newton, the calculus was geometrical, while Leibniz took it toward analysis. However, the concept of a derivative had been developed almost 1200 years before Newton and Leibniz by Bhaskara II, who provided differentiation of the trigonometric functions, for example (in modern notation), he established $\delta(\sin x) = \cos x \cdot \delta x$. Bhaskara II also gave a statement of Rolle's theorem, concluded that the derivative vanishes at a maxima, and introduced the concept of the instantaneous motion of a planet in his collection *Siddhanta Siromani*. Then, in the fourteenth century, Madhava of Sangamagramma invented the ideas underlying the infinite series expansions of functions, power series, the trigonometric series of sine, cosine, tangent, and arctangent (these series have been credited to Gregory, Taylor, and Newton), rational approximations of infinite series, tests of convergence of infinite series, the estimate of an error term, and early forms of differentiation and integration. Madhava fully understood the limit nature of the infinite series. This step has been called the "decisive factor onward from the finite procedures of ancient mathematics to treat their limit-passage to infinity," which is in fact the kernel of modern classical analysis. Parameshvara Namboodri, a disciple of Madhava, stated an early version of the mean value theorem in his *Lilavathi Bhasya*. This is considered to be one of the most important results in differential calculus and was later essential in proving the fundamental theorem of calculus, which shows the inverse character of tangent and area problems. Torricelli was the first to understand the fundamental theorem of calculus geometrically, and this was extended by Gregory while Barrow established a more generalized version, and finally Newton completed the mathematical theory. Nilakanthan Somayaji, following the footsteps of Madhava and his father Parameshvara, provided a derivation and proof of the arctangent trigonometric series and gave the relationship between the power series of π and arctangent, namely,

$$\frac{\pi}{4} = 1 - \frac{1}{3} + \frac{1}{5} - \frac{1}{7} + \frac{1}{9} - \frac{1}{11} + \cdots,$$

which in the literature has been credited to Gregory and Leibniz. L'Hôpital is known as the author of the world's first text book on differential calculus, but Jyesthadevan wrote the calculus text *Yuktibhasa* in Malayalam (a regional language of the Indian state of Kerala) almost 150 years earlier. Calculus in Leibniz's sense (which became widely used partially because of his carefully selected notation) was extended by Leibniz himself, the Bernoulli brothers, Euler, Bolzano, Cauchy, Weierstrass, Riemann, and several other eighteenth and nineteenth century mathematicians. Their work made calculus one of the most powerful, supple, and practical tools of mathematics. In fact, calculus finds applications in every branch of the actuarial science, business, computer science, demography, economics, engineering, medicine, the physical sciences, statistics, and this list continues growing. Acknowledging the importance of calculus, von Neumann said that "[t]he calculus was the first achievement of modern mathematics and it is difficult to overestimate its importance. I think it defines more unequivocally than anything else the inception of modern mathematics, and the system of mathematical analysis, which is its logical development, still constitutes the greatest technical advance in exact thinking." Today, calculus, its advances and abstractions, and its applied branches such as differential equations and optimization have become a major part of mathematical education at all levels.

In mathematics a **set** is defined as a collection of objects or elements, and **set theory** is considered to be a branch of mathematical logic. It has been documented that Cantor introduced set theory to the mathematical world in 1874. His two major books on set theory were published in 1883 and 1895. For Cantor, set theory was a part of metaphysics. However, during 200–875 AD, the Jain School of Mathematics in India utilized the concept of sets. In their work, the Jains introduced several different types of sets, such as cosmological, philosophical, karmic, finite, infinite, transfinite, and variable sets. They called the largest set an omniscient set, and the set containing no elements was known as the null set. They also defined the concept of a union of sets and used the method of one-to-one correspondence for the comparison of transfinite sets. In order to determine the order of comparability of all sets, they considered fourteen types of monotone sequences. Thus, Cantor alone is not the founder of set theory and his major contribution was the mathematical systematization of set theory, now known as naive (non-axiomatic) set theory, and the modern understanding of infinity. He also used the method of one-to-one correspondence to show that there are more real numbers than integers and generated an infinity of infinities from the power set operation.

After Cantor's work, naive set theory became widespread; however, by 1908 it led to several paradoxes (also known as antinomies), the simplest and best known of which is Russell's paradox. This gave rise to numerous attempts at finding a solution and finally resulted in the three main philosophies or schools of thought, namely, the logicist, formalist, and intuitionist schools. The logicists considered these paradoxes to be common errors, caused by errant mathematicians and not by a faulty mathematics. The intuitionists, on the other hand, considered these paradoxes to be clear indications that classical mathematics itself is far from perfect and felt that mathematics had to be rebuilt from the bottom up. After the discovery of paradoxes in naive set theory numerous axiomatic systems were proposed, which culminated in the axiomatic set theories. Of these axiomatic systems, the most commonly used is due to Zermelo and Frankel (Abraham Halevi Frankel, 1891–1965), and this system of set theory is often referred to as the Zermelo–Frankel set theory, which includes the axiom of choice.

Today, set theory is woven into the fabric of modern mathematics. In fact, the language of set theory is used to precisely define nearly all mathematical objects. The works of Bolzano, Dedekind, Charles Sanders Peirce, Schröder, Peano, Bendixson, Hausdorff, Zermelo, Russell, Sierpiński, Steinhaus, Kuratowski, Tarski, von Neumann, Kolmogorov, Hurewicz, Gödel, Pontryagin, Van Orman Quine, Smullyan, Paul Cohen, and others have made set theory and its many interrelated subfields (such as combinatorial set theory, descriptive set theory, fuzzy set theory, set-theoretic topology, combinatorial set theory, computable set theory, and category theory) one of the most active fields of mathematical research. One of the most exciting aspects of set theory is that its elements can be studied informally and intuitively, and so it is being taught in primary schools using Venn diagrams.

After Pythagoras several different definitions of mathematics have been proposed. Each one tries to define mathematics with a specific context in mind. For example:

Mathematics is like draughts (checkers) in being suitable for the young, not too difficult, amusing, and without peril of the state. (Plato)

Mathematics is the study of "quantity." (Aristotle)

Mathematics is the door and key to the sciences. (Roger Bacon, 1214–1294)

Mathematics is the science of order and measure. (Descartes)

Mathematics—the unshaken foundation of science, and the plentiful fountain of advantage to human affairs. (Barrow)

Mathematics is concerned only with the enumeration and comparison of relations. (Gauss)

Mathematics is the science of what is clear by itself. (Jacobi)

Mathematics is the science which draws necessary conclusions. (Benjamin Peirce, 1809–1880)

Mathematics seems to endow one with something like a new sense. (Charles Robert Darwin, 1809–1882)

Mathematics is the work of the human mind, which is destined rather to study than to know, to seek the truth rather than to find it. (Galöis)

Mathematics is not (as some dictionaries today still assert) merely "the science of measurement and number," but, more broadly, any study consisting of symbols along with precise rules of operation upon those symbols, the rules being subject only to the requirement of inner consistency. (Boole)

"Do you know what a mathematician is" Lord Kelvin (1824–1907) once asked a class. He stepped to the board and wrote

$$\int_{-\infty}^{\infty} e^{-x^2} dx = \sqrt{\pi}.$$

Putting his finger on what he had written, he turned to the class. "A mathematician is one to whom that is as obvious as that twice two makes four is to you."

Mathematics is a branch of logic. (Dedekind)

Mathematics is the science of quantity. (Charles Sanders Peirce)

Mathematics is the science of self-evident things. (Felix Klein)

Mathematics is the development "of all types of formal, necessary and deductive reasoning". (Whitehead)

Mathematics is a game played according to certain rules with meaningless marks on paper. (Hilbert)

Mathematics is a subject identical with logic. (Russell)

The whole of mathematics is nothing more than a refinement of everyday thinking. (Einstein)

Mathematics is the art of problem solving. (Polya)

Mathematics is a development of thought that had its beginning with the origin of man and culture a million years or so ago. To be sure, little progress was made during hundreds of thousands of years. (Leslie Alvin White, 1900–1975)

Mathematics is a spirit of rationality. It is this spirit that challenges, simulates, invigorates and drives human minds to exercise themselves to the fullest. It is this spirit that seeks to influence decisively the physical, normal and social life of man, that seeks to answer the problems posed by our very existence, that strives to understand and control nature and that exerts itself to explore and establish the deepest and utmost implications of knowledge already obtained. (Morris Kline)

Mathematics today is the instrument by which the subtle and new phenomena of nature that we are discovering can be understood and coordinated into a unified whole. In this some of the most advanced and newest branches of mathematics have to be employed and contact with an active school of mathematics is therefore great asset for theoretical physicists. (Homi Jehangir Bhabha, FRS, 1909–1966)

Mathematics is like a chest of tools, before studying the tools in detail, a good workman should know the object of each, when it is used and what it is used for. (Walter Warwick Sawyer, 1911–2008)

Mathematics has also been defined as follows: mathematics is a human creation; mathematics is a natural part of man's cultural heritage; mathematics is the accumulation of human wisdom in an effort to understand and harness the physical, social, and economic worlds; mathematics is a language and a language can be learned only by continuously using it; mathematics is a tool that ideally permits mediocre minds to solve complicated problems expeditiously; mathematics is the study of quantity, structure, space, and change; mathematics is the measurement, properties, and relationships of quantities and sets using numbers and symbols; mathematics is something that man himself creates, and the type of mathematics he works out is just as much a function of the cultural demands of the time as any of his other adaptive mechanisms; mathematics is the science which uses easy words for hard ideas; and mathematics seeks regularities and pattern in behavior, motion, number, or shape, or even in the substrata of chaos. A variety of quips and clichés can also define mathematics: "It's an art," "it's a science—in fact, it's the queen and servant of science," "it's what I use when I balance my checkbook," "a game that we play with rules we're not quite sure of," and the apologetic favorite "something I was never good at," and the list can go on and on.

About mathematics several interesting positive views also have been proposed. Perhaps from these we can find some more definitions of mathematics.

Mathematics gives life to her own discoveries; she awakens the mind and purifies the intellect; she brings light to our intrinsic ideas, she abolishes the oblivion and ignorance which are ours by birth. (Proclus Diadochus, 410–485 AD)

Neglect of mathematics works injury to all knowledge, since he who is ignorant of it cannot know the other sciences or the things of this world. (Roger Bacon)

Nature is written in mathematical language. (Galileo)

Mathematicians are like Frenchmen; whatever you say to them they translate into their own language, and forthwith it is something entirely different. (Johann Wolfgang von Goethe, 1749–1832)

Mathematics is one of the oldest of sciences; it is also one of the most active; for its strength is the vigor of perpetual youth. (Andrew Russell Forsyth, 1858–1942)

In mathematics I can report no deficiency, except it be that men do not sufficiently understand the excellent use of the Pure Mathematics. (Francis Bacon, 1909–1992)

Yet another positive view of mathematics and hence of mathematicians follows. Mathematicians are of two types: the 'we can' men believe (possibly subconsciously) that mathematics is a purely human invention; the 'there exists' men believe that mathematics has an extra-human *existence* of its own, and that 'we' merely come upon the *eternal truths* of mathematics in our journey through life, in much the same way that a man taking a walk in a city comes across a number of streets with whose planning he had nothing whatever to do. (Bell, Men of Mathematics)

We also encounter some negative views about mathematics, such as the idea that the mathematics that is certain does not refer to reality and mathematics that refers to reality is not certain. Sir William Hamilton (1788–1856), the famed Scottish philosopher, logician, and meta-physicist, viewed mathematics in a way that may be construed as a cruel attack on mathematics and hence on mathematicians: "Mathematics freeze and parch the mind," "an excessive study of mathematics absolutely incapacitates the mind for those intellectual energies which philosophy and life require," "mathematics cannot conduce to logical habits at all," "in mathematics dullness is thus elevated into talent, and talent degraded into incapacity," "mathematics may distort, but can never rectify, the mind." (Bell, Men of Mathematics). This view may have arisen because a mathematician who is deeply engrossed in his work so that his mind is tuned to only mathematics has his natural (common human) behavior replaced by his mathematical behavior. In other words, his whole being becomes that of a complete mathematician, devoid of the natural human thoughts that he would experience in any other situation.

The following parable of "The Blind Men and the Elephant" (a Story from the Buddhist Sutra) is relevant to our attempt to define mathematics. Several prominent citizens were engaged in a hot argument about God and the different religions and could not come to an agreement. So they approached Lord Gautama Buddha (1887–1807 BC) to find out what exactly God looks like. Buddha asked his disciple to get a large majestic elephant and four blind men. He then brought the four blind men to the elephant and told them to find out what the elephant would "look" like. The first blind man touched the elephant's leg and reported that it "looked" like a pillar. The second blind man touched its tummy and said that an elephant was an inverted ceiling. The third blind man touched the elephant's ear and said that it was a piece of cloth. The fourth blind man held on to the tail and described the elephant as a piece of rope. And all of them ran into a hot argument about the "appearance" of an elephant. The Buddha asked the citizens: "Each blind man had touched the elephant but each of them gives a different description of the animal. Which answer is right?" "All of them are right," was the reply. "Why? Because everyone can only see part of the elephant. They are not able to see the whole animal. The same applies to God and to religions. No one will see Him completely." By this story, Lord Buddha teaches that we should respect all legitimate religions and their beliefs.

This famous "blind men" episode is not meant to disrespect any mathematician. We state it only to point out that *any definition of mathematics, however elaborate or epigrammatic, will fail to lay bare its fundamental structure and the reasons for its universality* (Weyl). In fact, the definition of mathematics continues to change with time and innovation. However, mathematics rewards its creator with a strong sense of aesthetic satisfaction. It helps us understand man's place in the universe and enables us to find order in chaos. Under certain axioms, mathematics is the most absolute, ever lasting, precise, significant, and universal subject. It is perceived as the highest form of thought in the world of learning. No doubt, mathematics is one of the greatest creations of mankind—if it is not indeed the greatest. Mathematics will live forever.

Mathematical Science

Mathematical science is a term that refers to those academic disciplines that are mainly mathematical in nature but are not widely considered to be subfields of proper mathematics. While the characteristic features of mathematics are abstractness, precision, logical rigor, and the indisputable nature of its conclusions, mathematical science provides justification of new discoveries through the body of knowledge comprising mathematics. Such a justification, as and when found, is accepted and satisfying to the scientific community. Mathematical science pervades almost all of science, engineering, and technology. Its branches include astronomy, astrology, actuarial science, which began as a formal mathematical discipline during the late seventeenth century, and biometrics, which is estimated to be 31,000 years old and was used in Babylon as early as 500 BC. Mathematical science also includes computational science (developed during 2700–2300 BC in Sumer, while modern computational science rapidly grew into a major field of study and extensive use with the advent of modern computers in the 1940s), cryptography (the earliest known use was discovered in nonstandard hieroglyphs carved into mountains from the Old Kingdom of Egypt about 1900 BC), econometrics (the first known use of the term "econometrics" was by Powel Ciompa in 1910, although it has existed in the heart of economics since Xenophon's Oeconomicus of about 360 BC and Chanakya's Arthashastra during about 350–283 BC in India), and mathematical biology and biomathematics (established in the late eighteenth century). It also includes mathematical linguistics (a formal apparatus for describing the structure of natural languages and of some formal languages that began in the 1950s), mathematical psychology [which began during the early nineteenth century with Ernst Weber (1795–1878) and Gustav Fechner (1801–1887)], mathematical sociology (Nicolas Rashevsky, Anatol Rapoport, and others founded the discipline in the 1940s), mathematical statistics (came into existence along with statistics in around 1749), and operations research, both deterministic and stochastic, which began in the third century BC during the second Punic War. Mathematical science is also a part of population genetics (founded by Sewall Green Wright (1889–1988), John Burdon Sanderson Haldane (1892–1964), and Fisher in the early twentieth century), probability (while it existed in ancient and medieval times, it arose in the correspondence of Fermat and Pascal and also in a comprehensive treatment by Huygens in seventeenth century), statistical physics (which uses methods of probability theory and statistics), theoretical computer science (began long before the modern discipline of computer science that emerged by about mid-twentieth century; while formal algorithms have existed for millennia, it was Turing, Church, and Kleene who formalized the definition of algorithm in terms of computation), and theoretical physics, also called mathematical physics, which came into existence by 300 BC under the Pre-socratic philosophy and was continued by Plato and Aristotle (the modern era of theoretical physics seems to have begun with the Copernican paradigm shift in astronomy during the early sixteenth century). There remain only a few areas of human intellectual activity that have not been shaped significantly by the mathematical sciences. If the mathematical sciences (the body of knowledge that has been developed over the centuries for establishing and communicating truths about nature and the universe that makes use of the ever-evolving body of knowledge constituting mathematics) as well as the exploration of natural phenomena and the invention of machines and solution techniques is cut off, then we will simply collapse and return to the prehistoric era/stone age. The ideas and results of present-day mathematical science are studied almost exclusively in graduate departments and are used by researchers in a wide range of fields. **Mathematical methods**, which provide correct methods of solution for mathematical problems, exist in almost all fields of mathematics and constitute the essence of mathematical science. The domain of applications of a mathematical method is unlimited.

A **proof** is an explanatory note that the human senses can perceive and that leads the mind and intellect to establish the validity of a fact or argument. Proofs express subjective truth, not universal truth. Since humans have limited perceptive power, there are innumerable subtle assumptions that must be made during the process of proof which are beyond human intellect; therefore, any proof can never be exact. This is why we accept whatever truth can be established within our human limitations. Yet the great Sages and Rishis of history possessed extraordinarily subtle insight as well as unique experiences, which led them to establish certain truths. These are known as empirical theories, and are simply well-founded empirical generalizations or laws about the properties and behaviors of objects that are obtained by examining a large number of instances and seeing that they conform without exception to a single general pattern. These theories could never be questioned and have withstood the test of time. One accepts such theories because of his total faith in the wisdom of these sages, or because he believes this to be the primary (or the sole) path to knowledge. Before we can determine whether a proof is a mathematical proof, we need to understand a few mathematical concepts.

By a mathematical *statement* or *proposition*, we mean an unambiguous composition of words that is true or false. For example, ‘two plus two is four’ is a true statement, and ‘two plus three is seven’ is a false statement. However, $x - y = y - x$ is not a proposition, because the variables in this expression are not defined. If $x - y = y - x$ is asserted for all x, y real numbers, then this is clearly a false proposition; if $x - y = y - x$ is claimed for some real numbers, then this may be a true proposition. ‘Help me please’ and ‘your place or mine?’

are also not statements. Assertive statements that do not contradict themselves to the best of human knowledge are called *axioms* or *postulates*. The truth of axioms is based on our experience and our critical analysis, they are accepted and used without a doubt. Obvious or self-evident (this term is not very palatable to a strict mathematician) axioms are preferred. Some examples of axioms are: if x and y are real numbers, then $x + y$ and xy are real numbers; the sum of squares of two real numbers is nonnegative; for any given real number x there exists an integer n such that nx is greater than x, which is known as the *axiom of Archimedes*; perpendiculars dropped from two vertices of a triangle meet at a point; and opposite angles formed by two intersecting straight lines are equal. A mathematical theory is based on axioms, which are often accepted for the sake of studying the consequences that follow. A *theorem* consists of well-defined concepts and propositions $H_1, H_2, \ldots, H_n$ called *hypotheses*, and a proposition C called its *conclusion*. A theorem is true provided that its hypotheses imply its conclusion. A *mathematical proof/deductive formal proof/Greek proof* of a theorem consists of a sequence of propositions, ending with the conclusion C, that are considered *valid*. A valid proposition may simply be one of the hypotheses, or it may be derived or inferred from earlier propositions. A mathematical proof with a valid sequence of propositions is called a *valid proof*. If at least one of the propositions is invalid, then the argument is called a *fallacy*. A *lemma* is a simple theorem that is used to prove a more important theorem. Often, complicated proofs are easier to understand when they are established using a series of lemmas. A *corollary* is a proposition that can be established directly from a theorem that has been proved. An unproved statement based on guess or opinion, preferably based on some experience or other source of wisdom is called a *conjecture*. Once a conjecture is proved it becomes a theorem. Of all the known conjectures the most famous is the Goldbach conjecture: every even integer greater than 2 is the sum of two prime numbers. For example, $30 = 7 + 23 = 11 + 19 = 13 + 17$. This conjecture was made in the year 1742. Although it has been verified for all even numbers up to 10^{14}, it remains to be proved. Now consider the function $f(n) = n^2 - 79n + 1601$ for all natural numbers n. If we compute $f(n)$ for various values of n, we seem to always get a prime number: $f(1) = 1523$, $f(2) = 1447$, $f(39) = f(40) = 41$. Based on this experience we might make the conjecture that for each natural number n, $p(n) : n^2 - 79n + 1601$ is a prime number. However, this statement cannot be proved because it happens to be false. In fact, while it is true for $n = 1, 2, \ldots, 79$, for $n = 80$, $p(80) = 80^2 - 79 \times 80 + 1601 = 1681 = 41^2$. An example of this type which shows that the statement is false is called a *counterexample*. The ancient Chinese believed that $2^n - 2$ is not divisible by n when n is a composite number. As late as 1680, Leibniz also made this very statement. A single counterexample, namely, $n = 341$, was found in the nineteenth century that disproves this statement. Clearly, $n = 341 = 31 \times 11$ is composite, and $2^{341} - 2$ is divisible by 341. In fact, 341 is the least value of n for which the statement fails. Thus, one counterexample is sufficient to disprove a statement.

Although there are several different types of valid mathematical proofs, we will list and illustrate (through examples) only those that are commonly used. Previous mathematical historians have claimed that "The Hindus, like the Chinese, have pretended that they are the most ancient people on the face of the Earth, and that to them all sciences owe their creation. But it is probably that these pretensions have no foundation; and in fact no science or useful art (except a rather fantastic architecture and sculpture) can be definitely traced back to the inhabitants of the Indian peninsula prior to the Aryan invasion" (Walter William Rouse Ball, 1850–1925), "Oriental mathematics may be an interesting curiosity, but Greek mathematics is the real thing" (Hardy), "The Greeks are not clever schoolboys or scholarship candidates, but fellows at another college" (John Edensor Littlewood, 1885–1977), "Compared with what the Greeks achieved, the mathematics of Egypt and Babylonia is the scrawling of children just learning to write, as opposed to great literature. These civilizations barely recognized mathematics as a distinct discipline, so that for over a period of 4000 years hardly any progress was made in the subject" (Morris Kline, in the year 1962), and "Before the development of Greek culture the concept of deductive science was unknown to the Eastern people of antiquity. In the mathematical documents which have come down to us from these peoples, there are no theorems or demonstrations, and the fundamental concepts of deduction, definition, and axiom have not yet been formed. These fundamental concepts made their first appearance only with the Greek mathematics" (Árpád Szabó, 1913–2001, in the year 1967). From our previous considerations and the following illustrations it will become clear that these statements should be wiped off from the history of mathematics.

Direct Proof. In this form of proof, the hypotheses are shown to imply the conclusion directly. For example, we shall show that for every natural number n, $n^3 + n$ is even. Clearly, $n^3 + n = n(n^2 + 1)$. If n is even, then $n(n^2 + 1)$ is even. If n is odd, then n^2 is odd, and hence $n^2 + 1$ is even, and so $n(n^2 + 1)$ is even. Alternatively, we can consider the following two cases: (i) Suppose n is even. Then, $n = 2k$ for some natural number k, and so $n^3 + n = 8k^3 + 2k = 2(4k^3 + k)$, which is even. (ii) Suppose n is odd. Then, $n = 2k + 1$ for some natural number k, and so $n^3 + n = (8k^3 + 12k^2 + 6k + 1) + (2k + 1) = 2(4k^3 + 6k^2 + 4k + 1)$, which is even.

Aryabhatta, in his *Aryabhattiyam*, states and proves several geometric results. One of his theorems states: If in a circle, a chord CD and diameter AB intersect each other at right angles at E, then $AE \times EB = CE^2$ (Fig. 1). His proof is as follows: In ΔAEC and ΔCEB, we have

$$\angle CEA = \angle CEB = 90^\circ, \quad \angle CAE = 90^\circ - \angle ACE = \angle ECB,$$

and

$$\angle ACE = 90^\circ - \angle ECB = \angle EBC,$$

so ΔAEC and ΔCEB are similar. Thus, $AE/CE = CE/EB$, or $AE \times EB = CE^2$.

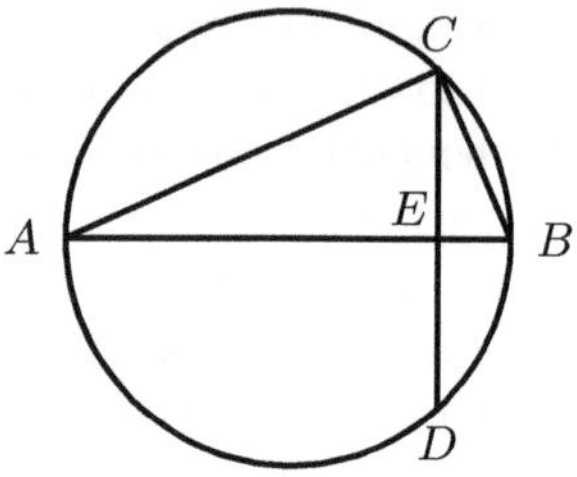

Fig. 1 Aryabhatta theorem

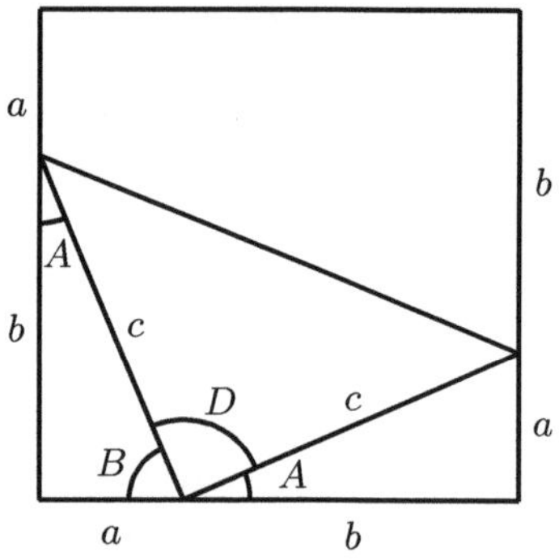

Fig. 2 Pythagoras theorem

As a further example of direct proof, consider the Pythagoras theorem. If a and b are the lengths of the two legs of a right triangle and c is the length of the hypothenuse, then $a^2 + b^2 = c^2$. The following direct proof was discovered by US Representative James A. Garfield (1831–1881) 5 years before he became the 20th President of the USA. The following figure shows three triangles forming half of a square with sides of length $a + b$ (Fig. 2). The angles A, B, and D satisfy the relations

$$A + B = 90^\circ \quad \text{and} \quad A + B + D = 180^\circ.$$

Thus, $D = 90^\circ$, and hence all the three triangles are right triangles. The area of the half square is

$$\frac{1}{2}(a + b)^2 = \frac{1}{2}(a^2 + 2ab + b^2),$$

while the equivalent total area of the three triangles is

$$\frac{1}{2}ab + \frac{1}{2}c^2 + \frac{1}{2}ab.$$

Equating these two expressions, we get

$$a^2 + 2ab + b^2 = ab + c^2 + ab \quad \text{or} \quad a^2 + b^2 = c^2.$$

Constructive Proof. In this form of proof, either a solution or an algorithm for determining a solution is specified. As we have mentioned above, the Sulvasutras contain an abundance of geometric constructions. We illustrate this method by showing that there is a positive integer that can be written as the sum of cubes of positive integers in two different ways. Indeed, after considerable computation we find that $1729 = 10^3 + 9^3 = 12^3 + 1^3$, which was a trivial exercise for the genius Ramanujan.

To calculate the value of $\sqrt{2}$ to specified degree of accuracy, the Babylonians in the eighteenth century BC used the following algorithm (which was found inscribed on a clay tablet, now catalogued as the Yale Babylonian Collection 7,289): Set $x_1 = 1$ and for $n \geq 1$,

$$x_{n+1} = \frac{1}{2}\left(x_n + \frac{2}{x_n}\right).$$

This is perhaps one of the oldest known algorithms, and is now obtained by applying Newton's method to the equation $x^2 - 2 = 0$; however, historians Otto Neugebauer (1899–1990) and Abraham Sachs (1915–1983) believe that the Babylonians obtained this algorithm based on the following principle: Suppose x is a guess which is too small (large), then $2/x$ will be a guess which is too large (small). Hence, their average $(x + 2/x)/2$ is a better approximation. Of course, the Babylonians did not know the convergence of this algorithm.

Nonconstructive Proof. In this form of proof we establish the existence of a solution, but give no indication as to how it might be found. As an example, we shall follow Dov Jarden to show that there exist irrational numbers a and b such that a^b is rational. Consider the irrational numbers $a = b = \sqrt{2}$. If the number $a^b = \sqrt{2}^{\sqrt{2}}$ is rational, we are done. If $\sqrt{2}^{\sqrt{2}}$ is irrational, we consider the numbers $a = \sqrt{2}^{\sqrt{2}}$ and $b = \sqrt{2}$ so that $a^b = \left(\sqrt{2}^{\sqrt{2}}\right)^{\sqrt{2}} = \sqrt{2}^{\sqrt{2}\sqrt{2}} = \sqrt{2}^2 = 2$ is rational. Note that in this proof we could not find irrational numbers a and b such that a^b is rational.

Proof by Contradiction (reductio ad absurdum) "is a wonderful logical tactic. By assuming the opposite of what we intend to show, we seem to be putting the eventual goal in jeopardy. Yet, in the end, calamity is averted." The British mathematician Hardy described proof by contradiction as "one of mathematicians' finest weapons. It is a far finer gambit than any chess gambit; a chess player may offer the sacrifice of a pawn or even a piece, but a mathematician offers the game." To illustrate the method of proof by contradiction we shall show that if n^2 is even for some integer n, then n itself is even. Suppose, for contradiction, that n is odd. Then $n = 2m + 1$ for some integer m. But, then $n^2 = 4m^2 + 4m + 1$ is odd, which contradicts the assumption that n^2 is even. Now we shall use the method of contradiction to show that $\sqrt{2}$ is irrational. The following proof was given by Euclid (some claim Pythagoras) and is probably the most famous nonexistence proof of all time. Suppose, for contradiction, that $\sqrt{2}$ is rational. Then, $\sqrt{2} = p/q$ where p

and q are natural numbers. By reducing the fraction if necessary we can assume that p and q have no common factors. In particular, p and q are not both even. Since $2 = p^2/q^2$ we have $p^2 = 2q^2$ and so p^2 is even. This implies that p is even. Hence, $p = 2k$, where k is some integer. But then, $(2k)^2 = 2q^2$ and therefore $q^2 = 2k^2$. Thus, q^2 and hence q is also even. This implies that both p and q are both even, which contradicts our earlier statement. Hence, $\sqrt{2}$ is irrational. As another example, we reproduce Euclid's finest proof, which shows that there are infinite number of primes. Suppose to the contrary that there are only finitely many prime numbers. Label them $p_1, p_2, p_3, p_4, \ldots$. As there are only a finite number of them, there must be a last one, say, p_n. Now let us consider the number $N = (p_1 \times p_2 \times p_3 \times \cdots \times p_n) + 1$. Surely, this number is larger than (and therefore not equal to) any one of $p_1, \ldots, p_n$. Also, none of $p_1, \ldots, p_n$ is a divisor of N. Hence, we have a number that is not a prime number, but has no divisor except 1 and itself and therefore is a prime number. This is a contradiction, and we conclude that our supposition of the finiteness of the number of primes is faulty.

Mathematical Induction (Domino Principle). According to Polya "induction is the process of discovering general laws by the observation and combination of particular instances. It is used in all of science, even in mathematics. Mathematical induction is used in mathematics alone to prove theorems of a certain kind." It is very likely that al–Karkhi was the first to use mathematical induction to validate his discoveries of the general binomial theorem and Pascal's triangle. This was followed by Maurolico, who used the method of mathematical induction to establish a variety of properties of integers. In 1838, De Morgan presented the first clear explanation of the method of mathematical induction, a term that he coined. Mathematical induction is used to establish the truth of an infinite list of propositions. This method can be described as follows: All the propositions $p(n), p(n+1), \ldots$, where $n \geq 0$ is an integer, are true provided (i) $p(n)$ is true and (ii) $p(k)$ implies $p(k+1)$ for all $k \geq n$. The first step (i) is called the *basis of induction*, whereas the second step (ii) is called the *inductive step*. As an illustration suppose we wish to prove that $1 + 2 + 3 + \cdots + n = n(n+1)/2$ for every positive integer n. The initial step of mathematical induction is to show that it is true for $n = 1$, that is, $1 = 1 \times (1+1)/2$, which is obvious. Then comes the induction step: Suppose the statement is true for a particular value of n, say, $n = k$, that is, $1 + 2 + 3 + \cdots + k = k(k+1)/2$. We must then show that it also holds for the next higher value of n, namely, $n = k + 1$. For this, it will suffice to note that

$$1 + 2 + 3 + \cdots + k + (k+1) = \frac{1}{2}k(k+1) + (k+1) = \frac{1}{2}(k+1)(k+2).$$

As an another example, we shall use mathematical induction to prove that any integer $n \geq 2$ can be expressed as a finite product of primes. Clearly, the statement is true for $n = 2$, because 2 itself is a prime. Now assume the statement is true for all integers between 2 and n, that is, any integer k with $2 \leq k \leq n$ can be expressed as a finite product of primes. If $n + 1$ is prime, the statement is true for $n + 1$. Otherwise, $n + 1$ has a prime divisor, that is, there are integers p and q with p prime and $2 \leq q \leq n$ such that $n + 1 = pq$. But by the induction hypothesis, q can be expressed as a finite product of primes and, therefore, so can $n + 1$.

For mathematical proofs several philosophical questions have been extensively addressed: Does every theorem have a proof? For the same theorem, why are several proofs useful? Is it the case that proof is mathematics, or mathematics is proof? What is the difference between a beautiful proof and an ugly proof? What is the role of proof in understanding the mathematics? Does the concept of a valid proof vary from culture to culture, as well as from age to age? This list continues. We conclude this section with the remark that like mathematics, a proof cannot be defined. We have seen over the course of 40 years of teaching that there is a constant decline in the appreciation of proofs; in fact, students try to avoid proofs and they are usually happy with numerical or computational examples. If this trend continues, in the near future the definition of proof will be very different.

Computing

Computing is a generic term implying any activity needing a computer or benefitting from it or creating it. It encompasses the design, development, and construction of hardware, firmware, and software systems. It involves structuring, processing, and managing different types of information, doing scientific/engineering work/research on and with it, making a computer system behave intelligently, and creating and using communications and entertainment media. Branches of computing include computer science, computational science, information technology, computer hardware and software engineering, and artificial intelligence.

The origin of computing dates back to the prehistoric era of mankind when a human being had the sense of comparing any two quantities to say/detect the one which is larger. The human hand is the original "calculating machine." It has been the tool most widely used at all times as an aid to counting calculation. Hands work well enough for representing numbers visually and immediately, but by its very nature, finger-counting cannot serve as a recording device. Hands, fingers (or finger knots), sand and pebble, chalk and slate, charcoal ink and palm leaves, pen and paper with or without the help of a table constitute the earliest tools for computing. Thus the history of computing is much older than the history of computing hardware, software, firmware, and modern computer technology, which started in the early twentieth century.

The most widely known computational tool in history was the *abacus*, which was probably invented in Babylon in 2400 BC. Its original form was lines drawn in sand with pebbles. Writing in the first century BC, Horace (65–8 BC) refers to the wax abacus, a board covered with a thin layer of black wax on which columns and figures were inscribed using a stylus. On the other hand, the Chinese Abacus (suan pan) was made of parallel grooves or rods, with buttons or pierced pellets that could be slid along these, that was a practical and formidable instrument which is still in common use throughout the Far East. But before they used their abacus, the Chinese used little ivory or bamboo sticks, called chou (literally, "calculating sticks") that they arranged on the squares of a tiled floor or on a table made like a chessboard.

Still used by abacists as a computation tool in a very limited way are *abaci* of a more modern design. The Chinese also employed counting boards as mechanical aids for calculation as early as the fourth century BC. Counting boards were similar in design to abaci and were usually made of wood or metal, and contained carved grooves in which mounted beads, pebbles, or metal discs were moved. Abaci differed from counting boards in that the beads on the abaci contained holes that enabled them to be held on a particular rod of the abacus. The Greeks invented a number of mechanical devices to assist with their problem solving. One of the most famous of these was the *Antikythera*, a machine designed to calculate astronomical positions that was built around 87 BC and lost a few years later, that has the oldest known complex gear mechanism and is sometimes called the first known analog computer (mechanical). Roman accountants used a real "pocket calculator" whose invention undoubtedly predates our era. Among the Romans, the word calculator meant, on the one hand, a "master of calculation" whose principal task was to teach the art of calculation to young people using a portable abacus with counters; and, on the other hand, the keeper of the accounts or the intendant in the important houses of the patricians, where he was also called dispensator.

An important requirement in computing is to store information for future use or further computation. In this context, the Incas success can be explained by their ingenious method of keeping accurate records by means of a highly elaborate and fairly complex system of knotted string. This device, called a *quipu* (an Inca word meaning "knot"), which was developed during third millennium BC, consisted of a main piece of cord about two feet long onto which thinner, colored strings were knotted in groups, these pendant strings themselves being knotted in various ways at regular intervals. Quipus are so simple and so valuable that they continued to be used for many centuries in Peru, Bolivia, and Ecuador. They served not only as a device for concrete numeration but also as a mnemotechnic tool for recoding numbers, maintaining administrative archives, and keeping count of contracts, calendars, and the like. For our purposes, these numbers on strings can be considered as a special form of written numbers.

Computational Science

Computational science or, equivalently, scientific computation or scientific computing involves the mathematical modelling of science and engineering problems, analyzing these models implemented on a computer, solving them using a computer (as opposed to using the analytic methods of applied mathematics), and applying quantitative analysis techniques. In practice, computational science is typically the application of a computer simulation (a useful part of mathematical modelling of many natural systems in areas such as physics, chemistry, biology, psychology, social science, economics, and engineering) and other forms of computation from numerical analysis and theoretical computer science to problems in various scientific disciplines. **Computational mathematics**, on the other hand, encompasses mathematical research in areas of science where computing plays the pivotal role, stressing numerical algorithms (in which most of the computational work, say, over 90 %, consists of arithmetic operations, like finding a root of a nonlinear equation using the Newton method), seminumerical algorithms (consisting of roughly 50 % numerical and 50 % nonnumerical operations, for instance, testing whether a given integer is prime or not), and nonnumerical algorithms (consisting of mostly non-arithmetic operations, such as searching and sorting). Computational mathematics emerged as a distinct field of applied mathematics in the mid-1950s. Currently, computational mathematics includes subjects such as numerical linear algebra, numerical solutions of ordinary and partial differential equations, discrete mathematics (for example, in its search for mathematical structures like groups), number theory (primarily testing and factorization), cryptography, and computational algebraic topology. Computation is prominent in research.

Computer science, unlike computational science, involves theory and laboratory experiments—the traditional forms of science and engineering. Computational science attempts to gain insight mainly through the mathematical models implemented on a computer. Thus, although computational science and computer science are very closely connected and even overlap (e.g., both involve the study of fundamental computational algorithms), they are significantly different. Computational science/mathematics may also be viewed as the interface between realistic mathematical models and scientific computing methods/algorithms. Scientists and engineers, belonging to various disciplines and whose main interest is not in computational science but instead the efficient and cost-effective nature of the solutions that it provides, develop computer programs (application software) that model systems under investigation and execute these programs with different sets of input data (parameters). In many cases, these models are very computationally intensive and are executed on a hyper-speed computing or distributed computing platform.

The pioneers of computer science and engineering had at the bottom of their mind the construction of a computing device, or simply a computer, that would be capable of performing computation, scientific or otherwise. These pioneers, the early computer scientists/technologists, contributed to both computer science (hardware/software) and computational science to varying degrees. Oughtred, after Napier's invention of logarithms and Gunter's creation of the logarithmic scales, invented two such scales in 1622 that were slid by one another to perform multiplication and division. Pascal invented the mechanical calculator in 1642. He was prompted to this invention when he saw the tedious arithmetical labor involved in his father's official work. His device was first called the Arithmetic Machine, then Pascal's Calculator, and later Pascaline. It could add and subtract two numbers directly, and it could multiply and divide by repetition. Pascal went through 50 prototypes before presenting his first machine to the public in 1645. Nine machines have survived that are now on display in European museums. In 1671, Leibniz embarked on inventing a machine that could add, subtract, multiply, and divide, which he gradually improved over the next couple of years. It was not considered a complete success because it did not fully mechanize the operation of carrying. In 1674, he anticipated a machine capable of performing some algebraic operations. In 1679, while pondering over his binary arithmetic, Leibniz imagined a machine in which binary numbers were represented by marbles, governed by some sort of punched cards. Modern electronic digital computers have replaced Leibniz's gravity-driven marbles with shift registers, voltage gradients, and electron pulses, but otherwise they run roughly as Leibniz had visualized. He also wondered if a machine could be built that could determine whether particular mathematical statements are true or false. Later, Babbage did pioneering work on the Difference Engine, a sophisticated calculator that could be used for the production of mathematical tables. In 1834, he also designed the Analytic Engine, which was the world's first mechanical computer. However, the machine was not built during Babbage's lifetime. Pehr Georg Scheutz (1785–1873), a Swedish lawyer, invented in 1837 and finalized in 1843 his Scheutzian calculation engine. This machine, which he constructed with his son Edvard Scheutz, was based on Babbage's Difference Engine. An improved model, roughly the size of a piano, was created in 1853 and demonstrated in 1855 in Paris at the World Fair. The machine was then sold to the British government in 1859. Scheutz created yet another machine in 1860, selling it to the USA. These devices were used for creating logarithmic tables. While Scheutz's machine was not perfect and could not produce complete tables, the Swedish scientist Martin Wiberg (1826–1905) reworked the construction from the ground up and in 1875 created a compact device, the size of a sewing machine, that could print complete logarithmic tables. The tables that it produced were published in English, French, and German. During 1935–1938, Konrad Zuse (1910–1995) built the world's first program-controlled computer, called Z1. Despite certain mechanical engineering problems it had all the basic ingredients of modern machines, using the binary system and today's standard separation of storage and control. Zuse's 1936 patent application (Z23139/GMD Nr. 005/021) also suggests a von Neumann architecture (which is the fundamental architecture used in modern computers, which von Neumann reinvented in 1945) with program and data modifiable in storage. In 1936, Turing published a key paper on computability in which he introduced the theoretical/hypothetical Turing machine, a mathematical machine that is capable of solving any conceivable mathematical problem that has an algorithm. The Church–Turing thesis states that anything that may be computed may be computed by a Turing machine. Turing also designed the automatic computing engine (ACE), which was the first complete design of a stored-program computer in Britain. The American physicist John Vincent Atanasoff (1903–1995) invented the first electronic digital computer, the Atanasoff–Berry computer, in 1939 at Iowa State College. This special-purpose computer was neither programmable nor Turing-complete. Challenges to his claim were resolved in 1973 (after 34 years) when the "Honeywell v. Sperry Rand" lawsuit ruled that Atanasoff was indeed the inventor of the computer. In 1941, Zuse built world's first fully functional programmable computer, Z3. Z3 pre-dates the Harvard Mark I (built by IBM between 1939 and 1944) and the Manchester Mark I. The Harvard Mark I computer (also known as the IBM automatic sequence controlled calculator (ASCC)), called simply the Mark I by Harvard University, was an electromechanical machine and the brainchild of Howard Hathaway Aiken (1900–1973), who conceived the idea of a general-purpose machine for scientific computation while working on his doctorate at Harvard. Mark I weighed 5 t, contained 500 miles of wiring, and accepted operations/instructions that were punched into a roll of paper tape. Its multiplication operation consisted of the addition of partial products, just like multiplication using Napier's bones. The machine was kept running 24 h a day and could handle in a day calculations that would have taken a contemporary computer 6 months.

The ENIAC (Electronic Numerical Integrator and Computer), the first general-purpose Turing-complete digital electronic computer, was built at the Moore School of Electrical Engineering of the University of Pennsylvania in Philadelphia during 1943–1945. Since it used electronic valves, it is considered to be a first generation machine. It began to operate in the summer of 1945, and the Moore School is now known as the birthplace of the computer industry. The logical design of the system was based on the ideas of John William Mauchly (1907–1980), an American physicist, while the hardware design and implementation was due to John Adam Presper Eckert Jr. (1919–1995), an American electrical engineer. ENIAC's construction was a great act of technological courage on the part of the designers as well as the Moore School. ENIAC contained 17,468 electronic valves (vacuum tubes), 7,200 crystal diodes, 1,500 relays, 70,000 resistors, 10,000 capacitors, and approximately 5 million hand-soldered joints. It weighed more than 60,000 lbs (27 t), was around $2.4 \times 0.9 \times 30$ m, took up 167 m^2, and consumed 150 kW of power. It cost almost $500,000 in 1945 (approximately $7 million today in 2014). An addition operation took 0.2 ms while a multiplication took 2.8 ms. The machine carried out calculations 1,000 times faster than its nearest rival, the Mark I; π was computed to 2,035 decimal places in 70 h. Its 17,468 vacuum tubes were a serious drawback, requiring too much power and producing too much heat. (An enduring, although questionable, legend is that every time the ENIAC was turned on, lights dimmed in all the houses in West Philadelphia.)

The EDVAC (electronic discrete variable automatic computer) was a direct successor of ENIAC. It was a binary serial computer with automatic addition, subtraction, and multiplication, as well as programmed division, and automatic checking with an ultrasonic serial memory of 1,000 44-bit words (later set to a 1,024-word re-circulating mercury-delay line memory). It had 4,000 tubes and 10,000 crystal diodes. The arithmetic unit did both fixed- and floating-point operations. Input and output were done via punched paper tape and IBM cards. It was delivered to the Ballistic Research Laboratories in Aberdeen, Maryland, in 1949 and underwent extensive tests. EDVAC became fully functional in 1951 and was used until 1962.

During 1847–1854, Boole formalized Boolean algebra, the foundation for digital logic (modern computing) and computer science. Boole and Babbage are among the fathers of computing. Shannon was the first person to see the applicability of Boolean algebra to the design of circuits and telephone routing switches. Shannon's later work at Bell Labs laid the foundation of modern information theory.

EDSAC 1 (electronic delay storage automatic calculator), a serial binary computer and the second usefully operational electronic digital stored-program computer, was a British machine. The computer was inspired by von Neumann's seminal First Draft of a Report on EDVAC of June 30, 1945, and was constructed by Sir Maurice Vincent Wilkes, FRS (1913–2010), and his team at the Cambridge University Mathematical Laboratory in England. It became operational in 1949 when it computed a table of squares and a list of prime numbers. The project was later supported by the British firm J. Lyons & Co., who were rewarded with the first commercially applied computer, LEO I, based on the EDSAC design. EDSAC 1 was replaced by EDSAC 2 in 1958, which stayed in use until 1965.

Many computer people left the Moore School around 1946. Eckert and Mauchly resigned and launched the first computer company, the Eckert–Mauchly Computer Corporation (later the name was applied to a division of Remington–Rand Company and successor organizations). Their product UNIVAC (universal automatic computer), built in 1951, was the first general-purpose computer for commercial use. This was preceded by the computer BINAC in 1949. UNIVAC needed only 5,000 vacuum tubes and relied on magnetic tape data storage instead of punched cards. The machine was initially installed at the United States Census Bureau in 1951, where it was used continuously over the next 12 years before its retirement to the Smithsonian Institution. UNIVAC received considerable publicity when the CBS network used it to provide early election-night predictions of the 1952 presidential race. Although the network's political pundits were skeptical of the machine's forecast, Dwight David Eisenhower (1890–1969) was elected President by almost exactly the same landslide voting margin that was predicted by UNIVAC. One of the system engineers contributing to the development of the UNIVAC was Hopper. Several versions of UNIVAC were developed that were widely used during the 1950s and 1960s.

There had been a surge of computing activities all over the world—in Western Europe and elsewhere. For instance, the first generation computer TIFRAC (Tata Institute of Fundamental Research Automatic Calculator) was developed in India at the Tata Institute of Fundamental Research (TIFR) in Mumbai. A TIFR Pilot Machine was developed in the 1950s that became functional in 1955 and was commissioned in November 1956. The enhanced TIFRAC computer became operational in 1957 and was commissioned in February of 1960. It performed computation for scientific research between the early 1960s and 1965. It included 2,700 vacuum tubes, 1,700 germanium diodes, and 12,500 resistors. It had $2,048$ 40-bit words of memory. Both machines were early adaptors of ferrite core memory. India's first second generation (transistor-based) computer, ISIJU–1, was designed and developed jointly by the Indian Statistical Institute, Kolkata, and Jadavpur University, Kolkata, during the early 1960s. Its logical design was given by Krishnamurthy and Murarka in 1962. ISIJU–1 became functional in 1965 and remained in use until 1975.

In 1958, under the leadership of Sergei Lvovich Sobolev (1908–1989) and Nikolay Brusentsov (born 1925), the world's first balanced ternary (using three-valued ternary logic instead of two-valued binary logic) computer, *Setun*, was developed and built at Moscow State University. Base 3 or more computers are not as stable as base 2 (binary) computers because of required stable and fast errorless switching problem (existing in silicon technology in the physical world). The Setun computer was built to fulfill the needs of Moscow State University and was manufactured at the Kazan Mathematical plant. Fifty computers were built until production was halted in 1965. In the period between 1965 and 1970, a regular binary computer replaced it at Moscow State University. Although this replacement binary computer performed equally well, it costed 2.5 times as much as the Setun. In 1970, a new ternary computer, the Setun-70, was designed.

Numerous computer programming languages have evolved from the early days of computing based on the requirements of the computational problems at hand. These formal languages were designed to be devoid of the ambiguities that exist in natural language. Backus and his team developed the first widely used high-level programming language, FORTRAN (FORmula TRANslation), in the mid-1950s. Since then there have been many versions of FORTRAN, including FORTRAN II (1958), FORTRAN IV, FORTRAN 66, FORTRAN 77, and FORTRAN 90. Backus is also credited for the Backus–Naur Form (BNF), a notation that is used worldwide to define formal language syntax. Besides FORTRAN, which is still in use, many other programming languages such as ALGOL (ALGOrithmic Language, developed in the late 1950s and 1960s) and its many variants such as ALGOL 60 and ALGOL 68, COBOL (Common Business Oriented Language, designed in the late 1950s), and PASCAL (named after Blaise Pascal, based upon ALGOL, developed in the early 1970s) were widely used when they were developed but now (2014) see limited use. Other programming languages that have been developed after the 1960s include BASIC (Beginner's All Purpose Symbolic Instruction Code, designed in 1963 as a teaching tool), the C language (which was introduced in the early 1970s and has became very popular in industry), C++ (an extension of C that was first released in 1983 and implemented in 1985), Miranda (developed in the mid-1980s), Java (which was developed in the early 1990s but released only in 1996), various functional programming languages (such as HOPE developed at Edinburgh University in the 1970s and REXX designed and developed at IBM in 1979), and logic programming languages. Others have become more or less history in computing, although many of their structures can be found in modern programming languages in some form or another.

During 1930–2000, the world experienced extensive and rapid refinements in (electronic) machine computation. Vannevar Bush (1890–1974), an American electrical engineer, inventor, and science administrator, was an analog computing pioneer. He introduced the Memex concept in 1930, which led to the development of Hypertext. An analog computer is based on realizing arithmetic operations by reading/measuring physical quantities such as current, voltage, and resistance in a carefully designed electrical circuit. Unlike a digital computer, analog computers are very limited in their accuracy. Analog computers might give at most 0.01 % accuracy and have very limited use as general-purpose computers. By the late 1950s, vacuum tubes were superseded by transistors, which generated little heat and had long service lives. The next decade saw the miniaturization of electronic circuitry; soon a million transistors were replaced by a single silicon chip. Computers accordingly became smaller, more powerful, and lowercost; instead of filling rooms, machines now sat on desks. Just as mechanical devices led to the Industrial Revolution, computing machines have become the engines that power the modern information revolution. Rapid advances in computer technology have led to an intriguing interplay between mathematician and machine. The computer has been an invaluable tool in furnishing counterexamples to outstanding conjectures or in verifying conjectures up to specific numerical bounds. There have been some remarkable successes with previously intractable problems. For example, π has been computed to 10 trillion digits, and the 48th Mersenne Prime, 2 raised to the 57,885,161 power minus 1 (which is 17,425,170 digits long), was discovered in 2013.

Sir Timothy John Berners-Lee (born 1955), an English computer scientist, was the inventor of the World Wide Web (www) that has revolutionized global communication. He proposed an information management system in March 1989, and he implemented the first successful communication between a hypertext transfer protocol (http) client and a server via the Internet in November of 1989. The invention of the World Wide Web was preceded by email communication, a very important development in computer communication. Raymond Samuel Tomlinson (born 1941), a programmer, implemented an email system in 1971 on the ARPANET (advanced research projects agency network). The ARPANET was one of the world's first operational packet switching networks that implemented TCP/IP (transmission control protocol/internet protocol), the basic language of the Internet. ARPANET was the progenitor of what was to become the global Internet. Email had been previously sent on other networks, such as AUTODIN. It was the first system able to send mail between users on different hosts connected to the ARPANET (previously, mail could only be sent to others who used the same computer). To achieve this, he used the sign to separate the user from their machine, which has been used in email addresses ever since. The first email sent by Tomlinson has not been preserved and had content that he describes as insignificant. The first message was sent between two machines that were literally side by side. The only physical connection they had was through the ARPANET. Tomlinson sent a number of test messages from one machine to the other, then he sent a message to the rest of his group explaining how to send messages over the network. The first use of network email announced its own existence.

In computational science, several breakthroughs were achieved. Some of them are as follows. James William Cooley (born 1926) and Tukey developed the Fast Fourier Transform (FFT) algorithm to compute discrete Fourier Transform and its inverse in 1965. Computing the DFT of n points in the naive way, using the definition, takes $O(n^2)$ arithmetic operations, while a FFT can compute the same DFT in only $O(n \log_2 n)$ operations. If $n = 2^{20}$, then n^2 will be 2^{40} and $n \log_2 n$ will be 20971520—a huge reduction in computation and hence in computational error. An improved version of the FFT, developed later, results in significantly less computation and also less error. Narendra Karmarkar developed a polynomial-time $O(n^{3.5})$ projective transformation iterative deterministic algorithm to compute a solution of a linear program $\text{Max}\, c^t x$ subject to $Ax \leq b,\ x \geq 0$

(null column vector) in 1984, where n is the number of variables. This algorithm obtains the next feasible direction toward optimality and scales back by a factor $0 < \alpha \leq 1$. It has undergone improvements over the years and a variant of it is now more widely used than the simplex algorithm (which ruled the linear optimization area for nearly four decades—1950s to 1980s) which is an exponential-time method in the worst case. In 2002, Manindra Agarwal, Neeraj Kayal, and Nitin Saxena (born 1981) of the Indian Institute of Technology, Kanpur, India, developed the AKS (Agarwal–Kayal–Saxena) primality test algorithm—a deterministic primality-proving method—which is general, polynomial time, deterministic, and unconditional all at the same time. Earlier algorithms developed over the centuries achieved at most three of these properties but not all four simultaneously. The algorithm is based upon the following theorem: An integer $n(\geq 2)$ is prime if and only if the polynomial congruence $(x-\alpha)^n \equiv (x^n-\alpha) \pmod{n}$ holds for all integers α coprime to n (or even just for some such integer α, in particular for $\alpha = 1$), where x is a free variable.

The foregoing (breakthrough) algorithms have had a tremendous positive impact on cost-effective computation in solving many real-world engineering and scientific problems. We emphasize the fact that these algorithms are the most-desired mathematically proved ones and are valid for any infinite precision (mathematical) computation that evidently includes finite precision computation which any man-made computer works on. We still have several outstanding problems for which a deterministic polynomial-time algorithm is yet to be developed. For example, the *Travelling Salesman Problem* (TSP—Given n cities and the distances between each pair of cities, find the shortest path that passes through each city only once and returns to the city of origin) and a verification (of its solution obtained by some other nondeterministic computational method).

Computers are everywhere, not just on our desks. Computers are hidden in our stereos and televisions, in our cars, in our wristwatches, and in our microwave ovens, and other household appliances. Computers make our telephone connections. Computers are used to make our airline reservations. Many of the images we see on the cinema screen are produced on a computer. In fact, the exponential growth that the world has seen during the last five decades in architecture, engineering, and technology, specifically information technology, has been made possible by fantastic progress in both computing devices (hardware) and software. Every 18 months processor speed doubles. Every 12 months bandwidth doubles, and every 9 months hard disk space doubles. Behind all this exponential growth is the computational mathematics that is superimposed on the hardware. All views—both negative and positive—of mathematics need to be modified when we land in the world of computational mathematics. Here we can see the real power of mathematics through its computational nature. In conclusion, the mathematical and computational sciences are closely connected and inseparable, and this union has presented mathematicians with new challenges, namely, finding hidden patterns. All of the science and engineering that we know today would be dwarfed without this computational arm of mathematics. At the root of science and technology is this arm. If we cut this arm, the whole of today's scientific world would simply collapse.

A **proof within the domain of computational mathematics** is a much easier task now than it was a couple of decades ago since it has only a finite number of distinct points, each corresponding to a distinct real (rational) number. The difference between computational mathematics and mathematics is precision that each has (implying word length). In mathematics, precision is always implicitly or explicitly assumed to be infinite and all real numbers—uncountably infinite and constituting a continuous (with no gap in between any two real numbers) space—are considered, while in computational mathematics, only negligibly few real numbers—countably finite and discrete (with gaps in between any two real numbers)—are included. Once we fix a reasonably (relative to engineering activities/computations) finite precision, any statement (lemma/theorem/corollary) can be checked for validity over the entire domain, which consists of only a finite number of points. Although by manual calculations or even by computations by a mid/(even) late twentieth century computer, it was not possible for a reasonable precision to check all possible cases of a statement (theorem/lemma/corollary) when such a number of cases are combinatorial and large. However, this is possible today, unlike in the past, due to the hyper-speed computing power (viz., 10 to the power 18 floating-point operations per second) available to us. Such a verification constitutes a computational proof. Intractability in most problems is nonexistent except in problems like the NP (nondeterministic polynomial-time)—hard problem in combinatorial optimization—the TSP of n cities—that demands centuries of computations even in the world's fastest computer provided the number of cities involved in the problem is not small. In fact, all of the $(n - 1)!$ possible paths must be evaluated to obtain the shortest route deterministically. If $n = 100$ (not an unreasonably large number, considering number of cities in the world), then we need to evaluate 99! paths (larger than an astronomical figure)! This is intractable even in the world's fastest computer. However, if $n = 10$, then the TSP can be deterministically solved in less than a minute on a modern laptop!

In spite of exponential growth in computing power, these problems will remain intractable for the foreseeable future. If the problem size is small, then such a combinatorial problem can be solved deterministically in a short time. The possible polynomial-time evolutionary approaches are used for these intractable problems to obtain a solution that cannot be shown to be the absolute best, yet such a solution could serve our purposes in practice.

However, any mathematical proof of a statement, when considered in the realm of computational mathematics, is the most desired proof (e.g., the Karmarkar projective transformation algorithm and the AKS prime-testing algorithm), because then we do not have to bother to check its validity through a computational proof (i.e., checking each and every point in the solution space, viz., n-dimensional domain). This would be superfluous, although this verification can be done for a reasonably finite precision. The Karmarkar and AKS algorithms have been mathematically proved to be deterministic and polynomial time (not combinatorial or exponential); therefore, these algorithms and their variations can safely be used and their computational complexities can accurately be spelled out. One and all (including mathematicians) are fully satisfied with the results of these algorithms and can accept the results with their eyes closed. A computational proof (subject to a specified precision) is satisfactory for an engineer or scientist, although such a proof is not satisfactory for a mathematician (who attributes little importance to any computational proof).

Computers do not use the mathematical real number system but instead a simulation of it—called the floating-point/fixed-point number system. Hence errors in numerical computation—both rounding and truncation—are intimately connected to computational mathematics and can never be gotten rid off. In general, the output of any arithmetic operation (add/subtract/multiply/divide) will involve error. Even the input data on which the operations are carried out are usually erroneous. The (non-discrete) quantities involved in computation are never known exactly. All that we can specify is an interval (preferably the narrowest possible) that contains the exact (error-free) real quantity; for example, a gallon of milk bought from a supermarket is not known to be exactly one gallon but it is the quantity of milk that lies between 1 gallon -10 ml and 1 gallon $+10$ ml, depending on the accuracy of the measuring device. Although there are many pitfalls in computational mathematics that do not exist in mathematics due to finite (often fixed) precision, the particular algorithm employed, and inherent error in the data that are input, which lead to erroneous results, obscurity, and instability, a computational proof is simple and straight-forward and can be exhaustively verified for the entire finite set of real numbers in computational mathematics. The use of methods like construction, deduction, induction, contradiction, and a combination of two or more of these mathematical methods is not essential, although not forbidden, in computational mathematics.

The four color conjecture (originally stated in 1852) states that, given any separation of a plane into contiguous regions, producing a figure called a map, at most four colors are needed to color the regions of the map so that no two adjacent regions have the same color. For example, in the following map we must make regions A, B, and C of different colors because they share pairwise boundaries, but then it would be impossible to color region D unless a fourth color was introduced (Fig. 3).

The four color conjecture attracted much interest during the last century, and several false proofs and counterexamples appeared. The five color theorem, which has a short elementary proof, states that five colors suffice to color a map and was proven in the 1890s. On the other hand, there are maps for which three colors are insufficient. So five colors are plenty but three are too few, and hence it all comes down to four. In 1976, the question of computer proofs burst dramatically upon the mathematical scene with the resolution of the four color conjecture. Kenneth Appel

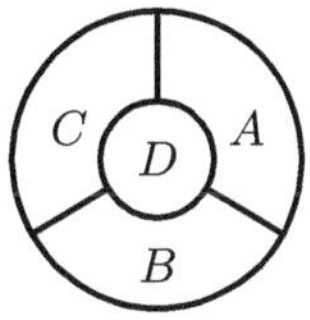

Fig. 3 Four color conjecture

(1932–2013) and Wolfgang Haken (born 1928) of the University of Illinois (USA) demonstrated that there is a particular set of 1,936 maps, each of which cannot be part of a smallest-sized counterexample to the four color theorem. This was followed by a special-purpose computer program, written to confirm that each of these maps has this property. Additionally, any map (including a counterexample) must have a portion that looks like one of these 1,936 maps. This demonstration required hundreds of pages of manual analysis. Appel and Haken concluded that no smallest counterexamples existed because any must contain, yet not contain, one of these 1,936 maps. This contradiction means there are no counterexamples at all and that the theorem is therefore true. Their proof was not accepted by all mathematicians because the computer-assisted proof was infeasible/intractable for a human to check by hand. Although the proof has gained wider acceptance, doubts continue to persist. As mathematician Ronald Lewis Graham (born 1935) asked when mulling over this complicated issue, "The real question is this: if no human being can ever hope to check a proof, is it really a proof?" At this point the question has no definite answer, although as computer proofs become more common, mathematicians will probably feel more comfortable with them. It is fair to say, however, that most mathematicians would breathe a sigh of relief if the four color theorem were established with a two-page proof—short, ingenious, and elegant—rather than with the brute-force machinations of a computer. Traditionalists long for the good old days of mathematics unplugged. Finally, we emphasize that the proofs of the Karmarkar and AKS algorithms were mathematical and not computer assisted, hence these are accepted by all mathematicians and others (including computational scientists) without any doubt.

A computer of today or of any future date, however powerful it might be, can definitely not prove all results that satisfy a mathematician because of infinity versus finiteness, as explained earlier. A proof in computational mathematics has the freedom to use mathematical proof methods (induction/contradiction/construction/deduction)—one or a combination of two or more—and computer assisted procedures. Subject to a certain limit of precision, and within the capability of the computer's power in terms of speed, storage, and bandwidth, any computational mathematical statement can be shown to be valid or not (very much unlike a mathematical statement). If we ask something beyond the foregoing limit and/or capability, then we transcend the domain of computational mathematics.

It is necessary for us to stress the fact that some of the contributors (who may be considered major contributors) have been named in this monograph while behind the scenes there have been and continue to be many other contributors, most of whom will not gain entry into the history of mathematics and computing. Our future generations may never know them and their invaluable contributions, yet they were truly real.

Creators

The title of any book, monograph, or article tells the reader broadly what he can expect to attain from the book by reading it. The title is essentially a shortest possible abstract of the book. By seeing it, the reader consciously, subconsciously, or unconsciously forms in his mind an expectation of the book's message. He then decides whether it is of interest to him. If it is, he then scans the bigger abstract/introduction and sometimes the headings of the contents of the book for further information and a better glimpse of what the book comprises. At this stage, the reader decides whether he should proceed further, into the whole book or into selected topics, or if he should simply leave the book aside without reading it. He may also scan the preface to learn how the authors orient/prepare the reader so that his inquisitiveness toward the book grows.

The reader will discover that the creators of the mathematical and computational sciences do not just contain the highlights of individuals' landmark contributions to the march of science and technology over the millennia. He will learn something much deeper from these diverse individuals: apparently, they are like any other learned human beings. Some were well to do, others were not. Some had to pass through several personal or family tragedies, while others had relatively smooth living. Some had to face social and environmental opposition and even persecution regarding their discoveries, political opinions, or religious beliefs; others received glorious appreciation and recognition in terms of rewards for their contributions. Some had long lives, others did not. Some enjoyed adventurous and eventful living while others preferred simple life. Some may be viewed as being eccentric, others may not. These creators are responsible for today's world, which may be considered highly developed compared to our world several millennia ago, but there is one thing that they have in common: this may be described as a kind of revelation through their intense concentration and highly focused minds. Even in the most adverse situations, such as utter poverty, environmental hostility, even death threats, they were completely engrossed in pursuing their goal of finding answers to the questions that they held in their minds. They used their minds as vital tools in pursuing

R.P. Agarwal, S.K. Sen, *Creators of Mathematical and Computational Sciences*,
DOI 10.1007/978-3-319-10870-4_2

relentlessly and ardently the answer to the queries encountered in connection with the physical environment or with abstract mathematical concepts or computational problems. Their focused minds were not limited to activity only during specific hours of the day. In fact, they remained intensely active at all times: while taking a lunch, a dinner, a walk, a bath, and even during their dreams while asleep.

Every reader of this book will be drawn automatically not only to learning individual contributions (including milestone discoveries) but also to meeting each individual character responsible for these contributions over the centuries. This makes the reading more interesting and effective. The reader will discover the fertile human-land nurtured by a number of factors that include these creators and their environment and society. Up from this land springs the much desired ideas, the revelations through the germinated seeds, viz., the creators—the scientists. The reader will be able to appreciate the environment of discovery, together with the individual personality, attitude, and focused mind, that brings forth innovation. We believe that this will have a profound positive impact on the reader's thought process and in turn on his character. In other words, the reader will be inclined to emulate the virtues of the scientists. This may go a long way toward molding and transforming him into one who will be a more valuable contributor to and potential innovator for our society.

Krishna Dwaipayana or Sage Veda Vyasa (Born 3374 BC)

Krishna Dwaipayana or Sage Veda Vyasa (born 3374 BC) along with his disciples Jaimani, Paila, Sumanthu, and Vaisampayana codified the four Vedas (Veda means wisdom, knowledge, or vision): *Rigveda* contains hymns and prayers to be recited during the performance of rituals and sacrifices; *Samaveda* contains melodies to be sung on suitable occasions; *Yajurveda* contains sacrificial formulas for ceremonial occasions; and *Atharvaveda* is a collection of magical formulas and spells. Each Veda consists of four parts—the Samhitas (hymns), the Brahmanas (rituals), the Aranyakas (theologies), and the Upanishads (philosophies). This compilation consisted of 1,131 Sakhas (recessions); however, only ten are available. Sage Vyasa is also the narrator of the epic *Mahabharata* (of which *The Bhagavad Gita* is a part) and is credited with writing most, if not all of the 18 (counted, but 20) Puranas: *Agni* (15,400 verses); *Bhagavata* (18,000 verses); *Bhavishya* (14,500 verses); *Brahma* (24,000 verses); *Brahmanda* (12,000 verses); *Brahmavaivarta* (18,000 verses); *Garuda* (19,000 verses); *Harivamsa* (16,000 verses); *Kurma* (17,000 verses); *Linga* (11,000 verses); *Markandeya* (9,000 verses); *Matsya* (14,000 verses); *Narada* (25,000 verses); *Padma* (55,000 verses); *Shiva* (24,000 verses); *Skanda* (81,100 verses); *Vamana* (10,000 verses); *Varaha* (10,000 verses); *Vayu* (24,000 verses); and *Vishnu* (23,000 verses). Arithmetic operations (Ganit) such as addition, subtraction, multiplication, fractions, squares, cubes, and roots are enumerated in the Vishnu Purana. In Rigveda the distances of the Moon and Sun from the Earth are given as 108 times the diameters of these heavenly bodies, respectively (the current

Fig. 1 Srichakra

approximate values are 110.6 for the Moon and 107.6 for the Sun). The vedic literature consists of these four Vedas; six Vedangas: Shiksha (phonetics and phonology), Kalpa (ritual), Vyakarana (grammar), Nirukta (etymology), Chandas (meter), Jyotisha (astronomy for calendar issues, such as auspicious days for performing sacrifices). It also includes the four Upangas: Nyaya (law of nature), Mimamsa (rituals and spiritual philosophy), Itihasa-Purana (Mahabharata is Itihasa, the creation and dissolution of Universe, evolution, cycle of time, theology, different incarnations of the Supreme Godhead), and Dharma Sastras (guidelines to live a Dharmic life, specific guidelines to Grhasthas, guidelines for performing rituals, and the mathematical principles for construction of various altars).

A hymn from Atharvaveda is dedicated to *Srichakra*, also known as *Sriyantra* (see Fig. 1). Srichakra belongs to the class of devices utilized in meditation, mainly by those belonging to the tantric tradition of Sanatana Dharma. Srichakra is also inscribed in several temples. We shall consider the mathematics of Srichakra. The diagram consists of nine interwoven isosceles triangles: four point upwards, representing Sakti, the primordial female essence of dynamic energy, and five point downwards, representing Siva, the primordial male essence of static wisdom. Thus the Srichakra also represents the union of Masculine and Feminine Divine. Because it is composed of nine triangles, it is also known as the Navayoni Chakra. The triangles are arranged in such a way that they produce 43 smaller triangles, at each center of the smallest of which there is a big dot, known as bindu, the junction point between the physical universe and its unmanifested source. These smaller triangles are expected to form the abodes of different devatas (symbolic of the entire cosmos or a womb of creation), whose names are sometimes inscribed in their respective places. This is surrounded by a lotus of eight petals, a lotus of sixteen petals, grilled in turn by three circles, all enclosed in a square with four doors, one on each side. The squares represent the boundaries within which the deities reside, protected from the chaos and disorder of the outside world. The mathematical interest in the Srichakra lies in the construction of the central nine triangles, which is a more difficult problem than might first appear. A line here may have three, four, five, or six intersections with other lines. The problem is to construct a Srichakra in which all the intersections are correct and the vertices of the largest triangles fall on the circumference of the enclosing circle. It is amazing how the ancient Indians achieved accurate constructions of increasingly complex versions of the Srichakra, including spherical ones with spherical triangles. There is a curious fact about all correctly constructed Srichakra, whether enclosed in circles or in squares; in all such cases the base angle of the largest triangle is about $51°30'$.

Sulbasutras (About 3200 BC)

Sulbasutras (about 3200 BC). The meaning of the word *sulv* is to measure, and geometry in ancient India came to be known by the name *sulba* or *sulva*. Sulbasutra means 'rule of chords,' which is another name for geometry. The Sulbasutras are parts of the larger corpus of texts called the Shrautasutras, considered to be appendices to the Vedas, which give rules for constructing altars. For a successful ritual sacrifice, the altar had to conform to very precise measurements, so mathematical accuracy was seen to be of the utmost importance. Only seven Sulbasutras are extant, named for the sages who wrote them: Apastamba, Baudhayana (born 3200 BC), Katyayana, Manava, Maitrayana, Varaha, and Vidhula. The four major Sulbasutras, which are mathematically the most significant, are those composed by Baudhayana, Manava, Apastamba, and Katyayana. In 1875, George Thibaut (1848–1914) translated a large portion of the Sulvasutras, which showed that the Indian priests possessed significant mathematical knowledge. Thibaut was a Sanskrit scholar and his principal objective was to make the mathematical knowledge of the Vedic Indians available to the learned world. After commenting that a good deal of Indian knowledge could be traced back to requirements of ritual, Thibaut adds that these facts have a double interest: In the first place, they are valuable for the history of the human mind in general. In the second place, they are important for the mental history of India and for answering the question of the originality of Indian science. For whatever is closely connected with the ancient Indian religion must have sprung up among the Indians themselves, unless positive evidence of the strongest kind points to the contrary conclusion. The sulbas contain a large number of geometric constructions for squares, rectangles, parallelograms, and trapezia. They describe how to construct: a square n times in area to a given square; a square of area equal to the sum of the two squares; a square whose whole area is equal to the difference of two squares; a square equal to a rectangle; a triangle equal to a rectangle; a triangle equal to a rhombus; and a square equal to the sum of two triangles or two pentagons. Several theorems are also proved in the sulbas:

the diagonal of a rectangle divides it into equal parts; the diagonals of a rectangle bisecting each other and opposite areas are equal; the perpendicular through the vertex of an isosceles triangle on the base divides the triangle into equal halves; a rectangle and a parallelogram on the same base and between the parallels are equal in area; and the diagonals of a rhombus bisect each other at right angles. Baudhayana contains one of the earliest references to what is known today as the Pythagorean theorem, with a convincing bona fide proof: the rope that is stretched across the diagonal of a square produces an area double the size of the original square. This is a special case of the Pythagorean theorem for a 45° right triangle. The Pythagorean triples $(3, 4, 5)$, $(5, 12, 13)$, $(8, 15, 17)$, and $(12, 35, 37)$ are found in Apastamba's rules for altar construction. The Katyayana, written later, gives a more general version of the Pythagorean theorem: the rope which is stretched along the length of the diagonal of a rectangle produces an area which the vertical and horizontal sides make together. In other words, the square of the hypotenuse (comes from Greek words hypo, meaning under, beneath, or down, and teinen meaning to stretch) equals the sum of the squares of the sides.

A remarkable approximation of $\sqrt{2}$ occurs in three of the sulbas, Baudhayana, Apastamba, and Katyayana, namely,

$$\sqrt{2} \simeq 1 + \frac{1}{3} + \frac{1}{3 \cdot 4} - \frac{1}{3 \cdot 4 \cdot 34}.$$

This gives $\sqrt{2} = 1.4142156\ldots$, whereas the exact value is $1.414213\ldots$. The approximation is therefore correct to five decimal places. The sulbas also provide methods for squaring a circle, or the attempt to find a square whose area is equal to that of a given circle, and also the reverse problem of turning a given square into a circle. This leads to the one of the most curious approximations

$$\pi \simeq 18(3 - 2\sqrt{2}) = \left(\frac{6}{2 + \sqrt{2}}\right)^2,$$

where π is the ratio of the circumference of a circle to its diameter (the Greek symbol π was used first by the Welshman William Jones in 1706). This is followed by a discussion about the irrationality of $\sqrt{2}$ and π. The sulbas also include the problem of solving quadratic equations of the form $ax^2 + bx + c = 0$; several examples of arithmetic and geometric progressions; a method for dividing a segment into seven equal parts; and solutions of first degree indeterminate equations.

Aryabhatta (Born 2765 BC)

Aryabhatta (born 2765 BC) is the first famous Indian mathematician and astronomer. It is believed that he was born in Patliputra in Magadha, modern Patna in Bihar. He taught astronomy and mathematics when he was 23 years of age, in 2742 BC. His astronomical knowledge was so advanced that he could claim that the Earth rotates on its own axis, the Earth moves round the Sun, and the Moon rotates round the Earth; incredibly he believed that the orbits of the planets (the word planet means wanderer in Greek) are ellipses. He talked about the positions of the planets in relation to their movement around the Sun, referring to the light of the planets and the Moon as a reflection from the Sun. He explained the eclipses of the Moon and the Sun, day and night, and the length of the year as exactly 365 days. He calculated (the word calculate is derived from the Latin word calculus, meaning pebble, and pebbles served as the counters on the Roman abacus) the circumference of the Earth as 24,835 miles, which is close to the modern day calculation of 24,900 miles. He gave the duration of the planetary revolutions during a *mahayuga* as 4.32 million years. He declared, "the spotless jewel of the knowledge which lay so long sunk in the ocean of knowledge, both true and false, has been raised by me, there from, using the boat of my intelligence." In his Aryabhattiyam, which consists of 108 verses and 13 introductory verses, and is divided into four padas, or chapters (written in the very terse style typical of sutra literature, in which each line is an aid to memory for a complex system), Aryabhatta included 33 verses giving 66 mathematical rules *ganita* of pure mathematics. He described various original ways to perform different mathematical operations, including square and cube roots and solutions of quadratic equations. He provided elegant results for the summation of series of squares and cubes. He made use of decimals, the zero (sunya), and the place-value system. A decimal system was already in place in India during the Harappan period, which emerged before 2600 BC along the Indus River valley, as indicated by an analysis of Harappan weights and measures. In fact, weights corresponding to ratios of 0.05, 0.1, 0.2, 0.5, 1, 2, 5, 10, 20, 50, 100, 200, and 500 have been identified. To find an approximate value of π, Aryabhatta gave the following prescription: Add 4 to 100, multiply by 8, and add to 62,000. This is "approximately" the circumference of a circle whose diameter is 20,000. This means that $\pi = 62{,}832/20{,}000 = 3.1416$. It is important to note that Aryabhatta used the word asanna (approaching) to mean that not only is this an approximation of π but also the value is *incommensurable* or *irrational*. In a single verse Aryabhata discussed the concept of sine using the name ardha-jya, meaning "half-chord." For simplicity, people started calling it jya. When Arabic writers translated his works from Sanskrit into Arabic, they referred it as jiba. However, vowels are omitted in Arabic writings, so it was abbreviated as jb. Later writers substituted it with jiab, meaning "cove" or "bay" (in Arabic, jiba has no meaning). Later in the twelfth century, when Gherardo of Cremona (around 1114–1187) translated these writings from Arabic into Latin, he replaced the Arabic jiab with its Latin counterpart, sinus, which became sine in English. He also used the kuttaka (pulverize) method to compute integer solutions to equations of the form $by = ax + c$ and $by = ax - c$, where a, b, and c are integers, a topic that has come to be known as diophantine equations. His method is essentially the same as what is now known as the Euclidean algorithm to find the highest common factor of a and b. It is also related to continued fractions. His other treatise, which has been lost, is *Aryabhatta-siddhanta*, which contained a description of several astronomical instruments: the gnomon (shankuyantra), a shadow instrument (chhaya-yantra), possibly angle-measuring devices, semicircular and circular (dhanur-yantra/chakra-yantra) devices, a cylindrical stick (yasti-yantra), an umbrella-shaped device called the chhatra-yantra, and water clocks of at least two types, bow shaped and cylindrical. Calendric calculations devised by Aryabhata have been in continuous use in India for the practical purposes of fixing the Panchangam (the Hindu calendar). In the Islamic world, they formed the basis of the Jalali calendar introduced in 1073 AD by a group of astronomers including Omar Khayyám, versions of which (modified in 1925) are the national calendars in use in Iran and Afghanistan today.

The legacy of this genius continues to baffle many mathematicians. India's first artificial satellite, Aryabhata, and the lunar crater Aryabhata are named in his honor. The Aryabhatta Research Institute of Observational Sciences (ARIES) near Nainital, India, conducts research in astronomy, astrophysics, and atmospheric sciences.

Great Pyramid at Gizeh (Erected Around 2600 BC)

Great Pyramid at Gizeh (erected around 2600 BC) built in Egypt by Khufu (2589–2566 BC), whom the Greeks called Cheops (flourished around 2680 BC), is one of the most massive buildings ever erected. It has at least twice the volume and thirty times the mass (the resistance an object offers to a change in its speed or direction of motion) of the Empire State Building in New York and is built from individual stones weighing up to 70 tons each. The slope of the face to the base (or the angle of inclination) of the Great Pyramid is $51°50'35''$. The same angle also appears in the ancient Hindu Srichakra. From the dimensions of the Great Pyramid it is possible to derive the two famous irrational numbers, namely, π (half the perimeter of the base of the pyramid divided by its height gives 3.14) and the 'golden number' or 'divine proportion' ϕ (the length of the face divided by the half length of the side of the square base gives 1.61806). According to Herodotus (around 484–425 BC), known as the "Father of History," four groups of a hundred thousand men labored 3 months each over 20 years to build this pyramid; however, calculations show that not more than 36,000 men could have worked on the pyramid at one time without bumping into one another.

Cuneiform Tablets (About 2000 BC)

Cuneiform Tablets (about 2000 BC). The Sumerians lived in the southern part of Mesopotamia (Iraq). The name Mesopotamia is purely descriptive, coming from the Greek words meaning "between the rivers." Their civilization was absorbed by the Semitic Babylonians around 2000 BC. The famous King and lawmaker Hammurabi (1792–1750 BC) was one of the outstanding figures of the first Babylonian dynasty established by the Semites, who became master of the region. Hammurabi extended Babylonian territory over the whole of Mesopotamia as far as the eastern parts of Syria. Babylonian writing is known as cuneiform, which means "wedge shaped" in Latin. In the last century almost 400,000 cuneiform tablets, generally the size of a hand, have been unearthed. A large collection of these tablets is available at the British Museum (the word "museum" is derived from the goddesses Muses: the nine sisters of the arts and sciences, from which "music" is also obtained) and at Yale, Columbia, and the University of Pennsylvania. Of these, about 400 tablets contained the remarkable mathematics of the Sumerians and Babylonians. They used a counting scale of 60, which is used today in measuring time and angles. In the ancient system of weights, a scale of 60 was endorsed by God himself: Lord Yahweh said: "... Twenty shekels, twenty-five shekels and fifteen shekels are to make one mina" (Ezekiel 45:9–12). They gave several tables of reciprocals, products, and squares and cubes of the numbers from 1 to 50. A tablet now in the Berlin Museum lists $n^2 + n^3$ for $n = 1, 2, \ldots, 20, 30, 40, 50$. The Babylonians used partially a positional number system. They were in full possession of the technique of handling quadratic equations. They solved linear and quadratic equations in two variables and even problems involving cubic and biquadratic equations. They formulated such problems only with specific numerical values for the coefficients, but their method leaves no doubt that they knew the general rule. The following is an example taken from a tablet: An area A, consisting of the sum of two squares, is 1,000. The side of one square is $2/3$ of the side of the other square, diminished by 10. What are the sides of the square? This leads to the equations $x^2 + y^2 = 1{,}000,\ y = (2/3)x - 10$, the solution of which can be found by solving the quadratic equation

$$\frac{13}{9}x^2 - \frac{40}{3}x - 900 = 0,$$

which has one positive solution $x = 30$. The Babylonians used a method of averaging to compute square roots quite accurately. In fact, a tablet from the Yale collection (No. 7289) gives the value of $\sqrt{2}$ as 1.414222 differing by about 0.000008 from the true value. Moreover, even the solution of exponential equations was carried out, as in the determination of the time required for a sum of money to accumulate to a desired amount at a given rate of interest, which was accomplished with the use of tables, as one might expect. In a tablet found in 1936 in Susa in Iraq, they sometimes used the value $3\frac{1}{8} = 3.125$ to approximate π, at other times they were satisfied with $\pi \simeq 3$. The later approximation is perhaps based on the following passage from the Old Testament of the Bible: "Also, he made a molten sea of ten cubits from brim to brim, round in compass, and five cubits the height thereof; and a line of thirty cubits did compass it round about" (I Kings 7:23 and 2 Chron. 4:2). This shows that the Jews did not pay much attention to geometry.

The Babylonians were aware of the Pythagorean theorem, especially noteworthy is their investigation of Pythagorean triples, established by the so-called Plimpton 322 tablet. In fact, the Babylonians were interested in a certain kind of Pythagorean triples that in modern terms can be stated as follows: Suppose u and v are relatively prime positive integers, that is, integers whose greatest common divisor is 1. Assume that not both are odd and that $u > v$. Then, if $a = 2uv$, $b = u^2 - v^2$, and $c = u^2 + v^2$, we have $\gcd(a, b, c) = 1$ and $a^2 + b^2 = c^2$. In Plimpton 322 the triple (13,500, 12,709, 18,541) is included, which is obtained by taking $u = 125$ and $v = 54$.

The Babylonians built pyramid-shaped *ziggurats*. The first story of a ziggurat might measure $n \times n \times 1$, the second story $(n-1) \times (n-1) \times 1$ and so on, with the top two stories measuring $2 \times 2 \times 1$ and $1 \times 1 \times 1$. The volume of such a ziggurat is $1^2 + 2^2 + \cdots + (n-1)^2 + n^2$, and the Babylonians knew that

$$1^2 + 2^2 + \cdots + (n-1)^2 + n^2 = \frac{n(n+1)(2n+1)}{6}.$$

According to the biblical story of the *Tower of Babel*, there was once an attempt to build a ziggurat 'with the top reaching heaven' (Genesis 11:4). They obviously believed that the infinite series $1^2 + 2^2 + 3^2 + \cdots$ converges.

The Babylonians also established some geometrical results, which include rules for finding areas of rectangles, right triangles, certain trapezoids, and volumes of prisms, right circular cylinders, and the frustum of a cone or square pyramid. Their cosmology was a mixture of science and religion. Like the Hindus, they believed that the Gods resided on the planets, from which it followed that their location would influence the actions of the Gods in human affairs. The Babylonians assumed that Earth was the center of the universe and everything revolved around it. This cosmology was later adopted by the early Christian thinkers and perpetuated for a 1,000 years after the fall of Rome (founded by Romulus and Remus about 750 BC). This "geocentric theory" was consistent with the Christian belief that the Son of God was born at the center of the universe.

Vardhamana Mahavira (Around 1894–1814 BC)

Vardhamana Mahavira (around 1894–1814 BC), who promulgated Jainism, was a contemporary of Gautama Buddha; consequently, Jainism is as old as Buddhism. His followers (the Jain School) until 875 AD considered mathematics to be an abstract discipline that should be cultivated for its own sake and not for fulfilling the needs of sacrificial ritual, as has been often suggested. Their main texts of special relevance are known as *Surya Prajnapti* (Sixth Century BC), *Jambu Dvipa Prajnapti*, *Sthananga Sutra* (Second Century BC), *Uttaradyyana Sutra*, *Bhagvati Sutra* (Third BC), and *Anuyoga Dvara Sutra* (Fifth Century AD). The Surya Pragyapti describes ellipses. The Sthananga Sutra gives a list of mathematical topics that were studied during the time of the Jains, namely, the theory of numbers, arithmetical operations, geometry, operations with fractions, simple equations, cubic equations, biquadratic equations, and especially permutations and combinations (vikalpa). Their major contributions can be summarized as follows:

Like the Vedic mathematicians, the idea of infinity is the essence of Jain Mathematics, which evolved from cosmology (time is eternal and without form, and the world is infinite and it was neither created nor destroyed). They devised a measure of time called a shirsa prahelika, which equalled $756 \times 10^{11} \times (8{,}400{,}000)^{28}$ days. Another example is a palya, which is the time taken to empty a cubic vessel of side one yojana (approximately 10 km) filled with the wool of new born lambs, if one strand is removed every century. The contemplation of such large numbers led the Jains to the concept of infinity. All numbers were classified into three groups, enumerable, innumerable, and infinite, each of which was in turn subdivided into three orders:

(a) Enumerable: lowest, intermediate, and highest;
(b) Innumerable: nearly innumerable, truly innumerable, and innumerably innumerable;
(c) Infinite: nearly infinite, truly infinite, and infinitely infinite.

The first group, the enumerable numbers, consisted of all the numbers from 2 (1 was ignored) to the highest. An idea of the 'highest' number is given by the following extract from the Anuyoga Dwara Sutra: *Consider a trough whose diameter is that of the Earth (100,000 yojana) and whose circumference is 316,227 yojana. Fill it up with white mustard seeds counting one after another. Similarly, fill up with mustard seeds other troughs of the sizes of the various lands and seas. Still the highest enumerable number has not been attained.* But once this number, call it N, is attained, infinity is reached via the following sequence of operations

$$N+1, N+2, \ldots, (N+1)^2-1$$
$$(N+1)^2, (N+2)^2, \ldots, (N+1)^4-1$$
$$(N+1)^4, (N+2)^4, \ldots, (N+1)^8-1$$

and so on. Five different kinds of infinities are recognized: infinite in one direction, infinite in two directions, infinite in area, infinite everywhere, and infinite perpetually. This was quite a revolutionary idea in more than one way. Thus the Jains were the first to discard the idea that all infinities are the same or equal; however, they did not include these in the rigorous realm of mathematics, which was not done until the late nineteenth century in Europe by Cantor.

A *permutation* is a particular way of ordering some or all of a given number of items. Therefore, the number of permutations which can be formed from a group of unlike items is given by the number of ways of arranging them. As an example, take the letters a, b, and c, and find the number of permutations that consist of two letters. Six arrangements are possible: ab, ac, ba, ca, bc, cb. Instead of listing all possible arrangements, we can work out the number of permutations by arguing as follows: The first letter in an arrangement can be any of the three, while the second must be either of the other two letters. Consequently, the number of permutations for two of a group of three letters is $3 \times 2 = 6$. The shorthand way of expressing this result is ${}_3P_2 = 6$. Similarly, a *combination* is a selection from some or all of a number of items, but unlike permutations, order is not taken into account. Therefore, the number of combinations that can be formed from a group of unlike items is given by the number of ways of selecting them. To take the same illustration as above, the number of combinations of two letters at a time from a, b, and c is three: ab, ac, bc. Again, instead of listing all possible combinations, we can work out how many there are as follows. In each combination, the first letter can be any of the three, but there is just one possibility for the second letter, so there are three possible combinations. (Although ab and ba are two different permutations, they amount to the same combination.) A shorthand way of expressing this result is ${}_3C_2 = 3$. In the Bhagvati Sutra, rules are given for the number of permutations and combinations of 1 selected from n, 2 from n, and 3 from n. Numbers are calculated in the cases where $n = 2, 3$, and 4. It then mentions that one can compute the numbers in the same way for large n. This Sutra also includes calculations of the groups that can be formed out of the five senses, and selections that can be made from a given number of men, women, and eunuchs.

The Anuyoga Dwara Sutra lists sequences of successive squares or square roots of numbers. Expressed in modern notation as operations performed on a certain number a, these sequences may be represented as

$$(a)^2, (a^2)^2, [(a^2)^2]^2, \cdots$$

$$\sqrt{a}, \sqrt{\sqrt{a}}, \sqrt{\sqrt{\sqrt{a}}}, \cdots.$$

In the same sutra, we come across the following statement on operations with power series or sequences. The first square root multiplied by the second square root is the cube of the second square root, and the second square root multiplied by the third square root is the cube of the third square root. Expressed in terms of a, this implies that

$$a^{1/2} \times a^{1/4} = (a^{1/4})^3 \quad \text{and} \quad a^{1/4} \times a^{1/8} = (a^{1/8})^3.$$

As a further illustration, the total population of the world is given as a "number obtained by multiplying the sixth square by the fifth square, of a number which can be divided by two 96 times." This gives $2^{64} \times 2^{32} = 2^{96}$, which in decimal form is a number of 29 digits.

This statement indicates that the laws of exponents,

$$a^m \times a^n = a^{m+n} \quad \text{and} \quad (a^m)^n = a^{mn},$$

were familiar to the Jains. From around the eighth century AD there is some interesting evidence in the *Dhavala* commentary by Virasenachary to suggest that the Jains may have developed the idea of logarithms (meaning ratio numbers, or reckoning numbers) to base 2, 3, and 4 without using them for any practical purposes. The terms *ardhacheda, trakeacheda,* and *caturarthacheda* of a quantity may be defined as the number of times the quantity can be divided by 2, 3, and 4, respectively, without leaving a remainder. For example, since $32 = 2^5$, the *ardhacheda* of 32 is 5. Or, in the language of modern mathematics, the *ardhacheda* of x is $\ln_2 x$, the *trakeacheda* of x is $\ln_3 x$, and so on.

It was around the second century AD when Jain mathematics had evolved the concept of sets. In Satkhandagama various sets were operated on with the help of logarithmic functions to base two, by squaring and extracting square roots, and by raising to finite or infinite powers. The general tendency was to repeat the operations a number of times in order to give rise to new sets.

The Jains' interest in sequences and progressions developed out of their philosophical theory of cosmological structures. Systematic representations of the cosmos constructed according to this theory contained innumerable concentric rings of alternate continents and oceans, the diameter of each ring being twice that of the previous one, so that if the smallest ring had a diameter of 1 unit, the next largest would have a diameter of 2 units, the next 2^2 units, and so on until the nth ring of diameter 2^{n-1} units. Arithmetic progressions were given the most detailed treatment. Separate formulas were worked out for finding the first term a, the common difference d, the number of elements n in the series, and the sum S of the terms. This was well explored in a Jain canonical text entitled, *Trilokaprajnapti*. One of its problems is to find the sum of a complicated series consisting of 49 terms made up of seven groups, each group itself forming a separate arithmetical progression and the terms of each group forming another.

The term *rajju* was used in two different senses by the Jain theorists. In cosmology, it was a frequently occurring measure of length, approximately 3.4×10^{21} km according to the Digambara school. But in a more general sense it was the term that the Jains used for geometry or mensuration (land measuring), in which they followed closely the Vedic Sulvasutras. Their notable contribution concerned measurements of the circle. In Jain cosmography, the center of Madhyaloka or the central part of the universe inhabited by humans consists of Jambudvipa (a large circular island) with a diameter of 100,000 yojana. While there are several estimates of the circumference of this island, including the rather crude 300,000 yojana, an interesting estimate mentioned in both the *Anuyoga Dwara Sutra* and the *Triloko Sara* is 316,227 yojana, 3 krosa, 128 danda, and $13\frac{1}{2}$ angula, where 1 yojana is about 10 km, 4 krosa = 1 jojana, 2,000 danda = 1 krosa, and 96 angula (literally a 'finger's breadth') = 1 danda. This result is consistent with taking the circumference given by $\sqrt{10}d$, where $d = 100{,}000$ jojana. The choice of the square root of 10 for π was quite convenient, since in Jain cosmology islands and oceans always had diameters measured in powers of 10.

Moscow and Rhind Papyruses (About 1850 and 1650 BC)

Moscow and Rhind Papyruses (about 1850 and 1650 BC). Ahmes (around 1680–1620 BC), more accurately Ahmose, was an Egyptian scribe. A surviving work of Ahmes is part of the *Rhind Mathematical Papyrus*, which is named after the Scottish Egyptologist Alexander Henry Rhind (1833–1863) who went to Thebes for health reasons, became interested in excavating, and purchased the papyrus in Egypt in 1858. It has belonged to the British Museum since 1863. When new, this papyrus was about 18 ft long and 13 in. wide. Ahmes states that he copied the papyrus from a now-lost Middle Kingdom original, dating around 2000 BC. This curious document, entitled *Directions for Knowing All Dark Things*, was deciphered by Eisenlohr in 1877, is a collection of problems in geometry and arithmetic, algebra, weights and measures, business, and recreational diversions. The 87 problems are presented with solutions, but often with no hint as to how they were obtained. It deals mainly with the reduction of fractions such as $2/(2n+1)$ to a sum of fractions each of whose numerators is unity, for example,

$$\frac{2}{29} = \frac{1}{24} + \frac{1}{58} + \frac{1}{174} + \frac{1}{232}.$$

Ahmes states without proof that a circular field with a diameter of 9 units is equal in area to a square with sides of 8 units, i.e., $\pi(9/2)^2 = 8^2$, which implies a value of π approximately equal to 3.16. Two problems deal with arithmetical progressions and seem to indicate that he knew how to sum such series. For example, Problem 40 concerns an arithmetic progression of five terms. It states: divide 100 loaves among 5 men so that the sum of the three largest shares is 7 times the sum of the two smallest $(x + (x+d) + (x+2d) + (x+3d) + (x+4d) = 100,\ 7[x + (x+d)] = (x+2d) + (x+3d) + (x+4d),\ x = 10/6,\ d = 55/6)$. The Rhind papyrus also contains problems dealing with pyramids, involving the idea of a *seqt*, which is the inverse of the gradient of the faces, that is, the cotangent of the angle they make with the horizontal line. Another important papyrus is the *Moscow Papyrus*, which was translated in 1930, contains 25 mathematical problems, and is about two centuries older than the Rhind Papyrus. It is also called the *Golenischev Mathematical Papyrus*, after its first owner Egyptologist Vladimir Goleniscev (1856–1947) who purchased it in 1892. It is now in the collection of the Pushkin State Museum of Fine Arts in Moscow. In this papyrus, most importantly, the solution to problem 14 indicates that the Egyptians knew the correct formula for obtaining the volume of a frustum, that is,

$$V = \frac{1}{3}h(a^2 + ab + b^2);$$

here, h is the altitude and a and b are the lengths of the sides of the square top and square base, respectively. Although most of the problems in these papyri are practical, according to Plato, the Egyptians held that mathematics had a divine source. In *Phaedrus* (around 444–393 BC) he remarks: "At the Egyptian city of Naucratis there was a famous old God whose name was Theuth; the bird which is called the Ibis was sacred to him, and he was the inventor of many arts, such as arithmetic and calculation and geometry and astronomy and draughts and dice, but his great discovery was the use of letters." The Berlin papyrus (about 1320 BC) and the Kahun papyrus (about 1700 BC) are also famous. Both of these provide examples of what might be interpreted as solutions to nonlinear problems.

Sage Lagadha (Before 1350 BC)

Sage Lagadha (before 1350 BC) is the author of Vedanga Jyotisha. This text is available in two recensions: one of 36 verses associated with the Rigveda and another of 45 verses associated with the Yajurveda. There are 29 verses in common. Vedanga Jyotisha describes rules for tracking the motions of the Sun and the Moon. The Vedanga Jyotisha includes the statement: "Just as the feathers of a peacock and the jewel-stone of a snake are placed at the highest point of the body (at the forehead), similarly, the position of Ganit (mathematics) is the highest among all branches of the Vedas and the Shastras."

Thales of Miletus (Around 625–545 BC)

Thales of Miletus (around 625–545 BC) was according to tradition (his whole figure is legendary), Greek's earliest mathematician, astronomer, and philosopher, who grew up along the eastern Aegean shores watching the swimming whales. The Greek historian Herodotus mentions that Thales was a Phoenician. He also states that the periodical inundations of the Nile (which swept away the landmarks in the valley of the river, and by altering its course increased or decreased the taxable value of the adjoining lands) rendered a tolerably accurate system of surveying indispensable and led to a systematic study of geometry by the Egyptian priests. Herodotus called Egypt "the Gift of the Nile." The fertile Nile Valley has been described as the world's largest oasis in the world's largest desert. According to the commentator Proclus Diadochus, "Thales was the first to go to Egypt and bring back to Greece this study; he discovered many propositions and disclosed the underlying principles of many others to his successors." He may have been influenced by Indian thought via Persia. Legend holds that it was Thales who first deductively proved: the angles at the base of an isosceles triangle are equal; if two straight lines cut one another, the vertically opposite angles are equal; a triangle is determined if its base and base angles are known, this he applied to find the distance of a ship at sea; the sides of equiangular triangles are proportional; a circle is bisected by its diameter; the angle subtended by a diameter of a circle at any point in the circumference is a right angle. He is also credited for showing that the sum of the angles of any triangle is 180°. Unfortunately, legend is all we have to rely on, for his actual proofs disappeared long ago. Still, the ancients held him in very high regard, classifying him as one of the "seven wise men of antiquity," others being: Pittacos of Mytilene (around 640–568 BC), Bias of Priene (about 600 BC), Solon (around 638–559 BC), Cleobulus of Lindos (about 600 BC), Myson of Chene, and Chilon of Sparta (about 600 BC). In this list of seven wise men, Thales is the only one who was a philosopher and a scientist. Of course, this list of wise men varies from one author to another. He is remembered by several apposite but probably apocryphal anecdotes. He astounded the Egyptians by calculating the height of a pyramid using proportionate right-angled triangles; diverted the river Halys, the frontier between Lydia and Persia, to enable a Lydian army under Croesus to cross; taught that a year contained about 365 days, and not 12 months of 30 days; believed that the Earth is a disc-like body floating on water; and correctly forecasted a plentiful olive crop 1 year after several bad crops, bought all olive presses around Miletus, and made a huge profit by renting them. He is believed to have predicted that a total solar eclipse would occur during May of 585 BC; however, it was not until long after the sixth century that Chaldean astronomers could give reasonably accurate predictions of eclipses of the Moon, though they were never able to predict that an eclipse of the Sun would be seen in a particular region. Thales was not the kindest of people. It is said that once when he was transporting some salt, which was loaded on mules, one of the animals slipped in a stream, dampened its load, and so caused some of the salt to dissolve. Finding its burden thus lightened it rolled over at the next ford to which it came. To break it of this trick, Thales loaded it with rags and sponges that absorbed the water, made the load heavier, and soon effectually cured it of its troublesome habit. Thales never married. When Solon asked why, Thales arranged a cruel ruse whereby a messenger brought Solon news of his son's death. According to Plutarch (around 46–120), Solon then "... began to beat his head and to do say all that is usual with men in transports of grief." But Thales took his hand and with a smile said, "These things, Solon, keep me from marriage and rearing children, which are too great for even your constancy to support, however, be not concerned at the report, for it is a fiction." Another favorite tale Plato told was that one night when Thales, while walking and stargazing, fell into a ditch, whereupon a pretty Thracian girl mocked him for trying to learn about the heavens while he could not see what was lying at his feet. The following theory of Thales' is silly: he believed that this globe of lands is sustained by water and is carried along like a boat, and on the occasions when the Earth is said to quake, it is fluctuating because of the movement of the water. It is no wonder, therefore, that there is abundant water for making the rivers flow since the entire world is floating. He introduced skepticism and criticism to Greek philosophy, which separates the Greek thinkers from those of earlier civilizations. His philosophy is called monism—the belief that everything is one. When Thales was asked what is most difficult, he said, "To know thyself." Asked what is most easy, he replied, "To give advice," and when asked what was the strangest thing he had ever seen, he answered "An aged tyrant." Aristotle records his teaching that "All things are full of Gods," and another ancient source attributes to him the statement, "The mind of the world is God and the whole is imbued with soul and full of spirits." Thales founded the Ionian School, which continued to flourish until about 400 BC. Anaximander and Anaximenes were his meritorious students. The importance of the Ionian School for philosophy and the philosophy of science is without dispute.

Anaximander of Miletus (Around 610–547 BC)

Anaximander of Miletus (around 610–547 BC) succeeded Thales to become the second master of the Ionian School. Little is known about his life and work; however, he is considered to be one of the earliest Greek thinkers of exact sciences. He claimed that nature is ruled by laws, just like human societies, and anything that disturbs the balance of nature does not last long. He tried to describe the mechanics of celestial bodies in relation to the Earth and presented a system where the celestial bodies turned at different distances. He regarded the Sun as a huge circular mass, 28 or 27 times as big as the Earth, with an outline similar to a fire-filled chariot wheel, on which appears a mouth in certain places and through which it exposes its fire. An eclipse occurs when the mouth from which comes the fire is closed. The Moon is also a circle, 19 times as big as the whole Earth and filled with fire like the Sun. Anaximander attributed some phenomena, such as thunder and lightning, to the intervention of elements rather than divine causes. He saw the sea as a remnant of the mass of humidity that once surrounded Earth. He explained rain as a product of the humidity pumped up from Earth by the Sun. He introduced into Greece the gnomon (for determining the solstices) and the sundial; however, its use, as well as the division of days into 12 parts, came from the Babylonians. He created a map of the known, inhabited world that contributed greatly to the advancement of geography, and probably inspired the Greek historian Hecataeus of Miletus (around 550–476 BC) to draw a more accurate version. He was also involved in the politics of Miletus as he was sent as a leader to one of its colonies. But his reputation is due mainly to his work on nature. He postulated that from the indefinite (or apeiron) comes the principle of beings, which themselves come from the heavens and the worlds. He asserted that the beginning or first principle is endless, unlimited mass, subject to neither old age nor decay that perpetually yields fresh materials for the series of beings which issue from it. He never defined this principle precisely, and it has generally been understood [e.g., by Aristotle and Saint Augustine (354–430)] as a sort of primal chaos. It embraced everything, the opposites of hot and cold, wet and dry, and directed the movement of things, by which there grew up a host of shapes and differences that are found in "all the worlds" (for he believed there were many). Anaximander explained how the four elements of ancient physics (air, Earth, water, and fire) are formed, and how Earth and terrestrial beings are formed through their interactions. He said that man himself and the animals had come into being by like transmutations. Mankind was supposed by Anaximander to have sprung from some other species of animal, probably aquatic. But as the measureless and endless had been the prime cause of motion into separate existences and individual forms according to the just award of destiny, these forms would at an appointed season suffer vengeance due to their earlier act of separation and return into the vague immensity from which they had issued. Thus the world and all definite existences contained in it will lose their independence and disappear in the "indeterminate." This perhaps presages the Heat Death of the Universe. The blazing orbs, which have drawn off from the cold Earth and water, are the temporary Gods of the world, clustering round the Earth, which to the ancient thinker is the central figure. Concerning Thales' water theory, he argued that water cannot embrace all of the opposites found in nature (for example, water can only be wet, never dry) and therefore cannot be the one primary substance, nor could any of the other candidates. For Anaximander, Gods were born, but the time is long between their birth and their death. Pythagoras was among his pupils.

Acharya Charak and Acharya Sudhrut (Before 600 BC)

Acharya Charak and Acharya Sudhrut (before 600 BC). Charak has been crowned the Father of Medicine. His *Charak Samhita* is considered to be an encyclopedia of Ayurveda, which means science of life. His principles, diagnoses, and cures have retained their potency and truth even after a couple of millennia. Charak revealed through his innate genius and inquiries the facts of human anatomy, embryology, pharmacology, blood circulation, and diseases like diabetes, tuberculosis, and heart disease. He described the medicinal qualities and functions of 100,000 herbal plants and emphasized the influence of diet and activity on mind and body. He proved the correlation of spirituality with physical health, which contributed greatly to the diagnostic and curative sciences. Sudhrut (Sushrut/Sushruta) is venerated as the father of plastic surgery and the science of anesthesia. He detailed the earliest surgical procedures in *Sushrut Samhita*, which is a unique encyclopedia of surgery. In it, he prescribed treatment for twelve types of fractures and six types of dislocations. Sushrut used 125 unique surgical instruments, mostly designed from the jaws of animals and birds, including scalpels, lancets, needles, and rectal speculums. He also described a number of stitching methods that used horse's hair and bark fibers as thread. In total he listed 300 types of operations. Sushrat Samhita mentions that the sage Sushruta (before 800 BC) performed cataract surgery. This encyclopedic text contains 184 chapters, and like Charak Samhita, describes 1,120 illnesses, 700 medicinal plants, 64 preparations from mineral sources, 57 preparations based on animal sources, and a detailed study on anatomy. In addition it places a high value on the well-being of children and expectant mothers. It goes into great detail regarding symptoms, first-aid measures, and long-term treatment, as well as classification of poisons and methods of poisoning. In this work there is also an interest in permutation and combination. Sushrut Samhita states that 63 combinations can be made out of six different rasa (tastes)—bitter, sour, salty, astringent, sweet, and hot—by taking the rasa one at a time, two at a time, three at a time, and so on. This solution of 63 combinations can be checked:

$$ {}_6C_1 + {}_6C_2 + {}_6C_3 + {}_6C_4 + {}_6C_5 + {}_6C_6 = 6 + 15 + 20 + 15 + 6 + 1 = 63. $$

Anaximenes of Miletus (Around 585–525 BC)

Anaximenes of Miletus (around 585–525 BC) was either a contemporary of Anaximander or his student. He regarded air (aer) as the origin and used the term 'air' as God. He posited that the first principle and basic form of matter was air, which could be transformed into other substances by a process of condensation and rarefaction. He stated: "Just as our soul, being air, holds us together, so do breath and air encompass the whole world." The Anaximenes crater on the Moon was named in his honor.

Pythagoras (Around 582–481 BC)

Pythagoras (around 582–481 BC) was born in the island of Samos, Greece, one of the most visited of the northeastern Aegean islands. In general people during his time were superstitious and had strong beliefs in spirits and supernatural forces/occurrences. Religious cults were popular in that era. His father Mnesarchus (an opulent merchant) and his mother Pythais wanted their son to receive the best possible education. During his childhood and boyhood days, his parents had an immense influence on his highly receptive mind. They provided the most vital building block of his multifaceted character, which has amazed people over more than two and a half millennia and continues to provoke thought among scientists, including mathematicians, regarding the brilliant character of this Greek philosopher-cum-scientist. Pythagoras learned a lot about science and philosophy from Pherecydes of Syros (about 600 BC), his first teacher. He remained in touch with him until Pherecydes' death. At the age of 18, shortly after the death of his father, Pythagoras went to Lesbos, a Greek island located in the northeastern Aegean Sea, where he worked and learned from Anaximander and Thales. Both personalities had a profound impact on the thinking process of young Pythagoras. On the recommendation of Thales, who himself visited Egypt, Pythagoras came to study with the priests at Memphis at the age of 23 and stayed there until he was 44. During these 21 years, among other things, he learned geometry. After leaving Egypt he traveled in Asia Minor, visited every place where he might learn something, and consulted with virtually every important figure. Pythagoras returned to Samos when he was about 55 and started a school there. Due to a lack of students, he moved to Croton in Southern Italy, where he started a school that was mainly involved in teaching and learning of mathematics, music, philosophy, astronomy, and their relationship with religion. Unlike in Samos, his birth place, it is said that in Croton around 600 of the most talented citizens of all ranks attended, especially those of the upper classes; even the women broke a law which forbade their going to public meetings and flocked to hear him. Among his most attentive auditors was Theano, whom he married at the age of 60. The couple had three children. Pythagoras regarded the bearing and raising of children as sacred

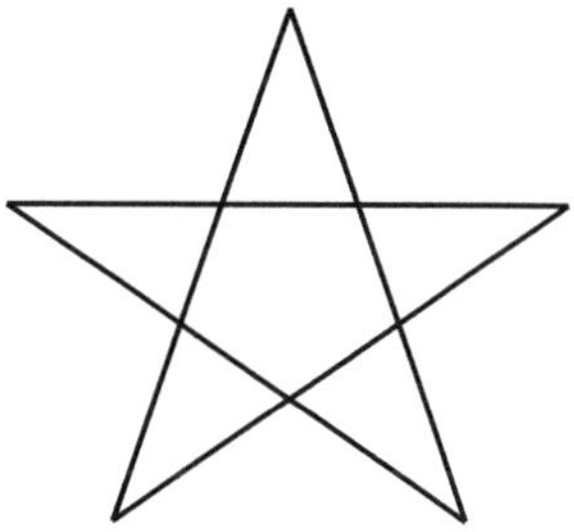

Fig. 2 Pentagram

responsibilities. He maintained that this was a duty of men and women to the Gods, so that their worship might continue into future generations. Theano wrote a biography of her husband, but unfortunately it is lost. The school in Croton reached its peak around 490 BC. Pythagoras taught the young disciples/students to respect their elders and to develop mentally through learning. He emphasized justice based on equality and encouraged calmness and gentleness. Pythagoreans were known for their mutual respect and devotion to each other. More than anybody else before him, Pythagoras taught spirituality combined with pursuit of scientific knowledge. Pythagoras was at the helm of a cult (a specific system of worship) known as the *secret brotherhood* that visualized numbers of everything and worshiped numbers and numerical relationships such as equations and inequalities. He believed that the principles of the universe and secrets of the cosmos, many of which are not perceptible to common human senses, are exposed by pure thought and reasoning through a process that can be expressed in terms of relationships involving numbers. The members of this cult, the Pythagoreans, were bound by oath not to reveal the teachings or secrets of the school. They were not allowed to eat meat or beans, which they believed was a necessary part of purification of the soul. For example, beans were taboo because they generate flatulence and are like the genitalia. The Pythagoreans believed that it is best to make love to women in winter, but not in the summer; that all disease is caused by indigestion; that one should eat raw food and drink only water; and that one must avoid wearing garments made of wool. The cult members divided the mathematical subjects that they dealt with into four disciplines: numbers absolute (arithmetic), numbers applied (music), magnitudes at rest (geometry), and magnitudes in motion (astronomy). This "quadrivium" was long considered to constitute a necessary and sufficient course of study for a liberal education. A distinctive badge of the brotherhood was the beautiful star pentagram (see Fig. 2)—a fit symbol of the mathematics that the school discovered. It was also the symbol of health.

The Pythagoreans gave 'divine significance' to most numbers up to 50. They considered even numbers feminine and related them to the evil, unlucky earthly nature; odd numbers were masculine and represented the good, lucky celestial nature. Each number was identified with some human attribute. One, the *monas*, stood for reason because it was unchangeable. It also stood for both male and female, odd and even, and was itself not a number. It represented the beginning and ending of all things, yet itself did not have a beginning or an ending, and it referred to the supreme God. Two stood for opinion and the process of creation: unity polarizing within itself becomes duality. Three, the triad, was the first true number, a principle of everything that is whole and perfect and all things with a beginning, middle, and end. It implied past, present, and future. Four represented justice because it is the first perfect square and the product of equals; four dots make a square and even today we speak of "a square deal." There are four seasons, four elements, four essential musical intervals, four kinds of planetary movements, four faculties (intelligence, reason, perception, and imagination), and the four mathematical sciences of the quadrivium (arithmetic, music, geometry, and astronomy). Five stood for marriage because it was the union of the first feminine and the first masculine numbers, $2 + 3 = 5$, and is manifested through the five essential figures, the tetrahedron, hexahedron, octahedron, dodecahedron, and icosahedron. Six, the hexad, was the first perfect number because it is a whole produced through its parts, the result of both the addition and the multiplication of its factors: $1 + 2 + 3 = 6$ and $1 \times 2 \times 3 = 6$, hence it reflects a state of health and balance. It is seen in the six extensions of geometrical forms and in the six directions of nature: north, south, east, west, up, and down. Like five it arises out of the first odd and even numbers, but by multiplication, 2×3, rather than by addition, and because of this it is associated with androgyny. The hexad is present in the arithmetic mean, $6, 9, 12$; the geometric mean, $3, 6, 12$; the harmonic mean, $3, 4, 6$; and is seen in music as $6, 8, 12$. The Pythagoreans praised the number six eulogistically, concluding that the universe is harmonized by it and from it comes wholeness, permanence, as well as perfect health. Seven, the heptad, cannot be generated by an operation on any other numbers, and hence expresses virginity and is symbolized by the virgin goddess Athena. Combined from the first triangular and the first square numbers, it is the number of the primary harmony $(3 : 4)$, the geometric proportion $(1, 2, 4)$, and the sides $(3, 4)$ around the right angle of the archetypal right-angle triangle. Since it cannot be divided by any number other than itself, it represents a fortress or acropolis (in Greek acro means top and polis stands for city-state). Seven brings order to nature. There are seven phases of the Moon and seven parts of the body: head, neck, torso, two arms, and two legs. The head has seven openings, the lyre has seven strings, and the lives of men and women have seven ages—infant, child, adolescent, young adult, adult, elder, and old person. There are seven wandering stars, or planets, from which the week and our names for the days of the week are derived. The octad, eight, was significant because it is the first cube $(2 \times 2 \times 2)$ and is thus associated with safety, steadfastness, and everything in the universe which is balanced and regulated. It is the source of all the musical ratios and is called "Embracer of Harmonies." It is also known as Eros, as it is a symbol for lasting friendship. The heavens are made up of nine spheres, the eighth of which encompasses the whole, introducing into the octad a notion of all-embracing presence. Nine was called "that which brings to fruition" because it completes the perfect 9 months before birth and is the number of the muses, particularly Terpsichore, the muse of dance and movement. The number ten, the decad, was the greatest of all: it contains in itself the first four integers—one, two, three, and four, $1 + 2 + 3 + 4 = 10$; it is the smallest integer n for which there

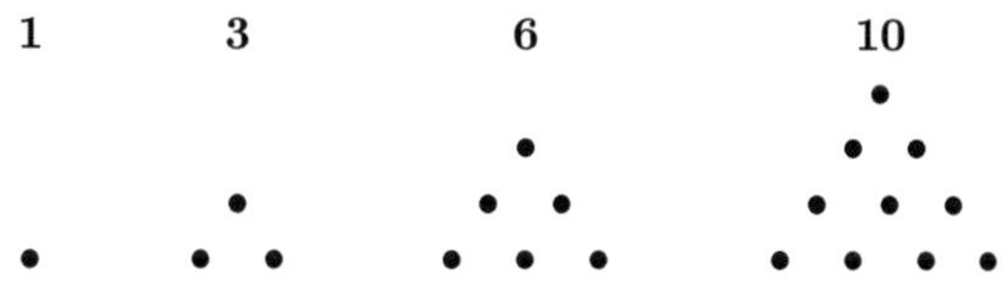

Fig. 3 Triangular numbers

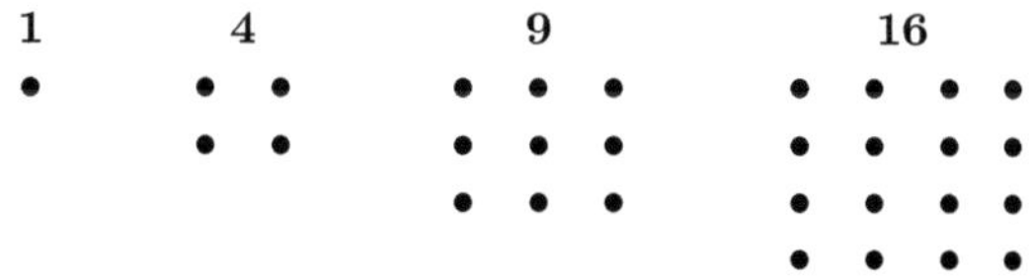

Fig. 4 Square arrays

are just as many primes between 1 and n as nonprimes, and it gives rise to the tetraktys. For the Pythagoreans, the tetraktys was the sum of the divine influences that hold the universe together or the sum of all the manifest laws of nature. They recognized tetraktys as fate, the universe, the heaven, and even God. The tetraktys was so revered by the members of the brotherhood that they shared the following oath, "I swear by him who has transmitted to our minds the holy tetraktys, the roots and source of ever-flowing nature."

The Pythagoreans pictured the integers as groups of points like constellations and called them *figurative numbers*. From such configurations one can read some remarkable number-theoretic laws. For example, Fig. 3 shows the Triangular Numbers. The rows of the triangles contain $1, 2, 3, 4, \ldots$ points, and the number of points in an n-rowed triangle is the sum of the first n positive integers, $1 + 2 + \cdots + n = n(n+1)/2$. For example, $1 + 2 = 3$, $1 + 2 + 3 = 6$, and $1+2+3+4 = 10$. In this way the Pythagoreans obtained the well-known sequence of Triangular Numbers: $1, 3, 6, 10, 15, 21, 28, \ldots$.

Even more remarkable laws can be read off from a square array (Fig. 4).
For example, $n^2 + (2n + 1) = (n + 1)^2$, and $1 + 3 + 5 + \cdots + (2n - 1) = n^2$.

Besides triangular and square numbers, Pythagoras spoke of cubic numbers, oblong numbers (having two long sides, two short sides, and four right angles), pentagonal numbers such as $1, 5, 12, 22, \ldots$, and spherical numbers. He also worked with prime numbers, perfect numbers [numbers equal to the sum of their divisors, such as 6 $(1+2+3 = 6)$], and amicable numbers like 284 and 220, where the sum of the divisors of each number equals the other number $(1 + 2 + 4 + 71 + 142 = 220$ and $1 + 2 + 4 + 5 + 10 + 11 + 20 + 22 + 44 + 55 + 110 = 284)$. According to a story, a person asked Pythagoras what a friend was, and he replied, "One who is another I, such as 220 and 284." The amicable numbers are supposed to have been known to the Hindus before the days of Pythagoras. The Pythagoreans saw them as good omens. The next pair of amicable numbers, 17,296 and 18,416, was not discovered until 1636. Today more than four hundred pairs are known, but the question of whether the number of amicable numbers is infinite still remains to be settled.

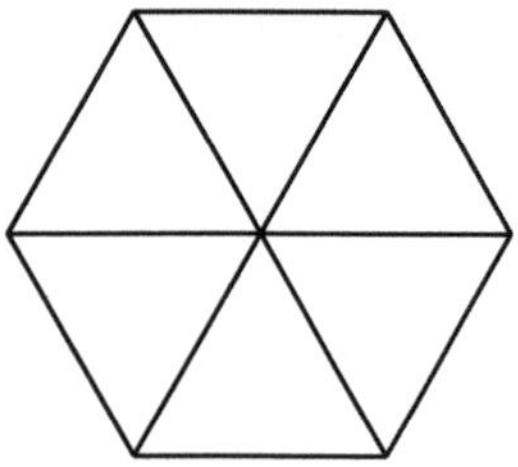

Fig. 5 Six equilateral triangles meeting at a point

The Pythagoreans related arithmetic to geometry; for example, to multiply 2×2, they constructed a square with each side equal to 2 units. The area of this square, 4, is equal to the product of its sides. The Pythagoreans noted that when we attempt to tile a floor with square tiles, we succeed because the meeting point of four right angled corners leaves no space, that is, four right angles add up to 360°. Their next observation was that six equilateral triangles (see Fig. 5) meeting at a point also leave no space.

In solid geometry there are only five regular figures whose plane faces are congruent, regular polygons. They are the tetrahedron (a pyramid with equilateral triangles as each of its four faces), the cube (a six-faced solid, each of whose faces is a square), the octahedron (with equilateral triangles as each of its eight faces), the dodecahedron (with regular pentagons as each of its twelve faces), and the icosahedron (a 20-faced solid with equilateral triangles as faces). The first three were known to the Egyptians; Pythagoras discovered the remaining two. The Pythagoreans dedicated the angles of an equilateral triangle to the Gods and the angles of a square to the Goddesses.

He probably learned the theorem of plane geometry in Egypt, the Pythagoras (or Pythagorean) theorem: For a right-angled triangle the square on the hypotenuse is equal to the sum of the squares on the other two sides, but the earliest known mention of Pythagoras' name in connection with the theorem occurred five centuries after his death, in the writings of Cicero and Plutarch. However, as we have noted, this theorem and its converse (which lead to Pythagorean triples) and their proofs were known several centuries before him. Interestingly, Denis Henrion (1580–1632), in 1615, comments: "Now it is said that this celebrated and very famous theorem was discovered by Pythagoras, who was so full of joy at his discovery that, as some say, he showed his gratitude to the Gods by sacrificing a Hecatomb of oxen. Others say he only sacrificed one ox, which is more likely than that he sacrificed a hundred, since this philosopher was very scrupulous about shedding the blood of animals." In any case, the Pythagorean theorem has probably been proven in more ways than any other theorem in mathematics. The book *Pythagorean Propositions* contains 370 different proofs of the Pythagorean theorem. Nicaragua issued a series of ten stamps commemorating mathematical formulas, including the Pythagorean theorem. The discovery of irrational numbers is also credited to Pythagoras; however as we have remarked, the irrationality of $\sqrt{2}$ and π was discussed several centuries before him.

The Pythagoreans also had a model of the universe, in which every sphere in the universe revolves around a Central Fire. From the Pythagoreans originated the doctrine of the "harmony of the spheres," a theory according to which the heavenly bodies emit constant tones, that correspond to their distances from the Earth. It was Pythagoras who first called the universe kosmos, a word whose Greek root implies both order and adornment.

It is strange that historians have misled the world by reporting that Pythagoras was a contemporary of Gautama Buddha, who actually lived during (1887–1807 BC). Some claim that Pythagoras traveled as far as India; however, he was certainly influenced by ideas from India, transmitted via Persia and Egypt. The book *India in Greece* published in 1852, in England, by the Greek historian Edward Pococke reports that Pythagoras, who taught Buddhist philosophy, was a great missionary. His name indicates his office and position; Pythagoras in English is equivalent to putha-gorus in Greek and Budha-guru in Sanskrit, which implies that he was a Buddhist spiritual leader. Specifically, Pythagoras believed that the soul is an eternal, self-moving number that passes from body to body through metempsychosis, or transmigration, and that after spiritual purification the soul will cease reincarnation and eventually unite with the Divine.

Pythagoras was able not only to recall past lives (in one he had been Euphorbus, a warrior who fought heroically in the Trojan War) but also to see into the souls of others. Once, upon seeing a dog being severely beaten, he rushed over and restrained the dog's owner, saying, "You must stop this. I know from the sound of his cries that within this animal is the soul of my late friend Abides." Another time, he identified the soul of the legendary Phrygian King Midas inhabiting the body of Myllias, a citizen of Kroton. At Pythagoras' urging, Myllias traveled to Asia to perform expiatory rites at Midas' tomb.

In the town of Tarentum he observed an ox in a pasture feeding on green beans. He advised the herdsman to tell his ox that it would be better to eat other kinds of food. The herdsman laughed, saying that he did not know the language of oxen, but if Pythagoras did, he was welcome to tell him so himself. Pythagoras approached the ox and whispered into his ear for a long time. The ox never again ate beans, and lived to a very old age near the temple of Hera in Tarentum, where he was treated as sacred.

It appears that Pythagoras was exiled from Croton and had to move to Tarentum where he stayed for about 16 years. He then moved again, to Metapontus, where he lived for four more years before he died at the age of about 99. Supposedly he was burned to death in his school, which was set on fire by the local Crotonians who regarded the weird and exclusive Pythagoreans with suspicion. Some say that Pythagoras managed to escape, only to be murdered several years later in another town. Herodotus considered him an "important sophist" (sophist means wisdom). He used to play the kithara, an ancient form of the guitar, and often sang as he played. He is reputed to have been able to soothe both animals and people with his music and should rightly be regarded as the founder of music therapy. To Pythagoras we owe the very word mathematics and its doubly twofold branches:

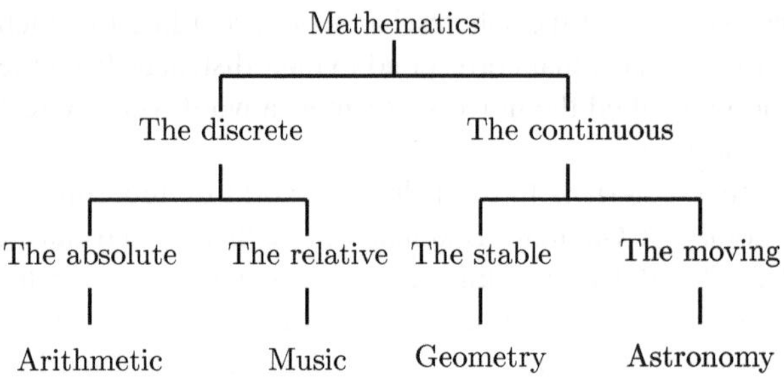

Theano of Crotona (About 546 BC)

Theano of Crotona (about 546 BC) was one of the few women in ancient mathematics and is also thought to have been a physician. Her father Pythonax of Crete (an Orphic philosopher and physician) was a great supporter of Pythagoras. She was first a student of Pythagoras and then became his wife. The couple had three children: two sons, Mnesarchus and Telauges (about 500 BC), and a daughter, Damo. She and her daughter carried on the Pythagorean School after the death of Pythagoras. She is believed to have written treatises on mathematics, physics, medicine, child psychology (derived from psyche which means mind or soul, and logos which stands for an account or discourse), marriage, sex, women, and ethics. In mathematics, she is mainly credited for her work on the Golden mean and Golden rectangle. A story is preserved that once, when asked how long it was before a woman becomes pure after intercourse, she replied, "The moment she leaves her own husband, she is pure. But she is never pure again, after she leaves any other." It is also reported that Theano advised a woman who was on her way to meet her husband to set aside her modesty with her clothes, and when she left him, to put them on again.

Pingala (About 500 BC)

Pingala (about 500 BC) was the author of the *Chhandah-shastra*, the Sanskrit book on meters, or long syllables. This Indian mathematician was from the region that is present day Kerala state in India. According to the Indian literary tradition, Pingala was the younger brother of the great grammarian Panini, the author of the famous treatise *Asthadhyayi* of the fifth century BC. Modern scholars have tended to place him two or three centuries later. Pingala presented the first known description of a binary numeral system. He described the binary numeral system in connection with the listing of Vedic meters with short and long syllables. His work also contains the basic ideas of *maatraameru* and *meruprastaara*, concepts that are inappropriately now known as Fibonacci numbers and Pascal's triangle, respectively. His book

Chhandah-shastra contains the first known use of zero, indicated by a dot (·). In this book he also explored the relationship between combinatorics and musical theory, which was later reproduced by Mersenne. Around the end of the tenth century, Halayudha (about 975) produced a commentary on Pingala's Chhandah-shastra in which he introduced a pictorial representation of different combinations of sounds, enabling them to be read off directly. Halayudha's meruprastaara (or pyramidal arrangement) provides the binomial expansion of $(a + b)^n$, where $n = 1, 2, 3, \ldots$.

Hippasus of Metapontum (About 500 BC)

Hippasus of Metapontum (about 500 BC) was a Greek Pythagorean philosopher. According to legend, Hippasus made his discovery of the existence of irrational numbers at sea and was thrown overboard by fanatic Pythagoreans because his result contradicted their doctrine: that all numbers can be expressed as the ratio of integers. This discovery of Hippasus is one of the most fundamental discoveries in the entire history of science.

Anaxagoras of Clazomanae (500–428 BC)

Anaxagoras of Clazomanae (500–428 BC) came to Athens from near Smyrna, where he taught the results of Ionian philosophy. He was a Pre-Socratic Greek philosopher, said to have neglected his possessions, which were considerable, in order to devote himself to science. In reply to the question, what was the object of being born, he remarked: "The investigation of the Sun, Moon and heaven." He was the first to explain that the Moon shines due to reflected light from the Sun, which explains the Moon's phases. He also said that the Moon had mountains and he believed that it was inhabited. Anaxagoras gave some scientific accounts of eclipses, meteors, rainbows, and the Sun, which he asserted was larger than the Peloponnesus. This opinion and various other physical phenomena that he tried to explain were supposed to have been the results of direct action by the Gods; this led him to prosecution for impiety. While in prison he wrote a treatise on the quadrature of the circle. (The general problem of squaring a figure came to be known as the *quadrature problem*). Anaxagoras is also famous for introducing the cosmological concept of Nous (mind) as an ordering force. According to him, there is no smallest among the small and no largest among the large; but always something still smaller and something still larger. The Anaxagoras crater on the Moon is named in his honor.

Zeno of Elea (Around 495–435 BC)

Zeno of Elea (around 495–435 BC) was a pre-Socratic Greek philosopher of southern Italy and a member of the Eleatic School founded by Parmenides (about 480 BC). It is only through the dialogues of Plato that we know a little about Zeno's life. According to Plutarch, Zeno attempted to kill the tyrant Demylus, and failing to do so, bit off his tongue with his own teeth and spit it on the tyrant's face. He was an original thinker and made several ingenious statements, known as Zeno's paradoxes. (The word paradox is derived from two words of Greek origin; para: meaning faulty, disordered, false, or abnormal; and doxa: meaning opinion. Thus a paradox is a false or faulty opinion.) We discuss here his famous four paradoxes (preserved by Aristotle), which challenged the accepted notions of space and time and the relation of the discrete to the continuous. Zeno encountered these concepts in various philosophical circles of his time, particularly among the Pythagoreans. These paradoxes arose because he was attempting to rationally understand the notions of infinity for the first time.

1. Achilles, the mighty and fast, cannot overtake the tortoise, the slowest creature on the Earth, if there is race between the two.
2. Two horses cannot run towards each other.
3. It is impossible to cross a field. It takes infinite time to accomplish the task.
4. An arrow in motion is actually at rest.

Achilles and the Tortoise. Suppose there is a race between Achilles and a tortoise. Achilles, the hero of Iliad, being several times faster than the tortoise, offers a small concession to the poor creature. Achilles stands a few steps behind the tortoise when the race starts (Fig. 6). Zeno claims that because of this, Achilles cannot win the race. To understand this suppose that Achilles is at point A and the tortoise is at point T, when the race starts. If Achilles is to overtake the tortoise, then he first has to reach the tortoise. For this to happen, both Achilles and the tortoise will have to be at the same point, say P, at the same time. In other words, Achilles will have to cover the length AP in the same amount of time that the tortoise requires to cover the length TP.

Since the line segment TP is a part of the line segment AP, and since the whole is always greater than its part (one of Euclid's self-evident truths), the segment AP has more points in it than the segment TP? Thus Achilles can win the race, only if he crosses more points than the tortoise in the same time interval. Is this possible? Zeno's argument is an emphatic "no," based on the following "self-evident" facts.

I. An instant is the last indestructible part of time. There is no fractional instant.
II. A point is the last indestructible part of a line. There is no fraction of a point.

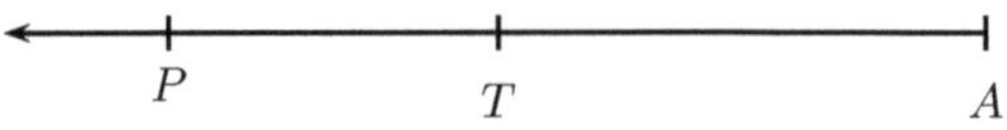

Fig. 6 Race between Achilles and a tortoise

If Achilles is faster than the tortoise; he will cover more points per instant than the tortoise does. Does he cover (say) two points while the tortoise covers just one, in an instant? If so, then the next question arises: how much time does he take to cross a point? Half an instant? Since a half-instant does not exist, Achilles must cover only a point in an instant. Similarly, the assumption that the tortoise, being slow, may take two instants to cross a point is absurd (the word absurd comes from the Latin absurdus meaning unmelodious or discordant). Thus both Achilles and the tortoise cover just a point per instant. Clearly, Achilles cannot win over the tortoise, as he is required to cover "more points" in the same "amount" of instants.

Horses Cannot Meet. If two horses approach each other, crossing a point per instant, then effectively they are crossing two points per instant. As examined above, this is impossible. Hence, the two horses running toward each other cannot meet.

No One Can Cross a Field. In order to cross a field, you will first have to cross half of it. But in order to cross the half, you will have to cross half of the half, i.e., $1/4$ of the distance. Repeating the argument, you will have to cover infinitely many distances $1/2, 1/4, 1/8, \ldots, 1/2^n, \ldots$. To cover any positive distance, you will need some positive amount of time, however small. Hence to cover infinitely many distances you will need infinite time. Clearly this is not possible for a mortal.

An Arrow in Motion is at Rest. Suppose an arrow leaves a bow and hits the target at 300 m in 10 s. If we assume (for simplicity) that the velocity is uniform, it is obvious that it will be at the distance of 150 m at the end of 5 s. More generally, given an instant of time, we will be able to state precisely the point at which the arrow will be found. Clearly, if we are able to locate the point precisely at each instant of time, the arrow must be static at that point and that instant. Thus, an arrow in flight is actually at rest.

Zeno's paradoxes confounded mathematicians for centuries, and it wasn't until Cantor's development (in the 1860s and 1870s) of the theory of infinite sets that these paradoxes could be fully resolved. Aristotle called him the inventor of the dialectic. Bertrand Russell described Zeno's paradoxes as "immeasurably subtle and profound." Around 435 BC he engaged in political activity in his native city, Elea, and he was put to death for plotting against the tyrant Nearchus. However, before dying, Zeno sank his teeth into Nearchus's ear.

Protagoras of Abdera (Around 484–414 BC)

Protagoras of Abdera (around 484–414 BC) was born in Abdera, the native city of Democritus. He was a pre-Socratic Greek philosopher and one of the best-known and most successful teachers of the Sophistic movement (the Sophist School started around 480 BC). He traveled throughout the Greek world and was a frequent visitor to Athens. He was a friend of Pericles and was said to have aided in framing the constitution for the colony of Thurii, which the Athenians established in southern

Italy in around 444 BC. Protagoras was a renowned teacher who addressed subjects related to virtue and political life. He was especially involved in the question of whether virtue could be taught. Plato said that Protagoras spent 40 years teaching and that he died at the age of 70. His most famous saying is: "Man is the measure of all things: of things which are, that they are, and of things which are not, that they are not." Protagoras was also a famous proponent of agnosticism. In his lost work, *On the Gods*, he wrote, "Concerning the Gods, I have no means of knowing whether they exist or not or of what sort they may be, because of the obscurity of the subject, and the brevity of human life." The main contribution of Protagoras was his method of finding the better of two arguments by discarding the less viable one. This is known as *Antilogies* and consists of two premises: the first is "Before any uncertainty two opposite theses can validly be confronted" and the second is its complement, the need to "strengthen the weaker argument." It seems all stories, such as an indictment against Protagoras by the Athenians, the burning of his books, and his death at sea are probably all fictitious. The Protagoras crater on the Moon was named in his honor.

Antiphon of Rhamnos (Around 480–411 BC)

Antiphon of Rhamnos (around 480–411 BC) was a sophist who attempted to find the area of a circle by considering it to be the limit of an inscribed regular polygon with an infinite number of sides.

Philolaus of Croton (Around 480–405 BC)

Philolaus of Croton (around 480–405 BC) was born in either Croton, Tarentum, or Heraclea. Philolaus was senior to Socrates (around 469–399) and Democritus, and a contemporary of Zeno of Elea, Melissus, and Thucydides. It is almost certain that he was in Croton during the persecution of the Pythagoreans. He was an immediate pupil and transcriber of Pythagoras, and after the death of his teacher great dissensions prevailed in the cities of lower Italy, where he taught Archytas. Philolaus wrote two books, *Bacchae* and *On Setting*, in Doric Greek. Only a few fragments from these books have survived. Plato used these works to compose his *Timaeus*. He believed that the universe is composed of four elements, both bounded and infinitely immense. According to Philolaus the world is one and was created from the Central Fire, which is equidistant from top and bottom of the universe. He supposed that a sphere of the fixed stars, the five planets, the Sun, Moon, and Earth, all moved round this Central Fire. To make the total a perfect ten, he added a tenth revolving body, called the Counter-Earth. He supposed the Sun to be a disk of glass that reflects the light of the universe. He made the lunar month consist of $29\frac{1}{2}$ days, the lunar year of 354 days, and the solar year of $365\frac{1}{2}$ days. Philolaus systematized Pythagoras' number theory. He stressed the importance of numerical groupings and the divine properties of numbers. According to him, "All things which can be known have number; for it is not possible that without number anything can be either conceived or known." Philolaus regarded the soul as a "mixture and harmony" of the bodily parts; he assumed a substantial soul, whose existence in the body is an exile on account of sin.

Hippocrates of Chios (About 470 BC)

Hippocrates of Chios (about 470 BC) began life as a merchant. In about 430 BC he came to Athens from Chios to open a school of geometry and began teaching, thus becoming one of the few individuals ever to enter the teaching profession for its financial rewards. He seems to have been well acquainted with Pythagorean philosophy. He wrote the first elementary textbook of geometry, which probably provided a foundation for Euclid's *Elements*. In this book, Hippocrates introduced the method of "reducing" one theorem to another, which, being proved, provides the solution to the original proposition; the *reductio ad absurdum* is an illustration of this method. He elaborated the geometry of the circle: proving, among other propositions, that similar segments of a circle contain equal angles; that the angle subtended by the chord of a circle is greater than, equal to, or less than a right angle as the segment of the circle containing it is less than, equal to, or greater than a semicircle. He also established the proposition that (similar) circles are to one another as the squares of their diameters, and that similar segments are as the square of their chord. This is equivalent to the discovery of the formula πr^2 for the area of a circle in terms of its radius. It means that a certain number π exists and is the same for all circles, although his method does not give the actual numerical value of π. It is thought that he reached these conclusions by regarding a circle as the limiting form of a regular polygon, either inscribed or circumscribed. In trying to square the circle, Hippocrates discovered that two Moon-shaped figures (lunes, bounded by pair of circular arcs) could be drawn whose areas were together equal to that of a right-angled triangle. Figure 7, with its three semicircles described on the respective sides of the triangle, illustrates this theorem. Hippocrates gave the first example of constructing a rectilinear (figures are called rectilinear if their sides, although numerous and meeting at all sorts of strange angles, are merely straight lines) area

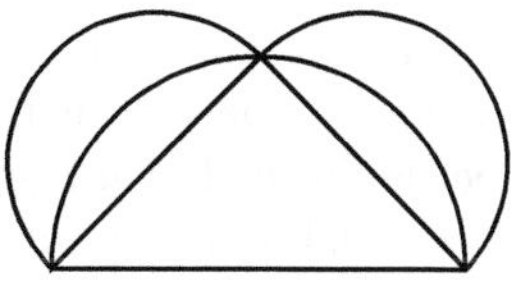

Fig. 7 Three semicircles on their respective sides of the triangle

equal to an area bounded by one or more curves *quadrature*. The culmination of attempts of this kind was the invention of the integral calculus by Archimedes.

In the twentieth century N.G. Tschebatorew and A.W. Dorodnow proved that there are only five types of squarable lunes, three of which were known to Hippocrates and the other two to Euler in 1771. Hippocrates was also able to duplicate the cube by finding two mean proportionals (Take $a = 1$ and $b = 2$ in $a : x = x : y = y : b$. Solve for x to get $x = \sqrt[3]{2}$.)

Hippias of Elis (About 460 BC)

Hippias of Elis (about 460 BC) was a Greek Sophist, a younger contemporary of Socrates. He is described as an expert arithmetician, but he is best known to us for his invention of a curve called the quadratrix, by means of which an angle can be trisected, or indeed divided in any given ratio. Hippias devised an instrument to construct this curve mechanically, but constructions that involved the use of any mathematical instruments except a ruler and a pair of compasses were objected to by Plato, and rejected by most geometricians of a subsequent date. (In the *Epic of Gilgamesh*, the narrator states that Gilgamesh was two-thirds God and one-third man. This is as impossible as trisecting an angle with a ruler and compasses—unless Gods and mortals are allowed to have an infinite amount of sex.) He lectured widely on mathematics and as well on poetry, grammar, history, politics, archaeology, and astronomy. Hippias was also a prolific writer, producing elegies, tragedies, and technical treatises in prose. His work on Homer was considered excellent. Nothing of his remains except a few fragments. Hippias has been pictured as an arrogant, boastful buffoon.

Democritus of Abdera (Around 460–362 BC)

Democritus of Abdera (around 460–362 BC) was born in Abdera in the north of Greece. We know very little about his life, except that his father was very wealthy, he traveled to Egypt, Persia, Babylon, India, Ethiopia, and throughout Greece, and Leucippus (about 440 BC, the founder of the atomism) was his teacher. Plato makes no reference to him, while Aristotle viewed him as a pre-Socratic and credited him with originating the theory of atomism. Democritus believed that space and the void had an equal right with reality to be existent. He considered the void a vacuum, an infinite space in which moved an infinite number of atoms (atoma). These atoms are eternal, invisible, and absolutely small, so small that their size cannot be diminished. They are absolutely full and incompressible, as they are without pores and fill the space they occupy entirely. They are homogeneous, differing only in shape, arrangement, position, and magnitude. Some matter is solid because its atoms have hooks that attach to each other; some are oily because they are made of very fine, small atoms which can easily slip past each other. In his own words, "By convention sweet, by convention bitter, by convention hot, by convention cold, by convention color: but in reality atoms and void." However, it is impossible to say which of these ideas are due to Democritus and which are from Leucippus. Assuming this as a basis for the physical world, which had always existed and would always exist, he explained that all changes in the world are due to changes in motion of the atoms, or changes in the way that they are packed together. He believed that nature behaves like a machine—a highly complex mechanism. There was no place in his theory for divine intervention. He wanted to remove the belief in Gods, which he believed were only introduced to explain phenomena for which no scientific explanation was then available. Democritus believed that the soul will either be disturbed, so that its motion affects the body in a violent way, or it will be at rest, in which case it regulates thoughts and actions harmoniously. Freedom from disturbance is the condition that causes human happiness, and this is the ethical goal. There are two forms of knowledge: legitimate and bastard. The bastard knowledge is concerned with perception through the senses: sight, hearing, smell, taste, and touch. The legitimate knowledge can be achieved through the intellect. According to Democritus, the knowledge of truth is difficult because perception through the senses is subjective. He proposed that the universe contains many worlds, some of them inhabited. Democritus extracted the essence of herbs and devoted his later life to researches into the properties of minerals and plants. He wrote almost 70 books on subjects such as ethics, physics, mathematics, music, literature, logic, and language. In mathematics, he wrote on numbers, geometry, tangencies, mappings, and irrationals. Democritus was the first to state that the volume of a cone is one-third of that of a cylinder having the same base and equal height, and that the volume of a pyramid is one-third of that of a prism having the same base and equal height. He enunciated these propositions 50 years before they were scientifically proved by Eudoxus. To reach his conclusions, Democritus thought of these solids as being built up of innumerable parallel layers. This foreshadows the great constructive work of Archimedes and, centuries later, that of Cavalieri and Newton. Plato is blamed for using his influence with the Romans to have all of Democritus' books burned for being "ugly and demeaning." When the early Christians burned down the library of Alexandria, the jewel of Egypt, all the remaining copies of his books were lost forever.

Democritus was highly esteemed by his fellow citizens because he foretold certain incidents that later proved to be true. There are many sayings associated with him, such as moderation increases enjoyment and makes pleasure even greater; the brave man is he who overcomes not only his enemies but also his pleasures; there are some men who are masters of cities but slaves to women; he who chooses the advantages of the soul chooses things more divine, but he who chooses those of the body, chooses things human. He referred to Egyptian mathematicians as 'rope stretchers' and used to say that "he prefers to discover a causality rather than become a King of Persia."

Bryson of Heraclea (About 450 BC)

Bryson of Heraclea (about 450 BC) was a student of Socrates. Bryson considered the circle squaring problem by comparing the circle to polygons inscribed within it.

Theodorus of Cyrene (Now Shabhat, Libya) (About 431 BC)

Theodorus of Cyrene (now Shabhat, Libya) (about 431 BC) was a Greek mathematician. He is said to have been Plato's teacher. Theodorus drew some figures to discover many irrationals, $\sqrt{2}, \sqrt{3}, \sqrt{5}, \sqrt{6}, \sqrt{7}, \sqrt{8}, \sqrt{10}, \sqrt{11}, \sqrt{12}, \sqrt{13}, \sqrt{14}, \sqrt{15}, \sqrt{17}$, 'at which point,' says Plato, "for some reason he stopped."

Archytas of Tarentum (Around 428–347 BC)

Archytas of Tarentum (around 428–347 BC) was an ancient Greek philosopher, mathematician, astronomer, statesman, and strategist. He was one of the first to write about mechanics, and he is said to have been very skilled in making toys and models, including a wooden dove that could fly, and a rattle which, as Aristotle says, "was useful to give to children to occupy them from breaking things about the house (for the young are incapable of keeping still)." Archytas lived at Tarentum in Southern Italy. He found time to take a considerable part in the public life of his city and is known for his enlightened attitude in his treatment of slaves and in the education of children. He was a Pythagorean, and was in touch with the philosophers of Athens, numbering Plato among his friends. On one occasion, he is said to have used his influence in high quarters to save the life of Plato. In music, Archytas gave the numerical ratios for the intervals of the tetrachord on three scales, the enharmonic, the chromatic, and the diatonic. He held that sound was due to impact and that higher tones correspond to quicker and lower tones to slower motion communicated to the air. Archytas introduced the concept of a harmonic mean, and according to Eutocius of Ascalon (around 480–540), solved the problem of doubling the cube with a geometric construction. The curve that he used in his solution is named after him. The crater Archytas on the Moon is named in his honor.

Plato of Athens (Around 427–347 BC)

Plato of Athens (around 427–347 BC) was one of the greatest Greek philosophers, a mechanician, a pupil of Socrates for 8 years, and Aristotle's teacher. Plato's teacher Socrates was not a mathematician; in fact, his metaphysical (the prefix meta means beyond and inclusive of all below) doubts have been recorded as "I cannot satisfy myself that, when one is added to one, the one to which the addition is made becomes two, or that the two units added together make two by reason of the addition. I cannot understand how when separated from the other, each of them was one and not two, and now, when they are brought together, the mere juxtaposition or meeting of them should be the cause of their becoming two." Socrates secured his firm belief in logical structure, employing classes with certain properties. Once we establish that something is a member of a class, we may assume that it has all of the properties of that class. After the execution of his teacher in 399 BC, Plato left Athens and spent some years traveling. During this time he studied mathematics. He visited Egypt with Eudoxus. Then Plato went to Cyrene, where he studied under Theodorus. Next he moved to Italy, where he became familiar with Archytas, then the head of the Pythagorean school, Eurytas of Metapontum (around 400 BC), and Timaeus of Locri (around 420–380 BC). He returned to Athens about the year 385 BC and formed a school for mathematics and other subjects related to philosophy. Because the school started its meetings in a park devoted to the local hero Academus, the school itself came to be known as "Plato's Academy," a name that came to be applied to all similar institutions. Later his academy was much like a modern university. There were grounds, buildings, students, and formal courses were taught by Plato and his aides. In less than 20 years his school became so famous that scholars and students flocked to it from every part of the Greek world. Within the mathematics "curriculum," disciplines included arithmetic, geometry, astronomy, and harmony (which survived to the Middle Ages as the quadrivium). The academy survived repeated invasions, outlived tyrant after tyrant, and saw two great civilizations fall—the Greek and the Roman—before its doors were finally closed in 529 AD by the Christian emperor Justinian (483–565) because it taught "pagan and perverse learning." In the first centuries of the Christian era, mathematics did not flourish. It was suspect because of its close connection with heathen philosophy. Many even considered it the work of the devil, since soothsayers and astrologers often called themselves mathematicians. "Let no man ignorant of mathematics enter here" is supposed to have been inscribed over the doors of the Academy. This motto, however, may have been originated by a twelfth-century monk who, generously and without regard for fact, first attributed it to Plato. There are no previous records that connect this phrase with the Academy. Another quotation ascribed to Plato is: "He is unworthy of the name of man who does not know that the diagonal of a square is incommensurable with its side."

Plato was not a mathematician, but was a strong advocate for all of mathematics. According to him the scientific aspect of mathematics, that is, its relation to the external world of reality, was of little importance in comparison with the study of mathematics for its own sake. ... mathematicians are really seeking to behold the things themselves, which can be seen only with the eye of the mind. In the construction of curves, he objected to the use of any instruments other than rulers and compasses. Plato affirmed the deductive organization of knowledge and was first to systematize the rules of rigorous demonstration. The Platonists (who believed that numbers are abstract, necessarily existing objects, independent of the human mind) are credited with discovery of two methods of proof, the method of *analysis*, and the method of *synthesis*. In analysis, one begins with a proposition to be proven, and deduces from the proposition other statements, until one arrives at a statement known to be valid. Synthesis is the procedure of reversing the steps, so the proposition may be proven. In connection with seismology he advocated that there are six types of motions; up and down, right and left, forward and back. Also, the Earth cannot possibly move in any of these directions, for on every side it is the lowest of all things in the world and therefore rests unmovable, having no cause for inclination more to one part than another. But some parts of the Earth, because of their lack of density, do jog and shake. Plato believed that the perfect ideals of physical objects are the reality. The world of ideals and relationships among them is permanent, ageless, incorruptible, and universal. Plato considered the nature of the four elements thought to compose the world: fire, air, water, and Earth. It is clear, said Plato, that these four elements are bodies and that all bodies are solids. The universe could have been created only out of perfect solids: tetrahedron, cube, octahedron, dodecahedron, and icosahedron. He thus advocated that fire is the shape of tetrahedron, Earth is the shape of cube, water is the shape of icosahedron, air is the shape of octahedron, and dodecahedron symbolizes the Universe. According to Proclus, 'Plato caused mathematics in general and geometry in particular to make a very great advance, by reason of his enthusiasm for them, which of course is obvious from the way he filled his books with mathematical illustrations, and everywhere tries to kindle admiration for these subjects, in those who make a pursuit of philosophy.' It is related that to the question, What does God do? Plato replied, "God always geometrizes." The knowledge at which geometry aims is the knowledge of the eternal.

Among Plato's pupils were the mathematicians Leodamas (about 380 BC), Neocleides, and Amyclas. Their school also included Leon (about 375 BC) and Theudius (about 350 BC), both of whom wrote textbooks on plane geometry, as well as Cyzicenus, Thasus, Hermotimus, Philippus, and Theaetetus (around 417–369 BC) who developed the theory of incommensurable magnitudes, studied the regular polyhedra, and worked on the theory of proportions, before dying from wounds in a battle in 369 BC. Plato died in 347 BC. He advocated that true knowledge could be acquired only through philosophical contemplation of abstract ideas and not through observation of the accidental and imperfect things in the real world.

Eudoxus of Cnidus (Around 400–347 BC)

Eudoxus of Cnidus (around 400–347 BC) was the most celebrated mathematician. Like Plato, he also went to Tarentum and studied under Archytas, then head of the Pythagoreans. He then traveled with Plato to Egypt and settled at Cyzicus, where he founded his school known as "The School of Eudoxus." Finally, he and his pupils moved to Athens, but he returned to Cnidus shortly before his death. He died while on a journey to Egypt. Eudoxus developed the theory of proportion, partly to place the doctrine of incommensurables (irrationals) upon a thoroughly sound basis. This task was done so well that it still continues, fresh as ever, after the great arithmetical reconstructions of Dedekind and Weierstrass during the nineteenth century. Specifically, he showed that the area of a circle is proportional to its diameter squared. He produced many theorems in plane geometry and furthered the logical organization of proof. Euclid documented most of this work in his fifth book. Eudoxus established the *method of exhaustions* fully, which depends on the proposition that *if from the greater of two unequal magnitudes there be taken more than its half, and from the remainder more than its half, and so on, there will at length remain a magnitude less than the least of the proposed magnitudes.* With the help of this proposition, the ancient geometers were able to avoid the use of infinitesimals. (Infinitesimal means something distinguishable from zero, yet which is exceedingly small—so minute indeed that no multiple of it can be made into a finite size). Eudoxus considered certain curves other than the circle. He explained the apparent motions of the planets as seen from the Earth. He also wrote a treatise on practical astronomy, in which he supposed a number of moving spheres to which the Sun, Moon, and stars were attached, and which by their rotation produced the effects observed. In all he required 27 spheres. Among the pupils of Eudoxus are Menaechmus, his brother Dinostratus (around 390–320 BC, who applied the quadratrix to the duplication of a cube and trisection of an angle problems), and Aristaeus of Croton (about 400 BC). The problem of duplication of a cube, that is, to find the edge of a cube whose volume is double that of a given cube, in the literature, is known as the Delian problem. According to a legend, Delians had consulted Plato on the subject. In one form of the story, which is related by Philoponus (about 490), it is asserted that in 430 BC the Athenians, when suffering from the plague of eruptive typhoid fever, consulted the oracle at Delos as to how they could stop it. Apollo replied that they must double the size of his altar which was in the form of a cube. To the unlearned supplicants nothing seemed more easy, and a new altar was constructed either having each of its edges double that of the old one (from which it follows that the volume was increased eightfold) or by placing a similar cubic altar next to the old one. Whereupon, according to the legend, the indignant God made the pestilence worse than before, and informed a fresh deputation that it was useless to trifle with him, as his new altar must be a cube and have a volume exactly double that of his old one. Suspecting a mystery, the Athenians applied to Plato, who referred them to the geometricians, and especially to Euclid, who had made a special study of the problem. The introduction of the names of Plato and Euclid is an obvious anachronism. Eratosthenes gives a somewhat similar account of its origin, but with King Minos as the propounder of the problem. In his version, King Minos was not satisfied with the size of the tomb built for his son and ordered that the tomb be doubled in size.

Xenocrates of Chalcedon (Around 396–314 BC)

Xenocrates of Chalcedon (around 396–314 BC) was the leader of the Platonic Academy during 339–314 BC. In mathematics, he is known for his book *Theory of Numbers* and several books on geometry. He used combinatorics to calculate the total number of syllables (1, 002, 000, 000, 000) that could be made from the letters of the alphabet. Xenocrates supported the idea of "indivisible lines" (and magnitudes) in order to counter Zeno's paradoxes. He was also a great philosopher.

Aristotle (Around 384–322 BC)

Aristotle (around 384–322 BC) was born in Stargirus in northern Greece. His father was the personal physician of the King of Macedonia. Because his father died when Aristotle was young, he could not follow the custom of following his father's profession. Aristotle became an orphan at a young age when his mother also died. His guardian, who raised him, taught him poetry, rhetoric, and Greek. At the age of 17, his guardian sent him to Athens to further his education. Aristotle joined Plato's Academy where for 20 years he attended Plato's lectures and later presented his own lectures on rhetoric. When Plato died in 347 BC (about), Aristotle was not chosen to succeed him because his views differed too much from Plato's. Instead, Aristotle joined the court of King Hermeas where he remained for 3 years and married the niece of the King. When the Persians defeated Hermeas, Aristotle moved to Mytilene and at the invitation of King Philip of Macedonia (382–336 BC), he tutored Alexander, Philip's son, who later became Alexander the Great (356–323 BC). Aristotle tutored Alexander for 5 years. After the death of King Philip, he returned to Athens and set up his own school, called the Lyceum. The school had a garden, a lecture room, and an altar to the Muses. Aristotle's followers were called the peripatetics, which means "to walk about," because Aristotle often walked around as he discussed philosophical questions. Aristotle taught at the Lyceum for 13 years where he lectured to his advanced students in the morning and gave popular lectures to a broad audience in the evening. When Alexander the Great died in 323 BC (about), a backlash against anything related to Alexander led to trumped-up charges of impiety against Aristotle. He fled to Chalcis to avoid prosecution. He only lived 1 year in Chalcis, dying of a stomach ailment in 322 BC (about). Aristotle wrote three types of work: those written for a popular audience, compilations of scientific facts, and systematic treatises. The systematic treatises included works on logic, philosophy, psychology, physics, and natural history. It was not until later in the twelfth century that Aristotle's original books on nature began to become available in Latin. His works on logic had been known earlier, and Aristotle was generally recognized as "the master of logic." But during the course of the twelfth century, Aristotle was transformed into the "master of those who know," and in particular a master of natural philosophy. He regarded logic as an independent subject that should precede science and mathematics. He contributed little to mathematics; however, his views on the nature of mathematics and its relations to the physical world were highly influential. According to him, mathematics originated because the priestly class in Egypt had the leisure needed for its study. He said, "mathematical sciences particularly exhibit order, symmetry, and limitation; and these are the greatest forms of the beautiful." Aristotle regarded the notion of definition as a significant aspect of argument. He made the distinction between potential infinity and actual infinity, and stated that only the former actually exists, in all regards. He wrote that a line segment is infinitely divisible, disagreeing with the views of pre-Socratics like Zeno and Democritus who believed that everything has an ultimate building block, or atom, which is indivisible. He maintained that "Truth is a remarkable thing. We cannot miss knowing some of it. But we cannot know it entirely." The supporters of Aristotle, called *scholastics*, held that the world is motionless. Aristotle proved the existence of God by considering the Earth stationary among the motion of a finite number of heavenly spheres. Aristotle observed that bugs appeared in spoiling meat and reasoned that life arose spontaneously from nonliving matter. To refute Aristotle's claim of spontaneous generation of life, Louis Pasteur (1822–1895) conducted an experiment in 1862 by isolating some meat broth in a sterile flask in order to demonstrate that the bugs Aristotle had observed grew from microscopic life forms too small to be seen.

Menaechmus of Proconnesus (or Alopeconnesus) (Around 375–325 BC)

Menaechmus of Proconnesus (or Alopeconnesus) (around 375–325 BC) succeeded Eudoxus as head of the school at Cyzicus, where he acquired a great reputation as a teacher of geometry and was (perhaps) for that reason appointed one of the tutors of Alexander the Great. In answer to his pupil's request to make his proofs shorter, Menaechmus made the well-known reply that "though in the country there are private and even royal roads, yet in geometry there is only one road for all." However, this quote is first attributed to Stobaeus, about 500 AD, and so whether Menaechmus really taught Alexander is uncertain. Menaechmus was the first to discuss the conic sections, which were long called the Menaechmian triads. He divided them into three classes, and investigated their properties, not by taking different plane sections of a fixed cone, but by keeping his plane fixed and cutting it by different cones. He showed that the section of a right cone by a plane perpendicular to a generator is an ellipse (application with deficiency), if the cone is acute angled; a parabola (exact application), if it is right angled; and a hyperbola (application with excess), if it is obtuse angled. He gave a mechanical construction for curves of each class. He acquainted himself with the fundamental properties of these curves. He also showed how these curves could be used in either of the two following ways to give a solution to the problem of duplication of a cube. In the first, he pointed out that two parabolas having a common vertex with their axes at right angles, such that the latus rectum of the one is double that of the other, will intersect at another point whose abscissa (or ordinate) will give a solution; for (using analysis) if the equations of the parabolas be $y^2 = 2ax$ and $x^2 = ay$, they intersect in a point whose abscissa is given by $x^3 = 2a^3$. It is probable that this method was suggested by the form in which Hippocrates had cast the problem; namely, to find x and y so that $a : x = x : y = y : 2a$, which implies $x^2 = ay$ and $y^2 = 2ax$. The second solution was given by Menaechmus as follows: Describe a parabola of latus rectum ℓ. Next describe a rectangular hyperbola, the length of whose real axis is 4ℓ, and having for its asymptotes the tangent at the vertex of the parabola and the axis of the parabola. Then the ordinate and the abscissa of the point of intersection of these curves are the mean proportionals between ℓ and 2ℓ. This is at once obvious by analysis. The curves are $x^2 = \ell y$ and $xy = 2\ell^2$. These intersect at a point determined by $x^3 = 2\ell^3$ and $y^3 = 4\ell^3$. Hence $\ell : x = x : y = y : 2\ell$. According to Proclus, Menaechmus was one of those who "made the whole of geometry more perfect." It is believed that he died in Cyzicus.

Aristaeus the Elder (Around 370–300 BC)

Aristaeus the Elder (around 370–300 BC) was a Greek mathematician and a contemporary of Euclid, though he was probably older. It is only from the work of Pappus that we know about Aristaeus and his work. He wrote on the five regular solids and on conic sections.

Callippus of Cyzicus (Around 370–300 BC)

Callippus of Cyzicus (around 370–300 BC) studied under Eudoxus and Aristotle. He found that Eudoxus' scheme of 27 spheres was insufficient to account for the planetary movements, and so he added seven more. He made measurements of the lengths of the seasons, finding them (starting with the spring equinox) to be 94 days, 92 days, 89 days, and 90 days. Callippus also measured the length of the year and constructed an accurate lunisolar calendar. The crater Calippus on the Moon is named after him.

Dicaearchus of Messana (Sicily) (Around 350–285 BC)

Dicaearchus of Messana (Sicily) (around 350–285 BC) spent the greater part of his life in Peloponnesus, Greece. He was a philosopher, cartographer, geographer, and mathematician. Dicaearchus was Aristotle's student in Lyceum and a friend of

Theophrastus (around 371–287 BC). Very little is known about his work, except that he made geometric constructions of a hyperbola and a parabola and worked mainly in the field of cartography, where he was among the first to use geographic coordinates.

Euclid of Alexandria (Around 325–265 BC)

Euclid of Alexandria (around 325–265 BC) taught and died at Alexandria in Egypt during the reign (323–283 BC) of Ptolemy I. We know very little about his life and character. However, it is believed that he was the son of Naucrates and the grandson of Zenarchus or of Berenice, that he was of Greek descent, and that he lived in Damascus although he had been born in Tyre. He may have received mathematical training from the students of Plato. In the Middle Ages, writers sometimes referred to him as *Euclid of Megara (around 435–365 BC)*, confusing him with a Greek Socratic philosopher. His masterpiece, the Elements, is divided into 13 books (each about the length of a modern chapter) and contains 465 propositions on plane and solid geometry and number theory. It is a misfortune that no copy of Euclid's Elements has been found that actually dates from the author's own time. Modern editions of the work are based on a revision that was prepared by the Greek commentator Theon of Alexandria (about 390), who lived almost 700 years after the time of Euclid. Theon's revision was, until the early nineteenth century, the oldest edition of the Elements known to us. In 1808, however, F. Peyrard found, in the Vatican library, a tenth-century copy of an edition of Euclid's Elements that predates Theon's recension. A study of this older edition indicates that the material of Euclid's original treatise undoubtedly underwent some editing in the subsequent revisions. The first Latin translations of the Elements were not made from the Greek but from the Arabic. It is known that there had been other Elements before Euclid's, and therefore Euclid's Elements is a highly successful compilation and systematic arrangement of works of earlier writers. However, Euclid is credited with being the first to describe mathematical systems from postulates and axioms. Recall that a *postulate* is something whose truth is assumed as part of the study of a science; and that an *axiom* is a principle that cannot be demonstrated, such as the notion that if equals are subtracted from equals then the remainders are equal. His geometry is systematic and coherent in its logical development. It is often said that, next to the Bible, Euclid's Elements may be the most translated, published, and studied of all the books produced in the western world. Euclid began Book I with a list of 23 definitions so the reader would know precisely what his terms meant. He introduced the *point* as "that which has no part" (one of his less illuminating definitions), an *equilateral triangle* as a triangle "which has its three sides equal"; and the *isosceles* triangle as a triangle "which has two of its sides alone equal." With these terms defined, Euclid presented five postulates (axioms) to serve as the foundations of his geometry, the starting points from which everything else would follow. These were given without proof or justification, they were simply to be accepted. Fortunately, such acceptance was not difficult because the postulates appeared to Euclid's contemporaries, and indeed to most of us today, as utterly innocuous. For example, the first three postulates are: one can draw a straight line from any point to any point; one can produce [extend] a finite straight line continuously in a straight line; and one can describe a circle with any center and distance. Books I, II, IV, and VI discuss lines, areas, and simple regular plane figures, and are mostly Pythagorean. Book III addresses circles and expounds Hippocrates; Book V elaborates the work of Eudoxus on proportion, which was essential to provide the properties of similar figures in Book VI; Books VII, VIII, and IX are arithmetical, giving an interesting account of the theory of numbers and are believed to be Pythagorean; Book X addresses the arithmetical side of the work of Eudoxus and a careful discussion of the method of exhaustions; Book XI is on elementary solid geometry; Book XII illustrates the method of exhaustions by formally proving Hippocrates' theorem for πr^2, the area of a circle; Book XIII gives and proves the constructions of the only possible five regular solids of Pythagoras and Plato. In Book I, propositions 47 and 48 provide remarkably clever proofs of the Pythagoras theorem and its converse. Book II is algebraic in substance but geometric in nature, so it can be called a treatise on geometric algebra. In addition to the Elements, four other works of Euclid have survived:

1. *Data* deals with the nature and implications of "given" information in geometrical problems; the subject matter is closely related to the first four books of the Elements.
2. *On Divisions of Figures*, which survives only partially in its Arabic translation, concerns the division of geometrical figures into two or more equal parts or into parts in given ratios. It is similar to a work by Heron of Alexandria, except Euclid's work which characteristically lacks any numerical calculations.
3. *Phaenomena* concerns the application of spherical geometry to problems of astronomy.
4. *Optics*, the earliest surviving Greek treatise on perspective, contains propositions on the apparent sizes and shapes of objects viewed from different distances and angles.

There are four other works which have been credited to Euclid but have been lost:

(i) *Conics* was a work on conic sections that was later extended by Apollonius into his famous work on the subject.
(ii) *Porisms* might have been an outgrowth of Euclid's work with conic sections, but the exact meaning of the title is controversial.
(iii) *Pseudaria*, or *Book of Fallacies*, was an elementary text about errors in reasoning.
(iv) *Surface Loci* concerned either loci (sets of points) on surfaces or loci which were themselves surfaces; under the latter interpretation, it has been hypothesized that the work might have dealt with quadric surfaces.

In the case of parallel lines Euclid assumed the *Parallel Postulate*: "If a straight line meet two straight lines, so as to make the two interior angles on the same side of it taken together less than two right angles, these straight lines, being continually produced, shall at length meet on that side on which are the angles which are less than two right angles." However, there are signs that he was not satisfied with this postulate, in the sense that he suspected it might not be necessary, and hundreds of later attempts by renowned mathematicians such as Proclus, al-Tusi, Wallis, the Italian Jesuit Girolamo Saccheri (1667–1733), Lambert, Legendre, and Eugenio Beltrami (1835–1900) to prove it or its equivalent, such as a line parallel to a given line has a constant distance from it (Proclus); there exist similar (but not equal) triangles, whose angles are equal but whose sides are unequal (Wallis); there exists at least one rectangle, a quadrangle whose angles are all right angles (Saccheri); a line perpendicular in one arm of an acute angle also intersects the other arm (Legendre); the sum of the angles of a triangle is equal to two right angles (Legendre); there exist triangles of arbitrarily large area (Gauss), and several others, all turned out to be inconclusive. Ignoring this postulate led to self-consistent non-euclidean geometry (this name is due to Gauss) in the nineteenth century. Euclid developed an algorithm for determining the *greatest common divisor* of two integers. In number theory, he succeeded in proving the existence of irrationals, especially demonstrating the irrationality of $\sqrt{2}$. In Proposition 20 of Book IX, he also devised an ingenious proof to show that there are infinite number of primes.

Euclid was careful to take nothing for granted. The Epicureans (founded around 307 BC) ridiculed him for proving so obvious a proposition as the one which states that the length of two sides of a triangle is greater than the third side. Apparently, Euclid did not stress applications. When a student asked what he would get by learning geometry, Euclid explained that knowledge was worth acquiring for its own sake and told his servant to give the student a coin "since he must make a profit from what he learns." Euclid's famous reply to King Ptolemy when asked how he could learn geometry easily—"In mathematics there are no short cuts, even for a king." He was modest and scrupulously fair, always ready to acknowledge the original work of others, and conspicuously kind and patient. The following problem that occurs in the Palatine Anthology (which contains material from 700 BC until 600 AD) has been attributed to Euclid: "A mule and a donkey were going to market laden with wheat. The mule said, 'If you gave me one measure I should carry twice as much as you, but if I gave you one we should bear equal burdens.' Tell me, learned geometrician, what were their burdens." $(7, 5)$.

Aristarchus of Samos (310–250 BC)

Aristarchus of Samos (310–250 BC) came between Euclid and Archimedes and was the first to assert that the Earth and the other planets (Venus, Mercury, Mars, Jupiter, and Saturn) revolved about the Sun, thus anticipating Copernicus' discovered by seventeen centuries. He gave important applications of mathematics to astronomy, particularly, the calculation of the ratio of the Sun's diameter to the Earth's diameter. He calculated that the distance from the Earth to the Sun is about 19 times greater than the distance from the Earth to the Moon.

Archimedes of Syracuse (287–212 BC)

Archimedes of Syracuse (287–212 BC) was a Greek mathematician and scientist. He stated that there are things which seem incredible to most men who have not studied mathematics, also, that the man who first states a theorem deserves as much credit as the man who first proves it. He was born in Syracuse, Sicily. Archimedes was the son of the astronomer Phidias and was possibly related to Hiero II (about 307–216 BC), King of Syracuse. In his youth he spent some time in Alexandria with the successors of Euclid. Most of the facts about his life come from the Roman biographer, Plutarch, who inserted a few tantalizing pages about him in the massive biography of the Roman soldier Marcellus (266–208 BC). In the words of one writer, "the account of Archimedes is slipped like a tissue-thin shaving of ham in a bull-choking sandwich." He was a man capable of intense concentration and when deep in thought was oblivious to everything around him. He would sit by the fire and draw diagrams in the ashes. After his bath, when he oiled his skin (as was the custom), he would become absorbed in working out problems, using his oiled skin as a slate and his fingernail as a stylus. When concentrating on a problem he thought of nothing else, forgetting to dress himself, to eat or sleep, and had to be carried by absolute violence to the table or tub.

Archimedes ranks with Newton and Gauss as one of the three greatest mathematicians who ever lived and is certainly the greatest mathematician of antiquity. His mathematical work is so modern in spirit and technique that it is barely distinguishable from that of a seventeenth-century mathematician. Among other mathematical achievements, Archimedes developed a general *method of exhaustion* for finding areas and volumes. The central idea of his method can be described briefly as follows: Given a region whose area is to be determined, we inscribe in it a polygonal region that approximates the given region and whose area we can easily compute. Then we choose another polygonal region that gives a better approximation, and we continue the process, taking polygons with more and more sides in an attempt to exhaust the given region. He used this method to find areas bounded by parabolas, spirals, volumes of cylinders, segments of spheres, and especially to approximate π and bounding its value between $22/7$ and $223/71$. It is almost certain that he was familiar with continued fractions. In spite of the limitations of the Greek numbering system, he devised methods for finding square roots. In *The Sand Reckoner*, Archimedes set out to calculate the number of grains of sand that the universe could contain. By doing so, he challenged the notion that the number of grains of sand was too large to be counted. He wrote: "There are some, King Gelon (around 230 BC), who think that the number of the sand is infinite in multitude: and I mean by the sand not only that which exists about Syracuse and the rest of Sicily but also that which is found in every region whether inhabited or uninhabited. And again, there are some who, without regarding it as infinite, yet think that no number has been named which is great enough to exceed its multitude." To solve the problem, Archimedes invented a method based on the Greek myriad (10,000) for representing numbers as large as 1 followed by 80 million billion zeros. He set the *Cattle Problem* to his friends in Alexandria. The problem dealt with eight herds, four of bulls and four of cows, according to their colors, white, black, yellow and dappled. Certain facts were stated, for example, the dappled bulls exceeded the yellow bulls in multitude by $(1/6 + 1/7)$ of the number of white bulls. The problem was to find the exact size of each herd. Algebraically, this leads to a set of seven equations with eight unknowns. There are an infinite number of solutions to these equations. He proved that the cubic equation $x^3 - ax^2 + (4/9)a^2b = 0$ can have a real and positive root only if a is greater than $3b$ (another very simple cubic equation occurred in the *Arithmetic* of Diophantus, after which no such equation is listed for almost 1,000 years in the European history). In mathematics he is best remembered for Archimedes' axiom, Archimedes' number, Archimedes' paradox, the Archimedean property, and the Archimedean solid; however, he was most proud of his discovery of the method for finding the volume of a sphere. He showed that the volume of a sphere is two-thirds the volume of the smallest cylinder that can contain it. At his request, the figure of a sphere and cylinder was engraved on his tombstone by Romans.

In addition to mathematics, Archimedes worked extensively in mechanics and hydrostatics. He was the first to construct a hydraulic organ. Nearly every school child knows Archimedes as the absent-minded scientist who, on realizing that a floating object displaces its weight of liquid, leaped from his bath and ran naked through the streets of Syracuse shouting, "Eureka, Eureka"! (meaning, "I have found it"!). Such stories are believed by common people because they like to believe that great men are ridiculous, a form of revenge for being forced to acknowledge that they are superior. Apocryphal or not, the tale does shed some light on Archimedes' personality. Archimedes actually created the discipline of hydrostatics and used it to find equilibrium positions for various floating bodies, laid down the fundamental postulates of mechanics, discovered the laws of levers, and calculated centers of gravity for various flat surfaces and solids. In the excitement of discovering the mathematical laws of the lever, he is said to have declared, "Give me a place to stand and I will move the Earth."

Although Archimedes was apparently more interested in pure mathematics than its applications, he was an engineering genius. He invented an instrument knows as the Archimedean sphere, which shows the movements of heavenly bodies. He designed the inclined screw, a strange device which consists of a tube, open at both ends, and bent into the form of a spiral like a corkscrew. If one end is immersed in water, and the axis of the instrument is inclined to the vertical at a sufficiently large angle, and the instrument is then turned, the water will flow along the tube and out at the other end. The device is known as the *Archimedean screw* and is in use to this day. Diodorus (flourished about 60 BC) reports that Archimedes invented it when he was in Egypt, to help the Egyptians raise the water which gathered in depressions after the flooding of the Nile. During the second Punic war (218–201 BC), when Syracuse was attacked by the Roman fleet under the command of Marcellus, it was reported by Plutarch that Archimedes' military inventions held the fleet at bay for 3 years. He invented super catapults that showered the Romans with rocks weighing a quarter ton or more, and fearsome mechanical devices with iron "beaks and claws" that reached over the city walls, grasped the ships, and spun them against the rocks. Archimedes may have used mirrors acting as a parabolic reflector to burn ships attacking Syracuse. After the first repulsion, Marcellus called Archimedes a "geometrical Briareus" (a hundred-armed mythological monster) who uses our ships like cups to ladle water from the sea. Eventually the Roman army was victorious, and contrary to Marcellus' specific orders, the 75-year-old Archimedes was killed by a Roman soldier. According to one report of the incident, the soldier cast a shadow across the sand in which Archimedes was working on a mathematical problem. When the annoyed Archimedes yelled, "Don't disturb my circles," the soldier flew into a rage and cut the old man down. Another version of Archimedes' death seems more probable: Archimedes was en route to meet Marcellus and carried with him various astronomical and mathematical instruments. Some soldiers mistook him for a rich citizen who might have something worth stealing, so they killed him. Marcellus regarded the killer as a murderer and treated Archimedes's surviving relatives with great honor and kindness.

With his death the Greek gift of mathematics passed into oblivion, not to be fully resurrected again until the sixteenth century. Unfortunately, there in no known accurate likeness or statue of this great man. The following verse depicts Jacobi's praise for Archimedes:

> To Archimedes came a youth eager for knowledge.
> Teach me, O Master, he said, that art divine
> Which has rendered so noble a service to the lore of the heavens,
> And back of Uranus yet another planet revealed.
> Truly, the sage replied, this art is divine as thou sayest,
> But divine it was ere it ever the Cosmos explored,
> Ere noble service it rendered the lore of the heavens
> And back of Uranus yet another planet revealed.
> What in the Cosmos thou seest is but the reflection of God,
> The God that reigns in Olympus is Number Eternal.

Galileo called him "divine Archimedes, superhuman Archimedes"; Sir William Rowan Hamilton remarked, "who would not rather have the fame of Archimedes than that of his conqueror Marcellus?"; Whitehead commented that "no Roman ever died in contemplation over a geometrical diagram"; Hardy said "Archimedes will be remembered when Aeschylus is forgotten, because languages die and mathematical ideas do not"; and the philosopher Francois Marie Arouet Voltaire (1694–1778) remarked "there was more imagination in the head of Archimedes than in that of Homer" (around 800–700 BC).

Ctesibius of Alexandria (Working 285–222 BC)

Ctesibius of Alexandria (working 285–222 BC) is considered to be the founder of the Alexandrian school of mathematics and engineering and was probably the first head of the Museum of Alexandria. Unfortunately, very little is known of his life and work, beyond the fact that he was the son of a barber from Aspondia, a suburb of Alexandria, and lived from about 285–222 BC. His work is chronicled by Vitruvius (around 75–15 BC), Athenaeus (flourished about 200), Philo of Byzantium (about 250 BC), Proclus, and Hero of Byzantium. Even though his work has not been fully studied, it is obvious that as an inventor and mathematician Ctesibius was second only to Archimedes in the world of ancient Greece. His work on the elasticity of air was extremely important, earning him the title of father of pneumatics, for the first treatises on the science of compressed air and its uses are his. Like all his other works, however, his *On Pneumatics* has not survived. His *Memorabilia*, a single compilation of his research, cited by Athenaeus, is also lost. In his study On Pneumatics, he proved that air is a material substance, and he devised many mechanisms operated by compressed air, beginning with a system of adjustable mirrors in his father's barber shop. He invented the piston pump, the pressure pump, and the double siphon; he perfected and multiplied the uses of the water clock in many different contexts; and he devised numerous types of catapults and other engines of war, many of which have been preserved. He also constructed mechanical figures operated by ratchet gears which he used to ornament his water clocks. Ctesibius is best known for three major inventions: the suction pump, the water clock, and the hydraulics, a musical instrument (the ancestor of the pipe organ), of which a single fine carved specimen has been discovered.

Eratosthenes of Rhodes (Around 276–194 BC)

Eratosthenes of Rhodes (around 276–194 BC) was born in Cyrene (now in Libya), a Greek colony west of Egypt. He studied under the supervision of Lysanias of Cyrene and the philosopher Ariston of Chios (flourished about 250 BC) who was a student of Zeno of Citium (around 334–262 BC, the founder of the Stoic school of philosophy. The Stoics believed strongly in duty and considered the universe to operate with rational principles), and the poet and scholar Callimachus (around 310–240 BC). Eratosthenes then spent some years studying at Plato's Academy in Athens. King Ptolemy II Philadelphus (around 283–246 BC, who married his own sister Arsinoe, around 305–248 BC; to marry one's own sister was a Pharaonic practice) invited him to Alexandria to tutor his son. Later King Ptolemy III Euergetes appointed Eratosthenes the chief librarian of the famous library at Alexandria, a central repository of ancient wisdom which held 700,000 volumes. Eratosthenes was an extremely versatile scholar, who wrote on mathematics, geography, astronomy, history, philosophy, and literary criticism. In mathematics, his method known as the Sieve of Eratosthenes was apparently the first methodical attempt to separate the primes from the composite numbers, and all subsequent tables of primes and of prime factors have been based on extensions of it. (The compilation of such a table involves a fantastic amount of work, which is not always rewarded. One table, published in 1776 at the expense of the Austrian imperial treasury, is reported to have had such a poor sale that the paper on which it was printed was confiscated and used in cartridges in war with Turkey.) Using the Sieve of Eratosthenes, we can find all the primes under 100 by eliminating after two every second number, after three every third number, and so on. This leaves us with the following numbers, which are all prime:

×	×	2	3	×	5	×	7	×	×
×	11	×	13	×	×	×	17	×	19
×	×	×	23	×	×	×	×	×	29
×	31	×	×	×	×	×	37	×	×
×	41	×	43	×	×	×	47	×	×
×	×	×	53	×	×	×	×	×	59
×	61	×	×	×	×	×	67	×	×
×	71	×	73	×	×	×	×	×	79
×	×	×	83	×	×	×	×	×	89
×	×	×	×	×	×	×	97	×	×

He is most noted for his chronology of ancient history and is credited by Cleomedes (about 20 AD) in *On the Circular Motions of the Celestial Bodies* with having calculated the Earth's circumference as 252,000 stadia (1 stadium = 559 ft) around 240 BC, using knowledge of the angle of elevation of the Sun at noon (the word noon comes from the word nones, the midday prayer service of medieval clergy) on the summer solstice in Alexandria and in the Elephantine Island near Syene (now Aswan, Egypt). Eratosthenes also measured the tilt of the Earth's axis by 23.5°, which gives us the seasons, the distance to the Sun as 804,000,000 stadia, and the distance to the Moon as 780,000 stadia. His three-volume *Geographica*, now lost except for fragments, was the first scientific attempt to put geographical studies on a sound mathematical basis. Eratosthenes earned the nickname "Beta" by ranking just below the best; he was looked upon as a second Plato. His writings include the poem Hermes, which put the fundamentals of astronomy into verse form, as well as literary works on the theater and on ethics which was a favorite topic of the Greeks. His name is given to a crater on the Moon, a period in the lunar geologic timescale, and a seamount in the eastern Mediterranean sea. He never married and in 194 BC became blind. It has been claimed that he committed suicide by starvation, a practice common in antiquity known as the "philosopher's death."

Apollonius of Perga (Around 262–200 BC)

Apollonius of Perga (around 262–200 BC) earned the title 'The Great Geometer.' Little is known of his life, but as a young man he came to Alexandria, where he studied under the followers of Euclid and later taught there. He traveled elsewhere, visiting Pergamum (today the town of Bergama in the province of Izmir in Turkey), where he met Eudemus of Rhodes (around 350–290 BC), a member of Aristotle's school who wrote histories of arithmetic, geometry, and astronomy, which is now lost. Apollonius wrote extensively, and many of his books are extant. His prefaces are admirable, showing how perfect was the style of the great mathematicians when they were free from the trammels of technical terminology. He speaks with evident pleasure of some of his results: 'the most and prettiest of these theorems are new.' What Euclid did for elementary geometry, Apollonius did for conic sections (intersections of a plane with a cone), addressing circles, ellipses, parabolas, and hyperbolas in eight books containing 487 propositions. In fact, he systemized and generalized the theory of conics as developed by his predecessors: Menaechmus, Euclid, Aristaeus, Archimedes, Conon of Samos (about 245 BC), and Nicoteles of Cyrene (about 250 BC). This work has influenced many later scholars including Ptolemy, Francesco Maurolico, Newton, and Descartes. Another great achievement of Apollonius was his complete solution of a problem about a circle satisfying three conditions. When a circle passes through a given point, or touches a given line, or touches a given circle, it is said to satisfy one condition. So the problem of Apollonius really involved nine cases, ranging from the description of a circle through three given points to that of a circle touching three given circles. The simplest of these cases were probably quite well known; in fact, one of them occurs in the Elements of Euclid. Pappus indicates six other works of Apollonius: Cutting of a Ratio, Cutting an Area, On Determinate Section, Tangencies, Plane Loci, and On Verging Constructions, each in two books. Hypsicles (around 190–120 BC) refers to a work by Apollonius comparing a dodecahedron and an icosahedron inscribed in the same sphere. Apollonius also wrote a work on the cylindrical helix and another on irrational numbers, which is mentioned by Proclus. Eutocius refers to a book entitled *Quick Delivery by Apollonius* in which he obtained an approximation for π, which was better than the approximation known to Archimedes. In *On the Burning Mirror*, he showed that parallel rays of light are not brought to a focus (which is Latin for hearthside or fireplace) by a spherical mirror (as had been previously thought) and discussed the focal properties of a parabolic mirror. Apollonius also composed eight excellent books on *The Elements of the Pyramid.* Ptolemy, in his book *Megale Syntaxis* (The Mathematical System), says Apollonius introduced the hypothesis (something believed to be true) of eccentric orbits, or equivalently, deferent and epicycles, to explain the apparent motion of the planets and the varying speed of the Moon. Apollonius' theorem demonstrates that the two models are equivalent given the right parameters. Apollonius also researched lunar theory. The Apollonius crater on the Moon was named in his honor. With the death of Apollonius, the golden age of Greek mathematics almost came to an end. In the second century AD, commentator Ptolemaus Chennus called him Epsilon (ϵ) as it resembles the Moon.

Diocles (Around 240–180 BC)

Diocles (around 240–180 BC) was a Greek mathematician and geometer, and a contemporary of Apollonius. He is believed to be the first to prove the focal property of a parabola. He is also known for the geometric curve *Cissoid of Diocles*, which he used for doubling the cube. Fragments of a work by Diocles titled *On Burning Mirrors* were preserved by Eutocius in his commentary of Archimedes' *On the Sphere and the Cylinder*. One of the fragments contains a solution using conic sections to solve the problem of dividing a sphere by a plane so that the resulting two volumes are in a given ratio. This was equivalent to solving a certain cubic equation. Another fragment uses the cissoid to find two mean proportionals.

Bakhshali Manuscript (About 200 BC)

Bakhshali Manuscript (about 200 BC) was found in 1881 in the village Bakhshali in Gandhara, near Peshawar, Northwest India (present-day Pakistan). The very word manuscript comes from the Latin words meaning 'written by hand.' It is written in an old form of Sanskrit on birch bark. Only about 70 mutilated birch barks still exist, the greater portion of the manuscript has been lost. This manuscript gives various algorithms and techniques for a variety of problems, such as computing square roots, dealing with negative numbers, and finding solutions of quadratic equations. It is one of the earliest evidences of Indian mathematics free from religious or metaphysical associations. The following problems are from this manuscript:

Three persons possess a certain amount of riches each. The riches of the first and the second taken together amount to 13; the riches of the second and the third taken together are 14; and the riches of the first and the third mixed are known to be 15. Tell me the riches of each. $(7, 6, 8)$.

A merchant pays duty on certain goods at three different places. At the first he gives $1/3$ of the goods, at the second $1/4$ (of the reminder), and at the third $1/5$ (of the remainder). The total duty is 24. What was the original amount of goods? (40).

Hipparchus (Around 190–120 BC)

Hipparchus (around 190–120 BC) was born in Nicaea in Bithynia, but spent much of his life in Rhodes. He is generally considered to be one of the most influential astronomers of antiquity. His work in astrology, geography, and mathematics, especially trigonometry (triangle measurement) and applied spherical trigonometry, is also remarkable. Hipparchus wrote at least 14 books, but his only preserved work is his commentary on the Phaenomena of Eudoxus, by Aratus (around 315–240 BC). Most of what is known about Hipparchus comes from Ptolemy's Megale Syntaxis, or Almagest (The very great or greatest), as it was named by the Arabs and then became known to medieval Europeans, with additional references to him by Pappus of Alexandria and Theon of Alexandria in their commentaries on the Almagest. Hipparchus is thought to be the first to calculate a heliocentric system, but he abandoned his work because the calculations showed the orbits were not perfectly circular, as was believed to be mandatory by the theology of the time. He may have been the first to develop a reliable method to predict solar eclipses. Hipparchus' other achievements include the discovery of Earth's precession, the compilation of the first comprehensive star catalog of the western world, and possibly the invention of the astrolabe, and the armillary sphere, which he used during the creation of much of the star catalogue. He is also credited with the invention or improvement of several astronomical instruments, which were used for a long time for naked-eye observations. In the second and third centuries, coins were made in his honor in Bithynia that bear his name and show him with a globe. Hipparchus was in the international news in 2005, when it was again proposed (as in 1898) that the data on the celestial globe of Hipparchus, or star catalog, may have been preserved in the only surviving large ancient celestial globe that depicts the constellations with moderate accuracy, the globe carried by the Farnese Atlas. The Astronomer's Monument at the Griffith Observatory in Los Angeles, California, USA, features a relief of Hipparchus as one of six of the greatest astronomers of all time and the only one from Antiquity.

Posidonius of Apamea (135–51 BC)

Posidonius of Apamea (135–51 BC) was born in Apameia in Syria in a Greek family. He went to Athens to complete his education, and there he studied under the Stoic philosopher Panaetius of Rhodes (around 185–109 BC). He settled in Rhodes after extensive travels. Posidonius gave new life to Stoicism by fortifying it with contemporary learning. Although all his writings have been lost, it is known that they were copious. He made contributions to Stoic physics and ethics, notably, the theory that a vital force emanating from the Sun permeated the world, as well as his doctrine of cosmic sympathy, through which man and all things in the universe are united. He calculated the circumference of the Earth, size and distance to the Moon, and the size and distance to the Sun. His other writings dealt with the natural sciences, mathematics, and military tactics. He had a strong influence on the Romans.

Bhaskara I (Before 123 BC)

Bhaskara I (before 123 BC) is the earliest known commentator of Aryabhatta's works. His exact time is not known, except that he was in between Aryabhatta and Varahamihira. Bhaskara I mentions the names of Latadeva, Nisanku, and Panduranga Swami as disciples of Aryabhatta. Moreover, he says that Aryabhatta's fame has crossed the bounds of the oceans and his works have led to accurate results, even after so much time. This shows that Bhaskara I lived much later than Aryabhatta. His works are *Mahabhaskariya*, *Aryabhatteeyabhashya*, and *Laghubhaskariya*. He was the first to write numbers in the Hindu–Arabic decimal system with a circle for the zero. Bhaskara I stated theorems about the solutions of today's so-called Pell (John Pell, 1610–1685) equations $Nx^2 + 1 = y^2$. His work Mahabhaskariya is divided into eight chapters, discussing topics such as the longitudes of the planets, association of the planets with each other and with the bright stars, the lunar crescent, solar and lunar eclipses, and the rising and setting of the planets. He gave a remarkable approximation formula for $\sin x$,

$$\sin x = \frac{16x(\pi - x)}{[5\pi^2 - 4x(\pi - x)]},$$

and the formula for the area of a triangle having sides a, b and c,

$$A = \sqrt{s(s-a)(s-b)(s-c)},$$

where $2s = a + b + c$. Ancient Indian mathematicians frequently used geometry to prove algebraic results. The following two interesting examples are due to Bhaskara I:

The two parallel faces of a figure resembling a drum-shaped musical instrument, called a Panava, are each 8 units, the ventral width 2 units, and the length between the faces is 16 units. Find the area of the Panava. (80 square units).

A hawk is sitting on a pole whose height is 18 units. A rat that had come a distance of 81 units out of its dwelling at the foot of the pole, while returning to its dwelling, is killed by the hawk. Find how far it had gone toward the hole, and also the horizontal motion of the hawk, the speeds of the rat and the hawk being the same. (42.5, 38.5).

Daivajna Varahamihira (Working 123 BC)

Daivajna Varahamihira (working 123 BC) was born in Kapitthaka or Ujjain, India, and was a Maga Brahmin. He was an astronomer, mathematician, and astrologer. His picture may be found in the Indian Parliament along with Aryabhata. He was one of the nine jewels (Navaratnas) of the court of legendary King Vikramaditya

(102 BC–18 AD). In 123 BC, Varahamihira wrote *Pancha-Siddhanta* (The Five Astronomical Canons), in which he codified the five existing Siddhantas, namely, Paulisa Siddhanta, Romaka Siddhanta, Vasishtha Siddhanta, Surya Siddhanta, and Paitamaha Siddhanta. The Panca-Siddhantika contains many examples of the use of a place-value number system. Varahamihira quotes a tribute paid to a Yavana (not Greek) astronomer by an early astronomer Garga, "the Yavanas are Mlechchas (people deviated from Sanatana Dharma) but amongst them this science of astronomy is duly established therefore they are honored as Rishis." This tribute to Yavanas has been misinterpreted to imply the influence of Greek astronomy on Indian astronomy in Garga's time. However, Garga's time was long before Varahamihira's time, and Yavanas were not Greeks at that time. In fact, there were Yavana kingdoms in South and Northwest of India before Maurya Asoka's time (1472–1436 BC). Varahamihira's other most important contribution is the encyclopedic *Brihat-Samhita* (The Great Compilation), which describes theoretical and predictive astrology. He also made some important mathematical discoveries, such as certain trigonometric formulas that correspond to $\sin x = \cos(\pi/2 - x)$, $\sin^2 x + \cos^2 x = 1$, and $(1 - \cos 2x)/2 = \sin^2 x$. He also developed new interpolation methods to produce sine tables that improved those of Aryabhata and constructed a table for the binomial coefficients

$$^{n}C_{r} = \frac{n(n-1)(n-2)\ldots(n-r+1)}{1.2.3.\cdots r}$$

known as Pascal's triangle and examined the pandiagonal magic square of order four. A magic square consists of a number of integers arranged in the form of a square so that the sum of the numbers in every row, in every column, and in each diagonal is the same, for example,

1	15	14	4
12	6	7	9
8	10	11	5
13	3	2	16

15	10	3	6
4	5	16	9
14	11	2	7
1	8	13	12

Magic squares have been considered strong talismans against evil, and possession of a magic square was thought to insure health and wealth. His son Prithuyasas also contributed to Hindu astrology; his book *Hora Saara* is famous on horoscopy. As stated in the verses in this work, Prithuyasas wanted to explain in a simplified manner, the difficult and brief principles of his father's works.

Thrasyllus of Mendes (Around 50 BC–36 AD)

Thrasyllus of Mendes (around 50 BC–36 AD), whose full name was Tiberius Claudius Thrasyllus, was an Egyptian astrologer, astronomer, grammarian, and mathematician who lived during the reign of the Emperor Tiberius (42 BC–37 AD), whom he served. He was renowned for the accuracy of his astrological predictions. He calculated the Exodus of the Israelites from Egypt to have taken place around 1690 BC. Today, he is most famous for having published one of the earliest editions of Plato's complete corpus and for his arrangement of those writings into tetralogies.

Brahmagupta (Born 30 BC)

Brahmagupta (born 30 BC) was born as a devout Hindu in Bhinmal city in the state of Rajasthan in northwest India. He wrote two treatises on mathematics and astronomy: the *Brahmasphutasiddhanta* (The Correctly Established Doctrine of Brahma), often translated as The Opening of the Universe, and the *Khandakhadyaka* (Edible Bite) which mostly expands on the work of Aryabhata. As a mathematician, he is considered to be the father of arithmetic, algebra, and numerical analysis. Most importantly, in Brahmasphutasiddhanta he treated zero as a number in its own right, stated rules for arithmetic on negative numbers and zero, and attempted to define division by zero. Particularly, he wrongly believed that $0/0$ was equal to 0. Such a belief may not be altogether always unjustified in a finite precision computing, e.g., obtaining a repeated zero of $f(x) = x^2 - 2x + 1$ by the second order Newton scheme where both $f'(x)$ and $f(x)$ tend to zero as x tends to 1 with, however, $f(x)$ tending to zero faster. He gave the sum of the squares of the first n natural numbers as $n(n+1)(2n+1)/6$, the sum of the cubes of the first n natural numbers as $(n(n+1)/2)^2$, the general solution of the quadratic equation $ax^2 + bx = c$, and a method for solving Diophantine equations of the second degree, such as $Nx^2 + 1 = y^2$ (Pell's equations). He also gave a formula for finding the area A (and the lengths of the diagonals) of any cyclic quadrilateral given its four sides a, b, c, d as

$$A = \sqrt{(s-a)(s-b)(s-c)(s-d)},$$

where $2s = a + b + c + d$. (Bhaskara's I formula is a special case of this formula, when one of the sides equals zero—this formula was rediscovered by Snell in 1619 AD). He described an inaccurate geometric construction for squaring the circle, which in modern terms would amount to $\pi = \sqrt{10}$, and the second order difference interpolation formula

$$f(x+nh) = f(x) + \frac{n}{2}[\Delta f(x-h) + \Delta f(x)] + \frac{n^2}{2}\Delta^2 f(x-h).$$

In astronomy, he invented methods for calculating the position of heavenly bodies over time (ephemerides), their rising and setting, conjunctions, and the procedure for calculating the dates for solar and lunar eclipses. He criticized the Puranic view that the Earth was flat or hollow like a bowl; instead, he observed that the Earth and heaven were spherical. However, he wrongly believed that the Earth did not move. He calculated the length of the solar year as 365 days, 6 h, 5 min, and 19 s, which is remarkably close to the actual value of 365 days, 5 h, 48 min, and about 45 s. He maintained that "all things fall to the Earth by a law of nature, for it is nature of the Earth to attract and keep things." Thus the roots of the law of gravitation were known almost 1,500 years before Newton's time. Later Bhaskara II called him *Ganita Chakra Chudamani*, which means "the gem in the circle of mathematicians." The following interesting problem is due to Brahmagupta.

Tell the number of elapsed days (residual degrees) for the time when four times the 12th part of the residual degrees increased by one, plus eight will be equal to the residual degree plus one. (11).

Geminus of Rhodes (Around 10 BC–60 AD)

Geminus of Rhodes (around 10 BC–60 AD) was a Greek astronomer and mathematician. Nothing is known about his life, except that he was a stoic philosopher and scholar. He is thought to have studied at the school in Rhodes and then later under Posidonius. During his career he wrote introductory works in mathematics and astronomy and is typically credited for the construction of the Antikythera mechanism, an ancient analog computer designed to calculate astronomical positions. It was recovered in 1900–1901 from the Antikythera wreck. He noted that the length of a day varied in different parts of the world. His book *Isagoge* was based on the work of Hipparchus. His mathematical work divided the field into pure and applied mathematics. Pure mathematics included number theory and the various properties of numbers. In applied mathematics he included surveying, musical harmony, optics, astronomy, mechanics, and accounting. He also wrote a commentary on Posidonius' work on Meteorology. Fragments of this commentary are preserved by Simplicius (about 520) in his commentary on Aristotle's physics. He gave an ingenious proof of the parallel postulate; however, it is false. The Geminus crater on the Earth's Moon is named for him.

Nicomachus of Gerasa (Around 60–120 AD)

Nicomachus of Gerasa (around 60–120 AD) lived near Jerusalem. He was a neo-Pythagorean, known for being one of the first thinkers to locate the natural numbers in the mind of God. According to him, "All things that have been arranged by nature according to a workmanlike plan appear, both individually and as a whole, as singled out and set in order by Foreknowledge and Reason, which created all according to Number, conceivable to mind only and therefore wholly immaterial; yet real; indeed, the really real, the eternal." In his book on number theory, the *Introductio arithmeticae*, Nicomachus considers the following infinite triangle, noting that the sum of the numbers in the nth row is n^3.

$$
\begin{array}{ccccccc}
 & & & 1 & & & \\
 & & 3 & & 5 & & \\
 & 7 & & 9 & & 11 & \\
13 & & 15 & & 17 & & 19
\end{array}
$$

This is in fact the first systematic work in which arithmetic is treated independent of geometry. Nicomachus said, "What is beautiful and definite and the object of knowledge is by nature prior to the indefinite and the incomprehensible and the ugly."

Menelaus of Alexandria (Around 70–140 AD)

Menelaus of Alexandria (around 70–140 AD) was a Greek or Egyptian mathematician and astronomer. He is known only through the writings of Pappus and Proclus. A conversation of his with Lucius, held in Rome, is recorded by Plutarch. Ptolemy mentions, in his work Almagest, astronomical observations made by Menelaus in Rome on the 14th of January, 98 AD. One of these observations was of the occultation of the star Beta Scorpii by the Moon. He wrote three books; however, *Sphaerica* is the only one that has survived, in an Arabic translation. This book deals with the geometry of the sphere and its application to astronomical measurements and calculations. He also introduced the concept of spherical triangles (figures formed by three great circle arcs, which he named "trilaterals") and proved Menelaus' theorem (an extension to spherical triangles of a previously known result). Menelaus also provided direct proofs of some of the Euclid's theorems, avoiding the method of reductio ad absurdum. A well-known theorem, dealing with the points in which a straight line drawn across a triangle meets the sides, still bears his name. His works were later translated by the sixteenth century astronomer and mathematician Francesco Maurolico.

Heron of Alexandria (About 75 AD)

Heron of Alexandria (about 75 AD) was probably an Egyptian, but definitely was a shrewd follower of Archimedes, bringing his mathematics to bear on engineering and surveying. Very little is known about his life, and even the century in which he lived is a subject of debate. In fact, many of his writings dealt with applications of mechanics, engineering, and measurements. He is also reputed to have invented a steam engine. One of his interesting theorems proves that when light from an object is broken by reflection on mirrors, the path of the ray between the object and one's eye is minimum. This is an instance of the *principle of least action*, which was formally adopted for optics and dynamics by Hamilton and was incorporated in the work of Einstein. We may regard Heron as the pioneer of Relativity. He explained, in his *Dioptra*, how to dig tunnels through mountains and how to measure the amount of water flowing from a spring. He is most remembered for his formula $A = \sqrt{s(s-a)(s-b)(s-c)}$, where A is the area of a triangle having sides a, b, and c, and $2s = a + b + c$. This formula appeared as Proposition 1.8 in his treatise *Metrica*, which had been lost for centuries until a fragment was discovered in 1894, followed by a complete copy in 1896. This formula is the same as that given by Bhaskara I, and is a particular case of the one given by Brahmagupta. He is also credited for giving the formula for the volume of the frustum of a cone as

$$V = \frac{1}{4}\pi h(r + R)^2,$$

where h is the height and r and R are the radii of the bases.

Zhang Heng (78–139 AD)

Zhang Heng (78–139 AD) was an astronomer, mathematician, inventor, geographer, cartographer, artist, poet, statesman, and literary scholar from Nanyang in central China's Henan Province. He lived during the Eastern Han Dynasty (25–220 AD). Zhang began his career by travelling extensively. In 111, he took a government post to record astronomical phenomena, forecast weather, and compile calendars. Based on this experience he wrote a summary of Chinese astronomical knowledge, including explanations of eclipses, the best star map to date, and various observational data. He proposed a theory of the universe that compared it to an egg. "The sky is like a hen's egg and is as round as a crossbow pellet. The Earth is like the yolk of the egg, lying alone at the center. The sky is large and the Earth is small." According to him, the universe originated from chaos. He said that the Sun, Moon, and planets were on the inside of the sphere and moved at different rates. He demonstrated that the Moon did not have independent light but that it merely reflected the light from the Sun. He showed that lunar eclipses occurred relatively frequently because the Moon moved faster within the sphere of the heavens. He drew a detailed map of the heavens. His chart showed 124 constellations consisting of a total of 2,500 stars, 320 of which were bright stars with names. "This is not including those observed by sailors," he wrote, "Of the small stars, there are eleven thousand five hundred and twenty." He is most famous in the West for his rotating celestial globe, and for inventing the first seismograph for measuring earthquakes in 132. His device was in the shape of a cylinder with eight dragon heads around the top, each with a ball in its mouth. Around the bottom were eight frogs, each directly under a dragon's head. When an earthquake occurred, a ball fell out of a dragon's mouth into a frog's mouth, making a noise. He is also known as a mathematician who proposed $\sqrt{10}$

(about 3.1623) for π, and was the inventor of the odometer, or "mileage cart." He excelled in writing and visual arts. In fact, he was also considered one of the four great painters of his era.

Alexandrian Claudius Ptolemaeus (Around 90–168 AD)

Alexandrian Claudius Ptolemaeus (around 90–168 AD), known in English as **Ptolemy**, was a Roman citizen of Egypt who wrote in ancient Greek. From the sketchy details of his life it is believed that he was born in the town of Ptolemais Hermiou in Thebaid and died in Alexandria around 168. He was a mathematician, geographer, astrologer, astronomer, and poet of a single epigram in the Greek Anthology, which was a collection of some 3,700 epigrams or short poems that was assembled some time in the fifth century by Metrodorus. He made a map of the ancient world in which he employed a coordinate system very similar to the latitude and longitude of today, though the formal study of coordinates had to wait another 1,500 years. One of his most important achievements was his geometric calculations of semichords. Ptolemy obtained, using chords of a circle and an inscribed 360-gon, the approximation $377/120$ for π. Having inherited the Babylonian and Greek view (following the lead of Hipparchus) that Earth is the center of the universe, Ptolemy wondered why at certain times of the year it appears that Mars is moving backward, in *retrograte motion*. Since the circle is clearly the most perfect geometric figure, ancient Greeks expected the universe to conform to their aesthetic taste and make all orbits perfectly circular. Ptolemy postulated a system wherein each body orbiting the Earth spins on a circle of its own, called an *epicycle*. Ptolemy required approximately 80 equations to describe, quite accurately, the locations of all the heavenly bodies of what we now call the solar system. He estimated that the Sun was at an average distance of 1210 Earth radii, while the radius of the sphere of the fixed stars was 20,000 times the radius of the Earth. Ptolemy's geocentric model went unchallenged until the middle of the sixteenth century and was defended by the Catholic Church for another few hundred years. A heliocentric (Sun at the center) system first appeared in print in 1548 in a work written by Copernicus. Kepler and Galileo then advocated this theory in the early seventeenth century, though Kepler replaced his circles with ellipses. Ptolemy's great works Almagest (originally the 'Mathematical Collection') on astronomy and *Geographia* on geography were translated by Islamic scholars and survived the ravages of history. Besides these works, he is also remembered for Ptolemy's Canon, the Ptolemy Cluster, Ptolemy's theorem, and Ptolemy's world map. The crater Ptolemaeus on the Moon and the crater Ptolemaeus on Mars are named after him.

Liu Hui (Around 220–280)

Liu Hui (around 220–280) was a Chinese mathematician who lived in the Wei Kingdom. Little is known about his life except that he wrote two works. The first was an extremely important commentary on the *Jiuzhang suanshu* or, as it is more commonly called, *Nine Chapters on the Mathematical Art*, which came into being in the Eastern Han Dynasty and is believed to have been originally written around 1000 BC. (It should be noted that very little is known about the mathematics 'art of calculation' of ancient China. In 213 BC, the Emperor Shi Huang of the Chin dynasty (221–206 BC) had all of the manuscripts of the kingdom burned.) The other was a much shorter work called *Haidao suanjing*, or *Sea Island Mathematical Manual*, which consisted of nine measurement problems, including the measurement of the heights of Chinese pagoda towers. Nine Chapters on the Mathematical Art consist of 246 questions with their solutions and procedures. Each of the nine chapters focuses on a separate part of daily mathematics. Many data recorded in this book were advanced for the world at that time, but some of the solutions lacked proof. In 263, Liu Hui annotated the book, giving full proof to the solutions and formulas in the book while correcting the errors. He expressed all of his mathematical results in the form of decimal fractions. Liu Hui also raised and defined a number of mathematical concepts, including area, power, and equation. By approximating circles by polygons, he approximated π as 3.141014 and suggested 3.14 as a practical approximation. He also provided commentary on the mathematical proof that is identical to the Pythagorean theorem, which Liu Hui knew as the Gougu theorem. He provided problems involving the building of canal and river dykes, giving results for total amount of materials used, the amount of labor needed, the amount of time needed for construction, etc. He computed the volume of various solids such as the prism, pyramid, tetrahedron, wedge, cylinder, cone, and the frustum of a cone. He also gave the formula for the volume of the truncated triangular prism, which has now been credited to Legendre. There is also evidence that he began to understand concepts associated with early work on differential and integral calculus. Liu Hui also showed how to solve n linear equations in n unknowns, with both positive and negative numbers, using a method which is essentially the same as Gaussian elimination. One system which he solved is the following: Three sheaves of a good harvest, two sheaves of a mediocre harvest, and one sheaf of a bad harvest yield a profit of 39 tous. Two sheaves of a good harvest, 3 sheaves of a mediocre harvest, and 1 sheaf of a bad harvest yield a profit of 34 tous. One sheaf of a good harvest, 2 sheaves of a mediocre harvest, and 3 sheaves of a bad harvest yield a profit of 26 tous. How much is the profit from one sheaf each of a good, mediocre, and bad harvest? Mathematically we need to solve the system

$$\begin{aligned} 3x + 2y + z &= 39 \\ 2x + 3y + z &= 34 \\ x + 2y + 3z &= 26. \end{aligned}$$

The Chinese interest in systems of linear equations was perhaps linked to their interest in magic squares. The square

4	9	2
3	5	7
8	1	6

was supposedly brought to humankind on the back of a tortoise from the River Lo in the days of Great Emperor Yu (Yii) (around 2200–2100 BC). A printed version of the *Nine Chapters* appeared in 1084 with each page carved on separate wooden blocks. (The oldest known book which was produced by the wood block technique is the Buddhist *Diamond Sutra* in the year 868; the entire Buddhist Canon which was printed during 972–983 required the engraving of 130,000 two-page blocks.) This work has also been recently translated into French. From his writing, it is clear that Liu Hui was a learned man, who not only had great expertise in mathematics but also was familiar with the literary and historical classics of China. The contributions of Liu Hiu have always aroused great interest in mathematicians and historians.

Sporus of Nicaea (Around 240–300)

Sporus of Nicaea (around 240–300) was probably born in Nicaea (Greek Nikaia). He studied with Philo of Gadara, and taught or was an older contemporary of Pappus. Sporus is known only through the writings of Pappus and Eutocius. Much of his work focused on squaring the circle and duplicating the cube, using approximations which are early examples of integration. He criticized the work of Hippias and Archimedes for not providing a more accurate approximation of π. He also wrote on mathematical geography, astronomy, and meteorology, and produced a critical edition of Aratus' *Phaenomena*. He died in the ancient district of Bithynia, (modern-day Iznik) in Bursa, in modern day Turkey.

Iamblichus (Around 245–325)

Iamblichus (around 245–325) was the chief representative of Syrian Neoplatonism, to whom we owe valuable work on the Pythagorean discoveries and doctrines. He also studied the properties of numbers. He enunciated the theorem that if a number that is equal to the sum of three integers of the form $3n, 3n-1, 3n-2$ is taken, and if the separate digits of this number are added, and if the separate digits of this result are added again, and so on, then the final result will be 6. For instance, the sum of 54, 53, and 52 is 159, the sum of the separate digits of 159 is 15, and the sum of the separate digits of 15 is 6. To anyone confined to the usual Greek numerical notation this must have been a difficult result to prove: possibly it was reached empirically.

Diophantus of Alexandria (About 250)

Diophantus of Alexandria (about 250) was either a Greek, a Hellenized Babylonian, an Egyptian, a Jew, or a Chaldean. He certainly had an un-Greek interest in something that was very like algebra. Little is known about him except the problems he proposed. His tombstone tells us all that is known about his personal life: "Here you see the tomb containing the remains of Diophantus, it is remarkable: artfully it tells the measures of his life. The sixth part of his life God granted for his youth. After a twelfth more his cheeks were bearded. After an additional seventh he kindled the light of marriage, and in the fifth year he accepted a son. Elas, a dear but unfortunate lad, half of his father he was and this was also the span a cruel fate granted it, and he controlled his grief in the remaining 4 years of his life. By this device of numbers tell us the extent of his life." If x is taken as the age of Diophantus at the time of his death, then the problem is to solve for x in the equation

$$\frac{x}{6} + \frac{x}{12} + \frac{x}{7} + 5 + \frac{x}{2} + 4 = x \quad (x = 84).$$

However, the accuracy of this information cannot be independently confirmed. He is celebrated for his writings on algebra and lived at the time of Pappus, or perhaps a little earlier. The chief surviving writings of Diophantus are six of the thirteen books forming the *Arithmetica* and fragments of his *Polygonal Numbers* and *Porisms*. Diophantus' work has a twofold importance: he made an essential improvement in mathematical notation, and at the same time added large installments to the scope of algebra as it then existed. His abiding work lies in the *Theory of Numbers* and of *Indeterminate Equations*. The method for solving the latter is now known as Diophantine analysis. Perhaps Diophantus was the first mathematician who recognized fractions as numbers; thus, he allowed positive rational numbers for the coefficients and solutions. He was more concerned with quadratic and higher types of equations, such as $x^4 + y^4 + z^4 = u^2$. He discovered four whole numbers x, y, z, u for which this equation is true. Certainly, Diophantus was not, as he has often been called, 'the father of algebra.' Nevertheless, his remarkable collection of indeterminate problems was not fully appreciated and further developed until much later. The following is a sample of beguiling, challenging problems from Arithmetica, which contains 189 problems with their solutions. Here, "number" means "positive rational number."

Find two square numbers such that their product added to either gives a square number. (Diophantus' answer: $(3/4)^2, (7/24)^2$.)

Find three numbers such that their sum is a square and the sum of any pair is a square. (Diophantus' answer: $80, 320, 41$.)

Find three numbers in arithmetic progression such that the sum of any pair is a square. (Diophantus' answer: $120\frac{1}{2}, 840\frac{1}{2}, 1560\frac{1}{2}$.)

Find two numbers such that their sum is equal to the sum of their cubes. (Diophantus' answer: $5/7, 8/7$.)

Find three numbers in geometric progression such that the difference of any two is a square number. (Diophantus' answer: $81/7, 144/7, 256/7$.)

Find a Pythagorean triangle in which the hypotenuse minus each of the legs is a cube. (Diophantus' answer: $40 - 96 - 104$.)

Pappus of Alexandria (Around 290–350)

Pappus of Alexandria (around 290–350) was born in Alexandria, Egypt, and was either Greek or a Hellenized Egyptian. The written records suggest that Pappus lived in Alexandria during the reign of Diocletian (284–305). He was a teacher, the author of commentaries on Euclid and Ptolemy, and author of a work on universal geography, much of which is now lost. Pappus' major work is *Synagoge*, or the Mathematical Collection, which is a compendium of mathematics of which eight volumes have survived. Book I is lost, and we conjecture that it covered arithmetic. Book II is partly lost and what remains deals with Apollonius' method of large numbers. Book III is divided into four parts: the first part includes the famous "Delian problem" of finding two mean proportionals between two given straight lines, and the second part provides a construction of the arithmetic, geometric, and harmonic means. The third part describes a collection of geometrical paradoxes, and the final part shows how each of the five regular polyhedra can be inscribed in a sphere. Book IV contains various theorems on circles, a study of various curves, and an account of the three classical problems of antiquity: the squaring of the circle, the duplication of a cube, and the trisection of an angle. It also contains a generalization of the theorem of Pythagoras on right-angled triangles. Book V deals with problems in geometry concerning the areas of different plane figures, the volumes of different solid figures that all have the same superficial area, a comparison of the five regular solids of Plato, and observations on the cells of a honeycomb (hexagonal figure occupying the least possible space). Book VI addresses problems in astronomy treated by Theodosius of Bithynia (around 160–100 BC), Autolycus of Pitane (360–290 BC), Aristarchus, and Euclid in his Optics and Phenomena. Book VII explains the terms analysis and synthesis and the distinction between theorem and problem. It presents an outline of the work of the ancient mathematicians Euclid, Apollonius, Aristaeus, and Eratosthenes, and discusses the 13 semiregular polyhedra. It also considers the subject of conic sections, completing the work of Apollonius. Finally, he proves a theorem on the creation of lines of harmonic progression, as well as a method for finding the volume generated by a surface rotated about a line and the area generated by a curve rotated about its center (Paul Guldin, 1577–1643 theorem). This was perhaps the earliest suggestion of a branch of mathematics called the *calculus of variations*. Book VIII deals with mechanics in both practical and theoretical terms. It also contains "Pappus'–Pascal's theorem" on conic sections, which includes the lineal pairs of a plane figure, which is an early example of the principle of duality. He is remembered for Pappus' centroid theorem, Pappus' chain, Pappus' harmonic theorem, Pappus' hexagon theorem, Pappus' trisection method, and for the focus and directrix of an ellipse.

Hypatia of Alexandria (370–415)

Hypatia of Alexandria (370–415) was the first recorded notable Greek woman mathematician and Neoplatonist philosopher. Her date of birth is highly debated. She was both just and chaste and remained a virgin. Although no images of Hypatia exist, nineteenth-century writers and artists envisioned her as an Athene-like beauty. She was the daughter of the mathematician and philosopher Theon, who was her teacher and the last fellow of the Museum of Alexandria (in Egypt). Alexandria at that time was the literary and scientific center of the world, containing numerous palaces, the Alexandrian Library and Museum, and influential schools of philosophy, rhetoric, and other branches of learning. She worked in Alexandria, but did not teach in the Museum, instead she received her pupils in her own home. People from several other cities came to study and learn from her. Hypatia wrote works on algebra, conic sections, and the construction of scientific instruments, and taught in the fields of astronomy and astrology. She edited the work On the Conics of Apollonius, which divided cones into different parts by a plane. This concept developed the ideas of hyperbolas, parabolas, and ellipses. Hypatia's influence on the book made its concepts easier to understand and ensured that the work would survive for many centuries. She became head of the Platonist school at Alexandria in about 400 AD. Hypatia actively stood for learning and science at a time in Western history when such activities were associated with paganism. According to Socrates Scholasticus (380–450), she paid the ultimate price for her beliefs—she was brutally murdered by religious zealots. Her death has been romanticized as follows: "On the fatal day, in the holy season of Lent, Hypatia was torn from her chariot, stripped naked, dragged to the church, and inhumanly butchered by the hands of Peter the Reader and a troop of savage and merciless fanatics: her flesh was scraped from her bones with sharp oyster-shells, and her quivering limbs were delivered to the flames." Her death led to the departure of many scholars from Alexandria and marked the beginning of the end of the great age of Greek mathematics. The Neoplatonic school she headed continued in Alexandria until the Arabs invaded in 642 AD. The library of Alexandria was burned by the Arab conquerors, its books were used as fuel for baths, and the works of Hypatia were destroyed. (According to a thirteenth century story, the commander, Amru, was willing to spare the library, but was dissuaded by the Caliph Omar I (579–644), who argued that if the books of the Greeks confirm what is written in the Koran (recitation, or lectionary), they are superfluous, if they contradict the Koran, they are dangerous. In either case, he claimed, they should be destroyed.) We know her writings today through the works of those who quoted her, even if unfavorably, and a few letters written to her by her contemporaries. Later, Descartes, Newton, and Leibniz expanded her work.

Sun Tsu (About 400)

Sun Tsu (about 400) was a Chinese mathematician who flourished between the third and fifth centuries. For the problem

divide by 3, the remainder is 2;
divide by 5, the remainder is 3;
divide by 7, the remainder is 2;
what will be the number?

he gave only one of the infinitely many solutions, but his method allowed him to give as many other solutions as he wanted. Problems of this kind are examples of what became universally known as the *Chinese Remainder Theorem* (used in multiple modulus residue arithmetic for error-free computation). In mathematical parlance, the problems can be stated as finding n, given its remainders of division by several numbers $m_1, m_2, \ldots, m_k$:

$$\begin{aligned} n &= n_1 \pmod{m_1} \\ n &= n_2 \pmod{m_2} \\ &\cdots \\ n &= n_k \pmod{m_k}. \end{aligned}$$

Sun Tsu was interested in astronomy and he investigated Diophantine equations during the development of a calendar. He also gave a formula for determining the sex of a fetus: Take 49, add the number of the month in which the woman will give birth, then subtract her age. From what now remains, subtract the heaven 1, subtract the Earth 2, subtract the man 3, subtract the four seasons 4, subtract the five elements 5, subtract six laws 6, subtract the seven stars 7, subtract the eight winds 8, subtract the nine provinces 9. If the remainder is odd, the child shall be a son; and if even, a daughter.

Tsu Ch'ung Chih (Zu Chongzhi) (429–500)

Tsu Ch'ung Chih (Zu Chongzhi) (429–500) was born in Jiankang (now Nanking, Kiangsu province), China. He was in the service of the Emperor Hsiao-wu of the Liu Sung dynasty (960–1279). He created various formulas that have been used throughout history. For example, in 475, he approximated π by $355/113$, or

3.1415926, which is correct to six decimal places. This remained the most reliable approximation of π for many centuries. Note that $\pi = 355/113$ can be obtained from the values given by Ptolemy and Archimedes:

$$\frac{355}{113} = \frac{377 - 22}{120 - 7}.$$

Chih, with his son Tsu Keng/Zu Gengzhi (450–520), wrote a mathematical text entitled *Zhui Shu* (Method of Interpolation), which is now lost. In 463 he created a calendar that was never used. He discovered a new way to keep time accurately using the shadow of the Sun. He calculated 1 year as 365.24281481 days, which is very close to the exact value of 365.24219878 days; he also gave the overlaps between Sun and Moon as 27.21223 days, which is very close to the exact value of 27.21222 days. In 478, he reinvented a mechanical South Pointing Chariot. The lunar crater Tsu Chung-Chi is named after him.

Anthemius of Tralles (Around 474–558)

Anthemius of Tralles (around 474–558) was a Greek Professor of Geometry in Constantinople (now Istanbul in Turkey) and architect. He came from a well-educated family and had two brothers who were doctors. One brother was a lawyer, and the other was described as a man of learning. As an architect, he is best known for replacing an old church in collaboration with Isidorus of Miletus at Constantinople in 532, by the order of Justinian. As a mathematician, he described the string construction of an ellipse using a string fixed at the two foci. His famous book, *On Burning Mirrors* (Ibn al-Haytham in Arabic), describes the focal properties of a parabola. Anthemius persecuted his neighbor and rival Zenon by reflecting Sunlight into his house. He also produced the impression of an earthquake in Zenon's house with the use of steam led under pressure through pipes connected to a boiler.

Anicius Manlius Severinus Boethius (Around 475–526)

Anicius Manlius Severinus Boethius (around 475–526) was born in Rome. He was well read in Greek literature and science. He was elected consul at a very early age and took advantage of his position to reform the coinage and to introduce the public use of Sundials, water clocks, etc. He reached the height of his prosperity in 522 when his two sons were inaugurated as consuls. His integrity and attempts to protect the provincials from abuse by the public officials earned him the hatred of the Court. King Theodoric sentenced him to death while absent from Rome. He was seized at Ticinum (now Pavia) and was tortured in the baptistry of the church by having a cord drawn round his head till the eyes were forced out of their sockets, after which he was beaten to death with clubs on October 23, 526. His death has been considered to mark the end of ancient mathematics in the Western Roman Empire. At a later time his merits were recognized, and tombs and statues were erected in his honor by the state. His *Geometry* consists of the enunciations of only the first book of Euclid and of a few selected propositions from the third and fourth books, with numerous practical applications (e.g., for finding areas). He added an appendix with proofs of the first three propositions to show that his work is accurate. His *Arithmetic* is based on the work of Nicomachus. He is memorably characterized as "the last of the Romans and the first of the Scholastics."

Sridhara (Before 486)

Sridhara (before 486) was perhaps born in Bengal or southern India. Bhaskara II referred to him as a distinguished mathematician and quoted his work in a number of places. Sridhara provided mathematical formulas for a variety of practical problems involving ratios, barter, simple interest, mixtures, purchase and sale, rates of travel, wages, and filling of cisterns. His *Patiganita* (Mathematics of Procedures) is considered to be an advanced mathematical work. In it, all of the algorithms for carrying out arithmetical operations are presented in verse form and no proofs are given. Some sections of this book are devoted to arithmetic and geometric progressions, including progressions with fractional numbers or terms, and formulas for the sums of certain finite series. His other famous works include *Ganitasara* (Essence of Mathematics) and *Ganitapanchavimashi* (Mathematics in 25 Verses). He gave the first correct formulas (in India) for the volume of a sphere and of a truncated cone.

Bhaskara II or Bhaskaracharya (Working 486)

Bhaskara II or Bhaskaracharya (working 486), the son of Chudamani Maheshvar, was born in Bijapur district, Karnataka, India. He was the head of the astronomical observatory at Ujjain. Bhaskaracharya's celebrated work Siddhanta Siromani (Crown of Treatises) consists of four parts, namely, *Leelavati, Bijaganitam, Grahaganitam*, and *Goladhyaya.* The first two deal exclusively with mathematics and the last two address astronomy. His popular text *Leelavati* (named after his daughter) was written in 486 AD and was translated into Persian in 1587. According to legend, Bhaskaracharya had calculated that his daughter Leelavati (the beautiful) might propitiously marry only at one particular hour on a given day. However, on that day just before the wedding the eager girl bent over the water clock and unknowingly dropped a pearl, which stopped the water clock. Before the mishap was noticed the auspicious hour had passed. To console the unhappy girl, the father gave her name to the book. Bhaskaracharya's contributions to mathematics include: a proof of the Pythagorean theorem; solutions of quadratic, cubic, and quartic indeterminate equations; solutions of indeterminate quadratic equations; integer solutions of linear and quadratic indeterminate equations; a cyclic Chakravala method for solving indeterminate equations, which was rediscovered by William Brouncker in 1657; solutions of Pell's equations; solutions of Diophantine equations of the second order, such as $61x^2 + 1 = y^2$ (this equation was posed in 1657 by Fermat, but its solution was unknown in Europe until the time of Euler in the eighteenth century); solutions of quadratic equations with more than one unknown, including negative and irrational solutions; a preliminary concept of infinitesimal calculus, along with notable contributions toward integral calculus; a concept of differential calculus, after discovering the derivative and differential coefficient; a statement of Rolle's theorem; derivatives of trigonometric functions and formulas; work on spherical trigonometry including a number of trigonometric results such as $\sin(a \pm b) = \sin a \cos b \pm \cos a \sin b$, and the following two remarkable identities:

$$\sqrt{a \pm \sqrt{b}} = \sqrt{[a + \sqrt{a^2 - b}]/2} \pm \sqrt{[a - \sqrt{a^2 - b}]/2}.$$

He conceived the modern mathematical convention that when a finite number is divided by zero, the result is infinity. He speculates the nature of the number $1/0$ by stating that it is "like the Infinite, Invariable God who suffers no change when old worlds are destroyed or new ones created, when innumerable species of creatures are born or as many perish." Bhaskaracharya postulated that the Earth had a gravitational force. In astronomy, his studies included Lunar and solar eclipses, latitudes of the planets, the Sunrise equation, the Moon's crescent, cosmography, and geography. Bhaskaracharya's greatness lies in making mathematics highly irresistible and attractive. We list here some of his stimulating problems.

A beautiful pearl necklace belonging to a young lady was torn in a love quarrel and the pearls were all scattered on the floor. One-third of the number of pearls was on the bed, one-fifth was under the bed, one-sixth was found by the pretty lady, one-tenth was collected by the lover, and six were left hanging on the thread. Tell me the total number of pearls in the necklace (30).

One person has three hundred coins and six horses. Another has ten horses (each) of similar value as well as a debt of one hundred coins. They are of equal worth. What is the price of a horse? (100).

The square-root of half of a swarm of bees went to a rose-bush to suck honey, followed by eight-ninths of the entire swarm. One lady bee was caught in a lotus flower which closed at night. She was humming in response to the humming call of a male bee. O lady, tell me the number of bees in the swarm (72).

A snake's hole is at the foot of a pillar which is 15 cubits high, and a peacock is perched on its summit. Seeing a snake at a distance of thrice the pillar's height, gliding toward his hole, he pounces obliquely upon him. Say quickly at how many cubits from the snake's hole do they meet, both proceeding an equal distance (20).

Inside a forest, a number of apes equal to the square of one-eighth of the total apes in a pack are playing noisy games. The remaining 12 apes, who are of a more serious disposition, are on a nearby hill and are irritated by the shrieks coming from the forest. What is the total number of apes in the pack? (16 and 48).

A single bee (out of a swarm of bees) was seen in the sky; $1/5$ of the remainder (of the swarm), and $1/4$ of the remainder (left thereafter) and again $1/3$ of the remainder (left thereafter), and a number of bees equal to the square-root of the numerical value of the swarm were seen in lotuses; two bees were on a mango tree. How many bees were there? (16).

One says, "Give me a hundred, friend, I shall then become twice as rich as you." The other replies, "If you give me ten, I shall be six times as rich as you." Tell me what is the amount of their (respective) capitals? (40,170)

What is that number, O learned man, which being multiplied by twelve and increased by the cube of the number is equal to six times the square of the number added with 35. (5)

What is that number which being multiplied by 200 and added to the square of the number and then multiplied by 2 and subtracted from the fourth power of the number will become one myriad less unity (9999)? Tell that number if thou be conversant with the operations of analysis. (11)

Flavius Magnus Aurelius Cassiodorus Senator (490–585)

Flavius Magnus Aurelius Cassiodorus Senator (490–585) was born in Scylletium (Italy). In Rome, he was a statesman and writer, serving in the administration of King Theodoric. He published two books, *De Institutione Divinarum Litterarum* and *De Artibus ac Disciplinis*, in which the grammar, logic, and rhetoric of the preliminary trivium (from which we get our word trivial), as well as arithmetic, geometry, music, and astronomy of the scientific quadrivium, were discussed. These books were used as standard texts during the Middle Ages.

Isidorus of Miletus (About 540)

Isidorus of Miletus (about 540) was one of the two Greek architects who designed the church of Hagia Sophia in Constantinople. He taught physics in Alexandria and then later at Constantinople and wrote a commentary on earlier books on building. He collected and publicized the writings of Eutocius, which were commentaries on the mathematics of Archimedes and Apollonius, and consequently helped to revive interest in their works. He was also an able mathematician, the T-square and string construction of a parabola and possibly the apocryphal Book XV of Euclid's Elements are credited to him.

Isidore of Seville (560–636)

Isidore of Seville (560–636) was born probably in Cartagena, Spain, and passed away in Seville, Spain, at the age of 76. He was an ancient Christian philosopher, an Archbishop of Seville, the compiler of an encyclopedia, and a visionary. Isidore studied in the Cathedral school of Seville. Although not much is known authentically, he spent enormous amount of time learning languages and then studied the literature diligently. He attempted to compile a summa (summary) of universal knowledge. The encyclopedia resulting from this tireless attempt combined all ancient and contemporary learning. It preserved many fragments of classical learning that would have otherwise been lost for good. His widely recognized works gave a new impetus to encyclopedic writing, which bore immense fruit in the coming centuries of the Middle Ages (fifth–fifteenth century). To recognize belatedly his contributions to the church, he was canonized a saint by the Roman Catholic Church in 1598, about 950 years after his death, by Pope Clement VIII (1536–1605). He was later declared a Doctor of the Church in 1722 by Pope Innocent XIII (1655–1724). In the latter half of the twentieth century, to recognize his vision/patronage, he was declared the patron saint of the Internet by the Vatican. He was also considered to be the patron saint of computers, their users/programmers, and their engineers/technicians. The University of Dayton, Ohio, USA, named their implementation of the Sakai Project in honor of Saint Isidore, in which a community of educators collaborate to create open software that advances learning, teaching, and research.

Virahanka (About 600 AD)

Virahanka (about 600 AD) was an Indian mathematician who showed how the Fibonacci sequence arises in the analysis of meters with long and short syllables. His work on prosody builds on the Chhanda-sutras of Pingala and was the basis for a twelfth century commentary by Gopala.

Anania Shirakatsi (610–685)

Anania Shirakatsi (610–685) was an Armenian scholar, mathematician, and geographer. His most famous works are *Geography Guide* and *Cosmography*. His greatest claim to fame is his discovery of the fact that the Earth is round and that there was more to the heavens and world than the standard Aristotelian belief purported at the time. The Anania Shirakatsi Medal is an Armenian State Award for scientists in the economics and natural sciences, engineers, and inventors. In 2005, the Central Bank of the Republic of Armenia issued an Anania Shirakatsi commemorative coin.

Venerable Bedé (673–735)

Venerable Bedé (673–735) was born in Jarrow, Northumbria, England, and died there. A lifelong monk, his works in theology, history, chronology, poetry, and biography have led him to be accepted as the greatest scholar of the early medieval era. He knew Greek, then a rare accomplishment in the West, and he received guidance from Pliny the Elder (23–79) and Isidore of Seville. His most famous work, *Historia ecclesiastica gentis Anglorum*, gained him the title "The Father of English History." In a treatise called *De loquela per gestum digitorum*, he explained a way of representing numbers up to 100,000 by using the hands.

Lalla (Around 720–790)

Lalla (around 720–790) was an Indian mathematician and astronomer who belonged to a family of astronomers. Although Lalla followed the traditions of Aryabhata, he did not believe in the rotation of the Earth. His most famous work, *Shishyadhividdhidatantra*, consisted of two volumes: *On the Computation of the Positions of the Planets* and *On the Sphere*. In the first volume he discussed the mean longitudes of the planets, true longitudes of the planets, the three problems of diurnal rotation, lunar eclipses, solar eclipses, syzygies, risings and settings, the shadow of the Moon, the Moon's crescent, conjunctions of the planets with each other, conjunctions of the planets with the fixed stars, and also the patas of the Moon and Sun. In the second volume, he examined: graphical representation, the celestial sphere, the principle of mean motion, the terrestrial sphere, motions and stations of the planets, geography, erroneous knowledge, instruments, and addressed some miscellaneous problems. He used the value of π as $62{,}832/20{,}000$, that is, $\pi = 3.1416$, which is correct to four decimal places. Lalla also wrote a commentary on *Khandakhadyaka*, a work of Brahmagupta, which has not survived. His most popular book on astronomy, *Jyotisaratnakosa*, was used in India for almost 300 years.

Alcuin of York (Around 735–804)

Alcuin of York (around 735–804) was born in Yorkshire and died in Tours. Most of his extant writings deal with theology or history. The textbook *Propositiones ad Acuendos Juvenes* is usually attributed to Alcuin. It contains about 53 mathematical word problems (with solutions) suitable for the instruction of young children. For example: If one hundred bushels of corn be distributed among one hundred people in such a manner that each man receives three bushels, each woman two, and each child half a bushel, how many men, women, and children were there? The general solution is $(20 - 3n)$ men, $5n$ women, and $(80 - 2n)$ children, where n may have any of the values $1, 2, 3, 4, 5, 6$. Alcuin only states the solution for $n = 3$; he gives as the answer 11 men, 15 women, and 74 children. This problem can also be found in Hindu, Chinese, and Arabic works. Other problems involve river crossings; three jealous husbands, each of whom can't let another man be alone with his wife; the problem of the wolf, goat, and cabbage; and the problem of the two adults and two children where the children weigh half as much as the adults. He explained the importance of the number six in the creation of the world by the fact that 6 is a perfect number. The second rebirth of the human race is linked with the deficient number eight, since Noah's ark contained eight human beings, from whom the whole human race was to spring. This shows that this second origin of mankind was less perfect than the first, which had been based on the number 6.

Masha'allah ibn Athari (740–815)

Masha'allah ibn Athari (740–815) was a Persian Jewish astrologer and astronomer from Basra (Iraq), who flourished under the Caliph (successor) al-Mansur (714–775). His name is usually Latinized as Messala or Messahalla. He wrote over 20 works on astrology; however, only a few have survived. He became an authority first in the Middle East and then in the West when horoscopic astrology was transmitted to Europe in the beginning of the twelfth century. The crater Messala on the Moon is named after him.

Abu Jafar Mohammed Ibn Musa al-Khwarizmi (Around 780–850)

Abu Jafar Mohammed Ibn Musa al-Khwarizmi (around 780–850) "Mohammed the father of Jafar and the son of Musa" was born to a Persian family perhaps in Khwarezm (Khiva), Uzbekistan; however, the epithet al-Qutrubbulli indicates he might be from Qutrubbull, a small town near Baghdad. Gerald James Toomer (born 1934) indicated that he was an adherent of the old Zoroastrian religion, but the pious preface to al-Khwarizmi's Algebra shows that he was an orthodox Muslim (true believer). In 786, Harun al-Rashid (766–809) became the fifth Caliph of the Abbasid dynasty, and ruled from his court in the capital city of Baghdad over the Islamic empire, which stretched from the Mediterranean to India. He brought culture to his court and tried to establish the intellectual disciplines, which at that time were not flourishing in the Arabic world. He had two sons, the younger was al-Amin (787–813) and the elder was al-Mamun (786–833). Harun died in 809 and there was an armed conflict between his sons. al-Mamun won the armed struggle, and al-Amin was defeated and killed in 813. al-Mamun became the new Caliph and ruled the empire from Baghdad, continuing the patronage of learning started by his father, who was said to have been inspired by a dream in which Aristotle appeared to him. al-Mamun founded an academy called the House of Wisdom (Bait al-hikma). He also built up a library of manuscripts, the first major library since the library at Alexandria, that collected important works from Byzantium. In addition to the House of Wisdom, al-Mamun set up observatories in which Muslim astronomers could build on the knowledge acquired by earlier people. In Baghdad, scholars encountered and built upon the ideas of ancient Greek and Indian mathematicians. al-Khwarizmi, al-Kindi, Banu Musa (around 800–860), Ibrahim al-Fazari (died 777), his son Mohammad al-Fazari (died 806), and Yaqub ibn Tariq, died about 796 AD) were scholars at the House of Wisdom in Baghdad. Their tasks there involved the translation of Greek and Sanskrit scientific manuscripts. They also studied and wrote on algebra, geometry, and astronomy. There, al-Khwarizmi encountered the Hindu place-value system based on the numerals $0, 1, 2, 3, 4, 5, 6, 7, 8, 9$, including the first use of zero as a place holder in positional base notation, and he wrote a treatise around 820 AD on what we call *Hindu–Arabic numerals*. The Arabic text is lost, but a Latin translation, *Algoritmi de numero Indorum* (that is, al-Khwarizmi on the Hindu Art of Reckoning), a name given to the work by Baldassarre Boncompagni (1821–1894) in 1857, changed much from al-Khwarizmi's original text, of which even the title is unknown. The French Minorite friar Alexander de Villa Dei, who taught in Paris around 1240, mentions an Indian King named Algor as the inventor of the new "art," which itself is called the *algorismus*. Thus the word "algorithm" was tortuously derived from al-Khwarizmi (Alchwarizmi, al-Karismi, Algoritmi, Algorismi, Algorithm) and has remained in use to this day in the sense of an arithmetic operation. This Latin translation was crucial in the introduction of Hindu–Arabic numerals to medieval Europe.

al-Khwarizmi dedicated two of his texts to the Caliph al-Mamun. These were his treatises on algebra and astronomy. The algebra treatise, *Hisab al-jabr w'al-muqabala*, was the most famous and important of all of al-Khwarizmi's works (the Arabic word al-jabr means setting of a broken bone/restoration/completion). al-Khwarizmi's original book on algebra is lost. It is the Latin translation of the title of this text, *Liber algebrae et almucabala*, in 1140 by the Englishman Robert of Chester (about 1150), and by the Spanish Jew John of Seville (about 1125) that gives us the word "algebra." This book was intended to be highly practical, and algebra was introduced to solve problems that were part of everyday life in the Islamic empire at that time, such as those that men constantly encounter in cases of inheritance, legacies, partition, lawsuits, and trade, and in all their dealings with one another, or where the measuring of lands, the digging of canals, geometrical computations, and other objects of various sorts and kinds are concerned. Early in the book al-Khwarizmi describes the natural numbers and then introduces the main topic of the first section of his book, namely, the solution of equations with a single unknown. His equations are linear or quadratic and are composed of units, roots, and squares. al-Khwarizmi's mathematics was done entirely in words, no symbols were used. A unique Arabic copy of this book is kept at Oxford and was translated in 1831 by F. Rosen. A Latin translation is kept in Cambridge. Western Europeans first learned about algebra from his works.

In 773, a man well versed in astronomy by the name of Kanaka from India brought with him the writings on astronomy, the *Siddhanta* of Brahmagupta. Around 820, Caliph al-Mansur asked al-Khwarizmi to translate this work from Sanskrit into Arabic, which became known as the *Sindhind.* It was promptly disseminated and induced Arab scholars to pursue their own investigations into astronomy. The original Arabic version of Sindhind is lost; however, a Latin translation by Adelard of Bath (around 1080–1152) in 1126, which is based on the revision by al-Majriti (about 950), has survived. The four surviving copies of the Latin translation are kept at the Bibliothéque Publique (Chartres), the Bibliothéque Mazarine (Paris), the Bibliotheca Nacional (Madrid), and the Bodleian Library (Oxford). This work consists of approximately 37 chapters and 116 tables on calendars; calculating true positions of the Sun, Moon and planets; tables of sines and tangents; spherical astronomy; astrological tables; parallax and eclipse calculations; and visibility of the Moon. al-Khwarizmi also wrote a longer version of Sindhind, which is also lost, but a Latin version has survived. Some historians assume that al-Khwarizmi was influenced by Ptolemy's work in casting tables in Sindhind; however, he did revise and update Ptolemy's Geography and authored several works on astronomy and astrology. The book *Kitab surat al-ard* (The Image of the Earth), which is based on Ptolemy's Geography, lists coordinates of latitudes and longitudes for 2,402 cities in order of "weather zones," mountains, seas, islands, geographical regions, and rivers, but neither the Arabic copy nor the Latin translation includes the map of the world.

A number of other works were also written by al-Khwarizmi on the astrolabe, sundial, and the Jewish calendar. Several Arabic manuscripts in Berlin, Istanbul, Tashkent, Cairo, and Paris contain further material that perhaps comes from al-Khwarizmi. He also wrote a political history containing horoscopes of prominent persons. According to one story, he was called to the bedside of a seriously ill caliph and was asked to cast his horoscope. He assured the patient that he was destined to live another 50 years, but caliph died within 10 days. A postage stamp was issued by the USSR in 1983 to commemorate the 1,200th anniversary of Muhammad al-Khwarizmi.

Ja'far ibn Muhammad Abu Ma'shar al-Balkhi (787–886)

Ja'far ibn Muhammad Abu Ma'shar al-Balkhi (787–886) was born in Balkh (Afghanistan) and died in al-Wasit (Iraq). He was a mathematician, astronomer, astrologer, and Islamic philosopher. He also wrote on ancient Persian history. His most important work is the Arabic treatise *Kitab al-mudkhal al-kabir ila 'ilm ahkam an-nujjum*, on astrology. Most of his works were later translated into Latin and were known in Europe, where he was called Albumasar. It has been reported that he went to Varanasi (India) and studied astronomy there for 10 years.

Govindasvāmi (or Govindasvāmin or Govindaswami) (Around 800–860)

Govindasvāmi (or Govindasvāmin or Govindaswami) (around 800–860) was an Indian astronomer–mathematician of the ninth century. His most famous treatise is a commentary on the Mahabhaskariya of Bhaskara I. In Govindaswami's commentary there appear many examples of using a place-value notation or, equivalently, positional notation in the Sanskrit system of numerals. One of the interesting aspects of the commentary is Govindaswami's construction of a sine table. Indian mathematicians as well as astronomers constructed sine tables with high precision so that the positions of the planets could be determined as accurately as possible. Govindaswami considered the sexagesimal (base 60) fractional parts of the 24 tabular sine differences from the Aryabhatiya. These led to more accurate sine values at intervals of $90°/24 = 3°45'$. In the commentary, Govindaswami found certain other empirical rules relating to computations of differences of the sine function in the argument range of 60–90°.

Abū-Yūsuf Ya'qūb ibn Ishāq al-Sabbah al-Kindī (Around 801–873)

Abū-Yūsuf Ya'qūb ibn Ishāq al-Sabbah al-Kindī (around 801–873), also known as **Alkindus**, was born in Kufah, Iraq, to an aristocratic family of the Kindah tribe that had migrated there from Yemen. His grandfather and father were the governors of Kufah. At that time, Kufah was a center for Arab culture and learning, and so al-Kindi received his preliminary education there. He then moved to Baghdad to complete his studies, and there he was patronized by the Caliph al-Mamun. al-Mamun appointed him, al-Khwarizmi, and the Banu Musa to the *House of Wisdom*, a recently established center for the translation of Greek and Sanskrit scientific manuscripts. In 833 al-Mamun died and was succeeded by al-Mutasim (around 794–842). al-Kindi continued to be in favor of al-Mutasim, and he employed him to tutor his son Ahmad. al-Mutasim died in 842 and was succeeded by al-Wathiq (816–847) who, in turn, was succeeded in 847 by al-Mutawakkil (822–861). He persecuted all nonorthodox and non-Muslim groups and destroyed most of the synagogues and churches in Baghdad. al-Kindi suffered a reversal of fortune when he was criticized for extolling the "intellect" as being the most immanent creation in proximity to God, which was commonly held to be the position of the angels. At one point he was beaten and his library temporarily confiscated. In 873, al-Kindi died in Baghdad as "a lonely man."

Almost unknown in the Western world, al-Kindi has an honored place in the Islamic world as a philosopher, scientist, astrologer, astronomer, chemist, mathematician, musician, physician, and physicist. In mathematics, he wrote many works on arithmetic that included manuscripts on Indian numbers, the harmony of numbers, lines, multiplication of numbers, the theory of parallels, relative quantities, measuring proportion and time, numerical procedures, and cancelation. He was a pioneer in cryptanalysis and cryptology and devised several new methods of breaking ciphers, including the frequency analysis method. He wrote on space and time, both of which he believed were finite, 'proving' his assertion with a paradox of the infinite. Using his mathematical and medical expertise, he was able to develop a scale that would allow doctors to quantify the potency of their medication. He also experimented with music therapy. al-Kindi described an early concept of relativity, which some see as a precursor to the later theory of relativity developed by Einstein. He was known as the philosopher of the Arabs in contrast to the later Islamic philosophers who, though Muslim, were not Arabs and often learned Arabic as a second language. His philosophical writings seem to be designed to show that he believed that the pursuit of philosophy is compatible with orthodox Islam. He believed that human beings are what they truly are in the soul, not in the body. al-Kindi wrote at least two hundred and sixty books, but only a few have survived in the form of Latin translations by Gherardo of Cremona and others have been rediscovered in Arabic manuscripts. Most importantly, twenty-four of his lost works were located in the mid-twentieth century in a Turkish library. Some consider him to be one of the twelve greatest minds of the Middle Ages. Some of his influential books are *An Introduction to Arithmetic, Manuscript on the use of Indian Numbers, Manuscript on Explanation of the Numbers Mentioned by Plato in his Politics, Manuscript on the Harmony of Numbers, Manuscript of Unity from the Point of View of Numbers, Manuscript on Elucidating the Implied Numbers, Manuscript on Prediction from the Point of View of Numbers, Manuscript on Lines and Multiplication with Numbers, Manuscript on Relative Quantity, Manuscript on Measuring of Proportions and Times, Manuscript on Numerical Procedures and Cancelation, Manuscript on the Body of the Universe is Necessarily Spherical, Manuscript on the Simple Elements and the Outermost Body are Spherical in Shape, Manuscript on Spherics, Manuscript on the Construction of an Azimuth on a Sphere, Manuscript on the Surface on the Water of the Sea is Spherical, Manuscript on how to Level a Sphere, and Manuscript on the Form of a Skeleton Sphere Representing the Relative Positions of the Ecliptic and other Celestial Circles.*

Mahavira (817–875)

Mahavira (817–875) was patronized by the great Rastrakuta King Amoghavarsa Nrpatunga (814–877). Amoghavarsa had become a Jain monk in the later part of his life. His capital was in Manyakheta in modern Karnataka. In keeping with the Jain tradition, Mahavira studied mathematics for its own sake and not in association with astronomy. His *Ganita Sara Samgraha* treats mathematical problems in a more simple and direct manner. In this work, Mahavira summarized and extended the works of Aryabhatta, Bhaskara, Brahmagupta, and Bhaskaracharya. He wrote,

"With the help of the accomplished holy sages, who are worthy to be worshiped by the Lords of the world, ..., I glean from the great ocean of the knowledge of numbers a little of its essence, in the manner in which gems are picked from the sea, gold from the stony rock and the pearl from the oyster shell; and I give out according to the power of my intelligence, the Ganita Sara Samgraha, a small work on arithmetic, which is not small in importance." This treatise contains: a naming scheme for numbers from 10 up to $10^2 4$, which are eka, dasha, ... mahakshobha; formulas for obtaining cubes of sums; techniques for least common denominators (LCM), which were not used in Europe before the fifteenth century; techniques for combinations ${}^nC_r = n(n-1)(n-2)\dots(n-r+1)/r!$, which were later invented in Europe in 1634 by Herrejoni; techniques for solving linear, quadratic, and higher order equations; arithmetic and geometric series; and techniques for calculating areas and volumes. He was the first person to mention that no real square roots of negative numbers can exist. The imaginary numbers were not identified until 1847 by Cauchy in Europe. According to Mahavira, "In all those transactions which relate to worldly, Vedic or other affairs, mathematics is employed. In the science of love, in economics, in music and in drama, in the art of cooking and similarly in medicine and in things like the knowledge of architecture, the science of computation is held in high esteem. Whatever is there in all the three worlds, which are possessed of moving and non-moving beings, all that indeed cannot exist without mathematics." The following interesting problems are due to Mahavira.

The price of nine citrons and seven fragrant wood apples taken together is 107; again the price of seven citrons and nine fragrant wood-apples taken together is 101. O mathematician, tell me quickly the price of a citron and of a fragrant wood-apple quite separately. (8,5).

A wizard having powers of mystic incantations and magical medicines seeing a cock fight going on, spoke privately to both the owners of the cocks. To one he said, "If your bird wins then you give me your stake money, but if you do not win, I shall give you two-thirds of that." Going to the other, he promised in the same way to give three-fourths. From both of them his gain would be only 12 gold pieces. Tell me, O ornament of the first-rate mathematician, the stake money of each of the cock owners. (42,40).

Three merchants begged money mutually from one another. The first on begging 4 from the second and 5 from the third became twice as rich as the others. The second on having 4 from the first and 6 from the third became thrice as rich. The third man on begging 5 from the first and 6 from the second became five times as rich as the others. O mathematician, if you know the *citra-kuttaka-misra* (name of the method for solving problems of this type given by Mahavira), tell me quickly what was the amount in the hand of each. (7, 8, 9).

A powerful, unvanquished, excellent black snake which is 80 angulas in length enters into a hole at the rate of $15/2$ angulas in $5/14$ of a day, and in the course of $1/4$ of a day its tail grows $11/4$ of an angula. O ornament of arithmeticians, tell me by what time this serpent enters fully into the hole? (8 days.)

Of a collection of mango fruits, the king took $1/6$, the queen $1/5$ of the remainder, and the three chief princes $1/4$, $1/3$, and $1/2$ of the successive remainders, and the youngest child took the remaining three mangoes. O you who are clever in miscellaneous problems on fractions, give out the measure of that collection of mangoes. (18).

One-fourth of a herd of camels was seen in the forest: twice the square root of that herd had gone to the mountain slopes; and 3 times five camels remained on the riverbank. What is the numerical measure of that herd of camels? (36).

Mahavira called the following products garland numbers, as they give the same numbers whether read from the left to right or from right to left.

$$\begin{aligned}
&139 \times 109 = 15151\\
&27994681 \times 441 = 12345654321\\
&12345679 \times 9 = 111111111\\
&333333666667 \times 33 = 11000011000011\\
&14287143 \times 7 = 100010001\\
&142857143 \times 7 = 1000000001\\
&152207 \times 73 = 11111111.
\end{aligned}$$

David Eugene Smith (1860–1944), an American mathematician, educator, and historian of mathematics, commented on the Ganita Sara Samgraha, "...oriental mathematics possesses a richness of imagination, an interest in problem solving and poetry, all of which are lacking in the treatises of the West, although abounding in the works of China and Japan." The Arabs were particularly attracted by the poetical, the rhetorical, and the picturesque approach in the Hindu treatment of the subject, more so than by the abstract approach. Mahavira is considered to be the last mathematician from the Jain school.

Thabit ibn Qurra (826–901)

Thabit ibn Qurra (826–901) was born in Harran (Mesopotamia) and died in Baghdad. As a mathematician and astronomer, he translated the works of Apollonius, Archimedes, Euclid, and Ptolemy, and wrote on the theory of numbers. He estimated the distance to the Sun and computed the length of the solar year. He solved a special case of the cubic equation by the geometric method, which al-Haitham had given particular attention to in the year 1000. This was the solution of cubic equations of the form $x^3 + a^2b = cx^2$ by finding the intersection of $x^2 = ay$ (a parabola) and $y(c - x) = ab$ (a hyperbola). He also worked on amicable numbers. In fact, the following remarkable formula is credited to him: If p, q, and r are prime numbers, and if they are of the form

$$p = 3 \cdot 2^n - 1, \quad q = 3 \cdot 2^{n-1} - 1, \quad r = 9 \cdot 2^{2n-1} - 1,$$

then p, q, and r are distinct primes, and $2^n pq$ and $2^n r$ are a pair of amicable numbers. For instance, for $n = 2$, $p = 11$, $q = 5$, $r = 71$ and since the pair of amicable numbers are $2^n pq$, $2^n r$, it follows that $2^2(11)(5) = 220$, and $2^2(71) = 284$. Therefore, 220 and 284 are amicable numbers. This formula was later proposed and used by Fermat. It is not known if Thabit's formula generates infinitely many amicable pairs, but it is known that there are some amicable pairs it does not generate, such as the pair 1184 and 1210, which was first discovered in 1866 by the 16-year-old B.N.I. Paganini. Thabit also gave a generalization of the theorem of Pythagoras.

Abū Kāmil, Shujā' ibn Aslam ibn Muḥammad Ibn Shujā' (850–930)

Abū Kāmil, Shujā' ibn Aslam ibn Muḥammad Ibn Shujā' (850–930) was an Egyptian Muslim mathematician during the Islamic Golden Age (around 850–1258). Abu Kamil Shuja is also known as al-Hasib al-Misri, meaning the 'calculator from Egypt.' Not much is known about his life. He contributed to algebra and geometry. His *Book of Algebra* contains a total of 69 problems. Kamil was probably the first mathematician who used irrational numbers as coefficients of an algebraic equation and accepted irrational numbers as solutions of the equation. Fibonacci employed Kamil's algorithms for the solutions of an algebraic equation of degree up to 8. This made Kamil one of the most important personalities who introduced algebra in Europe. Kamil solved the following system of nonlinear equations with three unknown variables, $x + y + z = 10$, $x^2 + y^2 = z^2$, $xz = y^2$. He wrote all problems rhetorically, lacking mathematical notation other than integers. He was one of the earliest mathematicians to recognize the contributions of al-Khwarizmi to algebra. He is often called "The Reckoner from Egypt."

Abu Abdallah Muhammad ibn Jabir ibn Sinan ar-Raqqi al-Harrani as-Sabi al-Battani (858–929)

Abu Abdallah Muhammad ibn Jabir ibn Sinan ar-Raqqi al-Harrani as-Sabi al-Battani (858–929) was born in Harran near Urfa (Turkey) and died in Damascus. He is considered to be the greatest Muslim astronomer and mathematician. al-Battani is mainly responsible for the modern concepts and notations of trigonometrical functions and identities. Various astrological writings, including a commentary on Ptolemy's *Tetrabiblon* ('four books'), are attributed to him, but his main work was an astronomical treatise with tables, *De Scientia* and *De Numeris Stellarum et Motibus* (About Science and Number of Stars and their Motion), which was extremely influential until the Renaissance. He made astronomical observations of remarkable range and accuracy throughout his life. His tables contain a catalog of fixed stars compiled during the year 800–801. He found that the longitude of the Sun's apogee had increased by 16° and 47 min since Ptolemy's planetary theory in 150 AD. This implies the discovery of the motion of the solar apsides. al-Battani determined several astronomical coefficients with great accuracy. He proved the possibility of annual eclipses of the Sun, and he did not believe in the trepidation of the equinoxes. al-Battani, who was called 'Albategnius' by the Latins, completed the introduction of functions *umbra extensa* and *umbra versa* (cotangents and tangents) and gave a table of cotangents in terms of degrees. He also wrote such books as 'The Book of the Science of the Ascensions of the Signs of the Zodiac in the Spaces Between the Quadrants of the Celestial Sphere,' 'A Letter on the Exact Determination of the Quantities of the Astrological Applications,' and 'Commentary on Ptolemy's Tetrabiblon.' His work, the *Zij*, influenced great European astronomers like Brahe and Kepler. Copernicus repeated what Al-Battani wrote nearly 700 years before him, as the Zij was translated into Latin thrice. al-Battani was called the Ptolemy of Baghdad. The crater Albategnius on the Moon is named after him.

Aboul Wéfá (940–997)

Aboul Wéfá (940–997) was born to a learned family in the province of Khorassan, in Northwest Persia. In 959 he moved to Baghdad and continued the tradition of his predecessors, combining original scientific work with commentary on the classics—the works of Euclid and Diophantus. He also wrote a commentary of al-Khwarizmi's algebra. None of these commentaries have been found. His textbook on practical arithmetic, *Book on What is Necessary from the Science of Arithmetic for Scribes and Businessmen*, written between 961 and 976, is famous. His large astronomical work, *Complete Book*, closely follows Ptolemy's Almagest, but is available only in part. Aboul Wéfá's achievements in the development of trigonometry are undoubted, specifically in the improvement of tables and in the means of solving problems of spherical trigonometry. He died in Baghdad.

Vijayanandi (Around 940–1010)

Vijayanandi (around 940–1010) was born in Benares (Varanasi), India. He was a mathematician and astronomer who made contributions to trigonometry. His most famous work was the *Karanatilaka*, which has survived only through the Arabic translation by al-Biruni. This work provides units of time measurement, mean and true longitudes of the Sun and Moon, the length of daylight, mean longitudes of the five planets, true longitudes of the five planets, the three problems of diurnal rotation, lunar eclipses, solar eclipses, the projection of eclipses, first visibility of the planets, conjunctions of the planets with each other and with fixed stars, the Moon's crescent, and the patas of the Moon and Sun.

Gerbert d'Aurillac, Pope Sylvester II (945–1003)

Gerbert d'Aurillac, Pope Sylvester II (945–1003) was born near Aurillac, Auvergen, France. He entered the monastery of St. Gerald of Aurillac in around 963, and in 967 Borrell II of Barcelona (947–992) visited the monastery, and the abbot asked the Count to take Gerbert with him so that he could study mathematics in Spain. He was in Rome in 971, where his main interest was in music and astronomy. He mastered all the branches of the trivium and quadrivium except logic; to learn this, he moved to Rheims where he studied at Archbishop Adalbero's school, which was the best available at that time. Here he was also invited to teach. Gerbert was especially famous for his construction of abaci and of terrestrial and celestial globes; he was accustomed to use the latter to illustrate his lectures. In 982 he received the abbey of Bobbio, and for the rest of his life he was involved in political affairs; he became Archbishop of Rheims in 991 and of Ravenna in 998; in 999 he was elected Pope, when he took the title of Sylvester II. As head of the Church, he at once commenced an appeal to Christendom to arm and defend the Holy Land, thus forestalling Peter the Hermit by a century. He died in Rome on May 12, 1003, before he had time to elaborate his plans. His library is still preserved in the Vatican. He made a clock which was long preserved at Magdeburg, and an organ worked by steam which was in Rheims even two centuries after his death. His mathematical works include a treatise on arithmetic entitled *De Numerorum Divisione*, and one on geometry, which so overawed his contemporaries that they believed he had a pact with the devil. Gerbert's own work on geometry is of unequal ability, including a few applications to land surveying and the determination of the heights of inaccessible objects, but much of it seems to be adapted from some Pythagorean textbook. However, he solved one problem, which is remarkable. His accomplishments corroborated the suspicions of some of his contemporaries that he had traded his soul to the devil.

al-Karkhi of Baghdad (953–1029)

al-Karkhi of Baghdad (953–1029) was an engineer and the most original writer of arithmetic. Three of his works are known: *al-Kafi fi al-Hisab* (Essentials of Arithmetic), *al-Fakhri fi'l-jabr wa'l-muqabala* (Glories of Algebra), which derives its name from al-Karkhi's friend, the grand vizier in Baghdad at that time; and *al-Badi' fi'l-hisab* (Wonderful Calculation). He used mathematical induction to validate his discoveries of the general binomial theorem and Pascal's triangle. He is also credited to discover the numerical positive solutions to equations of the form $ax^{2n} + bx^n = c$.

Abu-Ali al-Hassan ibn al-Hasan ibn al-Haitham (965–1039)

Abu-Ali al-Hassan ibn al-Hasan ibn al-Haitham (965–1039) was born in Basra (Iraq) and died in Cairo (Egypt). He was one of the most important Muslim mathematicians and one of the greatest investigators of optics of all times. A physician, he wrote commentaries on Aristotle and Claudius Galenus (129–217). His fame came from his treatise on optics, which became known to Kepler during the seventeenth century. This masterpiece, *Kitab al-Manazir* (Book of Mirrors), had a great influence on the training of later scientists in Western Europe. Ibn al-Haitham's writing reveals his precise development of the experimental facilities. His tables of corresponding angles of incidence and refraction of light passing from one medium to another show how he nearly discovered the law of the ratio of sines, for any given pair of media, which was later attributed to Snell. He investigated twilight, relating it to atmospheric refraction by estimating the Sun's height to be 19° below the horizon at the commencement of the phenomenon in the morning or at its termination in the evening. The figure generally accepted now is 18°. On this basis, Ibn al-Haitham estimated the height of the homogeneous atmosphere to be about 55 miles, a rather close approximation. He understood the laws of the formation of images in spherical and parabolic mirrors. He was also familiar with the reasons for spherical aberration and of magnification produced by lenses. He gave a much more sound theory of vision than that known to the Greeks, regarding the lens system of the eye itself as the sensitive part. Ibn al-Haitham was also able to solve a number of advanced questions in geometrical optics; for example, he solved the case of an aplanatic surface for reflection. During his later years, Ibn al-Haitham boasted that he could construct a machine that would control and regulate the course of the Nile River. He was accordingly summoned to Cairo by the Caliph Hakim (985–1021). Ibn al-Haitham somehow escaped by pretending to be insane. He had to maintain this hoax until the death of the Caliph. Later, he earned his living by writing mathematical books. Prior to his death in Cairo, al-Haitham issued a collection of problems similar to the *Data* of Euclid. He is known to have written nearly 200 works on mathematics, physics, astronomy, and medicine.

Abu Arrayhan Muhammad ibn Ahmad al-Biruni (973–1048)

Abu Arrayhan Muhammad ibn Ahmad al-Biruni (973–1048) was born in Khwarazm (Uzbekistan) and died in Ghazni (Afghanistan). He was among those who laid the foundation for modern trigonometry. As a philosopher, geographer, and astronomer al-Biruni was not only a mathematician but a physicist as well. He contributed to physics through studies in specific gravity and the origin of artesian wells. al-Biruni resided in India for nearly 13 years (1017–1030). He was amazingly well read, having knowledge of Sanskrit literature on topics such as astrology, astronomy, chronology, geography, grammar, mathematics, medicine, philosophy, religion, and weights and measures. He translated several of these works into Arabic. In his own work he described the religion and philosophy of India and its caste system and marriage customs. He characterized Hindu mathematics as a mixture of common pebbles and costly crystals. He also had a remarkable knowledge of the Greek sciences and literature. Taki Ed Din al-Hilali considers al-Biruni to be "one of the very greatest scientists of all time." Six hundred years before Galileo, al-Biruni had discussed the possibility of the Earth's rotation around its own axis. al-Biruni carried out geodesic measurements and determined the magnitude of the Earth's circumference in a most ingenious manner. With the aid of mathematics, he fixed the direction to Mecca in mosques all over the world. He produced 146 works in his lifetime (about 13,000 pages). He is remembered for his tongue of silver and gold.

Sripati (Around 1019–1066)

Sripati (around 1019–1066) was born in Rohinikhanda, Maharashtra (India). He was a follower of Lalla's teachings, and wrote on astrology, astronomy, and mathematics. Sripati's major works are: *Dhikotidakarana* (1039), a work of 20 verses on solar and lunar eclipses; *Dhruvamanasa* (1056), a work of 105 verses on calculating planetary longitudes, eclipses and planetary transits; *Siddhantasekhara*, a major work on astronomy in 19 chapters; and *Ganitatilaka*, an incomplete arithmetical treatise in 125 verses based on a work by Sridhara. In these works he provides the rules for the solution of simultaneous indeterminate equations of the first degree that are similar to those given by Brahmagupta. Sripati also wrote *Jyotisaratnamala*, which was an astrology text in 20 chapters based on the Jyotisaratnakosa of Lalla. He also wrote another book on astrology, *Jatakapaddhati*, which contains eight chapters.

Omar Khayyám (1048–1131)

Omar Khayyám (1048–1131) was born in Neyshapur (Iran). His father, Ibrahim, was perhaps a tentmaker (Khayyám means tentmaker). Omar moved to Samarkand and absorbed all that India and Greece had to teach. Afterward he moved to Bukhara and became established as one of the major mathematicians and astronomers of the medieval period. The Seljuk Sultan Jalal-al-Din Malik Shah (1055–1092) invited him to collaborate in devising a new calendar, the Jalali or Maliki. He measured the length of the year to be 365.24219858156 days, which is extremely accurate for his time. Omar wrote three books, on arithmetic, music, and algebra, all before he was 25 years old. In mathematics he is known for his commentary on Euclid's

Elements, the problems of irrational numbers and their relations to rational numbers, Euclid's parallel postulate, and a classification of equations with proofs of each, some algebraic but most geometric. Omar's most original work is his classification to 19 types of cubic equations, which, following Menaechmus, Archimedes, Thabit ibn Qurra, and al-Haitham, he solved by means of intersecting conic sections. Omar believed that for cubic equations arithmetic solutions were impossible. He also worked with the binomial theorem, anticipating Descartes and Newton by several centuries. Omar is known to the Western world as the author of *The Rubaiyat* (Persian poetry). We quote here one stanza from his work: "O thou, slave of the four (elements) and of the seven (spheres of the heavens), thou canst hardly move for the constraints put upon by these four and these seven. Drink wine, for I have told thee a thousand times: thou hast no hope of returning here, once thou art gone thou art gone forever." According to a story, when Omar was young, he and two fellow students, Nizam and Hassan, promised that if one of them became rich, he would share his wealth with the other two. Nizam did become rich, and it was due to Nizam's money that Omar could do mathematics. Omar's work on algebra was translated into French by F. Woepcke in 1851. According to al-Bayhaqi, one of his biographers, Omar Khayyám once read a book seven times to memorize it, and when he returned home, copied out the whole work from memory; a comparison with the original revealed very few errors. He died in Neyshapur. Khwajah Nizami of Samarkand, a pupil of his, related that he used to converse with his teacher in a garden, and that Omar once said that his tomb would be located in a spot where the north wind would scatter rose petals over it. A few years later, after the death of Omar, Khwajah Nizami visited Neyshapur and searched for Omar's grave. It was just outside the garden. Boughs of fruit trees hanging over the garden wall had dropped so many blossoms on the grave that the tombstone was completely hidden.

Rabbi Abraham Bar Hiyya ha-Nasi (1070–1136)

Rabbi Abraham Bar Hiyya ha-Nasi (1070–1136) was born in Catalonia (or perhaps in Barcelona). He was a Spanish Jewish mathematician and astronomer, also known as Savasorda (from the Arabic Sáhib al-Shurta). He is remembered for his role in the dissemination of the quadratic equation. ha-Nasi wrote several scientific works in the fields of astronomy, mathematics, land surveying, and calendar calculations, as well as two religious works, *Hegyon ha-Nefesh* on repentance and *Megillat ha-Megalleh* on redemption. Even these religious works have a scientific, philosophical flavor. Bar Hiyya wrote all his works in Hebrew. He also cooperated with Plato of Tivoli (around 1132–1146) in the translation of scientific works from Arabic into Latin.

Rabbi Abraham ben Meir ibn Ezra (Around 1090–1167)

Rabbi Abraham ben Meir ibn Ezra (around 1090–1167) was born at Tudela, Spain. He was one of the most esteemed biblical commentators among the Jews of the twelfth century. Abraham was also distinguished as a physician, mathematician, astronomer, poet, and grammarian. He left two original treatises on arithmetic, one of which concerns the properties of the numbers from 1 to 9. Abraham traveled extensively, lecturing before large audiences. His writings, some of which have been translated into Latin, are numerous, and evince originality, boldness, and independence. His style is pithy and often epigrammatic. He died in Rome. The crater Abenezra on the Moon is named in his honor.

Fibonacci (Leonardo of Pisa) (Around 1170–1250)

Fibonacci (Leonardo of Pisa) (around 1170–1250) or "son of Bonacci" was born in Pisa, Italy, and was reared as a teenager in Bougie on the Algerian coast where his father was a customs agent for Pisan merchants. Little is known about Leonardo, apart from an autobiographical précis in the introduction to his *Liber Abbaci*. Fibonacci learned basic computational arithmetic from a local teacher, an experience that motivated him to study mathematics in greater detail as he traveled about the Mediterranean on business. Returning to Italy, he completed the writing of Liber Abbaci (Book of the Abacus) in 1202. In this book he states that "The nine Indian numerals are ... with these nine and with the sign 0, which in Arabic is sifr, any desired number can be written" (Fibonacci learned about Indian numerals from his Arab teachers in North Africa). His *Practica geometria*, a collection of useful theorems from geometry and what would eventually be named trigonometry, appeared in 1220, which was followed 5 years later by *Liber quadratorum*, a work on indeterminate analysis. During his time, all works were hand written. His prowess in mathematics came to the attention of the Emperor Frederick II (1194–1250), who invited Leonardo to his court to engage in mathematical tournaments. When Liber Abbaci was written, Europe was ready for such a compendious volume on calculations of any kind in the fields of arithmetic, geometry, and algebra, as these subjects were becoming better known that time. Roman numerals were yielding to Hindu–Arabic numerals to such an extent that, although advocated by Leonardo at the beginning of his book with complete (but tedious!) instructions on how to use them in every arithmetic operation, their use in Florence (the cultural center of fourteenth-century Europe) was forbidden in 1299. A problem in Liber Abbaci led to the introduction of the Fibonacci sequence for which he is best remembered today: A man puts a pair of rabbits in a protected area. How many pairs of rabbits can be produced from that pair in a year if it is supposed that every month each pair begets a new pair which from the second month on becomes productive? The resulting sequence is $1, 1, 2, 3, 5, 8, 13, 21, 34, 55, \ldots$. This sequence, in which each number is the sum of the two preceding numbers, occurs in many places including Pascal's triangle, the binomial formula, probability, plants, nature, and on the piano keyboard, where one octave contains 2 black keys in one group, 3 black keys in another, 5 black keys all together, 8 white keys, and 13 keys total. A special number, closely related to the Fibonacci sequence, is called the *golden number* $\Phi = (1 + \sqrt{5})/2$. This number is obtained by taking the ratio of successive terms in the Fibonacci sequence, and it has some remarkable mathematical properties: its square is equal to itself plus one, while its reciprocal is itself minus one, and the product of π with the square root of ϕ is close to 4. The number ϕ is found in nature, art, architecture, poetry, music, and of course mathematics. Psychologists have shown that the golden ratio subconsciously affects many of our choices, such as where to sit as we enter a large auditorium, where to stand on a stage when we address an audience, and so on. Another interesting problem that appears in Liber Abbaci may have been suggested by a similar problem, number 79 in the Ahmes papyrus: Seven old women went to Rome; each woman had seven mules; each mule carried seven sacks; each sack contained seven loaves; and with each loaf were seven knives; each knife was put up in seven sheaths. Fibonacci also gave a very elegant proof of the identity

$$1^3 + 2^3 + \cdots + n^3 = (1 + 2 + \cdots + n)^2.$$

For this, like Pythagoreans, he used the patterns in the following triangular array

$$\begin{array}{rrrrrrrrrrrl}
 & & & & & & & & & & 1 & 2^0 \\
 & & & & & & & & & 1 & 1 & 2^1 \\
 & & & & & & & & 1 & 2 & 1 & 2^2 \\
 & & & & & & & 1 & 3 & 3 & 1 & 2^3 \\
 & & & & & & 1 & 4 & 6 & 4 & 1 & 2^4 \\
 & & & & & 1 & 5 & 10 & 10 & 5 & 1 & 2^5 \\
 & & & & 1 & 6 & 15 & 20 & 15 & 6 & 1 & 2^6 \\
 & & & 1 & 7 & 21 & 35 & 35 & 21 & 7 & 1 & 2^7 \\
 & & 1 & 8 & 28 & 56 & 70 & 56 & 28 & 8 & 1 & 2^8 \\
 & 1 & 9 & 36 & 84 & 126 & 126 & 84 & 36 & 9 & 1 & 2^9 \\
1 & 10 & 45 & 120 & 210 & 252 & 210 & 120 & 45 & 10 & 1 & 2^{10} \\
\cdot & \cdot & \cdot & \cdot & \cdot & \cdot & \cdot & \cdot & \cdot & \cdot & \cdot & \cdot
\end{array}$$

Unfortunately, besides spreading the use of the Hindu–Arabic numerals and his rabbit problem, Fibonacci's contributions to mathematics have been largely overlooked. However, after the Dark Ages (476–1000), he is considered the first to revive mathematics in Europe.

Vincent of Beauvais (Around 1190–1264)

Vincent of Beauvais (around 1190–1264) was a Dominican priest. He became lector and chaplain to the court of King Louis IX (1214–1270). Vincent wrote a huge encyclopedia, called *Speculum Maius* (Great Mirror), an 80-book compendium of knowledge that included human history from the creation to the time of Louis IX, natural history and science known to the West, and a compendium of European literature, law, politics, and economics. This work also contains passages from Euclid, Aristotle, and Boethius. Further, under the heading 'Algorismus' he gives a very clear explanation of the modern system of numerals, including the zero.

Li Chih (1192–1279)

Li Chih (1192–1279) also known as **Li Zhi** and **Li Ye** was born in Ta-hsing (now Beijing). He passed the civil service examination in 1230 and was the administrative prefect of Jun prefecture in Henan province until the Mongol invasion in 1233. Li was offered a government post by Khublai Khan (1216–1294) in 1260, but politely found an excuse to decline it. His *Ts'eyuan hai-ching* (Sea-Mirror of the Circle Measurements) written in 1248 includes 170 problems dealing with circles inscribed within or escribed without a right triangle, and with determining the relationship between the sides and the radii, some of the problems leading to equations of the fourth degree. Although Li does not explain how to solve these equations, it is often assumed that he used methods similar to Ruffini's (Paolo Ruffini, 1765–1822) rule. He spent his final years teaching at his home near Feng Lung mountain in Yuan, Hebei. Li told his son to burn all of his books except for Sea-Mirror of the Circle Measurements. Fortunately his other texts, both mathematical and literary, have survived.

Albertus Magnus (Around 1193–1280)

Albertus Magnus (around 1193–1280) was a German Catholic philosopher-saint of the Middle Ages with a profound knowledge of the natural sciences. He was born to the Count of Bollstädt in Lauingen, Bavaria, and passed away in Cologne. Albertus studied Aristotle's writings at Padua and theology in Bologna. He taught as a lecturer in Cologne, Germany. He commented on Aristotle's writings, thus making them accessible to a wider academic circle. In 1245, he went to Paris, received his Ph.D., and taught theology. He advocated for the peaceful coexistence of science and religion. He gave an enlightening commentary on the musical practice. The Catholic church honored him as a Doctor of the Church in 1931—a rare distinction bestowed only on 34 persons.

Nasir-al-Din al-Tusi (1201–1274)

Nasir-al-Din al-Tusi (1201–1274) was born in Tus, Khurasan (now Iran), where his father was a jurist in the 12th Imam School. Besides religion, he was also taught logic, physics, and metaphysics by his uncle. He also studied mathematics with other teachers, particularly algebra and geometry. Around the age of 22, al-Tusi joined the court of Nasir al-Din Muhtashim, the Muslim governor of Quhistan, where he was accepted into the Islamic community as a novice (mustajib). In 1256, when the Mongols conquered Almut, al-Tusi joined Hulegu as one of his ministers, and later on, as administrator of Auqaf. He was instrumental in the establishment and progress of the observatory at Maragha. al-Tusi made significant contributions in a large number of subjects and wrote about 64 treatises on subjects including geometry, algebra, arithmetic, trigonometry, spherical trigonometry (including six fundamental formulas for the solution of spherical right-angled triangles), medicine, metaphysics, logic, ethics, and Islamic scholastic philosophy. al-Tusi also wrote many commentaries on Greek texts. These included revised Arabic versions of works by Autolycus, Aristarchus, Euclid, Apollonius, Archimedes, Hypsicles, Theodosius, Menelaus, and Ptolemy. He pointed out several serious shortcomings in Ptolemy's astronomy and foreshadowed the later dissatisfaction with his system that culminated in the Copernican reforms. In addition he wrote poetry in Persian. al-Tusi's influence was felt widely in Renaissance Europe as late as 1651 and 1663; Wallis used his work on the Euclidean postulates. He died in Kadhimain, near Baghdad.

Ch'in Chiu-Shao (Around 1202–1261)

Ch'in Chiu-Shao (around 1202–1261) was a leading Chinese mathematician. In his book *Mathematical Treatise in Nine Sections* (1247), Ch'in solved polynomial equations up to tenth degree, particularly, he found a root of

$$x^4 - 763200x^2 + 40642560000 = 0,$$

by using the method *fan fa* which is now known as Horner's method (William George Horner 1786–1837, this method was also known to Viéte in 1600). In the preface, he mentions that he learned mathematics from a itinerant teacher. In this book he also developed the theory of indeterminate equations. One of his examples can be written as

$$x \equiv 32(\text{mod } 83) \equiv 70(\text{mod } 110) \equiv (30 \text{ mod } 131).$$

Like many traditional Chinese mathematical works, this book reflects a Confucian administrator's concern with calendaring and mensural and fiscal problems. Using the colors familiar on counting boards, Ch'in started the custom of printing negative numbers in black type and positive ones in red.

William of Moerbeke (1215–1286)

William of Moerbeke (1215–1286) was a Flemish medieval translator of philosophical, medical, and scientific texts from Greek to Latin. He was born in Moerbeke, Brabant, and died in Orvieto. William's translations were not only respected during the thirteenth century, but are respected even now by modern twentieth/twenty-first century scholars. He made the first known translation of Aristotle's complete works that include Aristotle's Metaphysics, thus contributing to the revival of the classic ideas of the European Renaissance. His translations were in line with Aristotle's thought and without excess. William also translated the mathematical theses of Heron of Alexandria and Archimedes, but he is better known for his translation of Elements of Theology by Proclus (1268), which was one of the basic texts behind the revival of the neo-Platonic philosophical currents in the thirteenth century.

Campanus of Novara (Around 1220–1296)

Campanus of Novara (around 1220–1296) was a mathematician, astronomer, astrologer, and physician, and was probably born in 1220 in Novara, Lombardy, northern Italy. He called himself Campanus Nouariensis in his writings, and contemporary documents referred to him as Magister Campanus and his full name as Magister Campanus Nouariensis. He is also referred to by other names, such as Campano da Novara and Giovanni Campano. Beginning in the sixteenth century, some authors used the forename Johannes Campanus or Iohannes Campanus. He is best known for his work on Euclid's Elements. In astronomy, he wrote *Theorica Planetarum*, in which he described the motions of the planets and their longitude. He also included instructions for building a planetary equatorium, with its geometrical description. Campanus tried to determine the time of each planet's retrograde motion. He also computed the distances to the planets and their sizes. This work has been called "the first detailed account of the Ptolemaic astronomical system ... to be written in the Latin-speaking West." He served as chaplain to Pope Urban IV (1195–1264), Pope Adrian V (around 1215–1276), Pope Nicholas IV (1227–1292), and Pope Boniface VIII (1235–1303). His contemporary, Roger Bacon, an English philosopher, wonderful teacher, and Franciscan friar, considered Campanus to be one of the greatest mathematicians of their time. Campanus' contributions to astrology were significant. A number of benefices were conferred on him, and Campanus was relatively wealthy at the time of his death at Viterbo, Italy. The crater Campanus on the Moon is named after him.

Jordanus Nemorarius (1225–1260)

Jordanus Nemorarius (1225–1260), also known as Jordanus de Nemore, was born in Borgentreich (near Warburg), Germany. He is considered to be one of the most prestigious natural philosophers of the thirteenth century. He wrote several books on arithmetic, algebra, geometry, and astronomy. His *Demonstratio de algorismo* gives a practical explanation of the Arabic number system and deals only with integers and their uses; *Demonstratio de minutiis* deals with fractions; *De elementis arithmeticae artis* is a theoretical work on arithmetic, which became the standard source of Middle Ages texts; *Liber phylotegni de triangulis* addresses geometry; *Demonstratio de plana spera* is a specialized work on geometry that studies stereographic projection; and the most impressive of all is the *De numeris datis*, which is the first advanced algebra to be written in Europe after Diophantus. On mathematical astronomy he wrote *Planisphaerium* and *Tractatus de Sphaera* and on statics *De ratione ponderis*. Jordanus used letters to represent numbers and stated general algebraic theorems. He is also credited with being the first to correctly formulate the law of the inclined plane. Jordanus visited the Holy Land and on his way back lost his life at sea.

Jacob ben Machir ibn Tibbon (Around 1236–1304)

Jacob ben Machir ibn Tibbon (around 1236–1304), a French Jewish astronomer, physician, and translator, was probably born in Marseilles. His work was utilized by Copernicus and Dante. As a highly respected physician, he served as regent of the faculty of medicine at the University of Montpellier. He translated from Arabic into Hebrew several scientific and philosophical works, such as the Elements of Euclid, treatises on the sphere by Menelaus of Alexandria, the treatise of Qusta ibn Luqam (died 912) on the armillary sphere, a treatise by Autolycus on the sphere in movement, and a compendium of the Almagest of Ptolemy. He is mainly remembered for his description of an astronomical instrument called the quadrant and his astronomical tables. He passed away at Montpellier in around 1304 at the age of about 68.

Yang Hui (1238–1298)

Yang Hui (1238–1298) was a Chinese mathematician from Qiantang (in modern Hangzhou), Zhejiang province, during the Song Dynasty (960–1279). He worked on magic squares, magic circles (arrangements of natural numbers around concentric and non-concentric circles), the binomial theorem, and is best known for presenting *Yang Hui's triangle*, which is the same as Pascal's triangle. He worked with decimal fractions and wrote them in a way that reminds us of our present methods. One of his problems leads to $24.68 \times 36.56 = 902.3008$. Yang Hui also considered quadratic equations with negative coefficients of x.

Maximus Planudes (1260–1330)

Maximus Planudes (1260–1330) was born at Nicomedia in Bithynia, but the greater part of his life was spent in Constantinople. He flourished during the reigns of Michael VIII Palaeologus (1223–1282) and Andronicus II (1259–1332). He wrote about the Indian system of numerals, which shows that this system was known in the Byzantine Empire by the thirteenth century, at about the same time that it gained acceptance in the West.

Richard of Wallingford (1292–1336)

Richard of Wallingford (1292–1336) was born in Berkshire (now Oxfordshire) in England. His father died when he was 10 years of age, and Richard was taken into the care of William de Kirkeby, the Prior of Wallingford. William sent his ward to study at Oxford. At the age of 23, Richard assumed the monastic habit at St. Albans Abbey in Hertfordshire, a ceremonial and nonmetropolitan county in the United Kingdom of Great Britain and Northern Ireland. After 3 years, he returned to Oxford and spent 9 years studying philosophy and theology, where he obtained a Bachelor of Divinity degree and was licensed to lecture in the sentences. In 1326, he became abbot of St. Albans. He made significant contributions to astronomy/astrology and horology while serving as abbot. He designed the astronomical clock as described in the Tractatus Horologii Astronomici in 1327. The clock was completed in 1356, 20 years after Richard's death, but was probably destroyed during Henry VIII's (1491–1547) reformation and the dissolution of St. Albans Abbey in 1539. Richard also designed and constructed a calculator, known as an equatorium, which he called Albion. This could be used for astronomical calculations such as lunar, solar, and planetary longitudes, and could predict eclipses. He published other works on trigonometry, celestial coordinates, astrology, and various religious works. He died of tuberculosis of the lymphatic glands at the age of 43, at St. Albans, Hertfordshire.

Chu Shih-Chieh (about 1300)

Chu Shih-Chieh (about 1300) was one of the greatest Chinese mathematicians of the Sung period (960–1279). We know very little about him, not even when he was born or when he died. He seems to have spent about 20 years as a wandering scholar, earning his living by teaching mathematics. He wrote two major influential texts: *Suan-hsüeh ch'i-meng* (Introduction to Mathematical Studies), which was lost until it reappeared in the nineteenth century; and *Ssu-yüan yü-chien* (Precious Mirror of the Four Elements). In these books, Chu offered the most accomplished presentation of the Chinese arithmetical-algebraic-computational method. He even extended the matrix solution of systems of linear algebraic equations to equations of higher degrees with several unknowns using methods that remind us of Sylvester in the nineteenth century. A diagram of Pascal's triangle also appears in the second book. For this diagram, he says that it belongs to his ancestors.

Ala al-Din Abu'l-Hasan Ali Ibn Ibrahim Ibn al-Shatir (1304–1375)

Ala al-Din Abu'l-Hasan Ali Ibn Ibrahim Ibn al-Shatir (1304–1375) was an astronomer, mathematician, and engineer. He was Muezzin at the great mosque Jami' al-Umawi in Damascus. He wrote a special treatise, *Rasd Ibn Shatir* (observatory of Ibn Shatir). He devised astronomical instruments and wrote various treatises explaining their structure and use. With regular and precise observations, Ibn al-Shatir investigated the motion of the celestial bodies and determined at Damascus the obliquity of the ecliptic path to be 23° 31 min in 1365; the correct value extrapolated from the present one is 23° 31 min and 19.8 s. He died in Damascus.

Nicole Oresme (1320–1382)

Nicole Oresme (1320–1382) was born in the village of Allemagne, France. Very little is known about his childhood. In 1348, he was a student of theology in Paris; in 1355, he received his Master's of Theology and the following year became grand master of the College of Navarre. In 1370, he was appointed chaplain to King Charles V (1338–1380) and advised him on financial matters, and in 1377, he was selected to be the bishop of Lisieux. Oresme was as an economist, mathematician, physicist, astronomer, philosopher, psychologist, musicologist, theologian, a competent translator, and one of the most original thinkers of the fourteenth century. He invented coordinate geometry before Descartes, finding the logical equivalence between tabulating values and graphing them. Oresme was very interested in limits, threshold values, and infinite series by means of geometric additions, which prepared the way for the infinitesimal calculus of Descartes and Galileo. He proved the divergence of the harmonic series $1+\frac{1}{2}+\frac{1}{3}+\cdots$ using the standard method still taught in calculus classes today. The divergence of this series was also proved later by Pietro Mengoli (1625–1686) and Jacob Bernoulli. He is credited with providing the rules of exponents. He was the first to prove Merton's theorem, namely, the distance traveled in a fixed time by a body moving under uniform acceleration is the same as if the body moved at a uniform speed equal to its speed at the midpoint of the time period. Oresme did not believe in Saxony's (Albert of Saxony around 1316–1390) generally accepted model of free fall, but preferred Aristotle's constant-acceleration model, the model that Galileo used 300 years later. He opposed the theory of a stationary Earth as proposed by Aristotle and taught the motion of the Earth 200 years before Copernicus; however, he later rejected his own ideas. He also wrote a work dealing with the nature of light, reflection of light, and the speed of light. Oresme defined the nature of infinity as follows: "And thus outside the sky there is a space that is empty and incorporeal, in a way other than that in which a space may be filled and corporeal, just as that duration called eternity is different from temporal duration, even if that duration were to be perpetual.... And that space I have referred to is infinite and indivisible and is God's immensity, and is God Himself, just as God's duration, called eternity, is infinite and indivisible and God Himself." Oresme died in 1382 in Lisieux, France.

Narayanan Pandit (1340–1400)

Narayanan Pandit (1340–1400) was one of the notable Kerala mathematicians. He wrote two works, an arithmetical treatise called *Ganita Kaumudi* and an algebraic treatise called *Bijganita Vatamsa*. Narayanan is also thought to be the author of an elaborate commentary of Bhaskara II's Leelavati, titled *Karmapradipika*. Although the Karmapradipika contains little original work, it contains seven different methods for squaring numbers, a contribution that is wholly original to Narayanan, as well as contributions to algebra and magic squares. His other major works contain a variety of mathematical developments, including a rule for calculating approximate values of square roots, investigations into the second order indeterminate equation (Pell's equations), solutions of indeterminate higher order equations, mathematical operations with zero, several geometrical rules, and a discussion of magic squares and similar figures. Narayanan also made some contributions to the ideas of differential calculus found in Bhaskara II's work. His work on cyclic quadrilaterals is also available.

Madhava of Sangamagramma (1340–1425)

Madhava of Sangamagramma (1340–1425) was born in Irinjalakkuda, near Cochin, Kerala, India, which was at that time known as Sangamagramma, where sangama means union and gramma stands for village. Very little is known about his life and his work has come to light only very recently. Although there is some evidence of mathematical activity in Kerala prior to Madhava, for example, the text *Sadratnamala* (about 1300), he is considered to be the founder of the Kerala school of astronomy and mathematics. Madhava was the first to invent the ideas underlying the infinite series expansions of functions, power series, trigonometric series of sine, cosine, tangent, and arctangent, rational approximations of infinite series, tests of convergence of infinite series, the estimate of an error term, early forms of differentiation and integration, and the analysis of infinite continued fractions. He fully understood the limit nature of the infinite series. This step has been called the "decisive factor onward from the finite procedures of ancient mathematics to treat their limit-passage to infinity," which is in fact the kernel of modern classical analysis. In Europe, the first such series were developed by James Gregory in 1667. George Joseph in *The Crest of the Peacock* suggests that Indian mathematical manuscripts may have been brought to Europe by Jesuit (Roman Catholic) priests such as the Italian Matteo Ricci (1552–1610, who adopted later a Chinese name Li Ma-dou) who spent 2 years in Kochi (Cochin) after being ordained in Goa in 1580. Kochi is only 70 km from Thrissur (Trichur), which was then the largest repository of astronomical documents. Whish and Hyne, two European mathematicians, obtained their copies of works by the Kerala mathematicians from Thrissur. It is not inconceivable that Jesuit monks may have taken copies to Pisa, where Galileo, Cavalieri, and Wallis spent time; or Padau, where James Gregory studied; or Paris, where Mersenne (who was in touch with Fermat and Pascal) acted as an agent for the transmission of mathematical ideas. He also gave many methods for calculating the circumference of a circle. A value of π correct to 13 decimal places, 3.1415926535898, is attributed to Madhava. However, the text *Sadratnamala*, usually considered to be prior to Madhava (some researchers have claimed that it was compiled by Madhava), gives the astonishingly accurate value of $\pi = 3.14159265358979324$, which is correct to 17 decimal places. He discovered the solutions of transcendental equations by iteration and found the approximation of transcendental numbers by continued fractions. In astronomy, Madhava discovered a procedure to determine the positions of the Moon every 36 min, and methods to estimate the motions of the planets. Madhava also extended some results found in earlier works, including those of Bhaskara. Madhava's works include *Golavada, Madhyamanayanaprakara, Mahajyanayanaprakara, Lagnaprakarana, Venvaroha, Sphutacandrapti, Aganita-grahacara*, and *Candravakyani*. Certainly, Madhava was one of the greatest mathematician–astronomers of the Middle Ages. His discoveries opened the doors to what has today come to be known as mathematical analysis.

The Kerala school of astronomy and mathematics flourished for at least two centuries beyond Madhava. Its active members were Parameshvara Namboodri, Nilakanthan, Jyesthadevan, Sankara Variar, Achyuta Pisharati, and Melpathur Narayana Bhattathiri. This school contributed much to linguistics (the relation between language and mathematics is an ancient Indian tradition), the ayurvedic and poetic traditions of Kerala, a formulation for the equation of the center of the planets, and a heliocentric model of the solar system.

Parameshvara Namboodri (Around 1370–1460)

Parameshvara Namboodri (around 1370–1460) was a direct disciple of Madhava. He belonged to the Alathur village situated on the bank of Bharathappuzha. He was the founder of the *Drigganita* system of astronomy and a prolific author of several important works. He made direct astronomical observations for 55 years before writing his famous work Drigganita. He also wrote commentaries on the works of Bhaskara I, Aryabhata, and Bhaskara II. His Lilavathi Bhasya, a commentary on Bhaskara II's Leelavati, contains one of his most important discoveries: an early version of the mean value theorem. This is considered to be one of the most important results in differential calculus and was later essential in proving the fundamental theorem of calculus. The *Siddhanta-Deepika* by Paramesvara is a commentary on Govindsvamin's commentary on Bhaskara I's Maha-bhaskareeya. This work contains some of his eclipse observations, including one made at Navakshethra in 1422 and two made at Gokarna, in 1425 and 1430. It also presents a mean value type formula for inverse interpolation of the sine function, a one-point iterative technique for calculating the sine of a given angle, and a more efficient approximation that works using a two-point iterative algorithm, which is essentially the same as the modern secant method. Parameshvara was also the first mathematician to give the radius of a circle with an inscribed cyclic quadrilateral, an expression that is normally attributed to Simon Antoine Jean L'Huilier (1750–1840). He is also credited for decimal floating point numbers.

Jemshid al-Kashi (Around 1380–1429)

Jemshid al-Kashi (around 1380–1429) was born in Kashan, Iran. He started his life in poverty, but devoted himself to astronomy and mathematics while moving from town to town. The condition of the region, and therefore his life, improved later. The first event in al-Kashi's life was his observation of an eclipse of the Moon, which he made in Kashan on June 2, 1406. In 1407, he wrote his first book, *Sullam al-sama* (The Stairway of Heaven), while he was very young. In this book he discussed how to determine the altitudes, sizes, and distances of the heavenly bodies. During 1410–1411, al-Kashi finished his second book, *Mukhtasar dar 'ilm-i hay'at* (Compendium of the Science of Astronomy), and dedicated it to a Timurid ruler of Iran. In about 1420, Ulugh Beg (1393–1449), then the ruler, founded a school for the study of theology and science at Samarkand. He invited al-Kashi and 60 other scholars to join him at this school. al-Kashi gladly moved to Samarkand to work with Ulugh Beg. al-Kashi's third important book, *Khaqani Zij*, on astronomical tables and based on the work by Nasir al-Tusi was completed at Samarkand. He dedicated this work to his new patron Ulugh Beg. This book contained a useful tool for calculating the coordinates of the heavens and helped astronomers measure distances, predict the motion of the Sun, Moon, and planets, and calculate longitudinal and latitudinal parallaxes. His other books include *Risala dar sharh-i alat-i rasd* (Treatise on the Explanation of Observational Instruments), *Nuzha al-hadaiq fi kayfiyya san'a al-ala almusamma bi tabaq al-manatiq* (The Method of Construction of the Instrument Called Plate of Heavens), *Risala al-muhitiyya* (Treatise on the Circumference), *The Key to Arithmetic*, and *The Treatise on the Chord and Sine*. In these works al-Kashi showed a great variety in numerical work, comparable to that reached by late sixteenth-century Europeans. He solved cubic equations by iteration (early version of Newton's method) and by trigonometric methods and knew the method now known as Horner's for the solution of general algebraic equations of higher orders. He gave the binomial formula for a general positive integer exponent and computed the value of π to 16 decimals. al-Kashi wrote many letters to his father about his life in Samarkand, which provide an unusual glimpse into the lives of his contemporary scholars and scientists. He also wrote about the observatory that was built at Samarkand in about 1424, which featured an immense astrolabe with a precision-cut marble arc, 62 yards long. After his death Ulugh Beg praised him as a remarkable scientist, one of the most famous in the area, who had a perfect command of the science of the ancients, who made original contributions in the fields of astronomy and mathematics, and who could solve the most difficult problems. With his death the account of Arabic mathematics was almost closed.

Nicholas of Cusa (1401–1464)

Nicholas of Cusa (1401–1464) was the son of a poor fisherman and is often referred to as Nicolaus Cusanus or Nicholas of Kues (Cusa was a Latin place name for a city on the Mosel). He was a German cardinal of the Roman Catholic Church, a philosopher, jurist, mathematician, and an astronomer. He received a doctorate in Canon law from the University of Padua in 1423. Most of his mathematical ideas can be found in his essays, *De Docta Ignorantia* (Of Learned Ignorance), *De Visione Dei* (Vision of God), and *On Conjectures*. He is widely considered to be one of the greatest geniuses and polymaths (a person whose expertise spans a significant number of different subject areas) of the fifteenth century. He made important contributions to the field of mathematics by developing the concepts of the infinitesimal and of relative motion. He approximated the value of π as $\frac{3}{4}(\sqrt{3}+\sqrt{6})$. His writings were essential for Leibniz's discovery of calculus as well as Cantor's later work on infinity; however, some of his mathematical deductions are incorrect. He said that if we can approach the Divine only through symbols, then it is most suitable that we use mathematical symbols, for these have an indestructible certainty. Copernicus, Galileo, and Kepler were aware of the writings of Cusanus. Kepler called Cusanus 'Divinely Inspired' in the first paragraph of his first published work. Predating Kepler, Cusanus said that no perfect circle can exist in the universe, thus opening the possibility for Kepler's model featuring elliptical orbits of the planets around the Sun. He also influenced Giordano Bruno (1548–1600) by denying the finiteness of the universe and the Earth's exceptional position in it (being not the center of the universe, and in that regard equal in rank with the other stars). In accordance with his wishes, his heart is within the chapel altar at the Cusanusstift in Kues.

George Pürbach (1423–1461)

George Pürbach (1423–1461), whose real surname is unknown, was born in Pürbach, a town within the confines of Bavaria and Austria. He was educated in Vienna and then visited several universities in Germany, France, and Italy. In 1450, he was appointed mathematical professor in Vienna, which he continued until his death at the age of 39. One of his most famous pupils is Regiomontanus. Pürbach wrote a work on planetary motions which was published in 1460; one on arithmetic, published in 1511; and a table of eclipses, published in 1514. He replaced Ptolemy's chords with the sines from Arabic mathematics and calculated tables of sines for every minute of arc for a radius of 600,000 units. This table, published in 1541, was the first transition from the duodecimal to the decimal system.

Johann Regiomontanus (1436–1476)

Johann Regiomontanus (1436–1476) was born in Königsberg, Germany. His real name was Johannes Müller. He studied mathematics at the University of Vienna under Georg von Peuerbach (1423–1461). He built astrolabes for Matthias Corvinus of Hungary and Cardinal Bessarion (1403–1472), and in 1465 a portable sundial for Pope Paul II (1417–1471). His first work consisted of an analysis of the Almagest. In this work, the functions sine and cosine were used and a table of natural sines was introduced. The book was published in Venice in 1496, after his death. After finishing this book, Regiomontanus wrote a work on astrology, which contains some astronomical tables and a table of natural tangents. This work was published in 1490. Regiomontanus left Vienna in 1461 and traveled for some time in Italy, Hungary, and Germany, and in 1471 moved to Nuremberg, where he established an observatory, opened a printing press, and lectured. He constructed a mechanical eagle that flapped its wings and saluted the Emperor Maximilian I (1459–1519) on his entry into the city. He then moved to Rome on an invitation from Pope Sixtus IV (1414–1484), who wanted him to reform the calendar. However, shortly after his arrival he died of the plague, but there was a rumor that he was assassinated. He was the first to study Greek mathematical works in order to make himself acquainted with their results and methods of reasoning. He was well read in the works of the Arab mathematicians. He compiled most of this study in his *De Triangulis Omnimodis* (On Triangles of all Kinds), which was completed in 1464; however it was published in 1533. Regiomontanus used algebra to find solutions of geometrical problems. He is considered to be one of the most prominent mathematicians of his generation. The crater Regiomontanus on the Moon is named after him.

Nilakanthan Somayaji (Around 1444–1544)

Nilakanthan Somayaji (around 1444–1544) was a disciple of Damodara (around 1410–1510) and the son of Parameshvara. He was a brahmin from Trkkantiyur in Ponnani taluk. His younger brother Sankara was also a scholar of astronomy. Nilakanthan's most notable work, *Tantrasangraha*, elaborates and extends the contributions of Madhava. Nilakanthan was also the author of *Aryabhatiya-Bhashya*, a commentary on the Aryabhatiya. Significantly, Nilakanthan's work includes inductive mathematical proofs, a derivation and proof of the arctangent trigonometric series, improvements and proofs of other infinite series expansions by Madhava, an improved series expansion of π that converges more rapidly, and the relationship between the power series of π and arctangent, namely,

$$\frac{\pi}{4} = 1 - \frac{1}{3} + \frac{1}{5} - \frac{1}{7} + \frac{1}{9} - \frac{1}{11} + \cdots . \tag{1}$$

Nilakanthan also gave sophisticated explanations of the irrationality of π, the correct formulation for the equation of the center of the planets, and a heliocentric model of the solar system.

Nicolas Chuquet (Around 1445–1488)

Nicolas Chuquet (around 1445–1488) was born in Paris. Not much is known about him except that his bachelor's degree was in medicine, though mathematics occupied much of his interest. He moved to Lyon at the age of about 35. Chuquet is famous for his work *Triparty en la science des nombres*, which was not published in his lifetime. Most of this work was copied without attribution by Estienne de la Roche (about 1480) in his 1520 textbook, *l'Arismetique*. In the 1870s, Aristide Marre discovered Chuquet's manuscript and published it in 1880. In this work, Chuquet covered the Hindu–Arabic numeration system, zero, positive and negative numbers, fractions, order, averages, perfect numbers, progressions, etc. While he called negative numbers 'absurd numbers,' he introduced our familiar numerical terms billion, trillion, quadrillion, etc. Chuquet's work with fractions led to his discovery of an iterative method involving interpolation that makes it possible to solve any algebraic problem having a rational solution. The following interesting problem is due to him:

A woman was carrying some eggs to market. On the way a man made her drop the eggs and had to pay her for them. But the woman did not know how many eggs she had, though she did know that if she counted them two by two, three by three, four by four, five by five, or six by six, she always had one egg left over. And when she counted them seven by seven there were no eggs left over. How many eggs did the woman have? (172,081)

Jacques Lefévre (1455–1536)

Jacques Lefévre (1455–1536) was born in Étaples and died at Nérac, a little man of poor and very humble parents. He studied and worked in the University of Paris. In 1492, he traveled in Italy, studying in Florence, Rome, and Venice, making himself familiar with the works of Aristotle and Plato. He edited the works of Euclid, Jordanus, and John Sacrobosco (around 1195–1256) and wrote commentaries of these works. In 1528, he completed a translation of the Old Testament, and in 1530, he completed translation of the entire Bible, which was the first in the French language.

Scipione del Ferro (1465–1526)

Scipione del Ferro (1465–1526) was born in Bologna, Italy. His father was employed in the paper making industry, which led to the invention of the printing press by Johannes Gutenberg (around 1400–1468) in the 1450s. It is interesting to note that it took more than 1,200 years for paper to make its way from China (Tshai Lun, around 50–121) to Europe (the time lag for gunpowder and the magnetic compass were four and two centuries, respectively). Del Ferro studied and in 1496 was appointed as a lecturer at the University of Bologna, where he taught Arithmetic and Geometry. He retained this post for the rest of his life, although during the last few years he also did some commercial work. There are no surviving scripts from del Ferro. This may be due to his resistance to publishing his work, or communicating to others, except to a very selected group of friends and students. However, he kept a notebook where he recorded all his important discoveries. After his death in 1526, this notebook was inherited by his son-in-law Hannival Nave, who was married to del Ferro's daughter, Filippa. Nave was also a mathematician, who replaced del Ferro at the University of Bologna in 1526. Nave had this notebook in 1543, which has not survived.

Mathematicians from del Ferro's time knew that the general cubic equation $x^3 + bx^2 + cx + d = 0$ can be simplified to one of three cases: $x^3 + mx = n$, $x^3 = mx + n$, and $x^3 + n = mx$, thus the term x^2 can always be removed by appropriate substitution. It was assumed that the coefficients m and n are positive. We recall that in the middle of the sixteenth century in Europe, negative numbers and zero were not in use; otherwise, all the three cases can be simplified to only one case: $x^3 + mx + n = 0$. Del Ferro's role in the history of mathematics is very important; he is credited for his role in algebraically solving one of the outstanding ancient problems of mathematics, namely, at least one of these cubic equations. There are conjectures as to whether del Ferro worked on this topic as a consequence of a trip that the Franciscan monk Lucas Pacioli (1446–1517) made to Bologna. Pacioli taught at the University of Bologna during 1501–1502 and discussed various mathematical topics with del Ferro. It is not certain whether they discussed cubic equations, but Pacioli included them in his famous treatise, *Summa de arithmetica, geometrica, proportioni et proportionalita*, which was published in 1494 in Venice. Cardano, in his book *Ars Magna*, published in 1545, states with great respect that it was del Ferro who was the first to solve the cubic equation, and that the solution he gives is del Ferro's method. Besides a few others, Antonio Maria Fiore (about 1506), who was a student of del Ferro, knew del Ferro's solution of cubic equations.

Very little is known about del Ferro's other works; however, he is also credited with rationalizing fractions, extending methods to rationalize fractions that had square roots in the denominator to fractions whose denominators were the sum of three cube roots, and examining which geometrical problems could be solved with a compass set in a fixed position. He died in 1526, in Bologna, Italy. It is likely that del Ferro would have attained more fame if his notebook had survived.

Nicolaus Copernicus (1473–1543)

Nicolaus Copernicus (1473–1543) was born in the city of Thorn (Torun) in Royal Prussia, part of the Kingdom of Poland, under the name Mikolaj Kopernik. He was the youngest of four children and never married. After the death of Nicolaus's father he was raised by a maternal uncle, who enabled him to enter the University of Krakow (now Jagiellonian University), which was then famous for its mathematics, philosophy, and astronomy curricula. He then studied liberal arts at Bologna, medicine at Padua, and law at the University of Ferrara, where in 1503 he obtained a doctorate in canon law. Then he returned to Poland and settled permanently at the cathedral in Frauenberg (Frombork), near his birthplace. Copernicus practiced medicine in addition to his ecclesiastical duties, wrote a treatise on monetary reform, and turned his attention to his favorite subject, astronomy. He challenged the geocentric model of Ptolemy and proposed the theory that the Earth has a daily motion about its axis and a yearly motion around a stationary Sun. His theory was based on the grounds that placing the Sun at the center of the solar system and assuming that Earth revolves about the Sun reduces the number of equations describing the motion of the planets from about 80 down to 30. Although Copernicus realized that his theory implied an enormous increase in the size of the universe, he declined to pronounce it infinite. His book *De revolutionibus orbium coelestium* appeared in 1543, after his death. The Vatican ignored the book as it only suggested that the mathematical model putting the Sun at the center makes more sense. He didn't assert that this is the way things are. Nicole Oresme also opposed the theory of a stationary Earth as proposed by Aristotle and advocated the motion of Earth some 200 years before Copernicus, though he eventually rejected his own ideas. The heliocentric model of Copernicus marked the beginning of the scientific revolution. In fact, as a consequence of this theory, Kepler determined the ellipticity of planetary orbits, Galileo formulated his new concept of motion, and Newton espoused his theory of universal gravitation.

Michael Stifel (1486–1567)

Michael Stifel (1486–1567) was born in Esslingen, Germany. He obtained his master's degree from the University of Wittenberg. He served in several different churches at different positions, but had to resign and flee every time due to bad circumstances. Stifel made the error of predicting the end of the world on October 3, 1533, and also used a clever rearrangement of the letters LEO DECIMVS to "prove" that Pope Leo X (1475–1521) was 666, the number of the beast given in the Book of Revelation. He was forced to take refuge in a prison after ruining the lives of many believing peasants who had abandoned work and property to accompany him to heaven. In the later part of his life, he lectured on mathematics and theology at the University of Königsberg and at the University of Jena. He invented logarithms independently of Napier, using a totally different approach. His most famous work is *Arithmetica Integra* which was published in 1554. This work contains binomial coefficients, multiplication by juxtaposition, the term "exponent," and the notation $+, -,$ and $\sqrt{}$. For irrational numbers, he wrote: "We are moved and compelled to assert that they are numbers, compelled that is, by the results which follow from their use. On the other hand ... just as an infinite number is not a number, so an irrational number is not a true number, but lies hidden in some sort of cloud of infinity."

Oronce Finé (1494–1555)

Oronce Finé (1494–1555) was born in Briançon (France). He was educated in Paris and obtained a degree in medicine in 1522. He spent time in prison in 1518 and again in 1524, probably for practicing judicial astrology. In 1531, he was appointed to the chair of mathematics at the Collége Royal, where he taught until his death. His most notable works are *Arithmetica Pratica*, printed in Paris in 1530 and reprinted in 1544, and the *Protomathesis* (in 15 books), published in 1532. He is also famous for trying to square the circle and was one of the first mathematicians to use decimal numbers.

Francesco Maurolico (1494–1575)

Francesco Maurolico (1494–1575) was born in Messina, Sicily, Italy, to a family of Greek origin that settled in Messina after the Fall of Constantinople in 1453. His father, Antonio, was a physician in Constantinople and later became Master of the Messina mint. In 1521, Maurolico was ordained a priest. In 1550, he entered the Benedictine Order and became a monk at the Monastero di Santa Maria del Parto á Castelbuono. Two years later, he was consecrated as abbot at the Cattedrale San Nicoló di Messina. He lived his whole life in Sicily except for short periods in Rome and Naples. Maurolico wrote important books on Greek mathematics, restored many ancient works from scant information, and translated many ancient texts such as those by Aristotle, Apollonius, Archimedes, Autolycus, Euclid, Menelaus, and Theodosius. He also worked on geometry, number theory, optics, conics, and mechanics. In his book *Arithmeticorum Libri Duo*, Maurolico presented a variety of properties of the integers together with proofs of these properties. For some of these proofs he used the method of mathematical induction. Maurolico gave methods for measuring the Earth in Cosmographia that were used in 1670 by Jean Picard (1620–1682) to measure the meridian. He made astronomical observations, in particular the supernova that appeared in Cassiopeia in 1572, which is now known as Tycho's supernova. He also made a map of Sicily, which was published in 1575. Maurolico is considered to be one of the great geometers of his time. He died in Messina. The lunar crater Maurolycus is named after him.

Niccoló Fontana (Tartaglia) (1500–1557)

Niccoló Fontana (Tartaglia) (1500–1557) was born in Brescia, Republic of Venice (now Italy), to a poor and very humble family. His father Michele used to ride his horse between Brescia and other towns in the district making deliveries. In 1505, Michele was murdered while out making deliveries, and Niccoló, his two siblings, and his mother suddenly plunged into poverty. In 1512, the French invaded Brescia during the War of the League of Cambrai, and by the end of battle, 46,000 residents were killed. Niccoló took refuge in the cathedral during the massacre along with his mother and younger sister, but a French soldier sliced Niccoló's jaw and palate and left him for dead. His mother discovered that he was still alive and nursed him back to health (according to legend, he survived only because a dog licked the horrible gash). This made normal speech impossible for Niccoló, prompting the nickname "Tartaglia" (stammerer). Tartaglia was self taught in mathematics. Lacking ordinary writing materials, he even used tombstones as slates. He had an extraordinary ability and was able to earn his living teaching at Verona and Venice. Tartaglia gradually acquired a reputation as a promising mathematician by participating successfully in a large number of debates. In the year 1535, Tartaglia announced that he could solve any equation of the type $x^3 + px^2 = q$. Antonio Maria Fiore (who knew del Ferro's solution) did not believe Tartaglia's claim and challenged him to the competition of solving various kinds of cubic equations at Venice. Eight days before the end of the competition, Tartaglia found a method to solve any cubic equation of the form $x^3 + px = q$ and solved all the 30 problems of his opponent in less than 2 h and won the competition. Tartaglia didn't take his prize for winning from Fiore, the honor of winning was enough. In 1539, Cardan learned of Tartaglia's discoveries and persuaded him to tell his secrets, which Cardan promised to keep confidential. With the use of Tartaglia's formula Cardan and Ferrari, his assistant, made remarkable progress. Tartaglia did not publish his formula, despite the fact that it had become well known. Tartaglia probably wished to keep his formula in reserve for upcoming debates. Cardan and Ferrari traveled to Bologna in 1543 and learned from Nave that it was del Ferro, not Tartaglia, who was the first to solve the cubic equation. Cardan felt that although he had sworn not to reveal Tartaglia's method, surely nothing prevented him from publishing del Ferro's formula. In 1545, Cardan published *Ars Magna*, which contained solutions to both the cubic and quartic equations and all of the additional work he had completed on Tartaglia's formula. The following year Tartaglia published a book, *New Problems and Inventions*, which stated his side of the story and his belief that Cardan had acted in bad faith, as well as malicious personal insults against Cardan. In response to this, Ferrari wrote to Tartaglia, berated him mercilessly and challenged him to a public debate. Cardan, however, had no intention of debating Tartaglia. In 1548, the contest between Tartaglia and Ferrari took place in the Garden of the Frati Zoccolanti, Milan. By the end of the first day, things had not gone in favor of Tartaglia, and he decided to leave Milan that night, resulting in Ferrari's victory.

Tartaglia is best known today for his conflicts with Cardan and for the Cardan–Tartaglia formula for solving cubic equations (since the publication of Ars Magna this result is referred to in the literature as Cardan's formula). In 1537, Tartaglia applied mathematics to the investigation of the paths of cannonballs for the first time; his work was later validated by Galileo's studies on falling bodies. He wrote a popular arithmetic text (which was translated into French by Guillaume Gosselin) and was the first Italian translator and publisher of Euclid's Elements in 1543. He also published Latin editions of Archimedes' works. Tartaglia gave a formula for the volume of a tetrahedron in terms of the distance values measured pairwise between its four corners, which is a generalization of Heron's formula for the area of a triangle. He is also known for having devised a method to obtain binomial coefficients. Tartaglia died in 1557, in poverty, in his house in the Calle del Sturion near the Rialto Bridge in Venice.

Jyesthadevan (Around 1500–1600)

Jyesthadevan (around 1500–1600) was a disciple of Damodara and the son of Parameshvara. He was a member of the Kerala School. His key work was the *Yuktibhasa*, the world's first Calculus text, which was written in Malayalam, a regional language of the Indian state of Kerala. It contained most of the developments of earlier Kerala School mathematicians, particularly those of Madhava. Similar to the work of Nilakanthan, it is unique in the history of Indian mathematics. It contains proofs of theorems, derivations of rules and series, a derivation and proof of the series of the arctangent function, and proofs of most of the mathematical theorems and infinite series that had been discovered by Madhava and other mathematicians of the Kerala School. It elaborates on the planetary theory later adopted by Brahe. He studied various topics found in many previous Indian works, including integer solutions of systems of first degree equations solved using the kuttaka method and rules for finding the sines and the cosines of the sum and difference of two angles. Jyesthadevan also gave the earliest statement of Wallis' theorem and geometrical derivations of infinite series.

Sankara Variar (Around 1500–1600)

Sankara Variar (around 1500–1600) was an astronomer–mathematician of the Kerala school of astronomy and mathematics. His family was employed as temple assistants in the Shiva temple at Trkkutaveli. He was a student of Nilakanthan and Jyesthadevan. Sankara's work includes *Kriyakramakari*, a lengthy prose commentary on Leelavati of Bhaskara II, and discussion of Madhava's work.

Girolamo Cardano (1501–1576)

Girolamo Cardano (1501–1576), better known as Cardan, was born in Pavia, Italy. He was the illegitimate child of Fazio Cardano (1444–1524) and Chiara Micheria, who was a widow with three children. Fazio was a learned jurist of Milan as well as a gifted mathematician. He was a close friend of Leonardo da Vinci (1452–1519), who studied mathematics under Pacioli and prepared a series of drawings of regular solids. In his autobiography, Cardan wrote that his mother Chiara attempted to abort him and that he was born "half-dead" and had to be revived in a bath of warm wine. He also wrote that according to the position of the stars and planets, he ought to have been born a monster. Shortly before his birth, Chiara had to move from Milan to Pavia to escape the plague; her three other children died. Chiara and Fazio married after many years. Cardan was a sickly child, and his hot-tempered parents whipped him for the slightest disobedience. Later he became his father's assistant and learned mathematics from him. In 1520, he entered the University of Pavia and later attended the University of Padua to study medicine despite his father's wish that he study law. His father died in 1524, leaving him a house and some inheritance, but by this time Cardan was in the middle of a campaign to become Rector of the university, which he won. At the university he picked up the habit of gambling and became addicted to shuffling cards, tossing dice, and playing chess. Often his understanding of the laws of probability helped him win. Cardan received his doctorate in medicine in 1525 and applied to join the College of Physicians in Milan. He had a reputation as a difficult man, whose unconventional, uncompromising opinions were aggressive, and therefore the College rejected his application on grounds of his illegitimate birth. Between 1526–1532, Cardan practiced medicine in Saccolongo, a small town near Padua, but was not very successful. In 1531, he married Lucia and had three sons, Giovanni, Giambatista, and Aldo, and a daughter, Clara. He moved to Gallarate near Milan in 1532 and again applied unsuccessfully to the College of Physicians. Unable to practice medicine, Cardan resumed gambling in 1533, but things went so badly that he was forced to pawn even his wife's jewelry. During 1534–1536, he taught mathematics in the Piattine schools of Milan. In 1536, he published his first book on medicine, *De malo recentiorum medicorum usu libellus*, in which Cardan attacked not only the College's medical ability but also their character. His third application, in 1539 to the College of Physicians in Milan, was accepted, likely due to the recommendation of the Milanese Senator Sfondrato whose child Cardan had saved from death. In the same year, his first mathematics book, *Practica arithmetice generalis et mensurandi singularis*, was published, which had great merit. In 1539, Cardan persuaded Tartaglia to demonstrate the method of solving cubic equations. Tartaglia revealed the secret to Cardano, who took the following solemn oath: "I swear to you by the Sacred Gospel, and on my faith as a gentleman, not only never to publish your discoveries, if you tell them to me, but I also promise and pledge my faith as a true Christian to put them down in cipher so that after my death no one shall be able to understand them." With Tartaglia's formula, Cardan and Ferrari, his assistants, worked together for 6 years. In 1545, Cardan published one of the best-known mathematics books, Ars Magna, which contained solutions to both the cubic and quartic equations and all of the additional work they had completed (which led to a decade-long fight between Cardan and Tartaglia). In this book, the square roots of negative numbers (complex numbers) were explicitly written for the first time, although he did not understand their properties. Cardan ended this book with this enthusiastic and rather touching statement, "Written in 5 years, may it last as many thousands." In 1543, he published a celebrated treatise on astrology, which is mostly based upon observations carefully collected by him that are seemingly well calculated to support his theories. It also contains evidence of his belief in dreams and omens. Cardan's wife Lucia died in 1546, but he was not greatly saddened because he was more interested in the fame he had achieved from his books, which were among the best sellers of the day. He even declined flattering offers from Pope Paul III (1468–1549) and the King of Denmark. In 1547, he was appointed Professor of Medicine at Pavia. In 1550, his encyclopedic work on the natural phenomena, *De Subtilitate Rerum*, was published. This work contains a wide variety of inventions, facts, and occult superstitions. He asserts that the inorganic realm of Nature be animated no less than the organic; all creation is progressive development; all animals were originally worms; and the inferior metals must be regarded as conatus naturae toward the production of gold. He believed that the animals were not created for the use of man, but exist for their own sake. A similar treatise, *De Varietate Rerum*, appeared in 1557. In 1552, he miraculously saved the life of John Hamilton (1512–1571), Archbishop of St. Andrews, Scotland, while the court physicians of the French King and the German emperor had failed. With this success, Cardan was considered to be the greatest physician in the world. He was immediately appointed Professor of Medicine at Pavia University, and with many wealthy patients he became a rich and successful man. He was the first to describe typhoid fever. In the history of deaf education, Cardan was one of the first to state that deaf people could learn without first learning how to speak.

When Cardan was at the height of his fame, he encountered what he called his "crowning misfortune." Cardan's son, Giambatista (who was qualified as a doctor in 1557) secretly married Brandonia di Seroni, a girl of indifferent character, who Cardan described as a worthless, shameless woman. Cardan continued to support his son financially and the young couple moved in with Brandonia's parents. However, the di Seronis were only interested in what they could extort from Giambatista and his wealthy father, while Brandonia publicly mocked her husband for not being the father of their three children. This taunting drove Giambatista to poison his wife. Following his arrest, he confessed to the crime. Cardan recruited the best lawyers, but at the trial the judge decreed that to save his son's life, Cardan must come to terms with the di Seronis. They demanded a sum that Cardan could never have paid. Giambatista was tortured in jail, his left hand was cut off, and he was executed on April 13, 1560. This was a blow from which Cardan never recovered. He could not forgive himself for failing to avert the disaster, and the terrible sufferings of his favorite son haunted him constantly. As the father of a convicted murderer, Cardan became a hated man. Realizing he had to move, Cardan applied for a professorship of medicine at Bologna and was appointed to the post in 1562. Cardan's time in Bologna was full of controversy. His reputation, in addition to his arrogant manner, ensured that he created many enemies. He humiliated a fellow Medical Professor in front of his students by pointing out errors in his lectures. After a few years, Cardan's colleagues tried to get the Senate to dismiss him by spreading rumors that his lectures were practically unattended. Cardan had further problems with his children. His son Aldo was a gambler and associated with individuals of dubious character. In 1569, Aldo gambled away all of his own clothes and possessions in addition to a considerable sum of his father's money. In an attempt to get money, Aldo broke into his father's house and stole a large amount of cash and jewelery. Cardan sadly reported Aldo to the authorities, and Aldo was banished from Bologna. In 1570, Cardan was imprisoned for a few months by the Inquisition. He was accused of heresy, particularly for having cast the horoscope of Christ, thus attributing the events of His life to the influence of the stars. He was forbidden to hold a university post and barred from further publication. This may have been a deliberate attempt on Cardan's part to gain notoriety; he wrote a whole chapter in his autobiography on wishing to "perpetrate his name" and thus gain a place in history. This is strange, for in all other respects Cardan gave the church his full support. Cardan went to Rome upon his release, where he was met with an unexpectedly warm reception. He was granted immediate membership of the College of Physicians and the Pope, who had now apparently forgiven Cardan, granted him a pension. It was in this period that his autobiography *De vita propria* was written, although it was not immediately published. It was published in Paris in 1643 and Amsterdam in 1654. Italian translations were published in Milan in 1821 and 1922, and in Turin in 1945. A German translation appeared in Jena in 1914, and a French translation in Paris in 1936.

Cardan also made important contributions to alchemy, probability, hydrodynamics, mechanics, geology, and music. His book *Liber de Ludo Aleae* (Games of Chance), which contains the first systematic treatment of probability as well as a section on effective cheating methods, was written in the 1560s but was published after his death in 1663. His chief claim to fame in mechanics was his affirmation of the impossibility of perpetual motion, except in heavenly bodies. He invented the combination lock, a gimbal consisting of three concentric rings that allowed a supported compass or gyroscope to rotate freely, and the Cardan shaft with universal joints, which allows the transmission of rotary motion at various angles and is used in vehicles to this day. Cardan wrote more than 200 works. Although many were burned, more than 100 manuscripts remained. His works were edited in ten volumes by Sponius (Lyons, 1663).

Cardan claimed that every important event in his life was presaged by strange dreams, crowing roosters, howling dogs, cawing ravens, sparks of fire, grunts, knocks, bangings, buzzings, etc. For example, one night he had a dream about a strange girl that he had never seen before. A week later, he claimed, he saw the same girl in the street and a few months later she became his wife. He died in 1576 in Rome on the day that he had (supposedly) astrologically predicted earlier; some suspect he may have committed suicide.

Robert Recorde (Around 1510–1558)

Robert Recorde (around 1510–1558) was born in Tenby, Wales. He entered the University of Oxford in about 1525 and was elected a fellow of All Souls College in 1531. From there, Recorde moved to the University of Cambridge and took a degree in medicine in 1545. Afterward, he returned to Oxford, where he publicly taught mathematics. It appears that Recorde later went to London and acted as physician to King Edward VI (1537–1553) and Queen Mary, to whom some of his books are dedicated. He died in the King's Bench prison at Southwark, where he was confined for debt, in 1558. Recorde published several works in mathematical subjects, chiefly in the form of dialogues between master and scholar. His book *The Grounde of Artes*, a popular arithmetic text that ran through 29 editions, was first published in 1542. Recorde's most cited work is *The Whetstone of Witte*, published in 1557. He is mainly remembered for introducing the signs $+$ for excess and $-$ for deficiency in 1540. These symbols were first used by Johann Widmann (1462–1500) of Eger (now in Czechoslovakia) in his *Mercantile Arithmetic*, published in 1489, and then by the Dutch mathematician Vander Hoecke in 1514, who also introduced the equal sign $=$ in 1557.

Gerhardus Mercator (1512–1594)

Gerhardus Mercator (1512–1594) was born in Rupelmonde, Flanders (now in Belgium). His father was a shoemaker, and Mercator was his seventh child. He went to school in Rupelmonde and studied Latin, religion, and arithmetic. By the time he was 7 years old he was able to speak and read Latin fluently. In the beginning, Mercator's aim was to have a career in the Church like his two eldest brothers. He was educated in 's-Hertogenbosch in the Netherlands by the famous humanist Macropedius. In 1530, he matriculated at the University of Louvain and obtained a master's degree in 1532 and chose not to proceed to a higher degree because he was completely disturbed by the discrepancies between Moses' version of the genesis of the world and the account given by Aristotle and the rest of the philosophers. He began to doubt the truth of all philosophers. Mercator traveled to a number of places and realized that geography might be the subject that can explain the structure of the world that God has created. He returned to Louvain in 1534, where he studied mathematics and its applications to geography and astronomy under Gemma Frisius (1506–1555). In 1535–1536, Mercator constructed a terrestrial globe with Gaspar Myrica and Gemma Frisius. His role in this project was not primarily as a cartographer, but as a highly skilled engraver of copper plates. The globe had been commissioned by the holy Roman Emperor Charles V. In 1537, they constructed a globe of the stars. Mercator's own independent mapmaking began when he produced a map of Palestine in 1537. The first map of the world, using Oronce Finé's projection, was produced by Mercator in 1538. This map is notable for giving North America its name. In 1540, Mercator's map of Flanders was commissioned for political purposes. He produced a new globe in 1541, which was the first to include rhumb lines. In 1544, Mercator was imprisoned in Rupelmonde Castle for 7 months for heresy. During these years, Mercator's main source of income came through his craftsmanship of mathematical instruments and giving mathematics tuition to the students at Louvain. In 1551, he produced a celestial globe, and in 1552, Mercator moved to Duisburg, Duchy of Cleves (now Germany), where he opened a cartographic workshop. In 1554, he completed his long project of producing a map of Europe. This map was drawn using a new projection devised by Johannes Stabius. This work established Mercator as the leading European mapmaker. During 1559–1562, he taught mathematics at the academic college of Duisburg. In 1564, he produced the map of Lorraine and the British Isles, which was used as a Catholic map against the Protestant Queen Elizabeth. The same year, he was appointed Court Cosmographer to the Duke Wilhelm of Cleve (1516–1592). In 1569, he used the *Mercator projection* for a wall map of the world, which has the property that lines of longitude, latitude, and rhumb lines all appear as straight lines. While this makes the map very useful for navigation, it distorts land masses; for example, Greenland looks as large as all of South America. He was the first to use the term *atlas* for a collection of maps. Mercator published corrected and updated versions of Ptolemy's maps in 1578 as the first part of his atlas. His atlas continued with a series of maps of France, Germany, and the Netherlands in 1585. Although the project was never completed, Mercator did publish a further series in 1589 that included maps of the Balkans (then called Sclavonia) and Greece. He suffered successive strokes in 1590, 1592, and 1593, and died in 1594 in Duisburg as a respected and wealthy citizen. He was buried in the city's main cathedral of Saint Salvatorus. Exhibits of his works can be seen in the Mercator treasury in that city. Some maps that were incomplete at the time of his death were completed and published by his son in 1595.

Pierre de la Ramée (1515–1572)

Pierre de la Ramée (1515–1572) was born in Cuts (near Noyon), Vermandois, France. His family was so poor that he had to take a job as a servant in the Collége de Navarre in order to be able to study there. He was very handsome and worked tirelessly. He slept only on straw and used to get up in the wee hours of the morning. His daily routine consisted of reading, writing, and thinking. He used to take a light meal in the morning, and in the evening he ate more. After supper he went for a walk for 2 or 3 h or talked with his friends. His health was good and he recovered from all of his illnesses through sobriety, abstinence, exercise, and above all by playing tennis, which was his usual form of amusement. Ramée made vigorous attacks on Aristotle's philosophy and Euclid's geometry. His works *Dialecticae Institutiones* (1543) and *Aristotelicae Animadversiones* (1543) were very influential. Besides philosophy, he also wrote several textbooks on arithmetic, algebra, geometry, astronomy, and physics. His teaching position was threatened, but through the efforts of Cardinal de Lorraine, Ramée was appointed to chair of rhetoric and philosophy at the Collége de France in 1551. In the religious wars of the period he sided with the reformers and fled to Germany (1568). He was killed in Paris during the St. Bartholomew's Day Massacre.

Jacques Pelletier (Peletier) du Mans (1517–1582)

Jacques Pelletier (Peletier) du Mans (1517–1582) was born in du Mans (France) to a bourgeois family. Peletier was a humanist, poet, and mathematician. In mathematics he is mainly remembered for his new naming convention for large numbers. He died in Paris.

Ludovico Ferrari (1522–1565)

Ludovico Ferrari (1522–1565) was born in Bologna. He began his career as a servant of Cardano and learned mathematics from him. He extended the work of Tartaglia and Cardano, solving the general fourth degree equation in 1540. His solution appears in the Ars Magna. (It is reported that in 1486, the Spanish mathematician Paolo Valmes was burned at the stake for claiming to have solved the quartic equation. Inquisitor General Tomás de Torquemada allegedly told him that it was the will of God that such a solution be inaccessible to human understanding.) Ferrari became rich in the service of the Cardinal Fernando Gonzalo. In 1565, he became a Professor of Mathematics at the University of Bologna, but retired later the same year due to ill health. Ferrari was murdered by his sister or her boyfriend. According to Cardano, Ferrari's sister refused to grieve at her brother's funeral, and having inherited his fortune, she remarried 2 weeks later. After she transferred all of her possessions to her new husband, he promptly left her and she died in poverty.

Rafael Bombelli (1526–1573)

Rafael Bombelli (1526–1573) was born in Bologna, Italy. His father was a wool merchant and his mother was the daughter of a tailor. They had six children, and Bombelli was the eldest son. He was taught privately by an engineer-architect, and so Bombelli turned to the same occupation and in addition became a mathematician. It is not known exactly how Bombelli learned leading mathematical works of that time, but he lived in the right part of Italy to be involved in the major events surrounding the solution of cubic and quartic equations. One of his main contributions was an outline of a series of five textbooks on algebra, but he could only complete the first three, which were published in 1572. The remaining two unfinished manuscripts were discovered in a library in Bologna by Ettore Bortolotti (1866–1947) in 1923 and were published in 1929. In his books, Bombelli introduced the rules for the multiplication of signed numbers, namely: plus times plus makes plus, minus times minus makes plus, plus times minus makes minus, and minus times plus makes minus. He even gave a geometric proof that minus time minus makes plus. Although complex numbers had been used long before Bombelli's book, he developed a calculus of operations with complex numbers. His rules, in our symbolism, are $(\pm 1)i = \pm i$, $(+i)(+i) = -1$, $(-i)(+i) = +1$, $(\pm 1)(-i) = \mp i$, $(+i)(-i) = +1$, and $(-i)(-i) = -1$. He also included examples involving addition and multiplication of complex numbers, such as $8i + (-5i) = +3i$ and

$$\left(\sqrt[3]{4+\sqrt{2i}}\right)\left(\sqrt[3]{3+\sqrt{8i}}\right) = \sqrt[3]{12 + 11\sqrt{2i} + 4i}.$$

Bombelli laid the cornerstone of the theory of complex numbers and showed that correct real solutions can be obtained from the Cardan-Tartaglia formula for the solution to a cubic equation even when the formula gives an expression involving the square roots of negative numbers. Although some historians of mathematics maintain that continued fractions were known to the Greeks, Bombelli used these for the first time to calculate square roots. Some results from Bombelli's incomplete Book IV are related to the geometrical procedures of Omar Khayyám. He died in 1573, probably in Rome, Italy.

Chitrabhanu (About 1530)

Chitrabhanu (about 1530) was a sixteenth century mathematician from Kerala who gave integer solutions to 21 types of systems of two simultaneous Diophantine equations in two unknowns. For each case, Chitrabhanu gave an explanation and justification of his rule as well as an example. Some of his explanations are algebraic, others are geometric.

Zuanne de Tonini da Coi (About 1530)

Zuanne de Tonini da Coi (about 1530) was a teacher in Brescia, a city and commune in the region of Lombardy in northern Italy. He was deeply interested in mathematics from the standpoint of problem solving. In 1530, he sent two cubic equations $x^3 + 3x^2 - 5 = 0$ and $x^3 + 6x^2 + 8x - 1000 = 0$ as a kind of challenge to Tartaglia. Tartaglia was unable to solve the equations for some time, but he did succeed in solving them. The procedure that was developed for the solution was an important step in solving the general problem of cubic equations.

Guillaume Gosselin (1536–1600)

Guillaume Gosselin (1536–1600) was born in 1536 in Caen (France), and perished there, probably during the tragic burning of his library. He worked for the restoration of Greek mathematics during the Renaissance. He also worked with Pelletier du Mans to promote the use of French in science. In 1577, Guillaume published *De Arte Magna* in Latin, and in 1578 a French translation entitled *Tartaglia Arithmetic*. De Arte Magna introduced the reader to the algebraic methods of his predecessors through the arithmetic problems of Diophantus that he discovered in the Latin translation by Guilielmus Xylander (1575). He summarized the rules that supported the conventional arithmetic, algebraic calculations on geometric progressions, extraction of roots, irrational expressions of computations, and notations of the objects of algebra, then the rules that resolved equations of first and second degrees with one unknown with numerical coefficients. This gave the resolution of systems with many unknowns, via linear combinations, where the unknowns were designated by two letters.

Christopher Clavius (1537–1612)

Christopher Clavius (1537–1612) was born in Bamberg, Germany. He did not contribute much to mathematics, but probably did more than any other German scholar of the century to promote knowledge of the subject. He was a gifted teacher and wrote highly esteemed textbooks on arithmetic and algebra. In 1574, he published an edition of Euclid's Elements that is very valuable. He also wrote on trigonometry and astronomy and played an important part in the Gregorian reform of the calendar, which was adopted in 1582. As a Jesuit, he brought honor to his order. In logic, Clavius' Law (inferring of the truth of a proposition from the inconsistency of its negation) is named after him. A large crater on the Moon has been named in his honor.

Ludolph van Ceulen (1539–1610)

Ludolph van Ceulen (1539–1610) was born in Hildesheim, Germany. Like many Germans during the Catholic Inquisition, he emigrated to the Netherlands. He taught fencing and mathematics in Delft until 1594, when he moved to Leiden and opened a fencing school. In 1600, he was appointed to the Engineering School at Leiden, where he spent the remainder of his life teaching mathematics, surveying, and fortification. He wrote several books, including *On The Circle* (1596), in which he published his geometric findings and an approximate value of π correct to 20 decimal places. Just before his death, Ludolph expanded his work and computed π correct to 35 decimal places. After his death, the *Ludolphine number*, 3.14159265358979323846264338327950288..., was engraved on his tombstone in Leiden. The tombstone was later lost, but was restored in 2000.

Francois Viéte (1540–1603)

Francois Viéte (1540–1603) was born in Fontenay-le-Comte, France. He is frequently called by his semi-Latin name of Vieta. He was educated in a cloister school and then at the University of Poitiers, where he graduated with a law degree in 1560. Like his father, he took up the practice of law in his home town and soon rose to prominence. Later, he served as royal counselor to Kings Henry III (1551–1589) and King Henry IV (1553–1610) of France. In his spare time he worked on mathematics and published his results at his own expense. He used letters and symbols in mathematics in the sixteenth century. Since then, algebra has become a generalized arithmetic. The letters, known as variables, are used to stand for numbers, which help greatly in solving complicated problems that involve tedious numerical calculations. He was given credit for finding a trigonometric solution of a general cubic equation in one variable. In relation to the three famous problems of antiquity, he showed that the problems of trisection of an angle and the duplication of a cube both depend upon the solution of cubic equations. He has been called the father of modern algebra and the foremost mathematician of the sixteenth century. Vieta's major publications are *Canon Mathematicus* in 1571, *In artem analyticam isagoge* in 1591, *Supplementum geometriae* in 1593, *De numerosa poteatatum resolutione* in 1600, and *De aequationum recognitione et emendatione*, which was published posthumously in 1615. In arithmetic he must be remembered for his plea for the use of decimal, rather than sexagesimal, fractions. In fact, in his 1571 book, he wrote "Sexagesimals and sixties are to be used sparingly or never in mathematics, and thousandths and thousands, hundredths and hundreds, tenths and tens, and similar progressions, ascending and descending, are to be used frequently or exclusively." In his 1593 book, he calculated π to 10 decimal places using a polygon of $6 \times 2^{16} = 393216$ sides. He also represented π as an infinite product

$$\frac{2}{\pi} = \cos\frac{\pi}{4}\cos\frac{\pi}{8}\cos\frac{\pi}{16}\cos\frac{\pi}{32}\cdots.$$

Vieta died in Paris in 1603, less than 3 months after he had received permission from King Henry IV to retire from public life.

Some entertaining anecdotes told about Vieta:

In 1593, the Belgian mathematician Adrianus van Roomen (1561–1615) proposed a challenge to all contemporary mathematicians to solve a certain 45th degree equation. The Dutch ambassador presented van Roomen's book to King Henry IV with the comment that at that time, there was no mathematician in France capable of solving this equation. The King summoned and showed the equation to Vieta, who immediately found one solution to the equation, and the next day presented 22 more. However, the negative roots escaped him. In return, Vieta challenged van Roomen to solve the problem of Apollonius, that is, to construct a circle tangent to three given circles. Van Roomen was unable to obtain a solution using Euclidean geometry. When he was shown the proposer's elegant solution, he immediately traveled to France to meet Vieta, and a warm friendship developed.

While working for King Henry III, Vieta successfully deciphered a Spanish code consisting of 500 characters. Philipp II (1527–1598) of Spain was so sure that his code was invulnerable that when he heard of this accomplishment, he complained to the Pope that the French were using sorcery against his country, contrary to good Christian morals.

Contemporary historian Jacques Auguste de Thou (1553–1617) said that when Vieta was absorbed with mathematics, he would closet himself in his study for days, sitting at his eating table without food or sleep, except what he could get leaning on his elbow, and neither moved nor sought refreshment at natural intervals.

Tycho Brahe (1546–1601), the Eldest Son of Otto Brahe (1518–1571)

Tycho Brahe (1546–1601), the eldest son of Otto Brahe (1518–1571) and Beatte Bille (1526–1605), both from noble Danish families, was born in Scania (Skane), now in Sweden, then in Denmark. He studied at the universities of Copenhagen and Leipzig, and then at Wittenberg, Rostock, and Basel. In 1566, when he was just 20, he lost part of his nose in a duel with another student in Wittenberg, and he wore a metal insert over his nose for the rest of his life. Tycho was an astronomer and an alchemist, and was known for his very accurate astronomical and planetary observations. His data were used by his assistant, Kepler, to derive the laws of planetary motion. The observatory Uraniburg, built by him in 1576 on a 2,000-acre island near Copenhagen, was the finest observatory in Europe. Tycho observed a new star in 1572 and a comet in 1577. He established the fact that these bodies were beyond the Moon and that therefore the heavens were not unchangeable, as Aristotle and other philosophers had believed. Tycho moved to Prague in 1599, taking his most portable instruments along. He suffered kidney failure and passed away on October 24, 1601, at the age of only 55. Before dying he said several times, "I hope that I will not appear to have died in vain."

Simon Stevin (1548–1620)

Simon Stevin (1548–1620) was born in Bruges, Flanders (now Belgium). Nothing is known about his early years of his education, although it is believed that he was brought up in the Calvinist tradition. Stevin worked for some time as a bookkeeper and cashier at a firm in Antwerp and then became an engineer in the army of Price Maurice of Nassau (1567–1625). While at this post he devised a system of sluices, which could flood the land as a defense should Holland be attacked. During 1571–1577, Stevin traveled extensively through Poland, Prussia, and Norway. After returning, he took a job as a clerk in the tax office at Brugge and in 1581 settled in the Dutch city of Leiden. Here, he attended the Latin school and then entered the University of Leiden. In 1583, Stevin graduated under the Latinized version of his name, Simon Stevinus. In 1585, Stevin published a 36-page booklet, *La Thiende* (The Tenth), which is an account of decimal fractions and their daily use. He compared fractions with an unknown island having beautiful fruits, pleasant plains, and precious minerals. From 1604, Stevin was quartermaster general of the army of the States General. He bought a house in The Hague in 1612 and married Catherine Krai soon afterward. They had four children. Stevin wrote 11 books, in which he made significant contributions to trigonometry, mechanics, architecture, musical theory, geography, fortification, and navigation. On imaginary numbers he wrote: "There are enough legitimate things to work on without need to get busy on uncertain matters." He is best known for his *Principles of Hydrostatics*, published in 1586. This book is considered to be the first substantial advancement in its subject beyond the work of Archimedes. In this book he discussed the triangle of forces, resolution of forces, stable and unstable equilibria, and pressure. Stevin died in 1620 in The Hague. Simon Stevin Square in Bruges contains his statue by Eugéne Simonis (1810–1893), which includes his inclined plane diagram.

John Napier (1550–1617)

John Napier (1550–1617) was the eighth Laird of Merchiston, born at Merchiston Castle, Edinburgh, Scotland, which is now a part of Napier University. His father, Sir Archibald Napier (1534–1608), served for more than 30 years as Master of the Mint. His mother was the sister of Adam Bothwell (around 1527–1593), first reformed Bishop of Orkney, who assisted at the marriage of his notorious kinsman, the Earl of Bothwell (1534–1578), to Queen Mary, and who also anointed and crowned the infant King James VI (1566–1625). He lost his mother when he was 13, and he was sent to the University of St. Andrews in the same year; however, he was a dropout and traveled to the European continent to continue his studies. In 1571, he returned to Gartnes, Scotland, where he built a castle. He had two children with his first wife and ten from his second. Napier lived as a landowner and devoted himself to running his estates. He experimented with fertilizers to improve his land and invented a hydraulic screw and revolving axle that could be used to remove water from flooded coal pits. His varied accomplishments earned him the nickname of "Marvelous Merchiston." Napier's study of mathematics was only a hobby. He invented a device called the *logarithm* (there is evidence that logarithms were known in the eighth century in India) that simplified arithmetic by replacing multiplication with addition. The equation that accomplished this is $\ln \alpha x = \ln \alpha + \ln x$. Napier spent the last 20 years of his life working on a table that he never finished, while the astronomer Brahe waited in vain for the information he needed to complete his calculations. The table *Mirifici logarithmorum canonis constructio* was completed 2 years after the deaths of Napier and Brahe by Napier's friend, Henry Briggs (1561–1630), in London. Logarithms became widely known as 'Briggs's logarithms,' and some older books on navigation still refer to them by this name. (Henry Savile (1549–1622), a classical scholar interested in mathematics, donated money to Oxford to fund two professorships, one in geometry and the other in astronomy. For his work on logarithms and other contributions to mathematics, Savile chose Briggs to become the first Savilian Professor of Geometry in 1619.) The number e, which is used now as a natural base of the logarithm, is often called the Napier number although Napier and Briggs did not work to any base. Laplace asserted that the invention of logarithms by Napier and Briggs "by shortening the labors increased many times the lives of the astronomers." His other mathematical contributions include a mnemonic for formulas used in solving spherical triangles, two formulas known as *Napier's analogies* used in solving spherical triangles, and an invention called *Napier's bones* (numbering rods were made of bone or ivory) that was used for mechanically multiplying, dividing, and finding square roots and cube roots. In principle this invention is the same as a method which had been long in use both in India and in Persia. Napier also found exponential expressions for trigonometric functions and introduced the decimal notation for fractions.

Napier took part in the religious and political controversies of the time. He was a staunch Protestant, and in 1593, he published the Plaine Discovery of the Whole Revelation of St. John. This book, a virulent attack on Catholicism that concluded the Pope was the Antichrist, gained him quite a reputation, not only within Scotland but also on the continent after the work was translated into Dutch, French, and German. To defend his faith and country, Napier designed various weapons. These included burning mirrors for setting enemy ships on fire, an underwater craft, and a tank-like vehicle. He also built and tested an experimental rapid-fire gun that he claimed would have killed 30,000 Turks without the loss of a single Christian.

Napier was commonly believed to be a magician and is thought to have dabbled in alchemy and necromancy. It was said that he would travel about with a black spider in a small box and that his black rooster was his kindred spirit. According to a story, one of Napier's servants once stole from him. In order to expose the thief, Napier told his servants to go, one by one, into a darkened room where they were to pat his psychic rooster. He assured them that the rooster would know from the patting who had been stealing and would then tell Napier. Unknown to the servants, Napier had smeared soot on the bird. The thief, who dared not touch it, was the only with clean hands. He believed that the end of the world would occur in 1688 or 1700. Napier used to frequently walk out in his nightgown and cap. These things, which to the vulgar appear rather odd, fixed on him the character of a warlock. He died in 1617 in Edinburgh, Scotland, and was buried in St. Cuthbert's Church. Napier crater on the Moon is named after him.

Thrikkandiyoor Achyuta Pisharati (1550–1621)

Thrikkandiyoor Achyuta Pisharati (1550–1621) was a student of Jyesthadevan and a member of the Kerala school of astronomy and mathematics. He was a renowned Sanskrit grammarian, astrologer, astronomer, and mathematician. He discovered the technique of 'reduction of the ecliptic.' He authored *Sphuta-nirnaya, Raasi-gola-sphuta-neeti* (raasi meaning zodiac, gola meaning sphere, and neeti roughly meaning rule), *Karanottama*, and a four-chapter treatise *Uparagakriyakrama* on lunar and solar eclipses.

Pietro Antonio Cataldi (1552–1626)

Pietro Antonio Cataldi (1552–1626) was born, studied, and died in Bologna (Italy). He began teaching mathematics in 1569 at the age of 17. Cataldi published more than 30 works on mathematics; however, he is primarily remembered for the *Trattato del modo brevissimo di trovar la radice quadra delli numeri*, published in 1613. This work represents a notable contribution to the development of continued fractions.

Jobst Bürgi (1552–1632)

Jobst Bürgi (1552–1632) was born in Lichtensteig (Switzerland) and died there 79 years later. He worked on astronomical clocks for German kings and on mathematics for himself. He invented logarithms independently of Napier. Bürgi's tables were not published until 1620, but they had been drawn up before 1610. The lunar crater Byrgius is named in his honor.

Melpathur Narayana Bhattathiri (1559–1632)

Melpathur Narayana Bhattathiri (1559–1632) was a student of Achyuta Pisharati and the last scholar from the Kerala school of astronomy and mathematics. His most important scholarly work, *Prkriya-sarvawom*, describes an axiomatic system that elaborates on the classical system of Panini. He is famous for his masterpiece poem *Narayaneeyam* in 1586, a devotional composition in praise of Sri Guruvayoorappan (Sri Krishna). It consists of 1,036 verses and gives a summary of 18,000 verses of the *Bhagavata Purana*. The day that Narayana dedicated his Narayaneeyam to Guruvayurappan is celebrated as "Narayaneeyam Dinam" every year at Guruvayur. It is said that by yogic strength he took the disease of his Guru Achyuta upon himself and relieved him of the pain of his paralysis.

Thomas Harriot (1560–1621)

Thomas Harriot (1560–1621) was born in Oxford. He received his Bachelor of Arts degree in 1579 from Oxford, and in 1584, he accompanied Sir Walter Raleigh (1552–1618) on his expedition to Virginia. This made him the first European mathematician to visit the Americas. Harriot used his knowledge of astronomy to provide navigational expertise. He was also involved in designing Raleigh's ships and served as his accountant. In 1588, he published his report of the New Found Land of Virginia, which described the resources that he found there. After returning to England, he became a prolific mathematician and astronomer to whom the theory of refraction is attributed. His algebra book, *Artis Analyticae Praxis* (1631), was published posthumously in Latin by his friend Walter Warner (1563–1643). Because of this work, Wallis accused Descartes of plagiarism. He wrote aa and aaa as a^2 and a^3, and introduced the signs $>$ and $<$ for strict inequalities. Harriot was the first person to make a drawing of the Moon with the use of a telescope (Greek words meaning 'far' and 'to look'), on July 26, 1609, over 4 months before Galileo. He is sometimes credited for the introduction of the potato to Great Britain and Ireland. He died in London from skin cancer. The Thomas Harriot College of Arts and Sciences at East Carolina University in Greenville is named in recognition of Harriot's scientific contributions to the New Found Land of Virginia.

Edward Wright (1561–1615)

Edward Wright (1561–1615) was born to Henry and Margaret Wright in Garveston village in Norfolk, East Anglia, England. Edward was educated at Gonville-and-Caius College, Cambridge, where he became a fellow during 1587–1596. He is widely known for his book *Certaine Errors in Navigation* (1599), which explained the mathematical basis of the Mercator projection. He set out a reference table (a table of values of the integral of the secant function) giving the linear scale multiplication factor as a function of latitude, computed for each minute of arc up to a latitude of 75°. This table was the essential step required to make navigational use and production of Mercator charts practical. Besides producing the Wright-Molyneux map (1599), he translated Napier's work on logarithms. This translation was published as *A Description of the Admirable Table of Logarithmes* (1616). Edward's contributions are enormous, specially considering his short life. He passed away at the age of 54.

Galileo Galilei (1564–1642)

Galileo Galilei (1564–1642) was an Italian pioneer of modern mathematics, physics, astronomy, and philosophy, who was closely associated with the scientific revolution. He excelled in many different endeavors, including lute playing and painting. He was born in Pisa in the same year as William Shakespeare (1564–1616) and the same year Michelangelo died. He was the first of the seven children of Vincenzo Galilei (1521–1591) and Guilia Ammannati (1538–1620). Vincenzo was a teacher of music and a fine lute player and carried out experiments on strings to support his musical theories. In 1581, Vincenzo published his *Dialogue on Ancient and Modern Music*, which exhibited great knowledge and laborious research. Galileo received some schooling in Pisa, and his father helped him with Greek and Latin. He was very fond of poetry in his youth and had a very strong memory for his favorite authors. He could recite whole works of Publius Vergilius Maro (Virgil, 70–19 BC), Quintus Horatius Flaccus (Horace), Lucius Annaeus Seneca (4 BC–65 AD), Francesco Petrarch (1304–1374), and Ariosto (1474–1533). Galileo was nicknamed 'the Wrangler.' He considered the priesthood as a young man, but at his father's urging he enrolled for a medical degree at the University of Pisa in 1581, which he did not complete. Instead, he began studying mathematics and natural philosophy. During 1582–1583, he took a course on Euclid's Elements from Ostilio Ricci (1540–1603), who was a former pupil of Tartaglia. He also studied the works of Euclid and Archimedes from the Italian translations that Tartaglia had made. In 1586, Galileo wrote his first scientific book, *La Balancitta*, which described Archimedes' method of finding the specific gravities of substances using a balance. In 1587, he worked on centers of gravity, which was a very popular topic among the Jesuit mathematicians. In 1588, Galileo gave a prestigious lecture on the dimensions and location of hell in Dante's Inferno at the Academy in Florence. In 1589, he secured the Chair of Mathematics in Pisa. At University of Pisa, Galileo was required to teach that the Sun and all the planets revolved around the Earth, which was the accepted theory during his time. In his early years at Pisa, he rediscovered (originally due to Charles Bouvelles, around 1470–1553) the curve "Helen of Geometers," or "the apple of discord," known as the cycloid. (A cycloid is the path traced out by a particle of dust on the rim of a wheel rolling along the ground, the shape of which is particularly graceful, and had suggested that arches of bridges should be built in this form.) He composed a severe critique of Tasso for the Florence Academy, in which he described his poetry as "pap for cats." During 1589–1592, he wrote *De Motu*, a series of essays on the theory of motion that he never published. These essays contain the important new idea that theories can be tested by conducting experiments. These essays also contain some conceptual errors. In 1592, Galileo moved to the University of Padua, where he taught geometry, mechanics, and astronomy until 1610. In this period, which he later described as the happiest time of his life, he made significant discoveries in both pure and applied sciences. He invented the Sector, a calculating instrument that has been used ever since by engineers and military officers. During 1595–1598, Galileo devised and improved a Geometric and Military Compass, suitable for use by gunners and surveyors. In 1593, Galileo constructed a thermometer, using the expansion and contraction of air in a bulb to move water in an attached tube. During 1602–1604, he corrected the errors he had made in *De Motu* and formulated the laws of falling bodies, especially on inclined planes, and the pendulum. He also discovered that a projectile follows a parabolic path. However, these famous results were not published for 35 years.

At the University of Padua he was exposed to Copernicus' theory that the Earth and all the other planets revolve around the Sun. In 1604, with the appearance of a New Star (now known as 'Kepler's supernova'), Galileo gave three public lectures against Aristotle's view of astronomy and natural philosophy. He used parallax arguments to prove that the New Star could not be close to the Earth, so all changes in the heavens cannot occur in the lunar region close to the Earth. In May of 1609, Galileo heard about a spyglass that a Dutchman had shown in Venice. By the end of 1609 Galileo, using his technical skills as a mathematician and craftsman, turned his own telescope on the night sky and began to make remarkable discoveries. His observations with this new telescope convinced him of the truth of Copernicus' Sun-centered, or heliocentric, theory. In 1610, Galileo made an account of his telescopic observations in a short book, *Starry Messenger*, which was published in Venice. Galileo had seen mountains on the Moon, rather than a perfect sphere as Aristotle had claimed. He observed the Milky Way, which was previously believed to be a nebula, and found it to be a multitude of stars packed so densely that they appeared to be clouds from Earth. He saw four small bodies orbiting Jupiter that he named Io, Europa, Callisto, and Ganymede, which later astronomers changed to *the Medicean stars*. However, the demonstration that a planet had smaller planets orbiting it was problematic for the orderly, comprehensive picture of the geocentric model of the universe. One month after his little book was published, Galileo resigned his post at Padua and became Chief Mathematician at the University of Pisa (without any teaching duties) and 'Mathematician and Philosopher' to the Grand Duke of Tuscany. In 1611, Galileo visited Rome to demonstrate his telescope to the influential philosophers and mathematicians of the Jesuit Collegio Romano. While in Rome he was made a member of the Accademia dei Lincei (founded in 1603 by the 18-year-old Marquis Frederico Cesi and three friends), which was especially important to Galileo, who signed his name afterward as 'Galileo Galilei Linceo.' In 1612, after a long series of observations, he gave accurate periods for *the Medicean stars*. He also observed the planet Neptune, but did not realize that it was a planet and took no particular interest in it.

In the same year, opposition arose to the heliocentric theory that Galileo supported. In 1614, from the pulpit of Santa Maria Novella, Father Tommaso Caccini (1574–1648) denounced Galileo's opinions on the motion of the Earth, judging them to be dangerous and close to heresy. Galileo went to Rome to defend himself against these accusations, but in 1616 Cardinal Roberto Bellarmino (1542–1621) personally handed Galileo an admonition forbidding him to advocate or teach Copernican astronomy. In 1618, Galileo made a major scientific error. When three comets appeared, he became involved in a controversy regarding the nature of comets, arguing that they were close to the Earth and caused by optical refraction. This was not acceptable to the Jesuits, and they considered Galileo to be a dangerous opponent. In 1622, Galileo wrote the book *Il Saggiatore* (The Assayer), which was approved and published in 1623. This book described Galileo's new scientific method and was written in the language of mathematics. Its characters are triangles, circles, and other geometric figures. In 1624, he developed the first known microscope. In 1630, he returned to Rome to apply for a license to print the *Dialogue Concerning the Two Chief World Systems* (Ptolemaic and Copernican), which was published in Florence in 1632 with formal authorization from the Inquisition and the Roman Catholic Church. This book took nearly 6 years to complete because of Galileo's poor health and takes the form of a dialogue between Salviati, who argues for the Copernican system, and Simplicio, who is an Aristotelian philosopher. Soon after the publication of this book the Inquisition banned its sale and ordered Galileo to appear before the Holy Office in Rome under suspicion of heresy. Because of the illness he was unable to travel to Rome until 1633. Galileo's adherence to experimental results and their honest interpretation led to his rejection of blind allegiance to authority, both philosophical and religious. The Inquisition found Galileo guilty and forced him to recant (publicly withdraw) his support of Copernicus. They forbade further publication of his work and condemned him to life imprisonment, but because of his advanced age allowed him serve his term under house arrest, first with the Archbishop of Siena and then at his villa in Arcetri outside of Florence. While Galileo was under house arrest, he wrote his finest work, *Discourses and Mathematical Demonstrations Concerning Two New Sciences*. This summarized the work he had done 40 years earlier, which is now called kinematics and strength of materials. This work was smuggled out of Italy and taken to Holland, where it was published.

Although Giambattista Benedetti (1530–1590) disproved the Aristotelian view that heavier objects fall faster than lighter objects in 1553, Galileo enunciated the correct mathematical law for the acceleration of falling bodies, which states that the distance traveled starting from rest is proportional to the square of time, and the law of time, which states that velocity is proportional to time. There is an apocryphal story that Galileo dropped a cannonball and a musket ball simultaneously from the leaning tower of Pisa to demonstrate that bodies fall at the same rate. He also concluded that objects retain their velocity unless a force acts upon them, contradicting the Aristotelian hypothesis that objects naturally slow down and stop unless a force acts upon them. Although al-Haytham, Jean Buridan (around 1300–1358), and Mo Tzu (around 470–391 BC) had proposed the same idea centuries earlier, Galileo was the first to express it mathematically. Galileo's *Principle of Inertia* states that "A body moving on a level surface will continue in the same direction at constant speed unless disturbed." This principle was incorporated into Newton's laws of motion as the first law.

Galileo is remembered for his physical theory that accounted for the tides based on the motion of the Earth, which was unfortunately a failure. Kepler correctly associated the Moon with an influence over the tides. A proper physical theory of the tides, however, was not available until Newton. Galileo described an experimental method to measure the speed of light. He is credited with being one of the first to understand sound frequency. He concluded that a waterfall breaks when the weight of the water becomes too great, which is the same reason that water pumps could only raise water to 34 ft. Galileo put forward the basic principle of relativity that the laws of physics are the same in any system that is moving at a constant speed in a straight line, regardless of the system's particular speed or direction. Therefore, there is no absolute motion or absolute rest. This principle provided the basic framework for Newton's laws of motion and is the infinite speed-of-light approximation to Einstein's special theory of relativity. In mathematics, Galileo's paradox, which was known earlier to Albert of Saxony, shows that there are as many perfect squares as there are whole numbers, even though most numbers are not perfect squares. Therefore, the part has the power of the whole. This paradox left no impression on his contemporaries, but played a significant role in Cantor's work 250 years later. He remarked that mathematics is the language of science, but important clues about the behavior of various equations can be obtained by observing physical processes.

Galileo had brilliant eyes, his hair was reddish, and he was naturally robust. Although a devout Roman Catholic, Galileo had a long term relationship with Maria Gamba, who was from Venice. They had three children out of wedlock, two daughters Virginia (1600–1634) and Livia (born 1601) and one son Vincenzio (born 1606). Because of their illegitimate birth the girls were never allowed to marry; however, Vincenzio was later legitimized and married. Galileo became blind at the age of 72 due to a combination of cataracts and glaucoma. In his last year, while totally blind, he designed an escapement mechanism for a pendulum clock. This was perhaps based on his knowledge of the pendulum, which he had understood since 1581. The legendary story is that while in the dome of the Cathedral of Pisa he noticed that a chandelier, now known as the "lamp of Galileo," swung with the same period as his pulse, regardless of its amplitude. He created sketches of various inventions, such as a candle and mirror combination that reflected light throughout a building, an automatic tomato picker, a pocket comb that doubled as an eating utensil, and what appears to be a ballpoint pen. He died at Arcetri in 1642, the year Isaac Newton was born. It was a sad end for so great a man to die condemned of heresy. Although his will indicated that he should be buried beside his father in the family tomb in the Basilica of Santa Croce, his relatives feared that this would provoke opposition from the Church. His body was concealed and placed in a fine tomb in the church in 1737 by the civil authorities, against the wishes of many in the Church. He was formally rehabilitated in 1741, when Pope Benedict XIV (1675–1758) authorized the publication of Galileo's complete scientific works, and in 1758, the general prohibition against heliocentrism was removed from the Index Librorum Prohibitorum. In 1992, 350 years after Galileo's death, Pope John Paul II (1920–2005) gave an address on behalf of the Catholic Church in which he admitted that errors had been made by the theological advisors in the case of Galileo. He declared Galileo's case closed, but did not admit that the Church itself had made mistakes. One Galileo's most telling quotes is "In questions of science, the authority of a thousand is not worth the humble reasoning of a single individual." A very similar twentieth century quote is attributed to Einstein, who in appreciation called Galileo the "father of modern science."

Johannes Kepler (1571–1630)

Johannes Kepler (1571–1630) was born near Stuttgart in Würtemberg. His father abandoned the home and later died abroad. His mother quarreled with all of her relations, including her son Johannes, who was glad to get away as soon as possible. He was a mathematician and astronomer who avidly accepted Copernicus' heliocentric theory and went so far as to defend it in public while still a student. Such rashness displeased his theology professors, who told him that he was unfit for the Protestant ministry. Thus Kepler came to teach high school mathematics and astronomy. He was as much a number mystic as a mathematician. In 1596, Kepler described the solar system with reference to the five regular polyhedrons. In a model, he fitted the five regular polyhedrons between the spheres of the six known planets. From the innermost planet out, the model gives the following sequence: Mercury, octahedron, Venus, icosahedron, Earth–Moon, dodecahedron, Mars, tetrahedron, Jupiter, cube, and Saturn. In 1598, Kepler had to flee to Prague because of the anti-Protestant sentiment during the Counter-Reformation. He was taken on as an assistant to Brahe. After his death in 1601, Kepler became an unpaid court mathematician to the Emperor Kaiser Rudolph II (1552–1612). Kepler formulated his three laws by studying many years' worth of data about the motion of the planets that had been gathered by Brahe.

1. Every planet describes an ellipse, with the Sun at one focus. The other focus is just a mathematical point at which nothing physical exists.

2. The radius vector from the Sun to a planet sweeps out equal areas in equal intervals of time.
3. The squares of the periodic times of planets are proportional to the cubes of the semimajor axes of the orbits of the planets.

It is amazing that he could come up with these laws just by staring at hundreds of pages of numerical data. Kepler could have simplified his task considerably by using the tables of logarithms that John Napier and his assistants were developing at the time. But Kepler could not understand Napier's mathematical justification for his tables, so he refused to use them. Later, Newton conceived that Kepler's laws could be derived, using calculus, from his inverse square law of gravitational attraction. In fact, this is one of the main reasons that Newton developed the calculus. The first law made a profound change in the scientific outlook on nature. From ancient times circular motion had reigned supreme, but now the circle was replaced by the ellipse. The second law is an early example of the infinitesimal calculus. The period of the Earth is 1 year; therefore, according to the third law a planet situated twice as far from the Sun would take nearly 3 years to complete its orbit. He was so overjoyed with the third law that he thanked God with these words:

"The wisdom of the Lord is infinite; so also are His glory and His power. Ye heavens, sing His praises! Sun, Moon, and planets glorify Him in your ineffable language! Celestial harmonies, all ye who comprehend His marvelous works, praise Him. And thou, my soul, praise thy Creator! It is by Him and in Him that all exists. That which we know best is comprised in Him, as well as in our vain science. To Him be praise, honor, and glory throughout eternity."

The modern derivations of Kepler's law show that they apply to any body driven by a force that obeys an inverse square law. For example, they apply to Halley's comet, the asteroid Icarus, the Moon's orbit about Earth, and the orbit of the spacecraft Apollo 8 about the Moon. In *Harmony of the Worlds* (1619), Kepler composed musical motifs for the harmony of the celestial spheres, an idea that goes back to Pythagoras. Mercury's motif, for example, runs up the scale and descends with an arpeggio. In contrast, Earth's motif is a two-note drone, mi, fa, mi ("Misery, famine, misery"). Kepler's mathematical work on the volume of a wine barrel was considered to be at the forefront of integral calculus and the calculation of volumes of solids of revolution. He also touched on the theories of recurring series and difference relations. He was the first to clearly enunciate the law of continuity, the continuous change of a mathematical entity from one state to another, by treating the parabola as the limiting case of either an ellipse or a hyperbola in which one of the two foci moves off to infinity. His personal life was full of misfortunes. When he was 4 years old, a case of smallpox impaired his eyesight. His favorite child died of smallpox. He married his first wife in part for her money and soon realized the error of his ways. When she died of madness, he decided to apply scientific methods to the selection of a second wife: He carefully analyzed and compared the virtues and defects of several ladies before selecting his second partner in matrimony. That marriage was also an unhappy one. He had an affectionate disposition, abundant energy, and methodical habits, with the intuition of a true genius and a readiness to look for new relations between familiar things. He died in 1630 in Ratisbon. The following two quotations are due to him: "Geometry is one and eternal shining in the mind of God. That share in it accorded to men is one of the reasons that Man is the image of God," and "I used to measure the Heavens, now I measure the shadows of Earth. The mind belonged to Heaven, the body's shadow lies here." He also said, "If there is anything that can bind the heavenly mind of man to this dreary exile of our earthly home and can reconcile us with our fate so that one can enjoy living—then it is verily the enjoyment of the mathematical sciences and astronomy."

William Oughtred (1575–1660)

William Oughtred (1575–1660) was born in Eton, Buckinghamshire (England), and was educated there and at King's College, Cambridge, where he became a fellow and worked for some time as a mathematical lecturer. He offered free mathematical tuition to his pupils, including Wallis and Christopher Wren (1632–1723), who is remembered as the architect of St. Paul's Cathedral and some 50 other churches. He is credited with the invention of the slide rule in 1622. His textbook *Clavis Mathematicae* (The Key to Mathematics) on arithmetic, published in 1631, contains practically all that was known of the subject. It was reprinted in several editions and was used by Wallis, Newton, and others. In this work he introduced the $\times$ symbol for multiplication, and used the double colon :: as a proportion sign. Oughtred also wrote a book in 1657 on trigonometry, in which he introduced the abbreviations sin and cos for the sine and cosine functions. He died at the age of 85 in Albury, Surrey. It has been falsely reported that the cause of his death was the excitement and delight that he experienced "at hearing the House of Commons had voted the King's return."

Willebrord Snell (1580–1626)

Willebrord Snell (1580–1626) was a Dutch astronomer and mathematician. At the age of 12 he had become acquainted with the standard mathematical works, and at the age of 22 he succeeded his father as Professor of Mathematics at Leiden. His fame rests mainly on his discovery of the law of refraction (Snell's law) in 1621, which played a significant role in the development of calculus and the wave theory of light. However, it is now known that this law was first discovered by Ibn Sahl (940–1000) in 984.

Johann Faulhaber (1580–1635)

Johann Faulhaber (1580–1635) was born in Ulm (Germany) and left his mortal body there at the age of 55. He learned weaving and started his career as a weaver before becoming a surveyor. However, his inclination toward mathematics led him to attend the school in Ulm, where he learned arithmetic and algebra. The sixteenth century mathematicians who were involved in algebra called themselves Cossists, or simply algebraist. The word cosa means "what" or "symbol" in Italian, referring to those who designate an unknown quantity (to be determined) using a symbol. Johann was possibly the first algebraist in Europe to introduce algebra to fourth/higher degree polynomial equations. Johann computed the sums of powers of integers. Jacob Bernoulli referred to Johann in his combinatorial mathematical paper on Art of Conjecturing (*Ars Conjectandi*) published in 1713 in Basel, Switzerland. Johann also collaborated with Kepler and deeply influenced Descartes. Perhaps his most important contribution was determining the sums of powers of integers, which was published in *Academia Algebra*. He also contributed by explaining logarithms.

Edmund Gunter (1581–1626)

Edmund Gunter (1581–1626) was born in Hertfordshire and educated at Westminster School. In 1599, he was elected a student of Christ Church, Oxford, in 1614 he became a preacher, and in 1615, he received the degree of bachelor in divinity. Throughout his life, Gunter was interested in mathematics with real-world applications. In 1619, he was appointed professor of astronomy in Gresham College, London. He held this post until his death. He is remembered for a seven-place table of the common logarithms of the sine and tangent of angles for intervals of a minute arc, as well as Gunter's chain, Gunter's quadrant, and Gunter's scale (lines, or rules). He coined the words cosine and cotangent.

Claude Gaspard Bachet de Méziriac (1581–1638)

Claude Gaspard Bachet de Méziriac (1581–1638) was born in Bourg-en-Bresse. He wrote *Problémes plaisants et délectables* in 1612 and its enlarged edition in 1624, which contains interesting tricks and questions, many of which are quoted by Walter William Rouse Ball (1850–1925) in *Mathematical Recreations and Essays*. He also wrote *Les éléments arithmétiques*, which exists in manuscript form, and made a translation of the Arithmetic of Diophantus. It was this translation in which Fermat wrote his famous margin note claiming that he had a proof of his famous last theorem. Bachet was the earliest writer to discuss the solution of indeterminate equations by means of continued fractions. He is also credited with the construction of magic squares. He was elected to the French Academy in 1635. The following moral problem is from his work:

There are fifteen Christians (C) and fifteen Turks (T) aboard a ship that runs into a storm. The captain says that the only way to save the vessel is to lighten her by throwing half of the passengers into the sea. The choice must clearly be made by lot, and the method agreed upon is that all the passengers shall stand in a line (circle), and the captain shall then count along the line in nines and throw every ninth passenger into the sea, until there are only 15 passengers left. However, since he himself was a Christian, the captain wants to save the Christians. How can he arrange the passengers so that it is always a Turk and never a Christian that is in the ninth place? (CCCCTTTTTCCTCCCTCTTCCTTTCTTCCT). (The same problem also appears earlier in the work of Chuquet, who divides the passengers into Christians and Jews.)

Pére Marin Mersenne (1588–1648)

Pére Marin Mersenne (1588–1648) was born in Maine, France, into a family of laborers and attended the College of Mans and the Jesuit College at La Fléche. He continued his education at the Sorbonne, studying theology from 1609 to 1611. He joined the religious order of the Minims in 1611, a group whose name comes from the word *minimi* (the members of this group considered themselves the least religious order). Besides prayer, the members of this group devoted their energy to scholarship and study. In 1612, he became a priest at the Palace Royale in Paris, and between 1614 and 1618 he taught philosophy at the Minim Convent at Nevers. He returned to Paris in 1619, where his cell in the Minims de l'Annociade became a meeting place of French scientists, philosophers, and mathematicians, including Fermat and Pascal. Mersenne corresponded extensively with scholars throughout Europe, serving as a clearinghouse for mathematical and scientific knowledge, a function later served by mathematical journals (and today also by the Internet). Mersenne wrote books covering mechanics, mathematical physics, music, and acoustics. He studied prime numbers and tried unsuccessfully to construct a formula representing all primes. In 1644, Mersenne claimed that $2^p - 1$ is prime for $p = 2, 3, 5, 7, 13, 17, 19, 31, 67, 127, 257$ but is composite for all other primes less than 257. It took over 300 years to determine that Mersenne's claim was wrong five times. Specifically, $2^p - 1$ is not prime for $p = 67$ and $p = 257$ but is prime for $p = 61,\ p = 87$, and $p = 107$. He is often referred to as "the father of acoustics" for his work in that field. It is also noteworthy that Mersenne defended two of the most famous men of his time, Descartes and Galileo, from religious critics. He also helped expose alchemists and astrologers as frauds.

Gerard Desargues (1591–1661)

Gerard Desargues (1591–1661) was born in Lyons. He was an architect, engineer, and was a technical advisor to the French government. He met Descartes at the siege of La Rochelle in 1628 and in the 1630s was a member of the Parisian group that included Descartes, Pascal, Fermat, and Mersenne. According to him, on every straight line there is one and only one point at infinity, and this point is common to all lines parallel to the given straight line and to no other; furthermore, two lines are parallel if they meet only at infinity. In 1636 and 1639 he published a treatise and proposals in which he invented *Projective Geometry*. This geometry had been practiced at least one century before by the artists to paint pictures that conveyed the illusion of three-dimensional height, width, and depth. Leonardo da Vinci's *Last Supper* is one of the best-known examples of the use of projective geometry or perspective. Unfortunately, his work was too difficult for his time; his terms were obscure and he did not use Cartesian symbolism. Desargues' geometric innovations were hidden for nearly 200 years. A manuscript by Desargues turned up in 1845, about 30 years after Poncelet had rediscovered projective geometry. Desargues' theorem states that in a plane, if two triangles are placed so that lines joining corresponding vertices meet in a point, then corresponding sides, when extended, will meet in three collinear points. (Collinear points are points lying on the same line.) However, this theorem appears nowhere in the work of Desargues, but rather in the work of Abraham Bosse (1602–1676).

Albert Girard (1595–1632)

Albert Girard (1595–1632) was born in Saint-Mihiel, France. He studied at the University of Leiden. He published a short treatise on trigonometry in 1626 that contains the first use of the abbreviations sin, cos, and tan. Girard was the first mathematician to offer a formula for the area of a triangle inscribed on a sphere; this result is called Girard's theorem. In 1627, he published an edition of Samuel Marolois' (1572–1627) Geometry with several additions. Girard had some early thoughts on the fundamental theorem of algebra, which he stated in *Invention Nouvelle en l'algÃ¨bre* (1629) published in Amsterdam. This work also contains the earliest use of brackets, a geometrical interpretation of the negative sign, the distinct recomposition of imaginary roots, and Newton's rule for finding the sum of like powers of the roots of a polynomial. Unfortunately, his work was not recognized during his lifetime. He died in Leiden.

René Descartes (1596–1650)

René Descartes (1596–1650) was born at La Haye, a small town about 200 miles southwest of Paris in Touraine, a region and former province of France. He was the son of a minor nobleman and belonged to a family that had produced a number of learned men. His mother died of tuberculosis shortly after he was born. He inherited the disease from his mother, and as an infant he was so pale and weak that the physician offered no hope of saving him. Afraid that he might lose him, his father kept careful watch over his son's delicate health. Descartes was a thoughtful child who asked so many questions that his father called him "my little philosopher." At the age of eight he was enrolled in the Jesuit College of La Fléche in Anjou, where he remained for 8 years to be educated in traditional Aristotelian philosophy. Besides the usual classical studies he received instruction in mathematics and Scholastic philosophy, which attempted to use human reason to understand Christian doctrine. In those 8 years, his teachers at the Jesuit College, recognizing his physical weakness and mental alertness, allowed him to remain in bed beyond rising hours. They noted that he used the time meditating and devouring one classic after another while the other boys recited in the classroom. It was a habit that he kept all his life. Indeed, when he visited Pascal in 1647, Descartes stated that he could only do good work in mathematics and maintain his health when no one was allowed to get him up in the morning. Upon his graduation from the college, Descartes went to Paris and fell in with a group of wealthy young men who spent their time gambling. He soon became tired of this life and hid himself away in the suburbs of Paris, in order to spend the next 2 years in mathematical investigation with Father Mersenne. During this period, Descartes entered the University of Poitiers, where he earned a degree in civil and canon law in 1616. After a brief probe into the pleasures of Paris he became a military engineer, first for the Dutch Prince of Nassau and then for the German Duke of Bavaria. After the wars, he returned to Paris where he stalked the city as an eccentric, wearing a sword in his belt and a plumed hat, and dabbled in the study of human physiology, philosophy, glaciers, meteors, and rainbows. He eventually moved to Holland, and during 1629–1633 he wrote *Le Monde*, which embodied an attempt to give a physical theory of the universe; however, the incomplete manuscript was not published until 1664. In 1637, he published his *Discourse on Method.* The Discourse, which contained important mathematical work and three essays on Meteors, Dioptrics, and Geometry, produced an immense sensation and his name became known throughout Europe, even though it was intentionally made difficult in order to evade popularity. In the very last line of Geometry, Descartes wrote "I hope that posterity will judge me kindly, not only as to things which I have explained, but also as to those which I have intentionally omitted so as to leave to others the pleasure of discovery." The English philosopher John Stuart Mill (1806–1873) called this "the greatest single step ever made in the progress of the exact sciences." In 1641, he published a work called *Meditations on First Philosophy*, in which he explained at some length his views of philosophy as sketched out in the

Discourse. In 1644, he issued the *Principia Philosophiae*, the greater part of which was devoted to physical science, especially the laws of motion and the theory of vortices. He laid down ten laws of nature, the first two being almost identical to the first two laws of motion given by Newton; the remaining eight laws are inaccurate. According to him, the nature of matter is uniform though there are three forms of it. He stated that the Sun is the center of an immense whirlpool of this matter, in which the planets float and are swept round like straws in a whirlpool of water. Each planet is thought to be at the center of a secondary whirlpool in which its satellites are carried. These secondary whirlpools are supposed to produce variations of density in the surrounding medium that constitutes the primary whirlpool, and so cause the planets to move in ellipses and not in circles. In 1647, he received a pension from the French court in honor of his discoveries. In 1649, Queen Christina (1626–1689, headstrong girl, interested in all aspects of learning) of Sweden summoned Descartes, against his wishes, to give her instruction in philosophy. The Queen, a rugged, tireless woman, wished to be instructed three times a week and made Descartes come for her lessons at five o'clock in the morning. Having long been accustomed to lying late in bed, Descartes was made ill by the cold of the winter mornings and died of pneumonia soon after, on February 11, 1650. He was buried in Stockholm and efforts to bring his remains to France failed. However, after 17 years his bones, except for those of his right hand, were returned to France and reinterred in Paris at what is now the Panthéon. The bones of the right hand were kept as a souvenir by the French Treasurer General, who made the negotiations.

Descartes is credited for the rectangular coordinate system, which led to the development of the entirely new field of analytic geometry, otherwise known as Cartesian or coordinate geometry. This concept is so common today that we often use rectangular coordinates without realizing it. For example, to locate a city on a detailed state map, we use a key that designates its location by a letter and a number. We read across from left to right to find the letter of the grid in which the city lies, and then read up and down to find the number. This is the same idea used to locate a point in the Cartesian plane. A story suggesting how Descartes may have come up with the idea of using rectangular coordinates in geometry follows: While watching a fly crawling about on the ceiling near a corner of his room, Descartes postulated that the path of the fly on the ceiling could be described only if one knew the relationship connecting the fly's distances from two adjacent walls. Although this story is apocryphal, Descartes' invention gave Newton, Leibniz, Euler, and the Bernoullis the weapon that Archimedes and Fermat lacked, limiting the articulation of their profound and far-reaching thoughts. He was the first to use the letters of the alphabet to represent unknown numbers and the first to write x^2 instead of xx. An interesting curve, known as the Cartesian oval, which has far-reaching research in geometry and analysis, was discovered by Descartes.

Descartes was a small man with a large head, projecting brow, prominent nose, and black hair that came down to his eyebrows. His voice was feeble. His disposition was cold and selfish. He never married, but had a daughter with a servant girl, Helen, who died young. He is regarded as a genius of the first magnitude. He was one of the most important and influential thinkers in human history and is sometimes called the founder of modern philosophy. In Descartes' words, men would soon be "the lords and possessors of nature." He is often referred to for his dictum, "I think therefore I am." Legend has it that he seated himself in a famous Parisian restaurant and the waiter asked him if he wanted a drink before dinner. He said "I think not" and promptly disappeared. This was the answer that he found to the question "Is there anything I can know with absolute certainty?" He realized that the fact that he was always thinking was beyond doubt and so he equated thinking with Being, that is to say, identity (I am) with thinking. Instead of the ultimate truth, he had found the root of the ego, but he didn't know that. Almost 300 years later another famous philosopher, Jean-Paul Sarte (1905–1980), looked at Descartes's statement very deeply and realized that the consciousness that says 'I am' is not the consciousness that thinks. This means that when you are aware that you are thinking, that very awareness is not a part of thinking. It is a different dimension of consciousness, and it is this awareness that says "I am." If there were nothing but thought in you, you wouldn't even know you were thinking. You would be like a dreamer who doesn't know he is dreaming. You would be as identified with every thought as the dreamer is with every image in his dream. When you know you are dreaming, you are awake within the dream, and another dimension of consciousness has come in. According to him, a philosopher is someone who knows something about knowing that nobody else knows so well. He said that perfect numbers, like perfect men, are very rare. To Descartes, the entire physical universe is a great machine operating according to laws that may be discovered by human reason, particularly mathematical reasoning.

In medicine, Descartes accepted and promoted William Harvey's (1578–1657) theory of the circulation of blood. He created a hypothetical body, which was the idealization of the human body, just as the forms studied in geometry are idealized forms that do not exist in nature. The following account of the Dark Ages was made by Jacobi in his address on Descartes: "History knew a midnight, which we may estimate at about the year 1000 AD, when the human race had lost the arts and sciences even to the memory. The last twilight of paganism was gone, and yet the new day had not begun. Whatever was left of culture in the world was found only with the Saracens, and a Pope eager to learn studied in disguise at their universities, and so became the wonder of the West. At last Christendom, tired of praying to the dead bones of the martyrs, flocked to the tomb of the Savior Himself, only to find for a second time that the grave was empty and that Christ had risen from the dead. Then mankind too rose from the dead. It returned to the activities and the business of life; there was a feverish revival in the arts and in the crafts. The cities flourished, a new citizenry was founded. Cimabue (1240–1302) rediscovered the extinct art of painting; Dante, that of poetry. Then it was, also, that great courageous spirits like Peter Abelard (1079–1142) and Saint Thomas Aquinas (1225–1274) dared to introduce into Catholicism the concepts of Aristotelian logic, and thus founded scholastic philosophy. But when the Church took the sciences under her wing, she demanded that the forms in which they moved be subjected to the same unconditioned faith in authority as were her own laws. And so it happened that scholasticism, far from freeing the human spirit, enchained it for many centuries to come, until the very possibility of free scientific research came to be doubted.

At last, however, here too daylight broke, and mankind, reassured, determined to take advantage of its gifts and to create a knowledge of nature based on independent thought. The dawn of this day in history is known as the Renaissance or the Revival of Learning."

Edmund Wingate (1596–1656)

Edmund Wingate (1596–1656) was born in Yorkshire. He matriculated from Queen's College, Oxford, in 1610 at the age of 14, received his B.A. in 1614 at the age of 18, and was admitted to Gray's Inn. Before 1624, Edmund went to Paris where he taught English to Henrietta Maria (1609–1669), the young princess of France. In England he learned the rule of proportion (the logarithmic scale), which he communicated to mathematicians in Paris. He was one of the first to publish, in the 1620s, on the principle of the slide rule, and was later the author of several popular expository works. Edmund was also a Member of Parliament during the Interregnum, the period of parliamentary and military rule by the Lord Protector Oliver Cromwell (1599–1658) under the Commonwealth of England after the English Civil War. It began with the overthrow and execution of Charles I in January 1649 and ended with the restoration of King Charles II (1630–1685) on May 29, 1660. Edmund passed away on Gray's Inn Lane and was buried in St. Andrew's, Holborn, at the age of 60.

Bonaventura Francesco Cavalieri (1598–1647)

Bonaventura Francesco Cavalieri (1598–1647) was born in Milan, Italy. In 1615, he joined the religious order Jesuati (not Jesuit) in Milan and the following year was transferred to the Jesuati monastery in Pisa. Here he studied philosophy and theology, and was taught geometry by Benedetto Castelli (1578–1643), who was a lecturer of mathematics at the University of Pisa. Cavalieri spent 4 years in Pisa and became an accomplished mathematician and a loyal disciple of Galileo. In 1620, Cavalieri was recalled to Milan, where he became a deacon and assistant to Cardinal Federico Borromeo (1564–1631). He lectured on theology there until 1623 and then became prior of St. Peter in Lodi. In 1626, he was appointed to be prior of the monastery of the Jesuati in Parma. In 1629, with the recommendation of Galileo, he secured the Chair of Mathematics at the University of Bologna. Cavalieri remained in this position until his death in 1647.

Cavalieri was one of the most influential mathematicians of his time. He published 11 books on mathematics, involving conic sections, trigonometry, optics, astronomy, and astrology. He was largely responsible for the early introduction of logarithms into Italy. He developed a general rule for the focal length of lenses and described a reflecting telescope. He developed a "method of the indivisibles," which he and his contemporaries used to determine areas and volumes, and the positions of centers of mass. This was a significant step on the way to modern infinitesimal calculus. Cavalieri also constructed a hydraulic pump for his monastery. The lunar crater Cavalerius is named for him.

Antoine de Lalouvére (1600–1664)

Antoine de Lalouvére (1600–1664), a French mathematician and theologian, was born at Rieux (Languedoc), to a noble family. He started schooling with emphasis on religion. In 1620, at the age of 20, he entered the Jesuit order. As an ordained Jesuit who had taken the fourth vow, he would have had a full education within the order, including a Ph.D. in theology. His primary interest in science was mathematics, in which he excelled. He was appointed professor of theology, mathematics, humanities, rhetoric, and Hebrew in the Jesuit college at Toulouse, in southwestern France. Antoine's main mathematics book, published in 1651, was *the Quadrature circuli* on the quadrature of the circle. The book was dedicated to Louis Phillippe XIV (1638–1715), the King of France (including Navarre). In 1658, he was drawn into a dispute about cycloids with Pascal. Antoine's name is best known because of this dispute. He passed away in Toulouse at the age of 64.

Puthumana Somayaji (Around 1600–1740)

Puthumana Somayaji (around 1600–1740) was born in Puthumana Illam (Kerala). He wrote *Karana Padhhathi*, which is a comprehensive treatise on astronomy and offers a detailed discussion of the various trigonometric series.

Pierre de Fermat (1601–1665)

Pierre de Fermat (1601–1665) known as the "Prince of Amateurs" was born at Beaumont-de-Lomagne, 36 miles northwest of Toulouse, France. His father was a successful leather merchant. He received a Bachelor of Civil Laws degree from the University of Orleans in 1631 and subsequently held various government positions, including a post as councillor to the Toulouse parliament. Although he was apparently financially successful, confidential documents of that time suggest that his performance in office and as a jurist was poor. He was a skilled linguist and one of the seventeenth century's greatest mathematicians, but refused to publish his work and rarely wrote completed descriptions even for his personal use. In 1629, he invented analytic geometry, but most of the credit has gone to Descartes, who hurried to publish his own similar ideas in 1637. At this time, 13 years before

Newton was born, Fermat also discovered a method for drawing tangents to curves and finding maxima and minima that amounted to the elements of differential calculus. Newton acknowledged, in a letter that became known only in 1934, that some of his own ideas on this subject came directly from Fermat. He also found the areas under curves by calculating the limits of sums of rectangular areas (as we do today) and developed a method for finding the centroids of shapes bounded by curves in the plane. Formulas for calculating arc length and finding the area of a surface of revolution can also be found in his work. In 1640, Fermat showed that with the first 64 natural numbers, the number of different magic squares that can be formed is more than 1,004,144,995,344! In a series of letters written in 1654, Fermat and Pascal jointly laid the foundation of the theory of probability. His discovery in 1657 of the principle of least time and its connection with the refraction of light was the first step ever taken toward a coherent theory of optics. It was in the theory of numbers, however, that Fermat's genius shone most brilliantly, for it is doubtful that his insight into the properties of the familiar but mysterious positive integers has ever been equaled. We list a few of his many discoveries in this field:

1. *Fermat's two squares theorem*: Every prime number of the form $4n + 1$ can be written as the sum of two squares in one and only one way.
2. *Fermat's little theorem*: If p is any prime number and n is any positive integer, then p divides $n^p - n$. This theorem is often stated in a slightly different form: if the prime p does not divide n, then it divides $n^{p-1} - 1$.
3. *Fermat's last theorem*: The equation $a^n = b^n + c^n$ has no positive integer solutions for a, b, and c if $n > 2$.

Some particular cases of Fermat's little theorem were known to the ancient Chinese. Fermat communicated it, without proof, to his friend Bernard Frénicle de Bessy (around 1605–1675) in a letter dated 1640. The theorem was rediscovered about 40 years later by Leibniz with a claim of originality. Euler proved this theorem five times, each time generalizing the result a little more. The converse of this theorem is not generally true, for example, $341 = 11 \times 31$ and therefore 344 is not prime; however, 341 divides $N = 2^{340} - 1$ because one of the divisors of N is $2^{10} - 1 = 31 \times 33 = 3 \times 341$.

In his personal copy of Diophantus's Arithmetica, Fermat claimed that he could prove the "last theorem" that has tantalized mathematicians for over 300 years. Unfortunately, Fermat was not kind enough to leave us a proof. Indeed, he made the comment that he had discovered a "remarkable proof" of this fact but the margin of the book was too narrow for him to jot it down! As any school child knows, when $n = 2$, there are in fact infinitely many solutions. These solutions form "Pythagorean triplets," that is, numbers that represent the sides of a right-angled triangle. Some simple examples are $3^2 + 4^2 = 5^2$, $5^2 + 12^2 = 13^2$, and $8^2 + 15^2 = 17^2$. What Fermat claimed, however, was that such equations are not possible if the exponent or "power" is greater than 2. In other words, a cube cannot be written as the sum of two smaller cubes, a fourth power cannot be written as the sum of two fourth powers, and so on. Over the years, there have many attempts to prove this "Theorem," and it has been a happy hunting ground for enthusiastic mathematical amateurs. Many eminent mathematicians have also tried to prove this theorem. Some early progress was made by such prominent mathematicians as Euler and Gauss. The eighteenth and nineteenth century mathematicians were initially only able to do special cases, like when $n = 3, 5, 7, 11$, or 14. Nevertheless, the list of mathematicians who contributed to these results reads like a Who's Who of eighteenth and nineteenth century mathematics. Names like Dedekind, Kronecker, Ferdinand Gotthold Eisenstein (1823–1852) (whom Gauss ranked equal to Newton and Archimedes), and Kummer figure prominently. The last one, in particular, made great strides and showed, for the first time, how it was possible to prove the theorem for many exponents simultaneously, as opposed to the ad hoc methods that had been developed previously. Kummer and Dedekind are recognized as being the fathers of modern algebraic number theory (an area of mathematics that uses analysis to answer questions about number theory).

In 1908, a sensational announcement was made that a prize of 100,000 marks would be awarded for the complete solution of Fermat's problem. The funds for this prize, which was the largest ever offered in mathematics, were bequeathed by a German mathematician Paul Friedrich Wolfskehl (1856–1906) to the "Konigliche Gesellschaft der Wissenschaften in Göttingen" for this purpose. This announcement drew so much attention that during a brief span of 3 years (1908–1911) over a thousand printed papers and monographs containing supposed solutions reached the Committee. But alas, all were wrong. Since then, the number of papers submitted for this prize has become so large that they would fill a library. The Committee has very wisely included as one of the conditions that the article be printed, otherwise the number would have been still larger.

In the final leg of the last century, Andrew Wiles (born 1953), a Cambridge trained mathematician working at Princeton University, has finally proved the theorem. His work is the culmination of a deep and difficult branch of mathematics that started with an inspired idea by Gerhard Frey (born 1944) in the 1980s. Frey had the brainwave that another well-known number theory conjecture, the Shimura-Taniyama-Weil (Goro Shimura, born 1930, and Yutaka Taniyama, 1927–1958) conjecture, implied Fermat's Last Theorem. He could not quite prove this, but Ken Ribet (born 1948), a mathematician from Berkeley, succeeded in showing that Frey's idea was correct. Thus if one proved the Shimura-Taniyama-Weil conjecture then Fermat's Last Theorem would follow. This is essentially what Wiles did. The Shimura-Taniyama-Weil conjecture turns out to be stronger than what is necessary to prove Fermat's Last Theorem. Wiles proved a special case of the Shimura-Taniyama-Weil conjecture, which is enough to imply Fermat's Last Theorem. The Shimura-Taniyama-Weil conjecture had long been considered to be out of the reach of current mathematical techniques, so even this partial result was a tremendous success.

His method of proof is so complicated and difficult that his original proof, which was announced in 1993 and made headlines in several leading newspapers, was found to be incomplete. It is reassuring to note that even the greatest scientists are human after all and can make mistakes, which is a small comfort to us mere mortals. However, together with his student Richard Taylor (born 1962) from the

University of Cambridge, he was able to patch the gap in his proof in 1994. This amendment to the proof has been warmly received, because based on another work by Wiles' student Fred Diamond (born 1964), it now seems possible to prove an even more lucrative target, namely, the full Shimura-Taniyama-Weil conjecture. Wiles' manuscript has made the rounds of the leading mathematicians of the world. While the general consensus seems to be that it is correct, some mistakes have been found.

Fermat's Last Theorem may not seem to be a deeply earth-shattering result. Its importance lies in the fact that it has captured the imagination of some of the most brilliant minds over the last 300 years, and their attempts at solving this conundrum, no matter how incomplete or futile, have led to the development of some of the most important branches of modern mathematics. Pascal called him "the greatest mathematician in all of Europe" during the mid-seventeenth century. According to Bell, Fermat was a mathematician of the first rank, a man of unimpeachable honesty, and an arithmetician without any superior in history. However, Brahmagupta and Bhaskara II had addressed themselves to some of Fermat's problems long before they were thought of in the west, and had solved them thoroughly. They have not held a proper place in mathematical history or received credit for their problems and methods of solution. Andre Weil wrote in 1984, "What would have been Fermat's astonishment, if some missionary, just back from India, had told him that his problem had been successfully tackled by native mathematicians almost six centuries earlier."

Evangelista Torricelli (1608–1647)

Evangelista Torricelli (1608–1647) was an Italian physicist, mathematician, and a disciple of Galileo, whom he served as secretary. In addition to formulating the principle that "water in an open tank will flow out through a small hole in the bottom with the speed it would acquire in falling freely from the water level to the hole", he advanced the first correct ideas, which were narrowly missed by Galileo, about atmospheric pressure and the nature of vacuums, and invented the barometer as an application of his theories. Torricelli showed that although the area under the curve $y = 1/x$ from $x = 1$ to ∞ is infinite, the solid obtained by revolving this area about the x-axis has a finite volume, which is π.

Frans van Schooten (1615–1660)

Frans van Schooten (1615–1660) was born in Leiden, the Netherlands. His father taught him mathematics, French, and Latin before he enrolled at the University of Leiden in 1631. He thoroughly studied the work of Christoff Rudolff (1499–1545, who introduced the sign $\surd$ for the square root because it resembles a small r, for radix), Simon Stevin, van Ceulen, Girard, and Cavalieri. Van Schooten was also familiar with the work of Archimedes, Apollonius, and Pappus. He met Descartes in Leiden in 1632 and read the proofs of his Geometry. Van Schooten graduated with his Artium Liberalium Magister from the University of Leiden in 1635. He traveled to Paris and London in about 1637 and there met the leading mathematicians. Van Schooten returned to Leiden in 1643. In Paris he collected manuscripts of Viéte's work, and in Leiden he published his own works. He published the Latin edition of Descartes' Geometry. This expanded second edition was extremely influential. He also made some contributions to mathematics, especially in *Exercitationes mathematicae* (1657). Van Schooten tutored Christiann Huygens for a year. He also made a portrait of Descartes and had a wide correspondence with him. He is mainly credited for promoting Cartesian geometry. He died in Leiden.

John Wallis (1616–1703)

John Wallis (1616–1703) was born in Ashford, Kent. John's father died when he was about 6 years old. He was initially educated at a local Ashford school, but moved to James Movat's school in Tenterden in 1625 following an outbreak of plague. Wallis learned Latin, Greek, Hebrew, logic, and arithmetic during his early school years. In 1632, he enrolled at the University of Cambridge, where he received bachelor's and master's degrees in 1637 and 1640, respectively. In the same year he was ordained by the bishop of Winchester and appointed chaplain to Sir Richard Darley at Butterworth in Yorkshire. During 1643–1649, he served as a non-voting scribe at the Westminster Assembly. Wallis was elected to a fellowship at Queens' College, Cambridge in 1644, which he had to resign following his marriage in 1645 to Susanna Glyde (1619–1695). In 1649, he was appointed Savilian Professor of Geometry at the University of Oxford, where he worked for over 50 years until his death in 1703. Besides being professor, he was appointed as keeper of the University archives in 1657, was one of 12 Presbyterian representatives at the Savoy Conference in 1661, and led the formation of the Royal Society of London under a charter granted by King Charles II in 1662. Wallis was the most influential English mathematician before Newton. In 1655, Wallis published a treatise that defined conic sections analytically. This was the earliest book in which these curves were considered and defined as curves of second degree equations. In calculus, he extended the works of Kepler, Cavalieri, Descartes, Torricelli, and de Roberval, a personal enemy of Torricelli's that found the solution of the quadrature of the cycloid curve, which was not published until 1693. Wallis also calculated the area under the curve $y = a_0 + a_1x + \cdots + a_nx^n$. In his most famous work, *Arithmetica infinitorum*, which he published in 1656, he established the formula

$$\pi = 2 \cdot \frac{2}{1} \cdot \frac{2}{3} \cdot \frac{4}{3} \cdot \frac{4}{5} \cdot \frac{6}{5} \cdot \frac{6}{7} \cdot \frac{8}{7} \cdot \frac{8}{9} \cdots$$

without proof. In 1657, Wallis published the *Mathesis Universalis*, on algebra, arithmetic, and geometry, in which he used negative and fractional exponents and introduced the symbol ∞ for infinity. (This sign was used by the Romans to denote the number 1000, and it has been conjectured that this led to it being applied to represent any very large number. Voltaire described the ∞ as a 'love knot,' and he was skeptical about the sign making the idea of infinity any clearer.) In his *Tractatus de Sectionibus Conicis* (1659), he described the curves that are obtained as cross sections by cutting a cone with a plane as properties of algebraic coordinates. His *Mechanica, sive Tractatus de Motu* in 1669–1671 (three parts) refuted many of the errors regarding motion that had persisted since the time of Archimedes; he gave a more rigorous meaning to terms like force and momentum and assumed that the gravity of the Earth may be regarded as localized at its center. In his *Opera Mathematica I* (1695) Wallis introduced the term 'continued fraction.' He rejected the now common idea that a negative number is less than nothing, but accepted the view that it is something greater than infinity, showing that $-1 > \infty$. Wallis made contributions to the history of mathematics by restoring some ancient Greek texts, such as Ptolemy's Harmonics, Aristarchus' work on the magnitudes and distances of the Sun and Moon, and Archimedes' Sand Reckoner. During the English Civil War he was a Parliamentarian, and he put his mathematical talents to use in decoding enciphered letters. He had a great affinity for mental calculations. He slept badly and often did mental calculations as he lay awake in his bed. On the night of December 22, 1669, he occupied himself with finding the integral part of the square root of 3×10^{40} while in bed, and several hours afterward he wrote down the result from memory. Two months later, he was challenged to extract the square root of a number of 53 digits, which he performed mentally; 1 month later, he dictated the answer that he had not committed to writing. Besides his mathematical works, he wrote on theology, logic, and philosophy, and was the first to devise a system for teaching the deaf.

Wallis' life was embittered by quarrels with his contemporaries, including Huygens, Descartes, and the political philosopher Thomas Hobbes (1588–1679), which continued for over 20 years, ending only with Hobbes' death. Hobbes called Arithmetica infinitorum "a scab of symbols" and claimed to have squared the circle. It seems that for some individuals, quarrels give strength, encouragement, and mental satisfaction.

John Graunt (1620–1674)

John Graunt (1620–1674) was born in London. The study of statistics (taken from the Italian *statista*, meaning a person who deals with state affairs) was started by him. He collected and studied the death records of various cities in Britain and was fascinated by the patterns that he found in the whole population even though people died randomly. Graunt documented his findings in the book *Natural and Political Observations Made upon the Bills of Mortality* (1662). His work was appreciated by King Charles II, who helped Graunt become a member of the Royal Society (founded in 1662). In the later years of his life he lived in poverty, dying of jaundice and liver disease at the age of 53.

William Viscount Brouncker (1620–1684)

William Viscount Brouncker (1620–1684) was born in Castlelyons, Ireland. He entered Oxford University when he was 16 years old, where he studied mathematics, languages, and medicine. He received the degree of Doctor of Medicine in 1647. He was a founder and the second President of the Royal Society. In 1662, he became Chancellor to Queen Catherine, who was then chief of Saint Catherine's Hospital. Brouncker's contributions to mathematics are the reproduction of Brahmagupta's solution of a certain indeterminate equation, calculations of the lengths of the parabola and cycloid, quadrature of the hyperbola (which required approximation of the natural logarithm function by infinite series), and the study of generalized continued fractions, especially for π. Brouncker never married. He died in Westminster.

Nicolaus Mercator (Kauffmann) (1620–1687)

Nicolaus Mercator (Kauffmann) (1620–1687) was born in Holstein, lived in the Netherlands (1642–1648), lectured at the University of Copenhagen (1648–1654), lived in Paris (1655–1657), and taught mathematics in London (1658–1682). In 1668, he wrote a treatise entitled *Logarithmo-technica*, and discovered the series

$$\ln(1+x) = x - \frac{1}{2}x^2 + \frac{1}{3}x^3 - \frac{1}{4}x^4 + \cdots ;$$

however, the same series had been independently discovered earlier by Grégoire de Saint-Vincent (1584–1667). Mercator became a member of the Royal Society in 1666. He died in Versailles.

Blaise Pascal (1623–1662)

Blaise Pascal (1623–1662) was born in Clermont-Ferrand, and his family settled in Paris in 1631. His entire life was a constant struggle between one world and the other. As a child, Blaise was sickly, supposedly due to a witch's spell. At the age of one he suddenly became very weak and would scream hysterically at the sight of water or at seeing his parents embrace. This so-called "death spell" that had been put on him was transferred to a cat by an obliging witch, and little Blaise was spared. The cat fell out of the window and was killed on the spot, in accordance with the spell. His mother died when he was 3. His father, Étienne Pascal (1588–1651), due to his own view that "mathematics fills and satisfies the soul," forbade Blaise to study mathematics before the age of 15 so that he would not neglect Latin and the other languages. But due to Blaise's geometrical discoveries at the age of 12, which were performed in secret, Étienne changed his educational method and allowed his son to study mathematics. Blaise discovered that the sum of the angles of a triangle is a straight angle through a simple experiment that involved folding a paper triangle. Pascal's sister and primary biographer claimed that he independently discovered Euclid's first 32 propositions without ever reading a book on geometry. (However, it is generally agreed that this story is apocryphal). Nevertheless, by the time he was 16 years old, the precocious Pascal formulated one of the basic theorems of projective geometry, known as Pascal's mystic hexagon theorem: if a hexagon is inscribed in a conic, the points of intersection of the opposite sides will lie in a straight line. Pascal then published a highly respected essay on conic sections. Mersenne reported that in this treatise Pascal had deduced no fewer than 400 propositions on conic sections from the mystic hexagon theorem. Descartes, who read this essay, thought it was so brilliant that he could not believe that it was written by such a young man. Triangular arrays such as the following, in which consecutive odd integers are arranged

										1	1^3
									3	5	2^3
								7	9	11	3^3
							13	15	17	19	4^3
						21	23	25	27	29	5^3
					31	33	35	37	39	41	6^3
				43	45	47	49	51	53	55	7^3
			57	59	61	63	65	67	69	71	8^3
		73	75	77	79	81	83	85	87	89	9^3
	91	93	95	97	99	101	103	105	107	109	10^3
.	.	.	.	.	.	.	.	.	.	.	.

have been named after Pascal. However, he did not invent these; he only recognized the relationship between the triangular array and the binomial expansion. He also found connections between these arrays and combinatorial problems. It is believed that Pascal, along with Fermat, fathered the theory of probability. However, Levi ben Gerson (1288–1344) had developed some of the ideas used in probability theory hundreds of years before them. Pascal invented and constructed the first calculating machine Pascaline in 1642. In 1647, he proved that the vacuum existed, which was not well received by the world at that time because it contradicted Aristotle's claim that "Nature abhors a vacuum." Pascal's contributions to physics include the discovery that air pressure decreases with altitude as well as the principle of fluid pressure that bears his name. However, the originality of his work is questioned by some historians. Pascal enriched various mathematical disciplines with important results, but also had his own ideas on the nature and value of the "mathematical method." Through his investigations of proof in geometry and of "the difference between the spirit of geometry and the spirit of intuition" he built a bridge between the mathematical thought of the ancients and our modern concept of the nature of the exact sciences. Some time later, he came to believe that God's plan for him did not include mathematics and he dropped the subject entirely. However, while experiencing a particularly nagging toothache when he was 35, Pascal let his thoughts wander to mathematics and the pain disappeared. He took this as a heavenly sign and made a quick and intense return to mathematical research. Although it took him a week, Pascal managed to discover the fundamental properties of the cycloid curve. Bedridden by sickness many times, he had migraine headaches since his youth. Blaise lived in great pain caused by a malignant growth in his stomach, which killed him at 39 after it spread to his brain. Pascal's notes on Christianity were collected after his death in the *Pensees* (Thoughts). An interesting quotation included in these notes asks "is it probable that probability gives assurance? Nothing gives certainty but truth: nothing gives rest but the sincere search for truth." Pascal wrote this when he had gone to live at the Jansenist convent of Port-Royal after a carriage accident in 1654. He developed "Pascal's wager": If you bet that God exists and live accordingly, you will gain much even if God does not actually exist, and will benefit from his favor if he does; if you bet the opposite, then you will lose everything if he does exist. Pascal said "The mathematicians who are merely mathematicians reason correctly, but only when everything has been explained to them in terms of definitions and principles. Otherwise they are limited and insufferable, for they only reason correctly when they are dealing with very clear principles." He is also known for claiming that "Men never do evil so completely and cheerfully as when they do it from religious conviction." In the last years of his life he tore his flesh using a belt of spikes wrapped round his body, hitting it with his elbow every time a thought entered his mind that was not sufficiently pious.

Christiaan Huygens (1629–1695)

Christiaan Huygens (1629–1695) was born in The Hague (the Netherlands), to an important Dutch family. He was educated at home until he was 16 years old. Huygens' mathematical education was influenced by Descartes, who was a regular visitor. During 1645–1647, he studied law and mathematics at the University of Leiden, and from 1647 to 1649 he studied at the College of Orange at Breda. Through his father's acquaintance with Mersenne, a correspondence between Huygens and Mersenne developed. Huygens wrote his first book in 1657 on probability theory, *De Ratiociniis in Ludo Aleae* (On Reasoning in Games of Chance). This work mostly duplicated results that Cardano had arrived at a 100 years before, but since Huygens was the first to publish, it was as if Cardano's work had never existed. In the same year his invention of the pendulum clock was patented, which was a breakthrough in timekeeping. He formulated Newton's second law of motion in a quadratic form, and in 1659, he derived the now well-known formula for centripetal force, the force exerted by an object in circular motion. In his 1659 book *Systema Saturnium*, Huygens reported that he had discovered the rings encircling the planet Saturn using a 50 power refracting telescope that he designed himself. In this book he also addressed his observations about the Moon, the planets, and the Orion nebula. In the *Horologium Oscillatorium sive de motu pendulorum* (1673) he described the theory of pendulum motion. He showed the tautochrone property of the cycloid curve: that a bead sliding down this curve will reach the bottom in the same time, no matter where on the curve it begins. In the same work he also derived the law of centrifugal force for uniform circular motion. Huygens was the first to derive a formula for the period of an ideal mathematical pendulum (with a massless rod or cord):

$$T = 2\pi\sqrt{\frac{\ell}{g}}.$$

In 1675, Huygens patented a pocket watch. He invented many other devices, including a 31 tone to the octave keyboard instrument that made use of his discovery of 31 equal temperament. Huygens is best known for his argument that light consists of waves, expounded in his *Traité de la Lumiére* in 1678, now known as the Huygens-Fresnel (Augustin Jean Fresnel, 1788–1827) principle, which became instrumental for understanding wave-particle duality. He experimented with double refraction (birefringence) in Icelandic crystal (calcite) and explained it with his wave theory and polarized light. Huygens developed a balance spring watch more or less contemporaneously with Hooke, and controversy over who should be given credit for this important invention persisted for centuries. He also contributed many elegant results to the infinitesimal calculus.

Huygens became a member of The Royal Society in 1663. In the year 1666, Huygens moved to Paris where he held a position at the French Academy of Sciences (founded in 1666). In 1684, he published *Astroscopia Compendiaria* which presented his new aerial (tubeless) telescope. Huygens believed in the existence of extraterrestrial life. Prior to his death in The Hague in 1695, he completed a book entitled *Cosmotheoros* in which he discussed his thoughts on extraterrestrial life. He was of the opinion that life on other planets is pretty much similar to life on Earth.

Johann van Waveren Hudde (1629–1704)

Johann van Waveren Hudde (1629–1704), a Dutch mathematician, was born in Amsterdam. He made a remarkable contribution to analytic geometry and mathematical analysis. He can be considered to be the forerunner of Philippe de La

Hire as far as analytical geometry is concerned. He was a renowned administrator and served as mayor for about 30 years in his hometown. As an administrator, he consulted scientists such as Jan De Witt (1625–1672) and Huygens regarding maintenance of channels, which was one of his responsibilities in Holland. He passed away in his hometown at the age of 75.

Isaac Barrow (1630–1677)

Isaac Barrow (1630–1677) was born in London. His father was a linen draper by trade. Barrow started his schooling at Charterhouse, where he was so troublesome that his father was heard to pray that if it pleased God to take any of his children, he could best spare Isaac. He then attended Felstead school, where he learned Greek, Latin, Hebrew, and logic in preparation for University. Barrow entered Trinity College, Cambridge in 1643, where he distinguished himself as a classical scholar as well as a mathematician, earning his bachelor's degree in 1648. He was elected a fellow of the college in 1649 and received his master's degree in 1652 (the Ph.D. at Cambridge did not exist until 1919). Barrow then resided for a few years in college and became candidate for the Greek professorship at Cambridge, but in 1655 he was driven out during the persecution of the Independents. He spent the next 4 years traveling across France, Italy, Smyrna, and Constantinople, and after many adventures returned to England in 1660, where he was finally elected Professor of Greek at Cambridge. In 1662, he was made Professor of Geometry at Gresham College, and in 1663, he accepted the newly created Lucasian professorship of mathematics (after Henry Lucas, 1610–1663), a post he resigned in 1669. The claim is often made that Barrow resigned his Chair in favor of his pupil, Isaac Newton. In reality, Newton was not the pupil of Barrow, and while he appreciated Newton's mathematical genius and saw to it that Newton succeeded him, Barrow was more interested in advancing his own career. He was appointed chaplain to King Charles II in 1669 and in 1673 returned to Cambridge as Master of Trinity College, the office being a gift of the king's.

Barrow's earliest work in mathematics was a simplified edition of Euclid's Elements in Latin (1655) and in English (1660), which remained the standard textbook for half a century. In 1657, he published an edition of the Data. John Collins (1625–1683) published most of Barrow's lectures: *Lectiones Opticae* in 1669, which dealt with many problems connected with the reflection and refraction of light; *Lectiones Geometricae* in 1670, which represented the work that Barrow did while at Gresham; and *Lectiones Mathematicae* (posthumously) in 1683, which were delivered in 1664, 1665, and 1666 and were designed to revive interest in mathematics at Cambridge. Lectiones Mathematicae covered topics such as divisibility, congruence, equality, measurement, proportion, ratio, time, and space, and developed a method of determining tangents that prefigured the differential calculus developed by Newton, and recognized that the processes of integration and differentiation in calculus are inverse operations. His work in both optics and mathematics was soon overshadowed by Newton's publications. In 1675, he published an edition, with numerous comments, of the first four books of *On Conic Sections* by Apollonius and of the extant works of Archimedes and Theodosius.

Barrow was unmarried, a small man of immense physical strength, a vivacious personality, slovenly in dress, and an inveterate smoker. Once when traveling in the East, he saved the ship from capture by pirates by his own prowess. He died in 1677 from, according to John Aubrey (1626–1697), an overdose of opium, an addiction that he had acquired in Turkey. On May 20, 1663, Barrow became one of 150 scientists who were elected Fellows of the Royal Society (FRS). The lunar crater Barrow is named after him.

Robert Hooke (1635–1703)

Robert Hooke (1635–1703) was born in Freshwater, on the Isle of Wight, England, the son of a curate at All Saints Church. In his childhood, Hooke continually suffered from headaches and was not expected to reach adulthood. His parents gave up on his education, leaving him much to his own devices. At the age of 13, Hooke was able to enter Westminster School, boarding in the house of the headmaster Richard Busby (1606–1695), who was an outstanding teacher. Hooke had mastered the first six books of Euclid's Elements by the end of his first week at school. He applied geometry to mechanics and began to invent possible flying machines. He also learned to play the organ. In 1653, he entered Christ College, Oxford, where he won a chorister's place. In Oxford, Hooke was taught astronomy by Seth Ward (1617–1689) and impressed John Wilkins (1614–1672) with his knowledge of mechanics. Wilkins gave him a copy of his book Mathematical Magick. This book encouraged Hooke to continue to try to invent a flying machine. While at Oxford he met the chemist (and physicist) Robert Boyle (1627–1691) and gained employment as his assistant in 1655. It is believed that Hooke formally stated Boyle's Law, as Boyle was not a mathematician. In 1660, he discovered Hooke's Law of Elasticity, which describes the linear variation of tension with extension in an elastic spring. However, he did not announce the general law of elasticity until his lecture in 1678. Hooke's first publication was a pamphlet on capillary action in 1661. In 1662, he was appointed as Curator of Experiments of the newly formed Royal Society of London and took responsibility for demonstrating new experiments at the Society's weekly meetings. He became an FRS in 1663. Hooke published the book *Micrographia* in 1665, a best seller of its day, which contained a number of microscopic and telescopic observations and some original observations in biology. He coined the biological term *cell.*

The hand-crafted, leather and gold-tooled microscope that Hooke used to make the observations for Micrographia, originally made by Christopher Cock in London, is on display at the National Museum of Health and Medicine in Washington, DC. In 1665, he became Professor of Geometry at Gresham College, London. This position gave him rooms at the College, where he lived for the rest of his life. He was required to give one lecture each week in Latin and then repeat it in English, and was required to be unmarried but was permitted a housekeeper. Hooke also achieved fame as Surveyor to the City of London; as a chief assistant to Wren, he helped in rebuilding London after the Great Fire in 1666. He worked on designing the Monument, Royal Greenwich Observatory, and the infamous Bethlem Royal Hospital. Other buildings designed by Hooke include The Royal College of Physicians in 1679; Ragley Hall in Warwickshire; and the parish church at Willen, Milton Keynes (historical Buckinghamshire). In 1672, Hooke claimed that what was correct in Newton's theory of light and color was stolen from his own ideas in 1665. This marked the beginning of severe arguments between the two. The same year, he conjectured that the Earth moves in an ellipse round the Sun, and 6 years later he proposed the inverse square law of gravitation to explain planetary motions. However, these claims were rejected by Newton, and all references to Hooke's work were removed from the Principia.

Hooke was the greatest experimental scientist of the seventeenth century. He observed the plants, the animals, the farms, the rocks, the cliffs, the sea, and the beaches around him. He invented the anchor escapement and the balance spring, which made more accurate clocks possible; worked out the correct theory of combustion; invented or improved meteorological instruments such as the barometer, anemometer, and hygrometer; invented the universal joint, the iris diaphragm, and an early prototype of the respirator; and assisted Boyle in studying the physics of gases. Hooke also made important contributions to biology and had grasped the cardinal principle of paleontology. He, two and a half centuries before Charles Robert Darwin (1809–1882), had realized that the fossil record documents changes among the organisms on the planet, and species appear and become extinct throughout the history of life on Earth. Hooke's *Discourse of Earthquakes*, published 2 years after his death, shows that his geological reasoning had continued.

Hooke's health deteriorated over the last decade of his life. He was often troubled with headaches, giddiness, fainting, and a general decay all over his body that hindered his philosophical studies, yet he still read some lectures. He died in London in 1703. He amassed a sizeable sum of money during his career in London, which was found in his room at Gresham College after his death. It seems that no authenticated portrait of him survives today. This is due in part to the enmity of his famous, influential, and extremely vindictive colleague, Newton, who instigated the removal of Hooke's portrait from the Royal Society. It is often said that Hooke was a lean, bent, and ugly man, and so he may have been too embarrassed to sit for a portrait. Two craters, on the Moon and Mars, have been named in his honor.

James Gregory (1638–1675)

James Gregory (1638–1675) was born in Drumoak, Aberdeenshire, Scotland. James had two older brothers, Alexander (the eldest) and David. James was 10 years younger than David. James suffered for about 18 months from the quartan fever, which recurs at approximately 72-h intervals. James first learned mathematics from his mother, who taught him geometry, and when his father died in 1651 he was educated by his brother David. James studied Euclid's Elements, attended Grammar School, and then proceeded to Marischal College in Aberdeen, where he studied optics and the construction of telescopes. In 1663, encouraged by his brother David, he wrote the book *Optica Promota*. The book begins with 5 postulates, 37 definitions, and 59 theorems on the reflection and refraction of light. It then describes the compact reflecting telescope known by his name, the Gregorian telescope. This telescope design attracted the attention of several people including Moray, Newton, and Hooke, who eventually built the telescope. This book also described the method for using the transit of Venus to measure the distance of the Earth from the Sun, which was later advocated by Halley and adopted as the basis of the first effective measurement of the Astronomical Unit. During 1664–1668, Gregory was in Padua, Italy, where he worked closely with the philosopher Caddenhead and published two books, *Vera circuli et hyperbolae quadratura* in 1667 and *Geometriae pars universalis* in 1668. In the first book, he showed that the area of a circle can be obtained only in the form of an infinite convergent series, and therefore he inferred that the quadrature of the circle was impossible. In the second book he attempted to write calculus systematically, which perhaps made the basis of Newton's fluxions. In 1668, during Easter, he returned to London and thereafter was benefited through his friendship with Collins. In June of 1668, Gregory was elected an FRS. He presented many papers to the Society on a variety of topics including astronomy, gravitation, and mechanics. In July of 1668, Huygens published a review of *Geometriae pars universalis*, claiming that he had been the first to prove some of the results. Gregory was upset by Huygens' comments, which he took to mean that Huygens was accusing him of stealing his results without acknowledgement. This began an unfortunate dispute, particularly regarding priority, and as a consequence Gregory was very reluctant to publish his later mathematical discoveries. Robert Moray (1608–1673), then President of the Royal Society, attempted to arrange a meeting between Gregory and Huygens in Paris; however, it never occurred. Geometriae also contains series expansions of $\sin(x)$, $\cos(x)$, $\arcsin(x)$, and $\arccos(x)$; however, as we have seen earlier, these expansions were known to Madhava of India. Later he established that

$$\int \sec x dx = \ln(\sec x + \tan x) + c,$$

which solved a long standing problem in the construction of nautical tables. He published the *Exercitationes Geometricae* as a counterattack on Huygens. Although he did not disclose his methods in this small treatise, he discussed various series expansions, the integral of the logarithmic function, and several other ideas. Gregory arrived in St. Andrews late in 1668. Although he was not attached to a college, the Upper Room of the university library became his place of work. In early 1669, he married Mary Jamesome, who was a widow. They had two daughters and one son. Gregory anticipated Newton in discovering both the interpolation formula and the general binomial theorem as early as 1670. In early 1671, he discovered Taylor's theorem (published by Taylor in 1715); however, he did not publish because Collins informed him that Newton had already found a similar result. In 1671, he rediscovered Nilakanthan's arctangent series (1). In 1674, Gregory left St. Andrews for Edinburgh and became the first person to hold the Chair of Mathematics. However, he died suddenly 1 year after taking up the post. One night he was showing the Moons of Jupiter to his students with his telescope when he suffered a stroke, became blind, and died a few days later. A crater on the Moon is named for him.

Gregory's other contributions include: the solution of Kepler's famous problem of how to divide a semicircle with a straight line through a given point of the diameter in a given ratio (his method used Taylor series to the general cycloid); observing the splitting of sunlight into its component colors; one of the earliest examples of a comparison test for convergence, essentially giving Cauchy's ratio test, together with an understanding of the remainder; a definition of the integral which is essentially as general as that given by Riemann; his understanding of singular solutions to differential equations; his attempts to prove that π and e are not the solutions of algebraic equations; expressing the sum of the nth powers of the roots of an algebraic equation in terms of the coefficients; realizing that algebraic equations of degree greater than four could not be solved by radicals; and researches on Diophantine problems. The full brilliance of Gregory's discoveries became known only in the 1930s, when Turnbull examined his papers in the library in St. Andrews.

Philippe De la Hire (1640–1719)

Philippe De la Hire (1640–1719) was born in Paris, the son of Laurent de La Hire (1606–1656), a distinguished artist. In 1660, he went to Rome to study painting, but on his return to Paris devoted himself to the classics and to science. He showed a particular aptitude for mathematics and became a favorite pupil of Desargues. Philippe became a member of the French Academy of Sciences in 1678, in the astronomy section. During 1679–1682, he made several observations and measurements of the French coastline and in 1683 aided in mapping France by extending the Paris meridian to the north. In 1683, Philippe assumed the chair of mathematics at the Collége Royale. From 1687 onward he taught at the Académie d'Architecture. La Hire wrote on graphical methods in 1673; on conic sections in 1685; a treatise on epicycloids in 1694; and on conchoids in 1708. His works on conic sections and epicycloids were founded on the teachings of Desargues. He also translated the essay of Manuel Moschopulus (around 1350–1450) on magic squares and collected many of the theorems about them that were previously known; this was published in 1705. He also published a set of astronomical tables in 1702. La Hire's work also extended to descriptive zoology, the study of respiration, and physiological optics. Two of his sons rose to distinction, Gabriel-Philippe (1677–1719), in mathematics, and Jean-Nicolas (1685–1727), in botany. The mountain Mons La Hire on the Moon is named for him.

Seki Kowa (1642–1708)

Seki Kowa (1642–1708) also known as **Seki Takakazu** is generally regarded as the greatest Japanese mathematician of the Edo period. Not much is known about his personal life. It is believed that he was born in Fujioka, the second son in a samurai warrior family. Later in life he became the landlord of a '300 person' village. Kowa began writing mathematical texts when he was employed as an auditor by Shogun Tokugawa Lenobu (1662–1712). He was a prolific writer, and a number of his publications are either translations of mathematics from Chinese into Japanese or commentaries on certain works of well-known Chinese mathematicians. He spent many years studying thirteenth century Chinese calendars in order to replace the one used in Japan at that time. His interests in mathematics ranged from recreational mathematics, magic squares, and magic circles to solutions of higher order and indeterminate equations, conditions for the existence of positive and negative roots of polynomials, and continued fractions. He discovered determinants in 1683, 10 years before Leibniz, and the Bernoulli numbers 1 year before Bernoulli. He also calculated a value of π correct to ten decimal places by applying an ingenious extrapolation to a polygon with 2^{17} sides.

Sir Isaac Newton (1642–1727)

Sir Isaac Newton (1642–1727) is hailed as one of the greatest scientists and mathematicians of the English-speaking world. He had a more modest view of his monumental achievements: "... to myself I seem to have been only like a boy playing on the seashore, and diverting myself in now and then finding a smoother pebble or a prettier shell than ordinary, whilst the great ocean of truth lay all undiscovered before me." As he examined these shells, he discovered to his amazement more and more of the intricacies and beauties that lay within them, which otherwise would have remained locked to the outside world.

Newton was born prematurely on Christmas Day, 1642 (under the Julian calendar, which is equivalent to January 4, 1643 under the Gregorian calendar), in Lincolnshire, England, in the same year that Galileo died. He was so small and sickly that he could be put into a quart mug and his "poor little weak head" had to be supported by a special leather collar. At first it was thought that the baby would die before the day was out, but the frail infant survived and was christened Isaac, after his dead father. When Isaac was 3 years old, his mother Hannah Ayscough Newton (1623–1679) remarried with Barnabas Smith, the 63-year-old rector of a nearby village, and Isaac was left on the Woolsthorpe farm with his grandmother. At the age of 12 he was sent to school in nearby Grantham, where he boarded with a Mr. Clark and his wife. The couple had a stepdaughter of whom Isaac was very fond. She was his best friend and only playmate, and their friendship lasted even into old age. Isaac spent most of time writing poetry, drawing pictures, and constructing miniature machines such as clocks and windmills. In his windmill was a mouse, which he called the Miller. This mechanism was so arranged that the mouse could set the mill going at any moment. He made a waterclock 4 ft high. He liked to fly kites to which he attached lanterns that he set aloft at night, hoping to frighten people with his artificial "comets." He was often inattentive in school and was ranked among the bottom of his class. During these years, one of Newton's chief occupations was making sundials; several of these are still on the walls of the manor house at Woolsthorpe. Newton's maternal uncle, who himself was a graduate of Cambridge, later saw tremendous academic potential in the young lad and persuaded his mother to enroll him at Cambridge. So in 1661, at the age of 19, Newton entered Trinity College, Cambridge, as a freshman. Remarkably Cambridge, which has always been a great center of learning, deteriorated considerably during 1660s. Professors were appointed for political or ecclesiastical reasons, and for many scholarship was simply irrelevant. Some faculty members occupied their positions for half a century without having a single student, or writing a single book, or giving a single lecture! Some did not even live in the Cambridge area and visited the campus infrequently. Newton's initial interest was in chemistry, which remained with him throughout his life. In his first year at Trinity, he got hold of a copy of Euclid's Elements, but was disappointed to find it too simple. He called it a 'trifling book' and threw it aside (an act that he lived to repent). He then studied the works of Kepler, Wallis, Galileo, Fermat, Huygens, Descartes, and other mathematicians. To add to his mathematical exposure and training, Newton attended a series of lectures given by Isaac Barrow, the Lucasian Professor of Mathematics at Cambridge in 1663. Barrow not only recognized Newton's genius, but is believed to have handed Isaac his own chair at the age of 26. Newton's involvement in mathematics was very much a result of learning from his predecessors. In a letter to Robert Hooke, Newton said, "If I have seen farther than Descartes, it is because I have stood on the shoulders of giants."

From 1664 to 1666, a large part of Europe came under the severe attack of the bubonic plague, which has been named the *Great Plague*. It claimed almost one-third of Europe's population. As a result, Trinity College was closed and Newton had to return to his home at Lincolnshire to think and contemplate. It was during these 3 years that Newton made some of the greatest discoveries yet known to the world. Newton discovered: (1) The nature of colors. (2) The law of gravitation and the laws of mechanics. (3) The fluxional calculus.

He discovered that white light was composed of colored lights by breaking up a sunbeam and making the separate beams paint a rainbow ribbon of colors on a screen. Legend has it that one day Newton was sitting under an apple tree in deep contemplation, when suddenly he saw an apple fall onto the ground. In one version of this legend, the apple actually fell onto his head! This triggered his theory of gravitation: All bodies are attracted to other bodies with a force directly proportional to the product of their masses and inversely proportional to the square of their distance, or

$$F = \frac{kMm}{r^2},$$

where F is the force of attraction, M and m are the masses of the two bodies, r is the distance between them, and k is the same for all bodies. This formula was suggested to him by Hooke for planets. It is interesting to note that the scientific concept of gravity was known to Abu Ja'far al-Khazin (900–971), who rediscovered the Diophantus identity $(a^2+b^2)(c^2+d^2) = (ac \mp bd)^2 + (bc \pm ad)^2$ (this identity also appeared in Fibonacci's Liber Abbaci). This corrected the general belief that things were drawn to the Earth because it was the most important place in the universe, as Christ was born on Earth. Subsequently, he developed his three laws of motion (the "apple-head" tale is probably apocryphal):

1. If undisturbed, a moving body will continue to move in a straight line at a uniform speed.
2. A change in motion is proportional to the force causing the change and takes place in the direction in which the force is acting.
3. To every action there is always an equal and opposite reaction.

The first two of these laws were already stated by Descartes and Galileo.

Fluxional calculus deals with magnitude that is changing, such as distance, length, area, volume, pressure, and so forth. The rate at which distance changes is called *velocity*, and the rate at which velocity changes is called *acceleration*. Although the roots of this calculus can be found in the works of Napier, Kepler, Cavalieri, Pascal, Fermat, Wallis, and Barrow, the importance of Newton's invention in 1665–1666 was that it provided a general procedure for finding the rate of change of practically any function. This powerful and universal tool was soon applied with enormous success to almost every branch of science. In 1665, he discovered the binomial theorem (see James Gregory), and the general method of expression of algebraic functions in infinite series in 1670. He worked out the area of the hyperbola for the first time, and reveling in his triumph, carried out the computation to 52 decimal places. In 1672, based on the idea of Gregory, he invented a reflecting microscope. Newton also became a member of the Royal Society of London in 1672. Some years later he invented the sextant, which was rediscovered by John Hadley in 1731. Although Newton was very prolific in his scientific research, he was slow to publish his findings. It was Newton's good friend Halley who, being impressed with Newton's findings, persuaded him to publish his results. Halley even funded the entire publication of Newton's classic, *Philosophiae Naturalis Principia Mathematica*, or "mathematical principles of the philosophy of nature." This rare classical treatise was written in a language that is archaic, even unfamiliar, and was first put into print in 1687, with subsequent editions in 1713 and 1726. This book contains an exposition of physics, mathematics, and astronomy in the language of pure geometry. However, in its propositions there is no hint that the universe has a history and has passed through evolution. He regarded the universe as a machine, created by God in the beginning, that has run essentially unchanged forever. The creation of the Principia was a gift to the mathematical world. Lagrange described the Principia as the greatest production of the human mind, and Laplace considered the Principia to have a preeminence above all the other productions of human genius. In 1701, Newton resigned the Lucasian chair, and in 1703, he became the President of the Royal Society of London, a position that he held to his death. After Hooke (who Newton hated) died in 1703, Newton published his *Optics* in 1704, which had existed in manuscript form for about 30 years. The first account of the method of fluxions that Newton put into print appeared in Optics. This generated an unpleasant dispute between England and Germany during 1708–1716. In Germany, Leibniz developed the differential and integral calculus. The similarity between Leibniz's work and the fluxional calculus in Newton's Optics cast suspicion that Leibniz had plagiarized Newton's work. A Swiss mathematician living in England wrote a letter to the Royal Society with the implication that Leibniz was guilty of plagiarism. Upon hearing this, Liebniz protested vehemently and made the mistake of appealing to the Royal Society for justice. Newton, as President of the Society, assigned an "independent committee" that included only his friends. Newton wrote almost all of the articles defending himself, but published under his friends' names. The argument as to who first invented the calculus was so sour that it caused bitter tension between the two countries. Finally, Newton wrote and published the committee's protocol himself, officially accusing Leibniz of plagiarism. Still not satisfied, Newton anonymously published a copy of the protocol in the Royal Society's paper. After Leibniz's death, Newton allegedly said that it was a great satisfaction for him to "break Leibniz's heart." It is generally believed that Newton was the first inventor of the calculus but Leibniz was the first to publish it. On fluxions, Bishop George Berkeley (1685–1753) in his tract *The Analyst* wrote a knowledgeable and extremely witty attack: "And what are these fluxions? The velocities of evanescent increments. And what are these same evanescent increments? They are neither finite quantities, nor quantities infinitely small, nor yet nothing. May we not call them ghosts of departed quantities?" His criticisms were well founded and important, as they focused the attention of mathematicians on a logical clarification of the calculus. According to Berkeley our only knowledge of this world is what comes to us through our senses, and he went so far as to say that physical objects only exist relative to the mind. Newton's *Universalis Arithmetic* was published in 1707, and two more important works, on algebra and geometry, appeared at about the same time. Newton's method of interpolation appeared in *Methodus Differentialis* which was published in 1711. He is also remembered for solving Pappus' and the brachistochrone problems, finding orthogonal trajectories, providing the sum of the nth powers of the roots of an algebraic equation, suggesting a geometric method to duplicate a cube, providing a series for the computation of the value of π, and giving a rule by which the number of imaginary roots of an algebraic equations can be determined. When the Queen of Prussia asked Leibniz what he thought of Newton, he replied: "Taking mathematicians from the beginning of the world to the time when Newton lived, what he had done was much the better half."

In 1705, Newton was knighted by Queen Anne (1665–1714) in the Master's Lodge of Trinity College. He was the first scientist to receive this honor for his achievements. His Secretary had said that Newton rarely went to dine in the hall, and when he did, would often go in "with shoes down at heels, stocking untied, surplice on, and his head scarcely combed." He often forgot to eat his lunch, and when he did, he forgot that he had eaten it. He very rarely went visiting, and had few visitors. He never went to bed until two or three in the morning, and sometimes not until six. He never stayed in bed more than 4 or 5 h. When he got up in the morning, he often forgot to complete his dressing and sat on the edge of his bed in profound meditation for long periods, without moving. He often spent 18 out of 24 h writing. He spent virtually his entire life in Woolsthorpe, Cambridge, and London. He did not visit Oxford until he was 78, and there is no record of his having made any kind of tour, nor did he ever go out of England. An associate of many years recalled that there was but once that he saw Newton laugh. When Newton was asked how his genius differed from that of other men, he replied that he had a greater power of concentration. He could hold a problem in his mind with sustained intensity, until all the parts fell into place and the solution appeared to him. He was a very strange individual, tormented by paranoia, psychosis, and a strange relationship with women (especially his mother). He had dreadful dealings with many of his professional contemporaries. Newton's alchemical writings lead to the conclusion that he had tried to be a magician. His frantic alchemical research was largely concerned with trying to transmute base metals to gold. His lectures would last for half an hour unless there was no one at all in the audience; in that case he would stay only 15 min. Toward the end of his life, Newton devoted more time to theology than scientific research. His main manuscript on religion, *Observations Upon the Prophecies of Daniel, and the Apocalypse of John*, was published in 1733, after his death. His health was excellent until his last years; he never wore glasses and he lost only one tooth in all his life. At the age of 80, he felt terrible pain that was diagnosed as a kidney stone; however, in recent years this diagnosis has been discredited, and it is believed that he had angina pectoris. In 1727, at the old age of 85, Newton took his last breath and rested from his labors. Eight days later he was buried in Westminster Abbey with great honor. Voltaire, who attended Newton's funeral, said afterward, "I have seen a Professor of Mathematics, only because he was great in his vocation, buried like a king who had done good in his subjects." Newton's tombstone reads, "Mortals! Rejoice at so great an ornament to the human race." A few years later his heirs erected a monument to him that was placed in a favored spot of the Abbey, a spot that had been denied to high-ranking nobility. England had lost a great prince of mathematics forever, but his legacy lives on. Many years later, Laplace recorded his respect as follows: "Newton was the greatest genius that ever existed, and the most fortunate, for we cannot find more than once a system of the world to establish." Einstein has said that Newton "stands before us strong, certain, and alone: his joy in creation and his minute precision are evident in every word and in every figure." Nature to him was an open book, whose letters he could read without effort. His accomplishments were poetically expressed by Alexander Pope (1688–1744) in the lines,

> Nature and Nature's laws lay hid in night;
> God said, 'Let Newton be,' and all was light.

Aldous Huxley (1894–1963) said "If we evolved a race of Isaac Newtons, that would not be progress. For the price Newton had to pay for being a supreme intellect was that he was incapable of friendship, love, fatherhood, and many other desirable things. As a man he was a failure; as a monster he was superb."

Gottfried Wilhelm von Leibniz (1646–1716)

Gottfried Wilhelm von Leibniz (1646–1716) was a universal genius who won recognition in many fields: law, philosophy, religion, literature, politics, geology, metaphysics, alchemy, history, and mathematics. He was born in Leipzig, Germany. His father, a Professor of Moral Philosophy at the University of Leipzig, died when Leibniz was 6 years old. The precocious boy then gained access to his father's library and began reading voraciously on a wide range of subjects, a habit that he maintained throughout his life. At age 15 he entered the University of Leipzig as a law student, and in 1666, he applied for the degree of doctor of law, but was refused by the faculty on the grounds that he was too young. He left Leipzig and the very next year received a doctorate from the University of Altdorf (Nuremberg). Subsequently, Leibniz followed a career in law and international politics, serving as counsel to kings and princes. During his numerous foreign missions, Leibniz came in contact with outstanding mathematicians and scientists who stimulated his interest in mathematics—most notably, the physicist Huygens. Leibniz taught himself mathematics by reading papers and journals. As a result of this fragmented mathematical education, Leibniz often duplicated the results of others, and this ultimately led to a raging conflict over the invention of calculus—Leibniz or Newton? The argument over this question engulfed the scientific circles of England and Europe, with most scientists on the continent supporting Leibniz and those in England supporting Newton. The conflict was unfortunate, and both sides suffered in the end. The continent lost the benefit of Newton's discoveries in astronomy and physics for more than 50 years, and for a long period England became a second-rate country with regard to mathematics, hampered by Newton's inferior calculus notation. The fact is that both men discovered calculus independently. Newton looked at the problems of infinitesimal calculus from the point of view of physics, whereas Leibniz started from the tangent problem and created a practical calculus that is still in use today. Leibniz invented it 10 years after Newton in 1685, but he published his results 20 years before Newton published his own work on the subject. His most significant mathematical contribution, other than the calculus, was in logic. Leibniz introduced the word "coordinate."

Leibniz tried to reunite the Protestant and Catholic churches. According to Laplace, Leibniz saw the image of Creation in binary arithmetic. He imagined that Unity represented God, and Zero the void; that the Supreme Being drew all beings from the void, just as unity and zero express all numbers in the binary system of numeration. Leibniz communicated his idea to the Jesuit Grimaldi, who was the President of the Chinese tribunal for mathematics, with the hope that it would help convert the Emperor of China to Christianity, who was said to be very fond of the sciences. He believed that the sum of an infinite number of zeros was equal to $1/2$, and attempted to make this plausible by saying that it is the mathematical analog of the creation of the world out of nothing. Leibniz thought that i was a bizarre mix between existence and nonexistence, he likened i to the Holy Spirit; both have an ethereal and barely substantial existence. He became an expert in the Sanskrit language and the culture of China. The word 'function' was introduced into the vocabulary of mathematics by Leibniz and Jean Bernoulli in about 1700. He popularized the use of the integral symbol in calculus. In his early papers, Leibniz used the notation "omn" (an abbreviation for the Latin word "omnes") to denote integration. Then on October 29, 1675, he wrote, "It will be useful to write $\int$ for omn, thus $\int \ell$ for omn $\ell \cdots$". Two or three weeks later he refined the notation further and wrote $\int [\ ] dx$ rather than $\int$ alone. This notation is so useful and so powerful that its development by Leibniz must be regarded as a major milestone in the history of mathematics and science. Leibniz made important contributions to the theory of determinants and the calculus of finite differences. He also developed a method of calculating π without reference to a circle. In 1674, he rediscovered Nilakanthan's arctangent series (1). In 1696, Leibniz showed that the Bernoulli equation can be reduced to a linear equation by making the substitution $v = y^{1-n}$. He even invented a calculating machine that could perform the four operations and extract roots. With Huygens he developed the notion of kinetic energy. Leibniz was the founder of the Berlin Academy of Science and became its first President in 1700.

Leibniz never married. He was moderate in his habits, quick tempered but easily appeased, and charitable in his judgment of others' work. He is famous for his assertion that this is the "best of all possible universes." (Given that God created the universe, could he have failed to create the best?) Voltaire ridiculed this view in his *Candide*, but it is unlikely that Leibniz intended 'best' to mean 'most pleasant,' as it did in Voltaire's simple-minded interpretation. According to him, the universe is made up of simple substances, called 'monads,' which are capable of perception. Human souls are monads with memory and reason. The monads 'have no windows' in the sense that they have no direct interaction with the rest of the universe. They are related to each other only by a 'preestablished harmony' set up by God. At the end of the *Discourse on Metaphysics*, Leibniz claims that God wants a personal relationship with human beings. In 1710, he published *Theodicy*, which is considered to be the crown of his philosophical thinking. It became one of the most popular books of the eighteenth-century Enlightenment. In spite of his great achievements, Leibniz never received the honors showered on Newton and spent his final years as a lonely, embittered man. He died in Hanover in 1716 from a noxious potion he took during an attack of the gout. At his funeral there was one mourner, his Secretary. An eyewitness stated, "He was buried more like a robber than what he really was—an ornament of his country." The logarithmic spiral was inscribed on the lid of his casket, along with the phrase *Inclinata resurget* (Though bent, he will rise). He may be said to have lived not one life, but several.

John Flamsteed (1646–1719)

John Flamsteed (1646–1719) was born in Derby and died in Greenwich. After completing his basic education at Derby School, he studied mathematics and astronomy on his own during 1662–1669. He was ordained a deacon and was preparing to take up a living in Derbyshire when he was employed by King Charles II as Britain's first Royal Astronomer on March 4, 1675. The Royal Observatory at Greenwich was built for him and he began observing in 1676, but he had to fund and bring his own instruments. He was elected to the Royal Society in 1676, where he was a member of council during 1681–1684. In 1684, he was appointed priest to the parish of Burstow, Surrey. He held that office, as well as that of Astronomer Royal, until his death. His main work was collecting improved observations and position measurements of stars, which finally led to the compilation of a large catalog, *Historia Coelestis Britannica* (Flamsteed 1725), and an atlas of stars, *Atlas Coelestis* (Flamsteed 1729). His other notable work includes lunar theory, optics of telescopes, meteorological observations with barometers and thermometers, and longitude determination. Besides valuable work in astronomy, he invented the system (published in 1680) of drawing maps by projecting the surface of the sphere on an enveloping cone that can then be unwrapped. Flamsteed's bust is installed in the Museum of the Royal Greenwich Observatory. He is also remembered for his conflicts with Newton. His greatest enemy, Edmund Halley, succeeded him as the second Astronomer Royal. The crater Flamsteed on the Moon is named after him, and the asteroid 4987 Flamsteed is named in his honor. Several schools and colleges in Derbyshire also have been named after him.

Joseph Raphson (1648–1715)

Joseph Raphson (1648–1715) was an English mathematician. His most notable work is *Analysis Aequationum Universalis* (1690), which contained the method of approximating the roots of an equation. Although Newton knew about the method in 1671, it was published only in 1736. Raphson was a staunch supporter of Newton's claim of being the sole inventor of Calculus. He also translated Newton's Arithmetica Universalis into English. Raphson was made an FRS in 1689.

Ehrenfried Walter von Tschirnhaus (1651–1708)

Ehrenfried Walter von Tschirnhaus (1651–1708) was born in Kieslingswalde (Germany) now Slawnikowice (Poland). He was the youngest of seven children. When Tschirnhaus was 6 years old his mother died, and he was brought up by a loving stepmother. He was taught privately until the age of 15, and in 1666, he entered the Gymnasium in Görlitz. Tschirnhaus entered the University of Leiden in 1668 to study mathematics, philosophy, and medicine. In 1674, he began a long European tour and met Newton, Collins, Wallis, Leibniz (with whom he maintained a life-long correspondence), and Huygens. In mathematics he is remembered for the Tschirnhaus transformation (the word transformation describes changes of position, size, or shape) by which certain intermediate terms from a given algebraic equation can be removed. He is noted as the discoverer of caustics by reflection (catacaustics), which bears his name. He also solved the brachystochrone problem of Johann Bernoulli. He is considered the inventor of porcelain in Europe; however, porcelain had been produced in China long before it was made in Europe. In 1682, he became a member of the Académie Royale des Sciences (founded in 1666) in Paris. Tschirnhaus died in Dresden, Germany.

Michel Rolle (1652–1719)

Michel Rolle (1652–1719), a French mathematician and the son of a shopkeeper, received only an elementary education. He married early and as a young man struggled hard to support his family on the meager wages of a transcriber for notaries and attorneys. In spite of his financial problems and minimal education, Rolle studied algebra and Diophantine analysis (a branch of number theory) on his own. Rolle's fortune changed dramatically in 1682 when he published an elegant solution of a difficult, unsolved problem in Diophantine analysis. The public recognition of his achievement led to a patronage under minister Louvois, a job as an elementary mathematics teacher, and eventually a short-termed administrative post in the Ministry of War. In 1685, he joined the Academy des Sciences in a very low-level position for which he received no regular salary until 1699. He remained there until he died of apoplexy in 1719.

While Rolle's forté was always Diophantine analysis, his most important work was a book on the algebra of equations called *Traité d'algébre*, published in 1690. In this book, Rolle firmly established the notation $\sqrt[n]{a}$ (earlier written as $\sqrt{\textcircled{n}a}$) for the nth root of a, and proved a polynomial version of the theorem that today bears his name. (Rolle's Theorem was named by Giusto Bellavitis in 1846.) Ironically, Rolle was one of the most vocal early antagonists of calculus. He strove intently to demonstrate that it gave erroneous results and was based on unsound reasoning. He quarreled so vigorously on the subject that the Académie des Sciences was forced to intervene on several occasions. Among his several achievements, Rolle helped advance the currently accepted size order for negative numbers. Descartes, for example, viewed −2 as smaller than −5. Rolle preceded most of his contemporaries by adopting the current convention in 1691.

Jacob Bernoulli (1654–1705)

Jacob Bernoulli (1654–1705) was born in Basel, Switzerland. He was first of the eight prominent mathematicians in the Bernoulli family. He studied theology at the insistence of his father, but abandoned it as soon as possible in favor of his love for science. He taught himself the new calculus of Newton and Leibniz and was a Professor of Mathematics at Basel from 1687 until his death. He wrote on infinite series, studied many special curves, invented polar coordinates, developed logarithmic differentiation, and introduced the *Bernoulli numbers* that appear in the power series expansion of the function $\tan x$. A problem of interest in the history of mathematics is that of finding the *tautochrone*, the curve down which a particle will slide freely under gravity alone, reaching the bottom in the same time regardless of its starting point on the curve. This problem arose in the construction of a clock pendulum whose period is independent of the amplitude of its motion. The tautochrone was found by Huygens in 1673 by geometrical methods and later by Leibniz and Jacob Bernoulli using analytical arguments. Bernoulli's solution in 1690 was one of the first occasions in which a differential equation was explicitly solved. In his book *Ars Conjectandi* (The Art of Conjecturing), published posthumously in 1713, he formulated the basic principle in the theory of probability known as *Bernoulli's theorem* or the *law of large numbers*: if the probability of a certain event is p, and if n independent trials are made with k successes, then $k/n \to p$ as $n \to \infty$. At first sight this statement may seem trivial, but beneath its surface lies a tangled thicket of philosophical (and mathematical) problems that have been a source of controversy from Bernoulli's time to the present day. Among his many discoveries, the finest of all is the equiangular spiral. This curve is found in the tracery of the spider's web, in the shells upon the shore, and in the convolutions of the far-away nebulae. In geometry, it is mathematically related to the circle and in analysis to the logarithm. To Bernoulli this curve was the symbol of his life and faith; according to his wishes the spiral was engraved upon his tombstone with the words *Eadem mutata resurgo* (Though changed I shall arise the same).

Pierre Varignon (1654–1722)

Pierre Varignon (1654–1722) was born in Caen (France), and educated at the Jesuit College and the University in Caen, where he received his master's degree in 1682. Varignon became Professor of Mathematics at the Collége Mazarin in Paris in 1688, and the same year was elected to the Académie Royale des Sciences. In 1704, he was appointed departmental chair at Collége Mazarin and became Professor of Mathematics at the Collége Royal. He was elected to the Berlin Academy in 1713 and to the Royal Society in 1718. Varignon was a friend of Newton, Leibniz, and the Bernoulli family. He was the earliest and strongest French advocate of infinitesimal calculus and exposed the errors in Rolle's critique. Varignon recognized the importance of a test for the convergence of series, simplified the proofs of many propositions in mechanics, adapted Leibniz's calculus to the inertial mechanics of Newton's Principia, and treated mechanics in terms of the composition of forces. His other works include the application of differential calculus to fluid flow and to water clocks. He also created a mechanical explanation of gravitation and applied calculus to spring-driven clocks. Many of his works were published in Paris in 1725, 3 years after his death.

Edmund Halley (1656–1742)

Edmund Halley (1656–1742) was born in Haggerston, Shoreditch, London (England). His father came from a Derbyshire family and was a wealthy soapboiler in London; however, he lost most of his fortune in the great fire of London when Halley was 10 years old. Halley was first tutored privately at home, then at St. Paul's School, and then from 1673 at Queen's College, Oxford. At St. Paul's School he was equally distinguished for his classical and mathematical ability. He constructed dials, observed the change in the variation of the compass, and studied the heavens. Halley made important observations at Oxford, including an occultation of Mars by the Moon on August 21, 1676, which he published in the Philosophical Transactions of the Royal Society. He gave up his studies and in November of 1676, with the support of his father, Brouncker (who was the President of the Royal Society), and John Moore (1646–1714, who had been a major influence in the founding of the Royal Observatory), Halley sailed to St. Helena in the southern hemisphere to study the southern stars. During the voyage he improved the sextant, he collected a number of valuable facts related to the ocean and atmosphere, he noted the equatorial retardation of the pendulum, and on November 7, 1677 he observed a transit of Mercury, which suggested to him the important idea of employing similar phenomena for determining the Sun's distance. Halley returned to England in 1678 and published his catalogue of 341 southern hemisphere stars. This gave him a reputation as one of the leading astronomers and won him the title of "Southern Tycho." In December of 1678, he earned the Master of Arts degree, which was conferred on him at Oxford by King Charles II's command, and almost simultaneously he was elected as one of the Royal Society's youngest members. The fame and recognition that Halley so quickly achieved made some jealous, particularly John Flamsteed, who had praised him when he was a student. Six months later, the Royal Society sent Halley to Danzig to settle a long standing dispute between Robert Hooke and the astronomer Johannes Hevelius (1611–1687) regarding the respective merits of plain and telescopic sights. He then went on a continental tour, and observed the great comet of 1680 in Paris along with Giovanni Domenico Cassini (1625–1712). Halley spent much of 1681 in Italy. Halley married Mary Tooke in 1682 and settled in Islington. The couple lived harmoniously for 55 years and had three children. In March 1684 his father vanished and was found dead 5 weeks later. In Islington he was engaged chiefly in lunar observations with regard to the greatly desired method of finding longitude at sea; however, he was also busy with the momentous problem of gravity, namely, proving Kepler's laws of planetary motion. In August of 1684 he went to Cambridge to discuss this with Newton and found that Newton had already solved the problem, but had published nothing. Halley convinced him to write the *Philosophiae Naturalis*, which was published in 1687 at Halley's expense. In 1686, Halley published the second part of the results he had established in St. Helena, which included a chart on trade winds and monsoons. In this work he identified solar heating as the cause of atmospheric motions. He also established the relationship between barometric pressure and height above sea level. His charts were an important contribution to information visualization. In 1690, he built a diving bell, a device in which the atmosphere was replenished by way of weighted barrels of air sent down from the surface. In a demonstration, Halley and five companions dived to 60 ft in the River Thames, and remained there for over an hour and a half. In 1692, he presented the idea of a hollow Earth consisting of a shell about 500 miles thick, two inner concentric shells, and an innermost core, each with diameters approximately equal to those of the planets Venus, Mars, and Mercury, respectively. Atmospheres separate these shells, and each shell has its own magnetic poles. The spheres rotate at different speeds. Halley proposed this scheme in order to explain anomalous compass readings. He envisaged the atmosphere inside as luminous (and possibly inhabited) and speculated that escaping gas caused the Aurora Borealis. In 1693, he published an article on life annuities, which strongly influenced the development of actuarial science. During the same year he detected the "long inequality" of Jupiter and Saturn, and the acceleration of the Moon's mean motion. In 1696, he was appointed deputy comptroller of the mint at Chester, and in 1698, he was commissioned to be a captain of the "Paramour Pink" for making observations of the conditions of terrestrial magnetism. His findings were published in General Chart of the Variation of the Compass in 1701. This was the first such chart to be published and the first on which isogonic, or Halleyan, lines appeared. Immediately afterward he executed, by royal command, a careful survey of the tides and coasts of the British Channel. He produced a map based on this work in 1702. In 1703, Halley was appointed Savilian Professor of Geometry at Oxford University following the death of Wallis. In 1705, applying historical astronomy methods, he published Synopsis *Astronomia Cometicae*. In this work he predicted the return of the comet in 1758 that had already been seen earlier in 1456, 1531, 1607, and 1682. Although Halley had been dead for 17 years by this time, he achieved lasting fame when the comet (known as *Halley's Comet*) was observed on December 25, 1758. Halley's comet is due to return again in the year 2062. A recent study indicates that the comet has made about 2000 cycles so far with about the same number remaining before the Sun erodes it away completely. In 1710, he received an honorary degree of Doctor of Laws. In 1716, he suggested a high-precision measurement of the distance between the Earth and the Sun by timing the transit of Venus. In 1718, based on Ptolemy's catalogue, Halley deduced that the stars must have small motions of their own, and he was able to detect this proper motion in three stars. In 1720, Halley succeeded Flamsteed as Astronomer Royal, a position he held until his death. Although in his 64th year, he undertook this position to observe the Moon through an entire revolution of her nodes (18 years).

Halley worked for the Royal Society in various roles; he was editor of the Philosophical Transactions during 1685–1693 and Secretary during 1713–1721. He supported Newton in his controversy with Leibniz over who invented the calculus, serving as Secretary of a committee set up by the Royal Society to resolve the dispute. However, he went out of his way to make his own dispute with Flamsteed worse. In 1712, he arranged with Newton to publish Flamsteed's observations long before they were complete. To make matters worse, Halley wrote a preface, without Flamsteed's knowledge, in which he attacked Flamsteed for sluggishness, secretiveness, and lack of public spirit. He died in January 1742 in Greenwich, London. His tomb is in the old graveyard of St. Margaret's church, and his bust is in the museum of the Royal Greenwich Observatory. Besides Halley's comet he is remembered for Halley's crater on Mars, Halley's crater on the Moon, Halley's research station (Antarctica), Halley's method for the numerical solution of equations, and Halley's Street in Blackburn, Victoria, Australia.

Guillaume Francois Antoine de L'Hôpital (1661–1704)

Guillaume Francois Antoine de L'Hôpital (1661–1704) was a French mathematician. He was born to parents of the French high nobility and held the title of Marquis de Sainte-Mesme Comte d'Autrement. L'Hôpital showed mathematical talent quite early, and at the age of 15 solved a difficult problem about cycloids that had been posed by Pascal. As a young man he served briefly as a cavalry officer, but resigned because of nearsightedness. He gained fame as the author of the first textbook ever published on differential calculus, *L'Analyse des Infiniment Petites pour l'Intelligence des Lignes Courbes* (1696). L'Hôpital's rule appeared for the first time in that book. Actually, L'Hôpital's rule and most of the material in the calculus text were due to John Bernoulli, who was L'Hôpital's teacher. L'Hôpital dropped his plans for a book on integral calculus when Leibniz informed him that he intended to write such a text. L'Hôpital was apparently generous and personable, and his many contacts with major mathematicians provided the vehicle for disseminating major discoveries in calculus throughout Europe.

John Craig (1663–1731)

John Craig (1663–1731), a Scottish theologian who was fond of applying mathematics to divine matters, was born in Dumfries and lived in or near Cambridge. He was acquainted with the works of Newton and Leibniz on calculus. His major contribution is the book *Method of Determining the Quadratures of Figures Bounded by Curves and Straight Lines*, which was published in 1685 in London. In this book, John referred to Leibniz's work and used his differential notation. Leibniz reviewed John's book, in which John cited a paper by Leibniz that was actually written by von Tschirnhaus. Leibniz was then prompted to write an article to dispel any false impressions that readers might have had about his work.

Johann Bernoulli (1667–1748)

Johann Bernoulli (1667–1748) was the younger brother of Jacob. Like his brother, he made a false start in his career, studying medicine and taking a doctorate at Basel in 1694 with a thesis on muscle contraction. He became fascinated by calculus, quickly mastered it, and applied it to many problems in geometry, differential equations, and mechanics. In 1695, he was appointed Professor of Mathematics and Physics at Groningen in Holland, and on Jacob's death he succeeded his brother in professorship at Basel. The Bernoulli brothers sometimes worked on the same problems, which was unfortunate because of their jealous and touchy dispositions. For example, the question of the *catenary curve* that Jacob posed in 1690 follows: find the shape of the curve assumed by a chain, fixed at two points, and hanging under its own weight. Such a curve had been known for a long time; in fact, Galileo had assumed it to be a parabola. Jacob made no progress for a year and was chagrined to see Johann's correct solution. Occasionally, the friction between them flared up into a bitter and abusive public feud, as it did over the brachistochrone problem: to find the curve along which a particle will slide without friction in the minimum time from one given point to another, the second point being lower than the first but not directly beneath it. In 1696, Johann proposed the problem as a challenge to the mathematicians of Europe. It aroused great interest and was solved by Newton and Leibniz as well as the two Bernoullis. Johann's solution, which used the procedure called the *calculus of variations* for the first time, was more elegant. Jacob's, though rather clumsy and laborious, was more general. This situation started an acrimonious quarrel that dragged on for several years and was often conducted in rough language more suited to a street brawl than a scientific discussion. Johann appears to have been the more cantankerous of the two; much later, in a fit of jealous rage, he threw his own son out of the house for winning a prize from the French Academy that he coveted for himself. However, both the brothers had a hand in examining the divergence of the *harmonic series*, $1+1/2+1/3+\cdots$. It is interesting to note that Johann defended Leibniz's reputation in the controversy regarding the discovery of calculus.

William Whiston (1667–1752)

William Whiston (1667–1752) was born in Leicestershire, the fourth of nine children. After being privately educated until 17, he studied at Queen Elizabeth Grammar School for 2 years, and then entered Clare College, Cambridge. William attended Newton's lectures while at Cambridge and showed great promise in mathematics. He obtained his bachelor's degree in 1690, master's degree in 1693, and was elected Fellow in 1691 and probationary Senior Fellow in 1693. During 1694–1698, he was Chaplain to the bishop of Norwich. In 1696, he wrote the very popular treatise *A New Theory of the Earth*, in which he claimed that biblical stories like the creation and flood can be explained scientifically as descriptions of events with historical bases. He acted as Newton's deputy in the Lucasian Chair from 1699, in 1703 succeeded him as professor, and edited his *Arithmetica universalis* (1707; *Universal Arithmetic*). He was expelled in 1711, mainly for theological reasons. Afterward, he supported himself by giving public lectures on popular science. He was succeeded by Nicholas Saunderson (1682–1739), the blind mathematician who was born in Yorkshire in 1682, began lecturing at Cambridge on the principles of the Newtonian philosophy at the age of 25, wrote two long lasting large volumes on Algebra, and died at Christ's College, Cambridge, on April 19, 1739. Whiston died in London on August 22, 1752.

Abraham De Moivre (1667–1754)

Abraham De Moivre (1667–1754) was born in Vitry-le-Francois, Champagne, France. Although he was a French Protestant, De Moivre first attended the Catholic school of the Christian Brothers in Vitry, and when he was 11 years old went to the Protestant Academy at Sedan where he spent 4 years studying Greek. During 1682–1684, he studied logic at Saumur and then moved to Paris to study at the Collége de Harcourt. Until 1684 most of the mathematics texts he read were on his own. In 1685, following the revocation of the Edict of Nantes and the expulsion of the Huguenots, De Moivre was forced to seek asylum in England. After arriving in London he became a private tutor of mathematics, visiting his pupils and also teaching in the coffeehouses of London. He purchased a copy of Newton's Principia, cut up the pages so that he could carry a few with him at all times, and as he traveled from one pupil to the next he read them. He soon became distinguished among first-rate mathematicians. In fact, he became one of the intimate personal friends of Newton and was elected a Fellow of the Royal Society of London in 1697. De Moivre hoped for a Chair of Mathematics, but being a foreigner, he was discriminated against. In 1710, De Moivre was appointed to the Commission set up by the Royal Society to solve the Newton-Leibniz dispute concerning who had invented calculus first. No doubt, his appointment to this Commission was due to his friendship with Newton. The Royal Society knew the answer it wanted! It is interesting to note that De Moivre was given this important position, but he was not entitled to a university post. Although he published several papers in the Philosophical Transactions, he is best known for his memoir *Doctrine of Chances: A method of calculating the probabilities of events in play*, which was first printed in 1718 and dedicated to Newton. This work was reprinted in 1718, 1738, and 1756 with great alterations and improvements. The 1756 edition contains what is probably De Moivre's most significant contribution, namely, the approximation to the binomial distribution by the normal distribution in the case of a large number of trials. In 1722, he published his famous theorem $(\cos x + i \sin x)^n = \cos(nx) + i \sin(nx)$. De Moivre also published a *Treatise on Annuities* in 1724, which has passed through several revised and corrected editions in 1743, 1750, 1752, and 1756. In his *Miscellanea Analytica*, published in 1730, appears the formula for very large n,

$$n! \simeq (2\pi n)^{1/2} e^{-n} n^n,$$

which is known today as *Stirling's formula*. For example, $10! = 3{,}628{,}800$ and the formula gives 3,598,695.6. In 1733, De Moivre used this formula to derive the normal curve as an approximation to the binomial. He also pioneered the development of analytic geometry. It seems that Newton, in his later years, advised those who came to him with questions on mathematics, "Go to Mr. De Moivre; he knows these things better than I do."

De Moivre was unmarried and spent his later years in peaceful study. He was a steadfast Christian throughout his life. His main income was from tutoring mathematics, and he died in poverty. De Moivre, like Cardan, is famed for predicting the day of his own death. He found that he slept 15 min longer each night, and summing the arithmetic progression, calculated that he would die on November 27, 1754, the day that he would sleep all 24 h. Exactly 5 months before his death, De Moivre was elected as a foreign associate of the Paris Academy of Sciences.

Luigi Guido Grandi (1671–1742)

Luigi Guido Grandi (1671–1742) was born in Cremona, Italy. He became a member of the Order of the Camaldolese in 1687. In 1694, Grandi became a teacher of philosophy and theology at the Camaldolese monastery in Florence, and in 1714 became Professor of Mathematics at the University of Pisa. In mathematics, Grandi is best known for studying the rose curve ($r = \sin n\theta$) and other curves that resemble flowers. He considered the formula

$$\begin{aligned}\frac{1}{2} &= 1 - 1 + 1 - 1 + 1 - 1 + \cdots \\ &= (1-1) + (1-1) + (1-1) + \cdots \\ &= 0 + 0 + 0 + \cdots\end{aligned}$$

to be the symbol for Creation from Nothing. He obtained the result $1/2$ by considering the case of a father who bequeaths a gem to his two sons who each may keep the gem during alternating years. It then belongs to each son one half of the time. Grandi also wrote several works on geometry. His work in 1703 introduced Leibniz's calculus into Italy. His work on hydraulics, mechanics, and astronomy is also well recognized. Grandi's practical work on mechanics included experimenting with a steam engine. He died in Pisa.

Count Jacopo Francesco Riccati (1676–1754)

Count Jacopo Francesco Riccati (1676–1754) was an Italian savant who wrote on mathematics, physics, and philosophy. He was chiefly responsible for introducing the ideas of Newton to Italy. He was offered the presidency of the St. Petersburg Academy of Sciences (which in the eighteenth century became one of the main centers of mathematical investigations), but understandably he preferred the leisure and comfort of his aristocratic life in Italy to the administrative responsibilities in Russia. Though widely known in scientific circles of his time, he now survives only through the differential equation $y' = p(x) + q(x)y + r(x)y^2$, which bears his name. Even this was an accident of history, for Riccati merely discussed special cases of this equation without offering any solutions, and most of these special cases were successfully treated by various members of the Bernoulli family. This equation was also studied by his son Vincenzo Riccati (1707–1775) and Euler.

Jacques Cassini (1677–1756)

Jacques Cassini (1677–1756) was born in Paris at the Paris Observatory where his father, Giovanni Cassini, was head of the observatory. He started his education at home and then studied at the Collége Mazarin in Paris. In 1691, at the age of 14, Jacques defended his thesis on optics. At the age of 17 he was admitted to the membership of the French Academy of Sciences, in 1696 he was elected a Fellow of the Royal Society of London, and he became maítre des comptes in 1706. Jacques succeeded his father's position at the observatory in 1712. In 1713, he measured the arc of the meridian from Dunkirk to Perpignan, and published the results in a volume entitled *Traité de la grandeur et de la figure de la terre* in 1720. Jacques published the first tables of the satellites of Saturn in 1716. He was appointed an advocate in the Court of Justice in the same year, and in 1722 became a Councillor of State. Jacques also wrote *Eléments d'astronomie* in 1740. He is especially known for his study of the Moons of Jupiter and Saturn and his study of the structure of Saturn's rings. Jacques was involved in an accident, his carriage overturned, and he died due to his injuries in 1756. He was buried in the Church at Thury.

Jacob Hermann (1678–1733)

Jacob Hermann (1678–1733) was born and died in Basel. He received his initial mathematical training from Jacob Bernoulli and graduated with a degree in 1695. He was appointed to a chair in mathematics in Padua in 1707, but moved to Frankfurt an der Oder in 1713, and then St. Petersburg in 1724. He returned to Basel in 1731 to take a chair in ethics and natural law. Hermann worked on problems in classical mechanics. In 1729, he proclaimed that it was as easy to graph a locus in the polar coordinate system as it was to graph it in the Cartesian coordinate system. He appears to have been the first to show that the Laplace-Runge-Lenz vector is a constant of motion for particles acted upon by an inverse-square central force. He became a member of the Berlin Academy in 1701 and was elected to the Académie Royale des Sciences (Paris) in 1733, the year of his death. It is believed that Hermann was a distant relative of Euler.

Roger Cotes (1682–1716)

Roger Cotes (1682–1716) was born in Burbage, Leicestershire. He first attended Leicester School where his mathematical talent was recognized, and later studied at St Paul's School in London, and then at Trinity College, Cambridge. He obtained his undergraduate degree in 1702 and his master's in 1706. In 1707, he became a fellow of Trinity College, formed a school of physical sciences at Trinity in partnership with William Whiston, and was appointed the first Plumian Professor of Astronomy and Experimental Philosophy (founded to honor Thomas Plume, 1630–1704) at Cambridge University. During 1709–1713, Cotes and Newton worked together to prepare the second edition of Newton's *Principia*. They fully deduced, from Newton's laws of motion, the theory of the Moon, the equinoxes, and the orbits of comets. Cotes made substantial advances in theory of logarithms, numerical methods, computational tables, and integral calculus. He published only one scientific paper in his lifetime, entitled *Logometrica*, in which he successfully constructed the logarithmic spiral. He also invented the quadrature formulas, known as Newton–Cotes formulas, and was the first to introduce what is known today as Euler's formula. He died at the early age of 33 due to violent fever in Cambridge. On his death Newton remarked, "If he had lived we would have known something." After his death, many of Cotes' mathematical papers were hastily edited by Robert Smith (1689–1768) and published in a book, *Harmonia mensurarum*. Cotes' additional works were published later in Simpson's *The Doctrine and Application of Fluxions*.

Brook Taylor (1685–1731)

Brook Taylor (1685–1731) was a leading English mathematician. Taylor was born of well-to-do parents. Musicians and artists were entertained frequently in the Taylor home, which undoubtedly had a lasting influence on young Brook. In his later years, Taylor published a definitive work on the mathematical theory of perspective and obtained major mathematical results about the vibrations of strings. He was also one of the founders of the calculus of finite differences and was the first to recognize the existence of singular solutions of differential equations. Taylor's most productive period was from 1714 to 1719, during which time he wrote on a wide range of subjects: magnetism, capillary action, thermometers, perspective, and calculus. In his final years, Taylor devoted his writing efforts to religion and philosophy. According to Taylor, the results that bear his name were motivated by coffeehouse conversations about works of Newton on planetary motion and works of Halley on roots of polynomials. There also exists an unpublished work, *On Music*, that was intended to be part of a joint paper with Isaac Newton. The Taylor series, for which he is best known in mathematics, was known to James Gregory when Taylor was just a few years old. It is believed that Taylor was not aware of Gregory's work when he published his book *Methodus incrementorum directa et inversa* in 1715, which contained the Taylor series. Taylor's writing style was so terse and hard to understand that he never received credit for many of his innovations.

Taylor's life was scarred with unhappiness, illness, and tragedy. Because his first wife was not rich enough to suit his father, the two men argued bitterly and parted ways. His wife died in childbirth. After he remarried, his second wife also died in childbirth, though his daughter survived.

Gabriel Daniel Fahrenheit (1686–1736)

Gabriel Daniel Fahrenheit (1686–1736) was born in Danzig in the Polish-Lithuanian Commonwealth to a merchant's family. Fahrenheit's parents died in 1701 after consuming poisonous mushrooms, so he took business training as a merchant in Russia. Because of his interest in the natural sciences, he studied experimentation in that field, and after traveling around settled in 1717 in the Hague, the Netherlands. He learned glassblowing, and made barometers, altimeters, and thermometers. From 1718 onward, he gave lectures on chemistry in Amsterdam and became a member of the Royal Society in 1724. He is best known for inventing the alcohol thermometer in 1709, the mercury thermometer in 1714, and for developing the Fahrenheit temperature scale with $0°$ equal to the temperature of an equal mixture of ice and ammonium chloride, and $96°$ as the temperature of the human body. It is often said, erroneously, that Fahrenheit set $100°$ as the temperature of the human body. Except the USA, most countries have officially adopted the Celsius scale. The conversion formula for a temperature that is expressed in the Celsius representation is $F = (9C/5) + 32$. Fahrenheit died, unmarried, in the Netherlands, presumably in the Hague, where he was buried.

Christian Goldbach (1690–1764)

Christian Goldbach (1690–1764) was born in Königsberg, Prussia, the city known for its famous bridge problem. Goldbach was the son of a pastor. He became Professor of Mathematics and historian at the Academy in St. Petersburg in 1725. In 1728, Goldbach went to Moscow to tutor the son of the Tsar. He entered the world of politics when, in 1742, he became a staff member at the Russian Ministry of Foreign Affairs. Goldbach is known for his correspondence with eminent mathematicians, including Euler, Leibniz, and Nicholas Bernoulli (1695–1726, who died from a cold caught after plunging into the freezing Neva River). Goldbach is best known for his famous conjecture made in 1742 in a letter to Euler, that every even integer greater than 2 can be represented as the sum of two primes, for example $4 = 2 + 2,\ 28 = 23 + 5,\ 96 = 89 + 7$; this is still an open problem. In 1743, he observed that a polynomial $P_n(x) = a_0 + a_1x + a_2x^2 + \cdots + a_nx^n$ with integer coefficients $a_0, a_1, \ldots, a_n$ cannot represent primes only, that is, the integers $f(0), f(1), f(2), \ldots$ are not all prime. He also studied and proved some theorems on perfect powers and made several notable contributions to analysis. Goldbach died in 1764 in Moscow, Russia.

James Stirling (1692–1770)

James Stirling (1692–1770) was born at Garden, Stirlingshire (Scotland) to an aristocratic family. At the age of 18 he went to Oxford, where he was nominated one of Bishop Warner's exhibitioners at Balliol. In 1715, he was expelled from Oxford on account of his correspondence with members of the Keir and Garden families, who were known Jacobites. From Oxford he went to Venice and became a Professor of Mathematics, but fearing assassination after having discovered a trade secret of the glassmakers of Venice, he returned to London in 1725. In 1730, his most important work, *the Methodus differentialis, sive tractatus de summatione et interpolatione serierum infinitarum*, was published. In 1735, he communicated to the Royal Society the paper *On the Figure of the Earth, and on the Variation of the Force of Gravity at its Surface*. Later he published some papers on applied mathematical problems. In mathematics, he is mainly remembered for Stirling numbers of the first and second kinds, which arise in a variety of combinatoric problems.

Colin Maclaurin (1698–1746)

Colin Maclaurin (1698–1746) was a prominent Scottish mathematician and physicist. Maclaurin's father, a minister, died when the boy was only 6 months old, and his mother when he was 9 years old. He was then raised by an uncle who was also a minister. Maclaurin entered Glasgow University as a divinity student, but transferred to mathematics after 1 year. He received his master's degree at age 17, and in spite of his youth, began teaching at Marischal College in Aberdeen, Scotland. Maclaurin met Isaac Newton during a visit to London in 1719 and from then on he became Newton's disciple. During that period, some of Newton's analytic methods were bitterly attacked by major mathematicians, and much of Maclaurin's important mathematical work resulted from his efforts to defend Newton's ideas geometrically. Maclaurin introduced the method of generating conics and was the first to give a correct theory for distinguishing maxima and minima. He also demonstrated that a homogeneous rotating fluid mass revolves in an elliptic fashion. In calculus, Maclaurin's formula, which is a particular case of Taylor's formula and had appeared some 25 years earlier in the work of Stirling, is well known. He used this formula in his *Treatise on fluxions*, published in 1742. This treatise was the first systematic formulation of Newton's methods and was so carefully written that it was the standard of mathematical rigor in calculus until the work of Cauchy in 1821. He was the original discoverer of the rule for solving systems of equations that we now call Cramer's rule. Maclaurin was an outstanding experimentalist. He devised numerous ingenious mechanical devices, made important astronomical observations, performed actuarial computations for insurance societies, and helped to improve maps of the islands around Scotland. Maclaurin took an active part in opposing the march of the Young Pretender in 1745 at the head of a great Highland army, which overran the country and finally seized Edinburgh. He escaped, but the hardships of trench warfare and the subsequent flight to York proved fatal.

Daniel Bernoulli (1700–1782)

Daniel Bernoulli (1700–1782) was the son of Johann. He studied medicine, like his father, and took a degree with a thesis on the action of the lungs. Like his father, he soon gave way to his inborn talent and became a Professor of Mathematics at St. Petersburg. In 1733, he returned to Basel worked as Professor of Botany, Anatomy, and then Physics. He had the distinction of winning the prize of the French Academy ten times, including the one that infuriated his father. Over the years, Daniel Bernoulli published many works on physics, probability, calculus, and differential equations. In his famous book *Hydrodynamics* he discussed fluid mechanics and gave the earliest treatment of the kinetic theory of gases. He is considered by many to have been the first genuine mathematical physicist. In 1760, Daniel Bernoulli published "On the Mortality Caused by Smallpox, and the Advantage of Inoculation." In it, he addressed the question of determining the number of years of life, on average, gained by inoculation. The following anecdote is told about him: Once when traveling as a young man Daniel introduced himself to an interesting stranger with whom he had been conversing: "I am Daniel Bernoulli." "And I," said the other sarcastically, "am Isaac Newton." This delighted Daniel to the end of his days as the sincerest tribute he had ever received.

Anders Celsius (1701–1744)

Anders Celsius (1701–1744) was born in Uppsala, Sweden. His grandfathers were both professors in Uppsala: Magnus Celsius (1621–1679) the mathematician, and Anders Spole (1630–1699) the astronomer. His father, Nils Celsius (1658–1724), was also Professor of Astronomy. Anders Celsius was very talented in mathematics from an early age. Like his father, he became Professor of Astronomy at Uppsala University in 1730 and remained there until his death of tuberculosis in 1744. During 1732–1735, he visited almost all of the notable European observatories, where he worked with many of the leading eighteenth century astronomers. In 1733, at Nuremberg, he published a collection of 316 observations of the aurora borealis made by himself and others over the period 1716–1732. Anders Celsius founded the Uppsala Astronomical Observatory in 1741, and in 1742, he proposed the Celsius temperature scale in a paper to the Royal Swedish Academy of Sciences. In his scale, temperature 0 represented the boiling point of water, and 100 the freezing point. The scale was turned around after his death. In the late 1950s the name 'centigrade,' which was used for Celsius temperature, became obsolete. 'Centigrade' was replaced by 'degrees Celsius.' The conversion formula for a temperature that is expressed in the Fahrenheit scale is $C = 5(F - 32)/9$. In astronomy, Anders Celsius made observations of eclipses and various astronomical objects. He is also remembered as a writer of poetry and popular science. Anders Celsius was a very active supporter of introducing the Gregorian calendar in Sweden, but he was not successful until 1753, almost 10 years after his death, when the Julian calendar was modified by dropping 11 days. His grave is next to his grandfather's, Magnus Celsius, in the church at Gamla Uppsala. The Celsius crater on the Moon is named after him.

Gabriel Cramer (1704–1752)

Gabriel Cramer (1704–1752) was born in Geneva, Switzerland. He was the son of physician Jean Cramer and Anne Mallet Cramer. Amazingly, he completed his doctorate at the age of 18 and was appointed co-Chair of Mathematics at the Académie de Calvin in Geneva at 20. As a part of his appointment, Cramer did lots of traveling in Europe and visited several leading mathematicians, including Johann Bernoulli, Euler, Halley, de Moivre, Stirling, Fontenelle, Pierre Louis Moreau de Maupertuis (1698–1759), Clairaut, and Comte de Buffon (1707–1783, who proposed an estimate of about 75,000 years for the age of the Earth, instead of the commonly accepted figure of approximately 6,000 years). In 1728, he proposed a solution to the St. Petersburg Paradox that came very close to the concept of expected utility theory given 10 years later by Daniel Bernoulli. In 1730, during a competition, the Paris Academy judged Cramer's entry second best, and the prize went to Johann Bernoulli. In 1734, he was appointed to be the Chair of Mathematics.

In 1742, Cramer published Johann Bernoulli's complete works in four volumes, and in two volumes in 1744. In 1745, Cramer and Johann Castillon (1704–1791) jointly published the correspondence between Johann Bernoulli and Leibniz. He edited a five-volume work by Christian Wolff (1675–1715), first published during 1732–1741, with a new edition during 1743–1752. In 1750, Cramer published his major mathematical work *Introduction á l'analyse des lignes courbes algébriques.* In it he presented a method to solve linear systems of equations, which is known today as *Cramer's Rule*; however, variations of this rule were already fairly well known. Cramer published articles on geometry, philosophy, the history of mathematics, the date of Easter, and the aurora borealis. He wrote an article on law where he applied probability to demonstrate the significance of having independent testimony from two or three witnesses rather than a single witness. Cramer's name is also attached to the Castillon–Cramer problem, which inscribes a triangle in a circle so that it passes through three given points. He is also known for Cramer's paradox. Cramer also served as a member of the Council of Two Hundred in 1734 and the Council of Seventy in 1749. In these councils he used his broad mathematical and scientific knowledge to tackle problems involving artillery, fortification, reconstruction of buildings, and excavation. He became an FRS in 1749. Cramer remained single and died in Bagnols-sur-Céze in 1752.

Émilie, Marquise du Châtelet–Laumont (1706–1749)

Émilie, Marquise du Châtelet–Laumont (1706–1749) was born in Paris into an elite aristocracy. Émilie was taught at home by excellent tutors and governesses, and by the age of 12 she had began learning Latin, Italian, and English. She studied Virgil, Tasso, Milton, Horace, and Cicero, and learned to ride and fence, but her true love was mathematics. In 1725, at the age of 19, Émilie married Florent Claude Chastellet, a military man. They had a daughter and a son in 1726 and 1727 respectively, as well as a boy in 1733 who did not live long. After the death of their third child, Émilie and Florent separated. In 1730, Florent was made a regimental colonel, and so he had to spend a significant amount of time with his troops. During this period, Émilie returned to the whirlwind of Paris high society, gambling, socializing, and enjoying more freedom. She surrounded herself with many men, including Duc de Richelieu, Maupetuis, and Voltaire (who once remarked that if God did not exist, it would be necessary to invent him). The friendship and affair with Voltaire blossomed for the remainder of her life. In 1734, Voltaire and Émilie moved to Cirey, a city near the Belgian border, where they set up a well-equipped physics laboratory and spent their days studying and writing. In this laboratory, all of the experiments that led to her Newtonian work were conducted. She and Voltaire independently competed in 1738 for a prize offered by the French Academy on the subject of fire. Although du Châtelet did not win, her dissertation was published by the academy in 1744. In 1740, she published a comprehensive textbook, *Institutions of physics*, expounding in part some ideas of Leibniz and a work on Vital Forces.

In this book, with respect to metaphysics she claimed, "it is certain that there are a number of points in metaphysics which lend themselves to demonstrations just as rigorous as the demonstrations of geometry, even if they are different in kind." During the last 4 years of her life, she translated Newton's Principia from Latin into French and added an algebraic commentary, most of which was based on her work with the scientist Clairaut. She also added a section on tides, which she summarized from a prize essay by Daniel Bernoulli. This has been the only French translation of Newton's Principia to date, which was finally published in 1759 and remains her best-known work. In 1748, Émilie fell in love with the Marquis de Saint-Lambert while visiting at Luneville and became pregnant in January 1749. Hearing this, Voltaire joined Saint-Lambert in plotting to make it appear as if the baby she was carrying was that of her husband. On September 4, 1749, she gave birth to a daughter; however, 6 days later Émilie died, and in 1751, the child also passed away. Émilie is also remembered for the translation of Mandeville's *The Fable of the Bees*, three chapters of a *Grammaire raisonnee*, and *A Discourse on Happiness*, which was published well after her death.

Leonhard Euler (Pronounced "Oiler") (1707–1783)

Leonhard Euler (pronounced "oiler") (1707–1783) was probably the most prolific mathematician who ever lived. He was born in Basel, Switzerland, and was the son of a Protestant minister who himself studied mathematics. Euler's genius developed early. He attended the University of Basel, where by the age of 16 he obtained both a Bachelor of Arts and a master's degree in philosophy. While at Basel, Euler had the good fortune of being tutored 1 day a week in mathematics by a distinguished mathematician, Johann Bernoulli. Urged by his father, Euler then began to study theology. The lure of mathematics was too great, however, and by age 18 Euler had begun to do mathematical research. Nevertheless, the influence of his father and his theological studies remained, and throughout his life Euler was a deeply religious, unaffected person. At the age of 19, he transmitted two dissertations to the Paris Academy, one on the analysis of the optimum placement of masts on a ship and the other on the philosophy of sound. Remarkably, at this point in his life, Euler had never even seen an ocean-going vessel. These essays mark the beginning of his splendid career. At various times, Euler taught at St. Petersburg Academy of Sciences (in Russia), the University of Basel, and the Berlin Academy of Sciences. Euler's energy and capacity for work were boundless. His collected works form about 60–80 quarto sized volumes and it is believed that much of his work has been lost. What is particularly astonishing is that Euler became sightless in his right eye during the mid-1730s and was blind for the last 17 years of his life, yet this was one of his most productive periods! In 1771, his house caught fire and he was trapped, unable to escape through the smoke and flames that he could feel but not see. His servant, Peter Grimm, dashed into his room and carried him out to safety. Euler's manuscripts were also saved, but everything else burned. Euler's flawless memory was phenomenal. Early in his life he memorized the entire *Aeneid* by Virgil and at age 70 could not only recite the entire work but also state the first and last sentence on each page of the edition that he memorized. His ability to solve problems in his head was beyond belief. He worked out, in his head, major problems of lunar motion that had baffled Newton. He once did a complicated calculation in his head to settle an argument between two students whose computations differed in the 50th decimal place. He gave the impression of a small-town, middle-class minister rather than Europe's greatest mathematician. His personal life was as placid and uneventful as is possible for a man with 13 children, except that he married twice. Euler remained undisturbed and often rocked a baby with one hand while he worked out difficult problems with the other. He could be interrupted constantly and then easily proceed from where he had left off without losing either his train of thought or his temper. For Euler, his home was his joy. In his 77th year, on November 18, 1783, Euler was sitting at the table having tea with one of his grandchildren when he suffered a stroke. "I am dying," he cried, and minutes later he became unconscious. He died a few hours afterward. He was buried in St. Petersburg (now Leningrad).

Euler has had a deeper influence on the teaching of mathematics than any other man. This came mainly through his three treatises: *Introductio in Analysin Infinitorum* (1748), *Institutiones Calculi Differentialis* (1755), and *Institutiones Calculi Integralis* (1768–1794). It has been said that since 1748 all elementary and advanced calculus textbooks are essentially copies of Euler or copies of copies of Euler. In these treatises, Euler summed and codified the discoveries of his predecessors, adding many of his own ideas. He extended and perfected plane and solid geometry, introduced the analytic approach to trigonometry, and was responsible for the modern treatment of the functions $\ln x$, e^x, Beta, and Gamma. He also adopted fundamental notation like π, e, i, and linked them together in the astonishing relation $e^{\pi i} + 1 = 0$, about which Klein remarked that all analysis is centered. Every symbol has its history: the principal whole numbers 0 and 1; the chief mathematical relations $+$ and $=$; π, the discovery of Hippocrates; i, the sign for the 'impossible' square root of minus one; and e, the base of Napierian logarithms. In 2004, the readers of the journal *Physical World* ranked this equality as the most beautiful equation in science. It is a special case of his famous formula, $e^{i\theta} = \cos\theta + i\sin\theta$, that connects the exponential and trigonometric functions. He contributed many important ideas to differential equations. In fact, the various methods of reduction of order, the notion of an integrating factor, and power series solutions are all due to Euler. In addition, he gave the first systematic discussion of the calculus of variations. He was the first and greatest master of infinite series, infinite products, and continued fractions, and his works are full of striking discoveries in these fields. In number theory, he gave the first published proofs of both Fermat's theorem and Fermat's two squares theorem. He proved that every positive integer is a sum of four squares, investigated the law of quadratic reciprocity (which connects any two odd primes), and initiated the theory of partitions (the number of ways in which a number n can be expressed as a sum of smaller whole numbers). The theory of graphs originated with Euler, he developed this idea to solve a problem that was posed to him by the citizens of the Prussian city of Königsberg (now Kaliningrad in Russia). The city is divided by the Pregel River, which encloses an island. The problem was to determine whether it is possible to start at any point on the shore of the river or on the island, and walk over all of the bridges, once and only once, returning to the starting spot. In 1736, Euler showed that the walk was impossible by analyzing the graph. He is also known as the "Grandfather of Topology"; he founded the "network theory," which is one of the most practical forms of topology. (Topology is the geometry of distortion. It deals with fundamental geometric properties that are unaffected when we stretch, twist, or otherwise change an object's size and shape.) His introductory algebra text, originally written in German, is still read in its English translation. He made contributions to virtually every branch of mathematics as well as astronomy, physics, botany, and chemistry. He was also an expert in Oriental languages.

The following apocryphal story about Euler was told by de Morgan. In 1773, at the Catherine the Great's court, the French encyclopedist Denis Diderot (1713–1784) made himself a nuisance with his openly atheist views. Euler, a good Catholic, was called upon to silence him, so he told Diderot that he had a mathematical proof of the existence of God. This intrigued Diderot; in full view of the court, he asked Euler for his proof. "Sir," Euler said, "$(a + b_n)/n = x$, therefore God exists. Refute that"! Diderot (who knew mathematics well enough to write articles in his Encyclopedia about it) realized he had been made a fool of and eventually returned to his native France.

The French physicist Dominique-François-Jean Arago (1786–1853), speaking of Euler's incomparable mathematical facility, remarked that "He calculated without apparent effort, as men breathe, or as eagles sustain themselves in wind."

Thomas Simpson (1710–1761)

Thomas Simpson (1710–1761) was an English mathematician. His father was a weaver. He was trained to follow in his father's footsteps and had little formal education in his early life. His interest in science and mathematics was aroused in 1724, when he witnessed an eclipse of the Sun and received two books from a peddler, one on astronomy and the other on arithmetic. Simpson quickly absorbed the contents and soon became a successful fortune teller. At 19 his improved financial situation enabled him to give up weaving and marry his landlady, a 50-year old widow with two children. Then in 1733 some mysterious "unfortunate incident" forced him to move. He settled in Derby, where he taught in an evening school and worked at weaving during the day. In 1736, he moved to London and published his first mathematical work in a periodical called the *Ladies' Diary* (of which he later became editor). In 1737, he published a successful calculus textbook that enabled him to give up weaving completely and concentrate on textbook writing and teaching. His fortune improved further in 1740 when Robert Heath accused him of plagiarism. The publicity was marvelous, and Simpson proceeded to dash off a succession of best-selling textbooks, *Algebra* (ten editions plus translations), *Geometry* (12 editions plus translations), *Trigonometry* (five editions plus translations), and numerous others. It is interesting to note that Simpson did not discover the rule that bears his name; this was a well-known result by Simpson's time.

Ruđer Josip Bošković (Roger Joseph Boscovich) (1711–1787)

Ruđer Josip Bošković (Roger Joseph Boscovich) (1711–1787) was a physicist, astronomer, mathematician, philosopher, diplomat, poet, theologian, Jesuit priest, and polymath, born in the city of Dubrovnik in the Republic of Ragusa (now Croatia). He was the seventh child of the family and the second youngest. Roger left Dubrovnik for Rome at the age of 14, where two Jesuit priests introduced him to the Society of Jesus, renowned for its youth education. In 1731, he became a novice and studied mathematics and physics in the Collegium Romanum (Roman College). He was so outstanding in these subjects that he was straightway appointed as a professor of mathematics in the same college. Several years before this appointment, he was already well known for an elegant solution of the problem of determining the equator of the Sun and computing the period of its rotation by observation of the spots on its surface. Besides performing the assigned duties of a professor, Roger found time to explore various problems of physics. He published several dissertations involving the double refraction micrometer, the inequalities in terrestrial gravitation, the theory of comets, the transit of Mercury, the observation of the fixed stars, the aurora borealis (or northern lights, a spectacular natural phenomenon which can occasionally be seen in the night sky over Britain), the application of mathematics to the theory of the telescope, the limits of certainty in astronomical observations, the law of continuity, and problems of spherical trigonometry. In 1742, Pope Benedict XIV consulted Roger and other scientists as to the best means of securing the dome of Saint Peter's Basilica, in which a crack had been discovered. Roger's suggestion of placing five concentric iron bands was adopted. In 1753, he discovered the absence of atmosphere on the Moon. As a representative of the republic of Lucca, Roger went to Vienna in 1757 and succeeded in bringing about a satisfactory arrangement when a dispute cropped up between Francis the Grand Duke of Tuscany and the republic regarding the drainage of a lake. In 1761, he was elected as a member of the Russian Academy of Sciences. In 1764, Roger was appointed as the chair of mathematics at the University of Pavia. He was also the director of the observatory of Brera in Milan for 6 years. He spent 2 years (1783–1785) at Bassano keeping himself busy with the publication of his works pertaining to optics and astronomy. After visiting the convent of Vallombrosa for a few months, Roger went to Brera in 1786 and resumed his work, but he increasingly fell prey to ill health and frustration. He breathed his last at the age of 75 in Milan and was buried there in the church of St. Maria Podone.

Alexis Claude Clairaut (1713–1765)

Alexis Claude Clairaut (1713–1765) was born in Paris, France, where his father taught mathematics. Clairaut was a prodigy; he learned to read from Euclid's Elements, and by the age of ten he had mastered calculus and analytical geometry. At the age of 12 he wrote a memoir on four geometrical curves, and made such a rapid progress in the subject that in his 13th year he read before the Paris Academy. At 16 he published a book containing important additions to geometry, and as a result he became the youngest person ever elected to the Paris Academy at the age of 18. In 1731, he gave a demonstration of the fact noted by Newton that all curves of the third order are projections of one of five parabolas. In 1733, Clairaut wrote on the calculus of variations, and in the same year he published work on the geodesics of quadrics of rotation. The following year he studied differential equations, especially $y = px + f(p)$, $p = dy/dx$, that bear his name. In 1739 and 1740, he published further work on integral calculus and proved the existence of integrating factors for solving first order differential equations. In 1743, he proved Clairaut's theorem, which connects the gravity at points on the surface of a rotating ellipsoid with the compression and the centrifugal force at the equator. In 1749, Clairaut wrote a book on algebra, in which he took the subject up to the solution of equations of the fourth degree. He tried, with great success, to show why the introduction of algebraic notation was necessary and inevitable. This book was used for teaching in French schools for many years. Clairaut was the first scientist to recognize the necessity of better approximations in the calculus of perturbations. After Newton's death, exact observations proved that orbits of Jupiter, Saturn, and the Moon differ significantly from what had been calculated. A question was raised about whether this was related to inaccuracy of the calculus at the first approximation or whether an adjustment in the law of universal gravitation was needed. Clairaut made massive calculations and demonstrated that application of additional terms of expansion eliminated the indicated discrepancy. Euler's recommendation gave him the prize of the Petersburg Academy for that essay in 1750. This was followed in 1754 by some lunar tables. Subsequently, Clairaut wrote various papers on the orbit of the Moon and on the motion of comets as affected by the perturbation of the planets, particularly on the path of Halley's comet, which appeared only 1 month before the date he had predicted. Clairaut also made important contributions to the aberration of light. He was particularly interested in improving telescope design by using lenses made up of two different types of glass. He also published a geometry book in the year of his death in Paris. Clairaut was elected to the Royal Society of London, and the Academies of Berlin, St Petersburg, Bologna, and Uppsala.

Jean le Rond d'Alembert (1717–1783)

Jean le Rond d'Alembert (1717–1783) was the illegitimate son of Chevalier (Knight) Louis-Camus Destouches (1668–1726), an artillery general, and Madame de Tencin (1682–1749), a wealthy socialite who abandoned d'Alembert when he was an infant near the church of Jean le Rond. He is best known as a French physicist, mathematician, and man of letters. According to him, mathematics is the science that concerns itself with the properties of magnitude, insofar as they can be measured and calculated. In science he is remembered for *d'Alembert's principle* (that the internal actions and reactions of a system of rigid bodies in motion are in equilibrium) in mechanics which he proved at the age of 26, *d'Alembert's paradox* in hydrodynamics, and his solution of the wave equation. The wave equation was first derived and studied by d'Alembert in 1746. It also attracted the attention of Euler (1748), Daniel Bernoulli (1753), and Lagrange (1759). Solutions were obtained in several different forms, and the merits of, and relations among, these solutions were heatedly argued in a series of papers over the course of 25 years. The debated issues concerned the nature of a function and the kinds of functions that can be represented by trigonometric series. These questions were not resolved until the nineteenth century. He also made significant contributions in celestial mechanics and the theory of elasticity. In 1751, he divided classical mathematics into three main branches: pure, mixed, and physico-mathematics. Pure mathematics includes arithmetic and geometry; mixed mathematics embraces mechanics, geometric astronomy, optics, acoustics, pneumatics, and the art of conjecturing or the analysis of games of chance; and physico-mathematics applies mathematical calculation to experiment and seeks to deduce physical inferences whose certitude is close to geometric truth. The main work of d'Alembert's life was his collaboration with Diderot on the latter's famous *Encyclopédie*, which played a major role in the French Enlightenment by emphasizing science and literature and attacking the forces of reaction in church and state. He was a friend of Euler, Lagrange, and Laplace. d'Alembert never married, but when his longtime love Julie de Lespinasse (1732–1776) fell ill in 1765, he moved in to take care of her and stayed until her death in 1776. "Go ahead, faith will follow" were the encouraging words with which d'Alembert kept reinforcing the courage of his doubters. A famous remark made by him is "Algebra is generous: she often gives more than is asked of her."

Maria Gaetana Agnesi (1718–1799)

Maria Gaetana Agnesi (1718–1799) was born in Milan, Italy. Her father, Pietro di Agnesi, was a wealthy man who made money trading silk. Pietro had 21 children with his three wives, and Maria was the eldest of the children. Maria was a gifted scholar and linguist, she could speak French and Italian at the age of five, published an essay in Latin defending higher education for women at the age of nine (during the Middle Ages, under the influence of Christianity, many European countries were opposed to any form of higher education for females), and by thirteen she had mastered Greek, Hebrew, Spanish, German, Latin, and other languages. She was called the "Walking Polyglot." When she was 15 her father began to periodically assemble the most learned men of Bologna at his house, and Maria explained and defended various philosophical theses. At the age of 20 she wanted to enter a convent; however, her father refused to grant this wish, the meetings were discontinued, and she led a life of retirement, in which she devoted herself to the study of mathematics. In 1738, her father published 191 theses that she had defended in the meetings, under the title *Propositiones Philosophicae*. In 1748, Maria published two volumes of *Instituzioni Analitiche*, the first treated the analysis of finite quantities, and the second addressed the analysis of infinitesimals. This work was regarded as the best introduction to the extant works of Euler. A French translation of the second volume appeared at Paris in 1775, and an English translation of the whole work by John Colson (1680–1760) was published in 1801. Today, Agnesi is remembered chiefly for a bell-shaped curve called the *witch of Agnesi*. This name, found only in English texts, is the result of a mistranslation. Agnesi's own name for the curve was versiera or "turning curve"; however, Colson probably confused versiera with avversiera, which means "wife of the devil," and translated it into "witch." After the success of this work, Maria was elected to the Bologna Academy of Sciences, and in 1750, Pope Benedict XIV appointed her to the Chair of Mathematics and natural philosophy at the Bologna University. She was the second woman to be appointed professor at a university; however, after the death of her father in 1752 she gave up mathematics altogether. In 1762, when the University of Turin asked if she would serve as a reader for a paper on the calculus of variations, she declined the honor and missed the opportunity to examine the work of Lagrange.

Maria never married, educated her younger siblings, and was a very religious woman. She devoted the rest of her life to the poor, homeless, and sick, especially women. When the *Pio Instituto Trivulzo*, a home for the ill and infirm, was opened, Maria became its director. She took care of ill and dying women until her own death in 1799. She is considered to be the first important woman mathematician after Hypatia. A crater on Venus was named in her honor.

Johann Heinrich Lambert (1728–1777)

Johann Heinrich Lambert (1728–1777) was born in Mülhausen, Switzerland. His father was a small tailor, and so Lambert had to rely on his own efforts for his education. He first worked as a clerk in an ironworks, then in a newspaper office, and subsequently became a private tutor to a family, where he was allowed to use the library. During 1756–1758, he traveled to Germany, the Netherlands, France, and Italy and met several prominent mathematicians. In 1759, he moved to Augsburg and in 1764 re-moved to Berlin where he became editor of the Prussian astronomical almanac. He continued working in Berlin until his death in 1777. Lambert was the first to introduce hyperbolic functions into trigonometry. The expansion of π into a continued fraction was discovered by Lambert in 1761, and it is of great historical importance. In addition to his work on irrational numbers, he first proved the irrationality of π in 1768. He wrote landmark books on geometry, the theory of cartography, and perspective in art. His influential book on geometry foreshadowed the discovery of modern non-Euclidean geometry. Lambert is also credited for expressing Newton's second law of motion in the notation of the differential calculus. He is remembered for the Lambert quadrilateral, Lambert's cosine law, Lambertian reflectance, the Lambert cylindrical equal-area projection, the Lambert conformal conic projection, the Lambert azimuthal equal-area projection, Lambert's trinomial equation, and Lambert's W-function.

Benjamin Banneker (1731–1806)

Benjamin Banneker (1731–1806) spent the first half of his life tending a farm in Maryland. He gained a local reputation for his mechanical skills and his abilities in mathematical problem solving. In 1772, he acquired astronomy books from a neighbor and devoted himself to learning astronomy, observing the skies, and making calculations. In 1789, Banneker joined the team that surveyed what is now the District of Columbia. Banneker published almanacs yearly from 1792 to 1802. His almanacs contained the usual astronomical data and information about the weather and seasonal planting. He also wrote social commentary and made proposals for the establishment of a peace office in the President's Cabinet, for a department of the interior, and for a league of nations. He sent a copy of his almanac to Thomas Jefferson (1743–1826) along with an impassioned letter against slavery. Jefferson praised Banneker and sent the almanac to the French Academy to show what an American could achieve.

Alexandre-Théophile Vandermonde (1735–1796)

Alexandre-Théophile Vandermonde (1735–1796) was born and died in Paris. He was a sickly child, and so his physician father directed him to a career in music, but he later developed an interest in mathematics. His complete mathematical work consists of four papers published in 1771–1772. These papers include fundamental contributions to the roots of equations, the theory of determinants, and the *knight's tour problem*. In chess, the knight is the only piece that does not move in a straight line, and the problem is to determine whether a knight can visit each square of a chessboard by a sequence of knight's moves, landing on each square exactly once. Vandermonde's interest in mathematics lasted for only 2 years. Afterward, he published papers on harmony, experiments with cold, and the manufacture of steel. He also became interested in politics, joining the cause of the French revolution and holding several different positions in government. Although Vandermonde is best known for his determinant, it does not appear in any of his four papers. It is believed that someone mistakenly attributed this determinant to him.

Jean-Sylvain Bailly (1736–1793)

Jean-Sylvain Bailly (1736–1793) was born in Paris. As a boy, Jean wanted to pursue a career in the arts following his family tradition, but became strongly inclined to astronomy due to the influence of Abbé Nicolas Louis de Lacaille (1713–62), a French astronomer noted for his catalogue of nearly 10,000 southern stars. An outstanding student with extraordinary, endless patience and tolerance, Jean also accurately reduced Lacaille's observations to 515 unique stars. He also took part in the construction of an observatory at the Louvre. Due to his remarkable achievements, he was elected to the 31st seat of the French Academy of Sciences in 1763 and was made a foreign member of the Royal Swedish Academy of Sciences in 1778. He is noted for his computation of an orbit for the next appearance of Halley's Comet in 1759 and for his studies of the four known satellites of Jupiter. The lunar crater Bailly was named in his honor.

Joseph Louis Lagrange (1736–1813)

Joseph Louis Lagrange (1736–1813) was born in Turin (Italy) to a French–Italian family. Baptismal records list his name as Giuseppe Lodovico Lagrangia. He was the youngest of 11 children and the only one to survive beyond infancy. His father was a public official and wanted him to be a lawyer; however, Lagrange was attracted to mathematics and astronomy after reading a memoir by the astronomer Halley on the uses of algebra in optics. At age 16 he began to study mathematics on his own and by age 19 was appointed to a professorship at the Royal Artillery School in Turin. The following year, Lagrange sent Euler solutions to some famous problems using new methods that eventually blossomed into a branch of mathematics called calculus of variations. These methods and Lagrange's applications of them to problems in celestial mechanics, were so monumental that by age 25 he was regarded by many of his contemporaries as the greatest living mathematician. In 1758, Lagrange established a society that was subsequently incorporated as the *Turin Academy*. The first five volumes of its transactions include most of the early work of Lagrange on a wide variety of problems.

In 1776, with Euler's recommendation, Frederick the Great (a misogynist) invited Lagrange to come to Berlin, accompanying his invitation with a modest message that said "it is necessary that the greatest geometer of Europe should live near the greatest of kings." During his stay in Berlin, Lagrange distinguished himself not only in celestial mechanics but also in algebraic equations of the theory of numbers, feeding the stream of thought that later nourished Gauss and Abel. One of Lagrange's most famous works is a memoir, *Mécanique Analytique* in 1788, in which he reduced the theory of mechanics to a few general formulas from which all other necessary equations could be derived. In the preface he wrote "one will not find figures in this work, but only algebraic equations." After 20 years in Berlin, he moved to Paris at the invitation of Louis XVI (1754–1793). He was given apartments in the Louvre and was treated with great honor, even during the revolution. Napoleon Bonaparte (1769–1821), who believed that the advancement and perfection of mathematics are intimately connected with the prosperity of the state, was a great admirer of Lagrange and showered him with the titles of count, senator, and member of the Legion of Honor. He called him "the lofty pyramid of the mathematical sciences." Laplace presented Napoleon a copy of the *Mécanique* with high expectations, but Napoleon found an apparent oversight. "You have written this huge book on the system of the world without once mentioning the author of the universe." "Sire", Laplace retorted, "I had no need for that *hypothesis*." When Napoleon repeated this to Lagrange, the latter remarked "Ah, but that is a fine hypothesis. It explains so many things." The years Lagrange spent in Paris were devoted primarily to the didactic treatises that summarized his mathematical conceptions: *The Theory of Analytical Functions* in 1797 and its continuation *The Lectures of Function Calculations* in 1801. Among the enduring legacies of his work are Lagrange's equations of motion, generalized coordinates, the method of multipliers, his concept of potential energy, and the interpolating formula in 1795, which was the outgrowth of work on problems of expressing the roots of a polynomial as functions of its coefficients, and was published earlier by the British mathematician Edward Waring (1736–1798) in 1779. Some of his earlier work on the theory of equations led Galöis to his theory of groups. In fact, the important result in group theory that states that the order of a subgroup of a finite group G is a factor of the order of G is called Lagrange's theorem. His mathematical writing was always perfect in both form and matter, and he was very careful to explain his procedure so that one could understand the arguments easily. Lagrange headed the committee that led to the adoption of the metric system by many countries, in which the subdivision of money, weights, and measures is strictly based on the number ten. He also played an important role in verifying Newton's universal law of gravitation. It is an interesting fact that Lagrange's father speculated unsuccessfully in several financial ventures, and his family was forced to live quite modestly. Lagrange stated that if his family had money he would have made mathematics his vocation. In spite of his fame, Lagrange was always a shy, modest man. He married twice: first when he lived in Berlin, but his wife died in 1783 after a long illness; and then again in 1792 in Paris with the teenage girl Renće Le Monnier (nearly 40 years his junior) who he spent his quiet fruitful years with until he died in 1813. Two days before he died, Monge and other friends called, knowing that Lagrange was dying and that he wished to tell them something of his life. "I was very ill yesterday, my friends," he said. "I felt I was going to die; my body grew weaker little by little; my intellectual and physical faculties were extinguished insensibly; I observed the well-graduated progression of the diminution of my strength, and I came to the end without sorrow, without regrets, and by a very gentle decline. Oh, death is not to be dreaded, and when it comes without pain, it is a last function which is not unpleasant." On his death, he was buried with honor in the Pantheon. Lagrange is known for his quote "When we ask for advice, we seek an accomplice." He believed that a mathematician has not thoroughly understood his own work until he has made it so clear that he can go out and explain it effectively to the first man he meets on the street. According to him, in any serious and honest attempt to solve a mathematical problem there is a faithful look at truth.

Marquis de Condorcet (1743–1794)

Marquis de Condorcet (1743–1794), an eighteenth century French mathematician, Enlightenment philosopher, political scientist, and revolutionist, was born at Ribemont, Aisne, France. He lost his father at a very young age and was raised by his devout mother. He studied at the Jesuit College in Reims and at the College of Navarre in Paris, where he was known for his diverse talents. In 1759, at the age of 16, his analytical abilities were noticed by d'Alembert and Clairaut. Soon d'Alembert became his mentor, and at the age of 22 Condorcet wrote a treatise on the integral calculus that was widely appreciated. Condorcet worked with famous scientists such as Euler, and one of the Founding Fathers of the USA, Benjamin Franklin (1706–1790), who said: "What science can there be more noble, more excellent, more useful ... than mathematics." His most important contribution was in the development of theory of probability and the philosophy of mathematics. In 1785, he wrote an essay on the *Application of Analysis to the Probability of Majority Decisions*, which included (i) Condorcet's Jury Theorem, that is, if each member of a voting group is more likely than not to make a correct decision, the probability that the highest vote of the group is the correct decision increases as the number of members of the group increases, and (ii) Condorcet's Paradox, according to which it is possible for an elector to express a preference for A over B, B over C, and C over A, showing that majority preferences could become intransitive with more than two options. In 1786, at the age of 43, Condorcet married Sophie de Grouchy (1764–1822), an intelligent and highly educated girl more than 20 years younger than him. Condorcet strongly desired a rationalist reconstruction of the society, championed many liberal causes, and consequently took an active part in the French Revolution that swept France in 1789 like a tempest. He greeted the advent of democracy (rule by the people). Fearing for his life, he went into hiding. After a few months, in 1794, he was arrested near Paris and sent to prison, where after 3 days he died mysteriously; either he was murdered, committed suicide using a poison, or had a natural death at Bourg-la-Reine, France.

Gaspard Monge, Comte de Péluse (1746–1818)

Gaspard Monge, Comte de Péluse (1746–1818) was born at Beaune, Bourgogne, France. He was the son of a small peddler and attended the Oratorian College in Beaune, which was intended for young nobles and run by priests. In 1762, Monge went to Lyons where he continued his education at the Collége de la Trinité, and the next year, at the age of 17, he was put in charge of teaching a course in physics. He completed his education there in 1764 and returned to Beaune. He made a plan of the city of Beaune, inventing the methods of observation and constructing the necessary instruments; this plan was seen by a member of staff at the École Royale du Génie at Méziéres. He was so impressed by this work that in 1765, Monge was appointed to the École Royale du Génie as a draftsman. During this period, Monge came into contact with Charles Bossut (1730–1814) who was the Professor of Mathematics there. Bossut helped Monge with his work on the evolutes of curves of double curvature. Although his job did not require the use of his mathematical talent, Monge worked in his spare time at developing his own ideas of geometry. About a year later, Monge was asked to draw up a fortification plan that would prevent an enemy from either seeing or firing at a military position. He made full use of geometrical techniques to devise such a fortification, but he was strictly forbidden from communicating his findings to others. However, his mathematical abilities were recognized, and in 1768, Monge became Professor of Mathematics, succeeding Bossut, and in 1771, he became Professor of Physics at the École Royale du Génie. In 1780, he was appointed to a Chair of Hydraulics at the Lyceum in Paris. In 1781, he communicated his first paper to the French Academy. In this work he discussed the lines of curvature drawn on a surface, which led to the general differential equation. During 1785–1789, Monge undertook research in a wide range of scientific subjects, presenting papers to the Académie on the composition of nitrous acid, the generation of curved surfaces, finite difference equations, partial differential equations, double refraction, the structure of Iceland spar, the composition of iron, the action of electric sparks on carbon dioxide gas, capillary phenomena, the causes of certain meteorological phenomena, and a study in physiological optics. In 1792, upon the creation of an executive council by the Legislative Assembly, Monge accepted the office of Minister of the Marine and held this office for 8 months. In 1794, he was made a professor at the short-lived Normal School where he gave lectures on descriptive geometry, which forms the basis of his book *Application de l'analyse á la géométrie*, published in 1799. Descriptive geometry is a method for representing solids and other figures in ordinary three-dimensional space on one plane. Although there are many ways in which descriptive geometry can be developed or modified, they all go back to Monge. In 1796, Monge went to Italy on a commission to select the best art treasures for the conquerors and bring them to France. In 1798, he was sent to Rome on a political mission, and after completing it he joined Napoleon in Egypt. Monge returned to Paris in 1799 and took up his role as director of the École Polytechnique (created in 1794), a state-run institution of higher education and research in Palaiseau, Essonne,

France, near Paris. In 1799, Napoleon named Monge a senator on the Consulate for life. Over the next few years Monge continued his interest in mathematics, but his mathematical work mostly involved teaching and writing texts for the students at the École Polytechnique. During 1812–1815, he took active part in Napoleon's military campaign. After Napoleon was defeated at Waterloo, Monge feared for his life and fled France. In 1816, he returned to Paris. Two days after his return, he was stripped of all his honors, expelled from the services, harassed politically, and his life was continually threatened.

Monge is considered to be the father of differential geometry, which is roughly the study of properties of curves, surfaces, and so forth in the immediate neighborhood of a point, so that in powers higher than the second, distances can be neglected. In his descriptive geometry lay the nucleus of projective geometry, and his mastery of algebraic and analytical methods and their application to curves and surfaces contributed to analytical and differential geometry. Once Lagrange, appreciating the work of Monge, claimed that "with his application of analysis to geometry this devil of a man will make himself immortal!" Most of his miscellaneous papers were reproduced in the fourth edition of *Application de l'algébre á la géométrie*, which was published in 1819, a year after his death. On his death, the students of the École Polytechnique paid tribute to him despite the insistence of the French Government that none be paid. Monge was commemorated on the Eiffel Tower, and a Paris street is named after him.

John Playfair (1748–1819)

John Playfair (1748–1819) was born in Benvie (near Dundee), Scotland, where his father was Parish Minister. He was educated at home by his father until the age of 14, when he entered the University of St. Andrews to study for a general degree with the aim of entering the Church. Playfair obtained his Master of Arts degree in 1765. At the age of 18, he entered an 11 day contest for the Chair of Mathematics at Marischal College in Aberdeen. Although he was ranked third out of the six candidates, he proved his extraordinary talent and comprehensive knowledge of mathematics. In 1772, Playfair applied for the Chair of Natural Philosophy in the University of St. Andrews, which was left vacant after the death of his friend Professor Wilkie, but again another candidate was appointed. In 1773, he was ordained the Parish Minister of Liff and Benvie, succeeding his father. During this period Playfair continued his academic studies, and in 1779, his first mathematical paper, entitled "On the Arithmetic of Impossible Quantities," was published in the Philosophical Transactions. In 1782, he resigned his church position and tutored two sons of Ferguson of Raith (1723–1816) until 1787. This arrangement brought him closer to Edinburgh and he participated in the city's intellectual life. In 1783, Playfair became involved in the establishment of the Royal Society of Edinburgh, and became one of the original Fellows of the Society. In 1785, he was appointed Joint Professor of Mathematics in the University of

Edinburgh, a position that he held for 20 years. After 1787, Playfair published on various topics in the Transactions of the Royal Society of Edinburgh. In 1795, he wrote *Elements of Geometry*, in which he used algebraic notation to abbreviate the proofs. In this work, he provided an alternative to Euclid's parallel postulate: through a given point not on a given line, only one parallel can be drawn to the given line; however, this had already been given in the fifth century by Proclus. In early 1797, Playfair suffered a severe attack of rheumatism; however, he continued his work and wrote *An Analytical Treatise on the Conic Sections*. In 1802, he published his celebrated volume *Illustrations of the Huttonian Theory of the Earth*, which was a summary of the work of James Hutton (1726–1797). It was only through this work that Hutton's principle of uniformitarianism reached a wide audience. Despite his success as a mathematician, in 1805 he exchanged the Chair of Mathematics for the Chair of Natural Philosophy, leaving his former position to Dr. John Robison (1739–1805), whom he also succeeded as general Secretary to the Royal Society of Edinburgh. In 1807, he was elected an FRS. In 1812, he published the first volume of his lecture notes series *Outlines of Natural Philosophy*. It covered dynamics, mechanics, hydrostatics, hydraulics, aerostatics, and pneumatics. The second volume was devoted to astronomy, while a third volume, which was intended to complete the series and cover optics, electricity, and magnetism, was never completed. Playfair died in 1819 in Edinburgh, Scotland, due to bladder disease.

Playfair was admired by all men and beloved by all women, of whose virtues and intellect he was always a champion. Society felt happier and more respectable for his presence. In 1822, a collected edition of Playfair's works with a memoir by his eldest nephew, James George Playfair, appeared at Edinburgh in four volumes octavo. Craters on Mars and the Moon were named in his honor.

Pierre Simon de Laplace (1749–1827)

Pierre Simon de Laplace (1749–1827) was a French mathematician and theoretical astronomer who was so famous in his own time that he was known as the Newton of France. Laplace was born at Beaumont-en-Auge in Normandy. His father was a small cottager, or perhaps a farm laborer. Laplace intended to become a theologian, but his interest in mathematics was piqued by his instructors Christopher Gadbled (1734–1782) and Pierre le Canu at the provincial school in Caen. In 1768, when Laplace was leaving for Paris, Canu wrote him a letter of recommendation to give to d'Alembert; however, d'Alembert sent Laplace away with a problem and told him to come back in a week. Legend has it that Laplace solved it overnight. This impressed d'Alembert, who managed to get Laplace a mathematics teaching position at the Military School only a few days later. His main interests throughout his life were celestial mechanics, the theory of probability, and personal advancement. At the age of 24 he was already deeply engaged in the detailed application of Newton's law of gravitation to the solar system as a whole, in which the planets and their satellites are not governed by the Sun alone but interact with one another in a bewildering variety of ways. Even Newton had been of the opinion that divine intervention would occasionally be needed to prevent this complex mechanism from degenerating into chaos. Laplace decided to seek reassurance elsewhere and succeeded in proving that the ideal solar system of mathematics is a stable dynamic system that will endure unchange for a long time. This achievement was only part of a long series of triumphs recorded in his monumental treatise *Mécanique Céleste* (published in five volumes from 1799 to 1825), which summed up the work on gravitation done by several generations of illustrious mathematicians. Unfortunately for his reputation, he omitted all reference to the discoveries of his predecessors and contemporaries, and left it to be inferred that the ideas were entirely his own. The first four volumes were translated into English during 1814–1817 and published in 1829 at his own expense by the American Nathaniel Bowditch (1779–1838), who doubled its length by adding extensive commentary. Bowditch complained about Laplace's use of "It is easy to see...", for it invariably required several hours of hard work to see what Laplace claimed to be easy. (Bowditch also wrote a classical work on navigation, *New American Practical Navigator*, in 1802. These writings earned him membership of the Royal Society of London, and professorship in mathematics at Harvard, which he declined.) The principal legacy of the *Mécanique Céleste* to later generations lay in Laplace's development of potential theory, with its far-reaching implications in a dozen different branches of physical science, ranging from gravitation and fluid mechanics to electromagnetism and atomic physics. Even though he lifted the idea of the potential from Lagrange without acknowledgment, he exploited it so extensively that ever since his time the fundamental differential equation of potential theory has been known as Laplace's equation. However, this equation first appeared in 1752 in a paper by Euler on hydrodynamics.

His other masterpieces included the treatise *Théorie Analytique des Probabilités*, published in 1812, in which he incorporated his own discoveries in probability from the preceding 40 years. Again he failed to acknowledge the many ideas of others' that he mixed in with his own; regardless of this, his book is generally agreed to be the greatest contribution to this part of mathematics by any one man. In the introduction he says: "At bottom, the theory of probability is only common sense reduced to calculation." This may be so, but the following 700 pages of intricate analysis, in which he freely used Laplace transforms, generating functions, and many other highly nontrivial tools, has been said by some to surpass in complexity even the *Mécanique Céleste*.

After the French Revolution, Laplace's political talents and greed for position came to full flower. His countrymen speak ironically of his "suppleness" and "versatility" as a politician. What this really means is that each time there was a change of regime (and there were many), Laplace smoothly adapted himself by changing his principles—back and forth between fervent republicanism and fawning royalism—and each time he emerged with a better job and grander titles. He has been aptly compared to the apocryphal Vicar of Bray in English literature, who was twice a Catholic and twice a Protestant. The Vicar is said to have replied to the charge of being a turncoat, "Not so, neither, for if I changed my religion, I am sure I kept true to my principle, which is to live and die the Vicar of Bray."

To balance his faults, Laplace was always generous in giving assistance and encouragement to younger scientists. From time to time he helped forward the careers of men like the chemist Gay-Lussac, the traveler and naturalist Alexander von Humboldt (1769–1859), the physicist Poisson, and (appropriately) the young Cauchy, who was destined to become one of the chief architects of nineteenth century mathematics. Laplace's last words "what we know is so minute in comparison to what we don't know" reminds us of Plato's remarks on human knowledge: "In the final analysis, the theory of probability is only common sense expressed in numbers," and "All the effects of nature are only mathematical consequences of a small number of immutable laws."

Laplace once remarked that India gave us the ingenious method of expressing all numbers by means of ten symbols, each symbol receiving a value of position as well as an absolute value, a profound and important idea that appears so simple to us now that we ignore its true merit. But its very simplicity and the great ease with which it lends itself to all computation puts our arithmetic in the first rank of useful inventions, and we shall appreciate the grandeur of this achievement when we remember that it escaped the genius of Archimedes and Apollonius, two of the greatest men of antiquity. Laplace died in Paris exactly a century after the death of Newton, in the same month and year.

Lorenzo Mascheroni (1750–1800)

Lorenzo Mascheroni (1750–1800) was born near Bergamo, Lombardy (Italy). He became professor of mathematics at Pavia. In his *Geometria del Compasso* (1797) he proved that any geometrical construction that can be done with a compass and straightedge can also be done with compasses alone. However, this had been established in 1672 by the Dane Georg Mohr (1640–1697). This result is now known as the Mohr–Mascheroni Theorem. He is also known for the Euler–Mascheroni constant, denoted by γ. Lorenzo died in Paris.

Adrien–Marie Legendre (1752–1833)

Adrien-Marie Legendre (1752–1833) was born in Toulouse, France, to a wealthy family. He was given a quality education in mathematics and physics at the Collége Mazarin. During 1775–1780, out of interest, he taught with Laplace at ecole Militaire. In 1782, he won the prize on projectiles offered by the Berlin Academy. Legendre then studied the attraction of ellipsoids, leading to the introduction of what we call today the Legendre functions. Immediately afterward, he was appointed an adjoint member in the Académie des Sciences. In 1785, he published a paper on number theory that contains a number of important results, such as the law of quadratic reciprocity for residues, and the discovery that every arithmetic series with the first term coprime to the common difference contains an infinite number of primes. In 1787, Legendre became a member of the Anglo-French commission to make measurements of the Earth involving a triangulation survey between the Paris and Greenwich observatories. Due to his contribution to this project, he was elected to the Royal Society of London. In 1787, he made an important publication that contains Legendre's theorem on spherical triangles. In 1791, Legendre became a member of the committee of the Académie des Sciences, whose task was to standardize weights and measures to the metric system. In 1792, he supervised the major task of producing logarithmic and trigonometric tables. He had more than 70 assistants, and the work was completed in 1801. In 1794, Legendre published the influential elementary text *Elélments de géométrie*, which passed through many editions and was translated into several languages. In this text, he rearranged and simplified many of the propositions from Euclid's Elements. His 1796 conjecture of the prime number theorem was rigorously proved by Hadamard and Charles de la Vallée-Poussin (1866–1962) in 1898. His memoir, *Théorie des nombres*, was published in 1798 and appendices were added in 1816 and 1825; the third edition, issued in two volumes in 1830, includes the results of his various later papers and remains a standard work on the subject. In 1806, Legendre published *Nouvelles méthodes*, to determine the orbits of comets. His method involved three observations taken at equal intervals, assuming that the comet followed a parabolic path so that he ended up with more equations than unknowns. Legendre, independently of Gauss, gave the least squares method of fitting a curve to the data available, which he supplemented in 1810 and 1820. (The first significant mathematical contribution from the new country USA was the theoretical justification for the method of least squares by Robert Adrain (1775–1843), who came from Ireland.) Legendre's major work *Exercices de calcul intégral* appeared in three volumes in 1811, 1817, and 1819. In the first volume he introduced basic properties of elliptic integrals and the beta and gamma functions. Deeper results on the beta and gamma functions, along with applications to mechanics, the rotation of the Earth, the attraction of ellipsoids, and other problems appeared in the second volume. The third volume was mainly devoted to tables of elliptic integrals. His most fundamental work, *Traité des fonctions elliptiques*, appeared in two volumes in 1825 and 1826. A third volume was added a few weeks before his death and contains three memoirs on the research done by Abel and Jacobi. In 1830, he gave a proof of Fermat's last theorem for the exponent $n = 5$, which was also proved by Dirichlet in 1828. He is best remembered for the Gauss–Legendre algorithm, Legendre's constant, Legendre's equation, Legendre polynomials, Legendre's conjecture, and the Legendre transformation. The Legendre crater on the Moon is named after him.

Legendre had the misfortune of seeing most of his best work—in elliptic integrals, number theory, and the method of least squares—perfected by younger and abler men. For instance, he devoted 40 years to research on elliptic integrals, and his two-volume treatise on the subject had scarcely appeared in print when the discoveries of Abel and Jacobi revolutionized the field completely. Legendre lost his money during the French Revolution, so it was his various teaching positions and pensions that kept him at an acceptable standard of living. A mistake in office politics in 1824 led to the loss of his pension and he lived the rest of his years in poverty. He died in Paris in 1833.

Marc-Antoine Parseval des Chénes (1755–1836)

Marc-Antoine Parseval des Chénes (1755–1836) was born into a well-to-do family of high standing in Rosiéres-aux-Salines, France. Not much is known about his life. Marc is widely known for Parseval's theorem in the context of Fourier series. He opposed the French revolution and was imprisoned in 1792. He wrote poems against Napoleon's regime and, not surprisingly, had to flee France when Napoleon ordered his arrest. Marc succeeded in avoiding arrest and was able to return to Paris. He married at the age of 40, but separated soon afterward. Marc was nominated to the French Academy of Sciences five times during 1796–1828, but was never elected. His mathematical publications consisted of five papers published in 1806. He took his last breath in Paris, France.

Gaspard Clair François Marie Riche de Prony (1755–1839)

Gaspard Clair François Marie Riche de Prony (1755–1839) was born in Chamelet (France). In mathematics he is mainly remembered for producing logarithmic and trigonometric tables between 14 and 29 decimal places. With the assistance of Legendre, Lazare Carnot (1753–1823), other mathematicians, and between 70 and 80 assistants, the work was completed in 9 years. Prony died in Paris.

Sylvestre Francois Lacroix (1765–1843)

Sylvestre Francois Lacroix (1765–1843) was born in Paris to a poor family who provided a good education for their son through their hard work. Young Sylvestre had an affinity for mathematics. At the age of 14, he computed the motions of the planets. In 1782, at the age of 17, Sylvestre became a mathematics instructor at the École Gardes de Marine, Rochefort, France. After returning to Paris he taught astronomy and mathematics at the Lycée. He co-won the 1787 Grand Prix of the French Académie des Sciences, but was never awarded the prize. Starting in 1788 he taught mathematics, physics, and chemistry at the École d'Artillerie in Besancon. In 1793, he became examiner of the Artillery Corps, replacing Laplace. In 1794, he assisted his former teacher Monge in creating material for a course on descriptive geometry. In 1799, Sylvestre became professor at the École Polytechnique in Palaiseau, Essonne (near Paris). In the same year he became a member of the newly formed Institut National des Sciences et des Arts, now Institute de France.

He produced most of his texts for the purpose of improving his courses. In 1812, he began teaching at the Collége de France, a public higher education institution situated in Paris and was appointed chair of mathematics in 1815. During his career, Sylvestre produced several textbooks in mathematics. In 1812, Babbage set up "The Analytical Society for the Translation of Differential and Integral Calculus," and Sylvestre's books were translated into English in 1816 by George Peacock (1791–1858), an English mathematician. English translations of these books were used in British universities and beyond for over 50 years. Interestingly, Sylvestre was of the opinion that "algebra and geometry should be treated separately, as far apart as they can be, and that the results in each should serve for mutual clarification, corresponding, so to speak, to the text of a book and its translation." He passed away in Paris at the age of 78. The Sylvestre crater on the Moon was named for him.

Thomas Robert Malthus (1766–1834)

Thomas Robert Malthus (1766–1834) was a British demographer and political economist best known for his pessimistic but highly influential views. He was the first to observe that many biological populations increase at a rate proportional to the population. His first paper on populations appeared in 1798.

Jean-Robert Argand (1768–1822)

Jean-Robert Argand (1768–1822) was born in Geneva. He was self-taught, and mathematics was perhaps his hobby. In 1806, when managing a bookshop in Paris, he gave a geometrical representation of a complex number and applied it to show that every algebraic equation has a root. He not only admitted negative numbers but anticipated Descartes in the use of the symbols $+$ and $-$ [which were first used by Johannes Widman of Eger (1460–1500)] to indicate opposite directions on a line. He is also well known for Argand diagrams; however, a Norwegian surveyor, Caspar Wessel (1745–1818), invented these diagrams prior to Argand but did not receive credit since his work was published in a journal not ordinarily read by mathematicians. Moreover, Wessel wrote in Danish, so his audience was very restricted.

Jean Baptiste Joseph Fourier (1768–1830)

Jean Baptiste Joseph Fourier (1768–1830) was born at Auxerre in the Yonne département of France. His father was a tailor who remarried after the death of his first wife, with whom he had three children. Joseph was the ninth of the twelve children of the second marriage. Joseph's mother died when he was 9 years old, and his father died the following year. Right from the beginning Joseph showed talent in literature, and by 13 mathematics had become his real interest. By the age of 14 he had completed a study of the six volumes of Bézout's *Cours de mathématiques*. In 1783, he received first prize for his study of Bossut's Mécanique en général. In 1789, he visited Paris and read a paper on algebraic equations at the Académie Royale des Sciences. In 1790, he became a teacher at the Benedictine college, École Royale Militaire of Auxerre, where he had studied. In 1793, he became involved in politics and joined the local Revolutionary Committee. Fourier was arrested in 1794 during the French Revolution and was in serious danger of going to the guillotine. Fortunately for the world of mathematics, the political climate changed and Fourier was freed after Robespierre was beheaded. In 1795, Fourier started teaching at the École Polytechnique. In 1797, he succeeded Lagrange and was appointed as the Chair of Analysis and Mechanics. Fourier went with Napoleon on his Eastern expedition in 1798 and was made governor of Lower Egypt. He contributed several mathematical papers to the Egyptian Institute that Napoleon founded at Cairo. Fourier returned to France in 1801 and became prefect of the district of Isére in southeastern France, and in this capacity built the first real road from Grenoble to Turin. It was during this time in Grenoble that Fourier did his important mathematical work on the theory of the conduction of heat and made a systematic, although not rigorous, use of the series that now bears his name. He presented his basic papers on the subject to the Academy of Science of Paris in 1807 and 1811. However, these papers were criticized by the referees for lack of rigor and so were not published. According to Riemann, when Fourier presented his first paper to the Paris Academy in 1807, stating that an arbitrary function could be expressed as a Fourier series, Lagrange was so surprised that he denied the possibility in the most definite terms. However, to encourage Fourier, the Academy did award him a prize in 1812. In 1822, Fourier finally published his famous *Théorie Analytique de la Chaleur* (*The Analytic Theory of Heat*), in which he made extensive use of the Fourier series; however, he contributed nothing whatsoever to the mathematical theory of these series, which were well known much earlier to Euler, Daniel Bernoulli, Lagrange, and others. His results inspired a flood of important research that has continued to the present day. In fact, the Fourier series has proved to be highly valuable in various fields, such as acoustics, optics, electrodynamics, and thermodynamics. Harmonic analysis is the modern generalization of Fourier analysis, and wavelets are the latest implementation of these ideas. Fourier left unfinished work on determinate equations that was edited by Claude-Louis Navier (1785–1836) and published in 1831. Fourier is also credited for the discovery, in his essay in 1827, that gases in the atmosphere might increase the surface temperature of the Earth. This would later be called the greenhouse effect. He is also remembered for the Fourier transform, which is a linear operator that maps functions to other functions. It decomposes a function into a continuous spectrum of its frequency components, and the inverse transform synthesizes a function from its spectrum of frequency components. According to him, mathematics compares the most diverse phenomena and discovers the secret analogies that unite them. Fourier held the curious opinion that desert heat is the ideal environment for a healthy life and accordingly swathed himself like a mummy and lived in overheated room.

Johann Christian Martin Bartels (1769–1836)

Johann Christian Martin Bartels (1769–1836) was a German mathematician and an extraordinary teacher, born in Brunswick, Germany. In 1783, at the age of fourteen, Martin was employed as an elementary school teacher in the Katherinen-Volksschule situated near his home. Soon he began to teach a 6-year old Gauss. From 1791 Martin studied mathematics under Johann Friedrich Pfaff (1765–1825), a German mathematician in Helmstedt at the eastern edge of the German state of Lower Saxony, and Abraham Gotthelf Kästner (1719–1800), a German mathematician and epigrammatist in Göttingen. In the winter semester of 1793–1794 he studied experimental physics, astronomy, meteorology, and geology under Georg Christoph Lichtenberg (1742–1799), a German scientist, satirist, and Anglophile. In 1800, he worked as Professor of Mathematics in Reichenau, Switzerland. He joined the University of Kazan in 1808 as the chair of Mathematics. During his 12 year tenure he lectured on the history of mathematics, higher arithmetic, differential and integral calculus, analytical geometry and trigonometry, spherical trigonometry, analytical mechanics, and astronomy. In 1808, he taught Lobachevsky at Kazan University. When Lobachevsky was due to graduate, Martin spent 3 days lobbying the other professors to award him a master's degree. The university authorities did not want to give Lobachevsky a degree at all because of his poor behavior. Martin won the argument. Lobachevsky obtained a master's degree in 1811 and remained in Kazan to study with Martin. Lobachevsky took Martin's course on the history of mathematics, which prompted him to think about non-Euclidean geometry. In 1821, Martin moved to the University of Dorpat, now Tartu, Estonia, where he founded the Center for Differential Geometry. He remained at Dorpat until his death. He was elected to the St. Petersburg Academy of Sciences, and from 1826 he was also a corresponding member of the academy. He received high Russian honors. Most of Martin's contributions to mathematical research were made at the University of Kazan. However, he published some of his discoveries only after he moved to Dorpat. We only know about his discoveries through his students who included the results in their own work, acknowledging that their teacher Martin had given them in his lecture courses. One such result is the famous Frenet–Serret (Jean Frédéric Frenet, 1816–1900; Joseph Alfred Serret, 1819–1885) formula, discovered first by Martin. He introduced the method of moving trihedrons. Martin associated a trihedron to each point of a space curve, which was later called the Frenet trihedron. The world knows of these because they were published in a prize work by his student Karl Eduard Senff (1810–1849) in *Principal Theorems of the Theory of Curves and Surfaces* in 1831, with due acknowledgment to Martin. Frenet gave six of the formulas in 1847, and later Serret gave all nine. Martin corresponded with Gauss from the time he started working in Switzerland. Their correspondence continued through the years that he worked in Kazan and during the first few years that he was in Dorpat. After Gauss became famous, people used to say jokingly that Martin was the best mathematician in Germany because Gauss was the best mathematician in the world.

Robert Woodhouse (1773–1827)

Robert Woodhouse (1773–1827) was born in Norwich (England), the son of a linen draper. He was educated at Caius College, Cambridge. Woodhouse was elected an FRS in 1802. His first work, entitled *The Principles of Analytical Calculation*, was published at Cambridge in 1803. In this work he explained the differential notation and strongly recommended its employment. He did work in plane and spherical trigonometry in 1809, a historical treatise on the calculus of variations and isoperimetrical problems in 1810, on practical and descriptive astronomy in 1812, and on physical astronomy in 1818. He became the Lucasian Professor of Mathematics in 1820 and subsequently the Plumian professor in the university. As Plumian Professor he was responsible for installing and adjusting the transit instruments and clocks at the Cambridge Observatory. He held that position until his death in 1827. He was buried in Caius College Chapel.

Marie-Sophie Germain (1776–1831)

Marie-Sophie Germain (1776–1831), because she was a woman, was denied admission to the *École Polytechnic*, the French academy of mathematics and science. Her story illustrates the social attitude that the discipline of mathematics was no place for women. As a child, Germain had been fascinated by the mathematical works she found in her father's library. When she expressed an interest in studying the subject formally, her parents responded in horror. Not to be stopped, she obtained lecture notes from courses in which she had an interest, including one taught by Lagrange. Under the pen name M. LeBlanc, she submitted a paper on analysis to Lagrange, who was so impressed with the report that he wished to meet the author and personally congratulate 'him.' When he found out that the author was a woman he became a great help and encouragement to her. Lagrange introduced Germain to many of the French scientists of the time. In 1801, Germain wrote to Gauss to discuss Fermat's equation $x^n + y^n = z^n$. He commended her for showing "the noblest courage, quite extraordinary talents and a superior genius." Germain's interests included number theory and mathematical physics. She was the first to devise a formula describing elastic motion. The study of the equations for the elasticity of different materials aided the development of acoustical diaphragms in loudspeakers and telephones. In her memoir on elasticity, she said that "Algebra is but written geometry and geometry is but figured algebra." She would have received an honorary doctorate from the University of Göttingen based on Gauss' recommendation, but died before the honorary doctorate could be awarded. This would have been an honor for a woman in early nineteenth century Germany.

Jósef Maria Hoëne Wronski (1776–1853)

Jósef Maria Hoëne Wronski (1776–1853) was born in Poland, educated in Germany, and lived in France. Although Wronski's work was dismissed as rubbish for many years, and much of it was indeed erroneous, some of his ideas contained hidden brilliance and have survived. His noteworthy contribution to mathematics is his determinant. Wronski was more a philosopher than a mathematician, believing that absolute truth could be attained through mathematics. Wronski designed a caterpillar vehicle to compete with trains (though it was never manufactured) and did research on the famous problem of determining the longitude of a ship at sea. He spent several years trying to claim a prize for his work on the longitude problem from the British Board of Longitude, but like many other things in his life it also did not go well; his instruments were detained at Customs when he came to England, and he was in financial distress by the time they were returned to him. He finally addressed the Board, but they were unimpressed and did not award him the prize. He was a gifted but troubled man, and his life was marked by frequent heated disputes with other individuals and institutions. He eventually went insane.

Karl Friedrich Gauss (1777–1855)

Karl Friedrich Gauss (1777–1855) was one of the greatest mathematicians of all time. He stood with his contemporaries Immanuel Kant (1724–1804), Goethe, Ludwig van Beethoven (1770–1827), and Georg Wilhelm Friedrich Hegel (1770–1831). He was born in the city of Brunswick in northern Germany. He was a child prodigy; later in life he joked that he could operate with numbers even before he could talk. Gauss corrected an error in his father's payroll accounts at the age of three. His elementary school teacher asked the class to add up the numbers from 1 to 100, expecting to keep them busy for a long time. Young Gauss found the formula $1+2+\cdots+n = n(n+1)/2$ and instantly wrote down the correct answer. His father was a gardener and bricklayer without either the means or the inclination to help develop the talents of his son. Fortunately, Gauss' remarkable abilities in mental computation attracted the interest of several influential men in the community and eventually brought him to the attention of the Duke of Brunswick. The Duke was impressed with the boy and supported his further education, first at the Caroline College in Brunswick (1792–1795) and later at the University of Göttingen (1795–1798).

At the Caroline College, Gauss explored the work of Newton, Euler, and Lagrange. At the age of 14 or 15, he discovered the *prime number theorem*, "the number of primes $\leq n$ tends to ∞ as $n/\ln n$ as n tends to infinity," which was finally proved in 1896 by Hadamard and Vallée-Poussin, after great efforts by many mathematicians. He also invented the method of least squares for minimizing the errors inherent in observational data, which has broad applications in linear regression, signal processing, statistics, and curve fitting. Further, he conceived the Gaussian law of distribution in the theory of probability. At the university, when he was just 18, he discovered the law of quadratic reciprocity in number theory (which he called "the gem of arithmetic," and at least six separate proofs are found in his works) independently of Euler and made a wonderful geometric discovery. The ancient Greeks had known about ruler-and-compass constructions for regular polygons of 3, 4, 5, and 15 sides, and for all others obtainable from these by bisecting angles. But this was all, and there the matter rested for 2,000 years, until Gauss solved the problem completely. He proved that a regular polygon with n sides is constructible if and only if n is the product of a power of 2 and distinct prime numbers of the form $p_k = 2^{2^k} + 1$. In particular, when $k = 0, 1, 2, 3$, we see that each of the corresponding numbers $p_k = 3, 5, 17, 257$ are prime, so regular polygons with these numbers of sides are constructible. According to one story, Gauss approached his colleague Kästner with this discovery; however, he told Gauss that the discovery was useless since approximate constructions were "well known"; further, that the construction was impossible, so Gauss' proof had to be flawed; finally, that Gauss' method was something that he (Kästner) already knew about, so Gauss' discovery was unimportant. Despite Kästner's discouraging remarks, Gauss was so proud of this discovery that he requested that a 17 regular polygon be inscribed upon his tombstone (but, it was not). Gauss later toasted Kästner, who was an amateur poet, as the best poet among mathematicians and the best mathematician among poets. In his *Disquisitiones Arithmeticae*, published in 1798, he collected work done by his predecessors, enriching and blending it with his own into a unified whole. The book is regarded by many as the true beginning of the theory of numbers.

The fundamental theorem of algebra, which states that every algebraic equation with one unknown has a root, was first proved by Gauss in 1797 as part of his doctoral dissertation completed in 1799. This theorem had challenged the world's finest mathematicians, such as d'Alembert, Euler, Lagrange, and Laplace, for at least 200 years. Gauss' proof used advanced mathematical concepts from outside the field of algebra. To this day, no purely algebraic proof has been discovered. Gauss returned to the theorem many times and in 1849 published his fourth and last proof, in which he extended the coefficients of the unknown quantities to include complex numbers. Today, students often meet the fundamental theorem of algebra in a first course on complex variables, where it can be established as a consequence of some of the basic properties of complex analytic functions. In 1800, Gauss formulated the rule for determining the date of Easter, which enabled him to place his birthday on April 30. Although the concept of the determinant of a matrix is due to Seki Kowa

(1683), the term was first introduced by Gauss in 1801. He used this to "determine" properties of certain kinds of functions. Interestingly, the term matrix is derived from a Latin word for "womb" because it was viewed as a container of determinants.

In the last decades of the eighteenth century, many astronomers searched for a new planet between the orbits of Mars and Jupiter, where Bode's (Johann Elert Bode, 1747–1826) Law (1772) suggested that there ought to be one. The first and largest of the numerous minor planets known as asteroids was discovered by Giuseppe Piazzi (1746–1826) in that region in 1801, and was named Ceres. Unfortunately the tiny new planet was difficult to see under even the best of circumstances, and it was soon lost in the light of the sky near the Sun. The sparse observational data posed the problem of calculating the orbit with sufficient accuracy to locate Ceres again after it had moved away from the Sun. The astronomers of Europe attempted this task without success for many months. Finally, Gauss was attracted by the challenge; and with the aid of his method of least squares and the procedure that we now call *Gaussian elimination*, and his unparalleled skill at numerical computation, he determined the orbit. He told the astronomers where to look with their telescopes, and there it was. He had succeeded in rediscovering Ceres after all the experts had failed. This achievement brought him fame, an increase in his pension from the Duke, and in 1807 an appointment as Professor of Astronomy and first director of the new observatory at Göttingen. His treatise *Theoria Motus Corporum Coelestium*, published in 1809, remained the Bible of planetary astronomers for over a century. Its methods for dealing with perturbations later led to the discovery of Neptune. Gauss thought of astronomy as his profession and pure mathematics as his recreation, and from time to time he published a few of the fruits of his private research. His great work on the theta functions (1808), hypergeometric series (1812), general ellipsoids (1813), mechanical quadrature (1814), and continuous perturbations (1818) belongs to this period. This was typical Gaussian effort, packed with new ideas in analysis that have kept mathematicians busy ever since.

From the 1820s on, Gauss became actively interested in geodesy. In the paper *Disquisitiones generales circa superficies curvas* (1827) he founded the intrinsic differential geometry of general curved surfaces. In this work he introduced curvilinear coordinates, obtained the fundamental quadratic differential form, and formulated the concepts of Gaussian curvature and integral curvature. Extensions of these concepts opened the doors to Riemannian geometry, tensor analysis [developed by Elwin Bruno Christoffel, Gregorio Ricci-Curbastro (1853–1925), and Tullio Levi-Civita (1873–1941)], and the ideas of Einstein. Another great work from this period was his 1831 paper on biquadratic residues. Here he extended some of his early discoveries in number theory with the aid of a new method, his purely algebraic approach to complex numbers. He defined these numbers as ordered pairs of real numbers with suitable algebraic operations, and in so doing laid to rest the confusion that still surrounded the subject, and prepared the way for the later algebra and geometry of n-dimensional spaces. He defined complex integers (now called Gaussian integers) as complex numbers $a + ib$ with a and b being ordinary integers; he introduced a new concept of prime numbers, in which 3 remains prime but $5 = (1 + 2i)(1 - 2i)$ does not; and he proved the unique factorization theorem for these integers and primes. The ideas of this paper inaugurated algebraic number theory, which has grown steadily since. In 1831, he expressed his "horror of the actual infinite" as follows, "I protest against the use of infinite magnitude as something completed, which is never permissible in mathematics. Infinity is merely a way of speaking, the true meaning being a limit which certain ratios approach indefinitely close, while others are permitted to increase without restriction."

From the 1830s on, Gauss was increasingly occupied with physics, and he enriched every branch of the subject that he touched. In the theory of surface tension, he developed the fundamental idea of conservation of energy and solved the earliest problem in the calculus of variations involving a double integral with variable limits. In optics, he introduced the concept of the focal length of a system of lenses and invented the Gauss wide-angle lens (which is relatively free of chromatic aberration) for telescope and camera objectives. In collaboration with his friend and colleague Wilhelm Weber (1804–1891) he built and operated an iron-free magnetic observatory, founded the Magnetic Union for collecting and publishing observations from many places in the world, and invented the electromagnetic telegraph and the bifilar magnetometer. There are many references of his work in Maxwell's famous *Treatise on Electricity and Magnetism* (1873). In his preface, James Clerk Maxwell (1831–1879) says that Gauss "brought his powerful intellect to bear on the theory of magnetism and on the methods of observing it, and he not only added greatly to our knowledge of the theory of attractions, but reconstructed the whole of magnetic science as regards the instruments used, the methods of observation, and the calculation of results, so that his memoirs on Terrestrial Magnetism may be taken as models of physical research by all those who are engaged in the measurement of any of the forces in nature." In 1839, Gauss published his fundamental paper on the general theory of inverse square forces, which established potential theory as a coherent branch of mathematics. He had been thinking about these matters for many years; and among his discoveries were the divergence theorem (also called Gauss' theorem) of modern vector analysis, the basic mean value theorem for harmonic functions, and the very powerful statement that later became known as "Dirichlet's principle," which was finally proved by Hilbert in 1899.

His scientific diary, a little booklet of 19 pages, is one of the most precious documents in the history of mathematics. It was unknown until 1898, when it was found among family papers in the possession of one of Gauss' grandsons. It extends from 1796 to 1814 and consists of 146 very concise statements of the results of his investigations, which often occupied him for weeks or months. All of this material makes it abundantly clear that the ideas Gauss conceived and worked out in considerable detail, but kept to himself, would have made him the most unsurmountable mathematician of all time had he published them. For example, the theory of functions of a complex variable was one of the major accomplishments of nineteenth century mathematics, and the central facts of this discipline are Cauchy's integral theorem (1827) and the Taylor and Laurent (Pierre-Alphonse Laurent, 1813–1854) expansions of an analytic function (1831, 1843). In a letter written to his friend Bessel in 1811, Gauss explicitly states Cauchy's theorem and then remarks, "This is a very beautiful theorem whose fairly simple proof I will give on a suitable occasion. It is connected with other beautiful truths which are concerned with series expansions." Therefore, he knew Cauchy's theorem, and possibly both series expansions, many years earlier. However, for some reason the suitable occasion for publication did not come. This may be due to comments he made to Farkas (Wolfgang) Bolyai (1775–1852), a close friend from his university years with whom he maintained a lifelong correspondence: "It is not knowledge but the act of learning, not possession but the act of getting there, which grants the greatest enjoyment. When I have clarified and exhausted a subject, then I turn away from it in order to go into darkness again." As it was, Gauss wrote a great deal, but to publish every fundamental discovery he made in a form satisfactory to himself would have required several long lifetimes. Another prime example is non-Euclidean geometry. From the time of Euclid to the boyhood of Gauss, the postulates of Euclidean geometry were universally regarded as necessities of thought. Yet there was a flaw in the Euclidean structure that had long been a focus of attention: the so-called parallel postulate. This postulate was thought not to be independent of the others, and many had failed in their efforts to prove it as a theorem. Gauss joined in these efforts at the age of 15, and also failed. But his failure was different, for he soon came to the shattering conclusion, which had escaped all his predecessors, that the Euclidean form is not the only possible geometry. He worked intermittently on these ideas for many years, and by 1820 he was in full possession of the main theorems of non-Euclidian geometry (the term was coined by him). But he did not reveal his conclusions, and in 1829 and 1832, Lobachevsky and Johann Bolyai (son of Wolfgang) published their independent work on this subject. One reason for Gauss' silence in this case is that the intellectual climate in Germany at the time was totally dominated by the philosophy of Kant, and one of the basic tenets of this philosophy was the idea that Euclidian geometry is the only possible way of thinking about space. Gauss knew that this idea was totally false, but he valued his privacy and quiet life. In 1829, he wrote to Bessel: "I shall probably not put my very extensive investigations on this subject (the foundations of geometry) into publishable form for a long time, perhaps not in my lifetime, for I dread the shrieks we would hear from the Boeotians if I were to express myself fully on this matter." The same thing happened with his theory of elliptic functions (called 'elliptic' because these functions exist in the expression for finding the length of an arc of an ellipse), a very rich field of analysis that was launched primarily by Abel in 1827, and also by Jacobi in 1828–1829. Gauss had published nothing on this subject, and claimed nothing, so the mathematical world was filled with astonishment when it gradually became known that he had found many of the results published by Abel and Jacobi before they were born. Abel was spared this devastating knowledge by his early death in 1829, but Jacobi was compelled to swallow his disappointment and go on with his work. The facts became known partly through Jacobi himself. His attention was caught by a cryptic passage in the *Disquisitiones*, whose meaning can only be understood if one knows something about elliptic functions. He visited Gauss on several occasions to verify his suspicions and tell him about his most recent discoveries; each time, Gauss pulled a 30-year-old manuscript out of his desk and showed Jacobi what Jacobi had just shown him. The depth of Jacobi's chagrin is understandable. At this point in his life, Gauss was indifferent to fame and was actually pleased to be relieved of the burden of preparing the treatise on the subject, which he had planned for a long time. After a week's visit with Gauss in 1840, Jacobi wrote to his brother, "mathematics would be in a very different position if practical astronomy had not diverted this colossal genius from his glorious career." The Royal Society of Göttingen published a collection of his works in several volumes.

Gauss, despite an aversion to teaching, gave an annual course in the method from 1835 until he died on February 23, 1855, due to several heart attacks. He was buried at St. Albans Cemetery in Göttingen next to the unmarked grave of his mother. Gauss took complete care of his mother who became totally blind for the last 4 years of her life. The King of Hanover had a memorial coin struck with the dedication, "Georgius V Rex Hannoverae Mathematicorum Principi." Also, Sartorius von Walterhausen (1809–1876) said in his obituary "Gauss always strove to give his investigations the form of finished works of art. He did not rest until he had succeeded, and hence he never published a work until it had achieved the form he wanted. He used to say that when a fine building was finished, the scaffolding should no longer be visible."

Apart from science he had two wives and six children; his main interests were history and world literature, international politics, and public finance. He owned a large library of about 6,000 volumes in many languages, including Greek, Latin, English, French, Russian, Danish, and German. His acuteness in handling his own financial affairs is shown by the fact that although he started with virtually nothing, he left an estate over a hundred times as great as his average annual income during the last half of his life. Once, in 1828, he visited Berlin, and once, in 1854, he made a pilgrimage to be present at the opening of the railway from Hanover to Göttingen. He saw his first railway engine in 1836, but other than these quiet adventures, until the last year of his life he never slept under any other roof than that of his own observatory. He was the last complete mathematician and has been named "the prince of mathematics." Gauss once asserted that "mathematics is the queen of the sciences, and the theory of numbers is the queen of mathematics." His friend Sartorius von Waltershausen writes, "As he was in his youth, so he remained through his old age to his dying day, the unaffectedly simple Gauss. A small study, a little work table with a green cover, a standing-desk painted white, a narrow sopha and, after his 70th year, an arm chair, a shaded lamp, an unheated bedroom, plain food, a dressing gown and a velvet cap, these were so becoming all his needs." According to Kronecker, "The further elaboration and development of systematic arithmetic, like nearly everything else which the mathematics of our (nineteenth) century has produced in the way of original scientific ideas, is knit to Gauss."

The following anecdotes about Gauss are often told:

Alexander von Humboldt asked Laplace who was the greatest mathematician in Germany, and Laplace replied Johann Friedrich, "Pfaff." "But what about Gauss" the astonished von Humboldt asked, as he was backing Gauss for the position of director at the Göttingen observatory. "Oh," said Laplace, "Gauss is the greatest mathematician in the world."

The Paris Academy in 1816 proposed the proof (or disproof) of Fermat's Last Theorem as its prize problem for the period 1816–1818. Some tried to entice Gauss into competing, but Gauss resisted, replying "I am very much obliged for your news concerning the Paris prize. But I confess that Fermat's Theorem as an isolated proposition has very little interest for me, because I could easily lay down a multitude of such propositions, which one could neither prove nor dispose of."

Mary Fairfax Somerville (1780–1872)

Mary Fairfax Somerville (1780–1872) was born in Jedburgh, Scotland, as the fifth of seven children of Vice Admiral Sir William George Fairfax (1739–1813) and Margaret Charters Fairfax. Mary received a rather desultory education, and mastered algebra and Euclid in secret after she had left school, without any external help. In 1804, she married her distant cousin, the Russian Consul in London, Captain Samuel Greig, who died in 1807. They had two sons, one of them also died. After the death of Greig, she began to study astronomy and mathematics seriously. She began solving math problems posed by a mathematics journal, and in 1811 won a medal for a solution she submitted. She mastered J. Ferguson's astronomy and became a student of Isaac Newton's Principia, despite the fact that many of her family and friends disapproved. Mary frequently corresponded with Scotsman William Wallace (1768–1843), who at the time was mathematics master at a military college. She remarried in 1812 to another cousin, Dr. William Somerville (1771–1860), who was a surgeon. Dr. Somerville supported her studies, writing, and contact with scientists. They had three daughters and a son. Mary Somerville began publishing papers on her scientific research in 1826 and became the second woman scientist (after Caroline Herschel) to receive recognition in the UK. On the request of Lord Brougham (1778–1868) she translated the *Mécanique Céleste* of Laplace, which was published in 1831 with great acclaim under the title *The Mechanism of the Heavens*. This work contained full mathematical explanations and diagrams in order to render Laplace's difficult work comprehensible. It ran many printings and was used in British universities. Laplace paid her a compliment by stating that she was the only woman who understood his works. In recognition of her work, a portrait bust of her was commissioned by her admirers in the Royal Society and placed in the great hall. Mary Somerville's other works are *On the Connection of the Physical Sciences* (1834), *Physical Geography* (1848), and *Molecular and Microscopic Science* (1869). A statement in one of the editions of On the Connection of the Physical Science suggested some irregularities in the orbit of Uranus. This prompted John Couch Adams in England to search for the planet Neptune, for which he is credited as a codiscoverer.

In 1835, she and Caroline Herschel (1750–1848) became the first women members of the Royal Astronomical Society. In 1838, she and her husband moved permanently to Italy. Dr. Somerville died in 1860. In 1869, she completed her autobiography, parts of which were published by her daughter Martha after her death. In the same year she was elected to the American Philosophical Society. Although deaf and frail in her later years, she retained her mental faculties and even continued to, in her words, "read books on the higher algebra for 4 or 5 h in the morning, and even to solve problems" until her peaceful death at the age of 92. On her death a newspaper dubbed her the "Queen of Nineteenth Century Science."

Somerville College, Oxford University, was named after her. The term "scientist" was first coined by William Whewell (1794–1866) in an 1834 review of Somerville's On the Connexion of the Physical Sciences. Sommerville House, Burntisland was named after her, the building she lived in for some years of her life. Somerville Island off British Columbia near the border with Alaska was named after her by Sir William Edward Parry (1790–1855).

Siméon Denis Poisson (1781–1840)

Siméon Denis Poisson (1781–1840) was born in Pithviers, about 50 miles south of Paris. His father was a private soldier. Siméon had several brothers and sisters who did not survive. In 1798, he entered the École Polytechnique in Paris, and his abilities excited the interest of Lagrange and Laplace, whose friendship he retained to the end of their lives. In 1800, he published two memoirs, one on Étienne Bézout's (1730–1783) method of elimination, and the other on the number of integrals of an equation of finite differences, which were of such quality that he was allowed to graduate without taking the final examination. Immediately after finishing his course at the Polytechnique he was directly appointed Repetiteur there. He was made Professeur Suppléant in 1802, and professor in 1806, succeeding Fourier, whom Napoleon had sent to Grenoble. In addition to his professorship at the Polytechnique, in 1808 Poisson became an astronomer at Bureau des Longitudes. In 1809, he was appointed Chair of Mechanics in the newly opened Faculté des Sciences. He became member of the Institute in 1812, examiner at the military school at Saint-Cyr in 1815, and graduation examiner at the Polytechnique in 1816. In 1817, he married Nancy de Bardi, and they had several children. Poisson became councillor of the university in 1820 and geometer to the Bureau des Longitudes in succession to Laplace in 1827. The rest of his career, until his death in Sceaux near Paris, was completely occupied in writing and publishing his many works.

In 1802, Poisson started studying the problems related to ordinary and partial differential equations. In particular, he studied applications to a number of physical problems such as the pendulum in a resisting medium and the theory of sound. In 1808 and 1809, Poisson published three important papers with the Academy of Sciences. In *Sur les inégalités des moyens mouvements des planétes* he looked at the mathematical problems that Laplace and Lagrange had raised about perturbations of the planetary orbits. In *Sur le mouvement de rotation de la terre* and *Sur la variation des constantes arbitraires dans les questions de mécanique* he used

Lagrange's method of variation of arbitrary constants. In 1808, he published a new edition of Clairaut's *Théorie de la figure de la terre*. This was first published by Clairaut in 1743, confirming the Newton–Huygens belief that the Earth was flattened at the poles. In 1811, Poisson published his two-volume treatise *Traité de mécanique* which was an exceptionally clear treatment based on his course notes at the Polytechnique. In 1812, Poisson discovered that Laplace's equation is valid only outside of a solid. In 1813, Poisson studied the potential in the interior of attracting masses, producing results that find applications in electrostatics. He produced influential work on electricity and magnetism, which began a new branch of mathematical physics, followed by work on elastic surfaces. Papers followed on the velocity of sound in gasses, on the propagation of heat, and on elastic vibrations. In 1815, he published a work on heat that overlapped the work of Fourier. Poisson published a paper in 1821 on the liberation of the Moon and another in 1827 on the motion of the Earth about its center of gravity. His most important memoirs on the theory of attraction were published in 1829, *On the Attraction of Spheroids*, and in 1835, *On the Attraction of a Homogeneous Ellipsoid*. The Poisson distribution, which describes the probability that a random event will occur in a time or space interval under the condition that the probability of the event occurring is very small, but the number of trials is so large that the event actually occurs a few times, was published in 1838 in his *Research on the Probability of Judgments in Criminal and Civil Matters*. He also introduced the expressions 'law of large numbers' and 'law of small numbers,' which have applications in statistics and probability.

Poisson published between 300 and 400 mathematical works. He was awarded several honors, such as FRS in 1818, Fellow of the Royal Society of Edinburgh in 1820, and the Royal Society Copley Medal (first awarded in 1731 in honor of Godfrey Copley, 1653–1709) in 1832. However, he was not highly regarded by other French mathematicians either during his lifetime or after his death. In mathematics his name is given to Poisson's integral, Poisson's equation, Poisson brackets, Poisson's ratio, Poisson's constant, Poisson's summation formula, Poisson's process, Screened Poisson's equation, Poisson's kernel, Poisson's distribution, Poisson's regression, and Poisson's spot, among others.

According to Abel, Poisson was a short, plump man. His family tried to encourage him in many directions, from doctor to lawyer. At last he found his place as a scientist and produced over 300 works in his relatively short lifetime. Arago reported that Poisson frequently used to say "Life is good for only two things, discovering mathematics and teaching mathematics." He excelled in both pursuits.

Bernhard Placidus Johann Nepomuk Bolzano (1781–1848)

Bernhard Placidus Johann Nepomuk Bolzano (1781–1848) was born in Prague, Bohemia, known today as the Czech Republic. He was a Catholic priest who made many important contributions to mathematics in the first half of the nineteenth century. He was one of the first to recognize that many "obvious" statements about continuous functions require proof. His observations concerning continuity were published posthumously by a student he had befriended in 1850 in an important book, *Paradoxien des Unendlichen* (The Paradoxes of the Infinite). One of his results, now known as the Theorem of Bolzano states: Let f be continuous at each point of a closed interval $[a, b]$ and assume that $f(a)$ and $f(b)$ have opposite signs. Then there is at least one c in the open interval (a, b) such that $f(c) = 0$. He used what we now call a Cauchy sequence 4 years before Cauchy introduced it. In 1843, he produced a function that was continuous in an interval, but surprisingly had no derivative at any point in that interval. His function did not become known, and it is Weierstrass, about 40 years later, who is usually credited with the first example of a function of this kind. In analysis, Weierstrass proved the famous Bolzano–Weierstrass theorem in 1860 in his Berlin lectures. Bolzano's theories of infinity preceded Cantor's. Because of his great human compassion and willingness to stand firm in his beliefs, he was forced to retire from his position as Chair of Philosophy at the University of Prague.

Friedrich Wilhelm Bessel (1784–1846)

Friedrich Wilhelm Bessel (1784–1846) was born in Minden (Germany) and started his working life as a clerk aboard a ship. In 1806, he became an assistant in the observatory at Lilienthal, and he was appointed director of the Prussian Observatory at Königsberg in 1810 and held this position until his death. Bessel was an intimate friend of Gauss, with whom he corresponded for many years. He embarked on a career in business as a youth, but soon became interested in astronomy and mathematics. His study of planetary perturbations led him to make a systematic study of the solutions, known as Bessel functions, of Bessel's differential equation in 1824. He was the first man to determine the distance of a fixed star; his parallax measurement in 1838 yielded a distance for the star 61 Cygni of 11 light years, or about 360,000 times the diameter of the Earth's orbit. In 1844, he discovered that Sirius, the brightest star in the sky, has a traveling companion and is therefore what is now known as a binary star. This Companion of Sirius, with the size of a planet but the mass of a star, and consequently a density many thousands of times the density of water, is one of the most interesting objects in the universe. It was the first dead star to be discovered and occupies a special place in modern theories of stellar evolution. He is also credited for developing the notion of a function.

Charles Jules Brianchon (1785–1864)

Charles Jules Brianchon (1785–1864) was a French mathematical painter. As a 21-year-old student, he discovered that in any hexagon circumscribed about a conic section (such as a circle), the three lines that join opposite diagonals meet in a single point. He also made connections between his result and Pascal's theorem concerning the points of intersection of opposite sides of a hexagon inscribed in a conic section.

Jacques Phillipe Marie Binet (1786–1856)

Jacques Phillipe Marie Binet (1786–1856) was born in Rennes, Bretagne, France. He entered the École Polytechnique in 1804, and after graduating in 1806 he worked for the Department of Bridges and Roads of the French government. In 1807, Binet became a teacher at École Polytechnique and 1 year later was appointed to assist the Professor of Applied Analysis and Descriptive Geometry. In 1814, he became examiner of descriptive geometry, and in 1815, he succeeded Poisson in mechanics. In 1816, Binet became an inspector of studies at Ècole Polytechnique and in 1823 succeeded Jean Baptiste Delambre (1749–1822) in the position of Chair of Astronomy at the Collége de France. The French revolution of July 1830, when Louis-Philippe (1773–1850) was proclaimed King of France, was unfortunate for Binet. On November 13, 1830, Binet was dismissed as inspector of studies; however, he was allowed to keep his Chair of Astronomy for almost 30 years.

Binet wrote over 50 papers in physics and astronomy. In mathematics he made significant contributions to number theory, specifically the theory of the Euclidean algorithm. He discovered the multiplication rule for matrices in 1812, which was used by Eisenstein to simplify the process of making substitutions in linear systems and was formalized by Cayley. Binet investigated the foundation of matrix theory and gave Binet's formula expressing Fibonacci numbers in closed form, but the same result was known to Euler over a 100 years earlier. He is also remembered for Binet's theorem and the Binet–Cauchy identity. In 1821, he was made a Chevalier in the Légion d'Honneur and in 1843 was elected to the Académie des Sciences. He died in Paris, France, in 1856.

Georg Simon Ohm (1789–1854)

Georg Simon Ohm (1789–1854) was a German physicist. His father was a locksmith, while his mother was the daughter of a tailor. Georg was taught by his father, who held him to a high standard in mathematics, physics, chemistry, and philosophy. He entered Erlangen Gymnasium at the age of 11 and in 1805 was admitted to the University of Erlangen. While there, instead of concentrating on his studies, Georg spent most of his time dancing, ice skating, and playing billiards, and so his father forced him to leave the university after three semesters and sent him to Switzerland, where he became a mathematics teacher. In 1809, Ohm left his teaching post and became a private tutor so that he could continue studying, privately, the work of the leading French mathematicians Lagrange, Legendre, Laplace, Jean Baptiste Biot (1774–1862), and Poisson. In 1811, he returned to the University of Erlangen, received his doctorate, and immediately joined the staff as a mathematics lecturer. However, he was dissatisfied after just three semesters, and Ohm gave up his university post. In 1817, he joined the Jesuit Gymnasium of Cologne as a teacher of mathematics and physics, where he remained until 1825. During this period he made an independent study of Fourier's work and performed experiments related to electricity. In 1826, Ohm published two important papers in which he gave a mathematical description of conduction in circuits modelled on Fourier's study of heat conduction. His most significant contribution to science appeared in 1827 in his famous book *Die galvanische Kette, mathematisch bearbeitet*, which states "the voltage drop E_R across a resistor is proportional to the instantaneous current I, say, $E_R = RI$, where the constant of proportionality R is called the resistance of the resistor." However, his contemporaries did not believe this law, as it seemed too good to be true. Ohm was considered unreliable because of this and was so badly treated that he had to live for several years in obscurity and poverty. His law was finally recognized by the Royal Society, and he was awarded the Copley Medal in 1841. He became a foreign member of the Royal Society in 1842. Ohm also became a corresponding member of the Berlin and Turin academies, and a full member of the Bavarian Academy. In 1849, he took up a post in Munich as curator of the Bavarian Academy's physical cabinet and began to lecture at the University of Munich. In 1852, just 2 years before his death, he was appointed to the Chair of Physics at the University of Munich. One of his pupils in Cologne was Dirichlet, who later became one of the most eminent German mathematicians of the nineteenth century.

Jean-Victor Poncelet (1788–1867)

Jean-Victor Poncelet (1788–1867) was born in Metz, France. He studied under Monge at the École Polytechnique. In 1810, Poncelet entered the military engineering college at Metz, and after 2 years he was commissioned a lieutenant of engineers and served in Napoleon's invasion of Russia. He was captured at the Battle of Smolensk in 1812 and was imprisoned by the Russians at Saratov. One legend states that at first, Poncelet had only scraps of charcoal salvaged from the meager brazier that kept him from freezing to death, with which he drew diagrams on the wall of his cell. He was repatriated to France in 1814. Poncelet brought with him a Russian abacus, which for many years was regarded as a great curiosity of "barbaric" origin. (The word *abacus* comes from a Semitic word meaning "dust," suggesting one cleared the table by wiping off the dust. Various forms of abaci were known to Roman, Greek, Egyptian, Chinese, and Indian civilizations as early as the seventh century BC, and there is no reason to suspect a common origin.) During his imprisonment he studied projective geometry (which played an important role in the development of the theory of invariance) and started to investigate those properties that figures share with their shadows, drafting the book *Applications d'analyse et de géométrie* that was published after 50 years in two volumes during 1862–1864. He also studied conic sections, discovered circular points at infinity, and developed the principle of duality independent of Joseph Gergonne (1771–1859), who was a student of Monge. During 1815–1825, he was a military engineer at Metz and from 1825 to 1835 was Professor of Mechanics there. He applied mechanics to improve turbines and waterwheels, more than doubling the efficiency of the waterwheel. In his lectures, he coined the term fatigue to describe the failure of materials under repeated stress. In 1838, Poncelet became Professor of Mechanics at the Sorbonne, and from 1848 he held the rank of general, commanding the École Polytechnique. Poncelet retired from his administrative duties in 1850 in order to devote himself to mathematical research. His other notable books are *Traité des des figures* in 1822, *Cours de mécanique appliqué aux machines* in 1826, and *Introduction a la mécanique industrielle* in 1829. He is considered to be the father of projective geometry and is remembered for the Poncelet–Steiner theorem, Poncelet's porism, and Poncelet points. He died in Paris in 1867. A French unit of power, the Poncelet, was named after him.

Augustin-Louis Cauchy (1789–1857)

Augustin-Louis Cauchy (1789–1857) was born in Paris on August 21, 1789, during a period of upheaval in French history. His father was Louis-Francois (1760–1848), and his mother was Marie-Madeleine Desestre. Louis-Francois was a parliamentary lawyer, an accomplished biblical scholar, and a lieutenant of police in Paris when the Bastille fell. Marie was an excellent woman but was not very intelligent. Like her husband, she was a devoted Catholic. Cauchy was the eldest of six children. His childhood fell during the bloodiest period of the Revolution. He and his family had to live in the village of Arcueil, and food was scarce. As a result of childhood malnutrition, Cauchy had to watch his health closely all through his life. Cauchy's first teacher was his own father, who taught him and his five brothers and sisters by writing his own textbooks. In this way, Cauchy acquired fluency in both French and Latin. At Arcueil, two men who frequently visited Cauchy's father were Marquis Laplace and Count Claude-Louis Berthollet (1748–1822). Before long, Laplace discovered that Cauchy had a phenomenal mathematical talent. In 1800, Louis-Francois was elected Secretary of the Senate. Lagrange, then professor at the Polytechnic, frequently visited Louis-Francois to discuss business. On one occasion, when Laplace and several other notable men were present, Lagrange pointed to young Cauchy and said, "You see that little young man? Well! He will supplant all of us in so far as we are mathematicians." Believing that the little boy might burn himself out, Lagrange advised his father, "Don't let him touch a mathematical book till he is 17." On another occasion he said, "If you don't hasten to give Augustin a solid literary education his tastes will carry him away; he will be a great mathematician but he won't know how to write his own language." Louis-Francois took this advice to heart and gave his son a sound literary education before turning him lose on advanced mathematics. At the age of 13, Cauchy entered the Central School of the Pantheon. From the beginning, Cauchy was the star of the school, winning all of the most prestigious prizes in Greek and Latin compositions. In 1804, Cauchy left the school, and in that same year he received his first communion, a solemn and beautiful occasion in the life of a Catholic. For the next 10 months he studied mathematics intensely with a good tutor and in 1805, at the age of 16, he entered the Polytechnique. In 1807, Cauchy joined the civil engineering school, completing his training in 1810. In March of 1810, Cauchy joined the military at Cherboug. Besides his baggage, Cauchy carried only four books, the Mécanique céleste of Laplace, the Traité des fonctions analytiques of Lagrange, Thomas á Kempis' Invitation of Christ, and a copy of Virgil's works. It was Lagrange's treatise that inspired Cauchy to seek a theory of functions free from the glaring defects of Lagrange's theory. Cauchy stayed at Cherboug for 3 years. Outside of his heavy duties his time was well spent. He said "Work doesn't tire me; on the contrary it strengthens me and I am in perfect health." He found time for research. By December of 1810, he had begun "to go over again all the branches of mathematics, from Arithmetic to astronomy." On top of all this, he still found time to instruct others, and he assisted the mayor of Cherboug by conducting school examinations. Yet, he still found time for his hobbies. At the age of 27, Cauchy had raised himself to the front rank of living mathematicians. His only serious rival was Gauss, who was 12 years older. Gauss produced the fundamental theorem of algebra in 1797, while Cauchy's memoir on the definite integral with complex-number limits was not produced until 1814. This has probably hurt the Polytechnique's printing budget; Cauchy was known for the massive length of his works, which usually ranged from 80 to 300 pages. Shortly after 1816, Cauchy became a member of the Academy of Sciences. He was also made Professor of the Polytechnique. His mathematical activity was incredible; sometimes two full length papers would be presented before the Academy in the same week. He became better known than Gauss to the mathematicians of Europe. In the midst of all this work, Cauchy found time to do his courting. In 1818, he married Aloise de Bure. He wrote her a poem that ends

> I shall love you, my tender friend,
> Until the end of my days;
> And since there is another life
> Your Louis will love you always.

They later had two daughters. In 1830, when the revolution unseated Charles X (1757–1836), as a matter of principle, Cauchy gave up all his positions and went into exile. Cauchy first went to Switzerland, where he sought distraction in scientific conferences and research. After the revolution, Charles made Cauchy Professor of Mathematical Physics at Turin. He learned Italian and delivered his lectures in that language. In 1833, Cauchy was entrusted with the education of Charles' heir. He did not manage to escape from his pupil until 1838, when he was almost 50. In 1843, Guglielmo Libri (1803–1869) was given the chair of mathematics at the Collége de France instead of Cauchy. This was the same Libri who later fled France when it was discovered that he had stolen library books. During the last 19 years of his life he produced over 500 papers on all branches of mathematics, including mechanics, physics, and astronomy. His total output of 789 papers filled 24 large quarto volumes. In elementary mathematical courses his name is remembered in Cauchy's root test, Cauchy's ratio test, Cauchy's product, Cauchy's inequality, Cauchy's integral theorem, and the Cauchy–Riemann equations (which were obtained earlier by d'Alembert in 1752). In 1815, Cauchy published a landmark paper in which he gave the first systematic and modern treatment of determinants. It was in this paper that $\det(AB) = \det(A)\det(B)$ for two square matrices A, B of the same size was proved for the first time in its full generality. Special cases of the theorem had been stated and proved earlier, but it was Cauchy who made the final jump. In 1821, he defined the basic concept of the limit as follows: When the values successively attributed to a particular variable approach indefinitely a fixed value, so as to end by differing from it by as little as one wishes, this latter is called the limit of all the others. The concept of continuity is also due to Cauchy. He defined the derivative of $y = f(x)$, with respect to x, as the limit when $\Delta x \to 0$ of the difference quotient $\Delta y/\Delta x = (f(x+\Delta x) - f(x))/\Delta x$. While dealing with infinite series of functions on an interval, he made mistakes by not distinguishing between pointwise and uniform convergences. His most famous false theorem asserts that the sum-function $s(x) = f_0(x) + f_1(x) + \cdots$ of a convergent series of continuous functions $f_n(x)$ is also continuous. Cauchy's contribution to the theory of determinants is his most prolific. He introduced the word 'characteristic' to matrix theory by calling the equation $|A - \lambda I| = 0$ the 'characteristic equation.'

Cauchy died rather unexpectedly in his 68th year on May 25, 1857, in Sceaux. Hoping to alleviate his bronchial trouble, he retired to the country to recuperate only to be smitten with a fever that proved fatal. A few hours before his death he was talking animatedly with the Archbishop of Paris about the charitable works he had in view; charity was one of Cauchy's lifelong interests. His last words were addressed to the Archbishop: "Men pass away but their deeds abide." Long after his death, Cauchy was severely criticized for overproduction and hasty composition. Abel described him as "mad, infinitely Catholic, and bigoted." Some writers praise his teaching, but others say he rambled incoherently, and according to one report of his day, he once devoted an entire lecture to extracting the square root of 17 to ten decimal places by a method well known to his students. One year Cauchy started his calculus course with 30 students and all but one dropped out. In any event, Cauchy is undeniably one of the greatest minds in the history of mathematics.

August Ferdinand Möbius (1790–1868)

August Ferdinand Möbius (1790–1868) was born in Schulpforta, Saxony-Anhalt (Germany). Möbius was a geometer, topologist, number theorist, statistician, and astronomer. He was a pupil of Gauss and became Professor of Astronomy and Higher Mechanics at the University of Leipzig in 1816 and director of its observatory in 1848. In 1827, Möbius published the book *Der barycentrische Calcul*, in which he gave applications of vectors to geometry. In 1837, he published a work on statics in which he used the idea of resolving a vector into components. Möbius is best known for Möbius strips, made by putting a half twist in a paper band before joining the ends. This allows a fly to walk on both sides of the strip without going over the edge. However, this was independently discovered by a pupil of Gauss, Johann Benedict Listing (1808–1882, who used the word 'topology' for the first time in 1847) at around the same time. Möbius was the first to introduce homogeneous coordinates to projective geometry; Möbius transformations bear his name. In number theory, the Möbius transform, the Möbius function $\mu(n)$, and the Möbius inversion formula are well known. A new branch of chemistry, called chemical topology, studies the structures of chemical configurations. A recent advance in this area is the synthesis of the first molecular Möbius strip, which was formed by joining the ends of a double-stranded strip of carbon and oxygen atoms. Möbius died in Leipzig in 1868.

Charles Babbage (1791–1871)

Charles Babbage (1791–1871), a Fellow of the Royal Society of London, was an English mathematician, a mechanical engineer, a computer scientist/technologist, and a philosopher. He is best known for his creation of the first mechanical computer, called the "Analytical Engine," along with the concept of programming (committing an ordered set of instructions to a computer). Charles was born in London to a wealthy banker. Parts of Charles' incomplete machine are on display in the London Science Museum. A perfectly functioning difference/analytical engine was constructed from Charles' original plans in 1991, one hundred and 20 years after his death. The completed engine demonstrated that Charles' machine would have worked exactly in the way he wanted. Nine years later, the Science Museum developed the printer that Charles had designed for the engine. Charles attended King Edward VI's Grammar School in Totnes, South Devon, but left the school after a while due to poor health and then studied under private tutors. He then attended Holmwood academy in Baker Street, Enfield, Middlesex. The academy had a good library that inspired Charles to study mathematics. He was coached by two more tutors after he left the academy. Babbage contributed two long papers on the calculus of functions to the *Philosophical Transactions of the Royal Society* in 1815 and 1816. Charles was awarded the Gold Medal of the Royal Astronomical Society "for his invention of a *Difference Engine* for calculating mathematical and astronomical tables" in 1824. During 1828–1839, Charles was Lucasian Professor of Mathematics at the University of Cambridge, and was a founding member of the Royal Astronomical Society in 1820 and the Royal Statistical Society in 1834. Charles worked in cryptography. He broke Vigenére's autokey cipher, widely known as the undecipherable cipher. His discovery was used to aid English military campaigns and remained unpublished for several years. Consequently, credit for the discovery was instead given to Friedrich Kasiski (1805–1881), a Prussian infantry officer who made the same discovery some years after Charles. In 1838, Charles invented the pilot (also called a cow-catcher), the metal frame attached to the front of locomotives that clears the tracks of obstacles. The discoveries of Charles have been viewed by some as being influenced by Indian thought, in particular, Indian logic. Mary Everest Boole (1832–1916), a self-taught mathematician, claimed that Charles along with Sir Frederick William Herschel (1738–1822), an FRS and a German-born British astronomer, was introduced to Indian thought in the 1820s by her uncle, Colonel Sir George Everest (1790–1866), a Welsh surveyor, geographer, and Surveyor General of India during 1830–1843, who is most famous for surveying the highest peak of the Himalayas. Charles passed away at the age of 80 in Marylebone, London. His brain was preserved in alcohol, and then in 1908 it was dissected. Today, his brain is on view in the science museum in London.

Nicolai Ivanovich Lobachevsky (1792–1856)

Nicolai Ivanovich Lobachevsky (1792–1856) was born in Nizhny Novgorod, Russia. His father, who was a clerk in a land surveying office, died when Nikolai was 7 years old, and his mother moved to Kazan (western Russia on the edge of Siberia) along with her three sons. There the three brothers attended Kazan Gymnasium, which was financed by government scholarships. Nikolai graduated in 1807 and then entered Kazan University, which was founded in 1804, as a free student, and he remained there throughout his career. Although Nikolai enrolled as a medical student, he decided to study a broad scientific course involving mathematics and physics. During his studies he was influenced by Johann Bartels, a former teacher and friend of Gauss. In 1809, Nicolai was made *Kammerstudenten*, or honor student in mathematics, which raised his confidence and lowered his inhibitions. In 1811, Lobachevsky received a master's degree in physics and mathematics, and in 1814, he became a lecturer. By 1816, he had become an extraordinary professor, and in 1822 a full professor. He served in many administrative positions, such as Dean of the Mathematics and Physics Department during 1820–1825, Head Librarian during 1825–1835, and Rector of Kazan University from 1827 to 1846. While Lobachevsky was Rector, two natural disasters struck the university, a cholera epidemic in 1830 and a big fire in 1842, which he handled meticulously so that the university was minimally damaged. For his activities during the cholera epidemic he even received a message of thanks from the Emperor. In 1832, he married Moisieva and had seven children. He retired (or was dismissed) in 1846, after which his health rapidly deteriorated. During his last years, he started working on his farm about 50 miles from Kazan. There he raised prize sheep and discovered a new way of processing wool, for which he won a government citation. He died as a blind man in 1856 in Kazan.

For over 2,000 years mathematicians tried unsuccessfully to prove Euclid's parallel postulate from other axioms. Then, in the early eighteenth century, Saccheri moved in the right direction; however, he was so dismayed at his findings that he completely gave up in favor of accepting Euclid's parallel postulate. Saccheri published his results in a 101-page book, inappropriately entitled *Euclid Vindicated from All Defects*, which resembles the *Theorie der Parallellinien* written in 1766 by Lambert. While Saccheri was frightened without the parallel postulate, Lobachevsky bravely replaced it with the postulate that there are more parallel lines than one through any given point; a consequence of this is that the sum of angles in a triangle must be less than 180°. He reported this non-Euclidean geometry, now known as *hyperbolic geometry*, in 1826. His paper *A concise outline of the foundations of geometry* was published in 1829 by the Kazan Messenger, but was rejected by Ostrogradsky when it was submitted for publication by the St. Petersburg Academy of Sciences. Lobachevsky's great mathematical achievement was not recognized in his lifetime, since the prevailing Kantian philosophy refused to take it seriously. It was regarded instead as a logical curiosity; however, it turned out to be of fundamental importance in Einstein's theory of general relativity. As we have noted earlier, Gauss had anticipated non-Euclidean geometry but withheld it from publication. In 1834, Lobachevsky found a technique for the approximation of the roots of algebraic equations, which in the literature is better known as the Dandelin–Gräffe (Germinal Pierre Dandelin, 1794–1847; Karl Heinrich Gräffe, 1799–1873) method. Lobachevsky also published several books: *New Foundations of Geometry* during 1835–1838, *Geometrical Investigations on the Theory of Parallels* in 1840, and *Pangeometry* in 1855.

George Green (1793–1841)

George Green (1793–1841) grew up in Sneinton, England. His parents were reasonably prosperous in the bakery business. His father, also named George Green, later built the tallest, most powerful, and most modern windmill (for grinding grain) in all of Nottinghamshire. The manager of Green's mill was William Smith, and he had a daughter named Jane Smith. George Green never married Jane Smith, but together they had seven children. He was almost entirely self-taught, and made significant contributions to electricity and magnetism, fluid mechanics, and partial differential equations. His most important work was an essay on electricity and magnetism that was published privately in 1828. In this paper, Green was the first to recognize the importance of potential functions. He introduced the functions now known as Green's functions as a means of solving boundary value problems and developed the integral transformation theorems of which Green's theorem (in Russia, known as Ostrogradsky's theorem) in the plane is a particular case. However, these results did not become widely known until Green's essay was republished in the 1850s through the efforts of William Thomson/Lord Kelvin. Green's mathematics provided the foundation on which Thompson, Stokes, Maxwell, John William Strutt (Lord Rayleigh, 1842–1919, a British physicist who won the Nobel Prize in Physics in 1904 for his discovery of the inert gas argon), and others built the present-day theory of electromagnetism. The George Green Library at the University of Nottingham is named after him. In 1986, Green's mill was restored to working order. It now serves both as a working example of a nineteenth century mill and as a museum and science center dedicated to George Green. On a visit to Nottingham in 1930, Einstein commented that Green had been 20 years ahead of his time.

Michel Floréal Chasles (1793–1880)

Michel Floréal Chasles (1793–1880) was born into a well off Catholic family at Épernon in France as the son of Charles-Henri Chasles, a wood merchant who became president of the chamber of commerce in Chartres. Michel attended the Lycée Impérial for his secondary education. Then, in 1812 he entered the École Polytechnique in Paris and studied under Poisson. He fought in the War of the Sixth Coalition to defend Paris in 1814 at the age of 21. Once the war was over, he pursued his mathematical studies vigorously. His book on geometry, which was published in 1837, made him famous. He became professor at the École Polytechnique in 1841 at the age of 48 and then was awarded a chair at the Sorbonne in 1846. A second edition of his work was published in 1875 and was translated into German by Leonhard Sohncke (1842–1897), a mathematician, scientist, and professor of physics in Karlsruhe, Munich, and Jena. One of the results for which Chasles is well known is his enumeration of conics. Lee Koppelman wrote in *Biography in Dictionary of Scientific Biography* (New York 1970–1990): "Chasles published highly original work until his very last years. He never married, and his few interests outside his research, teaching, and the Academy, which he served on many commissions, seem to have been in charitable organizations." Michel became a corresponding member of the Académie des Sciences in 1839 and a full member in 1851. He was elected an FRS in 1854 and won its Copley Medal in 1865. In 1867, he was elected the first foreign member of the London Mathematical Society. He was also a member of academies in Brussels, Copenhagen, Naples, Stockholm, St. Petersburg, and the USA. He breathed his last in Paris. Chasles' name is one of 72 that appear on the Eiffel Tower.

Olinde Rodrigues (1794–1851)

Olinde Rodrigues (1794–1851) was born into a Portuguese Jewish family in Bordeaux, France. He was the eldest of eight children. The family moved to Paris in the late 1790s. It is not known how Olinde learned advanced mathematics; however, in 1816 he was awarded a doctorate in mathematics from the Faculty of Science of the University of Paris for a thesis that contains one of the two results for which he is known today, namely, the Rodrigues formula for Legendre polynomials. The other result that bears his name is Rodrigues' rotation formula for the rotation of vectors. Soon after graduation, Rodrigues became interested in the scientific organization of the society and mostly wrote on politics and social reform. He died in 1851 in Paris. The ship *Franconia*, which was built in 1872, was later renamed the Olinde Rodrigues.

Gabriel Lamé (1795–1870)

Gabriel Lamé (1795–1870) entered the École Polytechnique in 1813 and graduated in 1817. He continued his education at the École des Mines, graduating in 1820. Immediately afterward, he went to Russia, where he was appointed director of the School of Highways and Transportation in St. Petersburg. Not only did he teach but he also planned roads and bridges while in Russia. He returned to Paris in 1832, where he co-founded an engineering firm. However, he soon left the firm, accepting the Chair of Physics at the École Polytechnique, which he held until 1844. While at this position, he was active outside academia as an engineering consultant, serving as chief engineer of mines and participating in the building of railways. Lamé contributed original work to number theory, applied mathematics, and thermodynamics. His best known work involves the introduction of curvilinear coordinates. His work on number theory included proving Fermat's Last Theorem for $n = 7$, as well as providing the upper bound for the number of divisions used by the Euclidean algorithm. In Gauss' opinion, Lamé was the foremost French mathematician of his time. However, French mathematicians considered him too practical, whereas French scientists considered him too theoretical.

Jacob Steiner (1796–1863)

Jacob Steiner (1796–1863) was born in Utzenstorf, Switzerland. His father was a peasant, and therefore Steiner did not have any opportunity to learn to read and write until he was 14, and only at the age of 18 did he become a pupil of Johann Heinrich Pestalozzi (1746–1827). Steiner then studied at the Universities of Heidelberg and Berlin, supporting himself with a very modest income from tutoring. Steiner's *Systematische Entwicklungen*, published in 1832, immediately made his reputation. In this work he introduced what are now called the geometrical forms and established a one-to-one correspondence between their elements. Through the influence of Jacobi, he received an honorary degree from the Königsberg University, and in 1834, a new Chair of Geometry was founded for him at the University of Berlin, a post he held until his death. Steiner extended many of Poncelet's ideas in projective geometry and published his many contributions in *Crelle's Journal*. His name is met in many places in geometry such as the Steiner solution and generalization of the Malfatti problem, Steiner chains, Steiner's porism, Steiner surface, Steiner tree, Poncelet–Steiner theorem, Steiner–Lehmus theorem, and the Steiner points of the mystic hexagram configuration. Unfortunately, he never prepared his lectures, often failed to prove results, and at every such failure used to make some characteristic remark. He has been considered to be the greatest pure geometer since Apollonius. Steiner died in 1863 in Bern, Switzerland. His will established the Steiner Prize of the Berlin Academy.

Sankaran Varman (1800–1838)

Sankaran Varman (1800–1838) is described as a very intelligent man and acute astronomer–mathematician from Kerala. He wrote an astronomical treatise, *Sadratnamala*, which serves as a summary of most of the results of the Kerala School. Although this book was written at a time when western mathematics and astronomy had been introduced in India, it comprises 211 verses of different measures and abounds with fluxional forms and series.

Mikhail Vasilevich Ostrogradsky (1801–1862)

Mikhail Vasilevich Ostrogradsky (1801–1862) was born in Pashennaya (now Ukraine). He came from a prosperous, aristocratic, and conservative social background. He attended the Poltava Gymnasium secondary school and in 1816 entered the University of Kharkov to study physics and mathematics. In 1820, he took and passed the exams necessary for his degree, but he was asked to retake the exams for religious reasons. Ostrogradsky refused to be reexamined, and so never received his degree. During 1822–1826, he studied at the Sorbonne and at the Collége de France in Paris and met with Laplace, Legendre, Fourier, Poisson, and Cauchy. In this period Ostrogradsky published some papers in the Paris Academy, which later in 1832 he incorporated in a fundamental work on hydrodynamics. In 1828, he returned to St. Petersburg and presented three important papers on the theory of heat, double integrals, and potential theory to the Russian Academy of Sciences. Based on the work in these papers, Ostrogradsky was elected a member of the Academy of Sciences. From 1830, Ostrogradsky lectured at the Institute of Communication and from 1832 at the Pedagogical Institute as well. In 1840, he wrote on ballistics introducing the topic to Russia. From 1847, he was chief inspector for the teaching of mathematical sciences in military schools. He made various contributions to ordinary and partial differential equations, elasticity, calculus of variations, integration of rational functions, number theory, geometry, probability theory, algebra, and in the fields of mathematical physics and classical mechanics. In addition to over 80 research publications, he wrote many excellent textbooks and established the conditions that allowed Chebyshev's school to flourish in St. Petersburg. He is remembered for Green–Ostrogradsky equation, Hamilton–Ostrogradsky (variational) principle, Ostrogradsky formalism, Einstein–Ostrogradsky–Dirac Hamiltonian, Horowitz–Ostrogradsky method, and Jacobi–Ostrogradsky coordinates. He was the first mathematician to publish a proof of the divergence theorem; however, Gauss had already proved the theorem while working on the theory of gravitation but his notebooks were not published until many years later. Therefore, the divergence theorem is sometimes called Gauss' Theorem. Ostrogradsky did not appreciate the work on non-Euclidean geometry, and in 1823, he rejected Lobachevsky's work when it was submitted for publication in the St. Petersburg Academy of Sciences. Ostrogradsky is considered to be Euler's disciple and was a leading Russian mathematician of his time. He died in Poltava (now Ukraine).

Julius Plücker (1801–1868)

Julius Plücker (1801–1868), a German mathematician and physicist, was born in Elberfeld, a municipal subdivision of the German city of Wuppertal. He made significant contributions to the field of analytical geometry and was instrumental in the discovery of electrons through his pioneering research of cathode rays. He also extended the study of a superellipse called a Lamé curve, after Gabriel Lamé, which is a geometric figure defined in the rectangular coordinate system as the set of all points (x, y) with $|x/a|^n + |y/b|^n = 1$, where n, a, and b are positive numbers. Julius studied at the universities of Bonn, Heidelberg, and Berlin after graduating from Gymnasium at Düsseldorf. He shifted to Paris in 1823. In Paris, he attended courses on geometry at the University of Paris and came into close contact with French geometers belonging to the famous school founded by Monge. He came back to Bonn in 1825, and in 1828, he was made extraordinary professor of mathematics in the University of Bonn. In the same year he published the first volume of his *Analytisch-geometrische Entwicklungen*, which introduced the method of abridged notation. In 1831, he published the second volume, in which he established projective duality. In 1836, Plücker was made professor of physics at the University of Bonn. In 1858, after a year of working with vacuum tubes made by his Bonn colleague Johann Heinrich Wilhelm Geissler (1814–1879, an expert glassblower and physicist who is famous for his invention of the Geissler tube, made of glass and used as a low pressure gas-discharge tube), Julius published the first of his classical researches on the action of the magnet on electric discharge in rarefied gases. He found that the discharge caused a fluorescent glow to form on the glass walls of the vacuum tube and that the glow could be made to shift by applying an electromagnet to the tube, thereby creating a magnetic field. It was later shown that this glow is produced by cathode rays. Julius, first by himself and later with Johann Wilhelm Hittorf (1824–1914), a German physicist, made many important discoveries in the spectroscopy of gases. Plücker was the first to use the vacuum tube with the capillary part, that is, the Geissler tube, by means of which the luminous intensity of weak electric discharges was raised sufficiently to permit spectroscopic investigation. He anticipated Robert Wilhelm Eberhard Bunsen (1811–1899) and Kirchhoff by announcing that the spectra emitted were characteristic of the chemical substance that emitted them and by indicating the value of this discovery for chemical analysis. In 1865, Julius moved again to the field of geometry and invented what was known as line geometry in the nineteenth century. In projective geometry, Plücker coordinates refer to a set of homogeneous coordinates introduced initially to embed the set of lines in three dimensions as a quadric in five dimensions. This construction uses second order minors or, equivalently, the second exterior power of the underlying vector space of dimension 4. It is now part of the theory of Grassmannians, in which these coordinates apply in generality (k-dimensional subspaces of n-dimensional space). Julius was the recipient of the Copley Medal from the Royal Society in 1866. He continued to hold the chair of physics at the University of Bonn until his death at the age of 67.

Sir George Biddell Airy (1801–1892)

Sir George Biddell Airy (1801–1892) was born at Alnwick, England. He was educated first at elementary schools in Hereford and then at the Colchester Royal Grammar School. As a boy, Airy was notorious for his skill in designing peashooters. In 1819, he entered Trinity College and was elected scholar of Trinity in 1822. In the following year he graduated as *Senior Wrangler* (the student who earned the year's highest score in the mathematical tripos was given the title Senior Wrangler; the Mathematical Tripos was the most intense mathematical challenge ever devised: 4 days of tests, up to 10 h a day) and obtained first Smith's prize. In 1824, he was elected a fellow of Trinity, and in 1826, he was appointed Lucasian Professor of Mathematics. In 1828, he was elected Plumian Professor of Astronomy and Director of the new Cambridge Observatory. Before these appointments, he had contributed at least three important memoirs to the Philosophical Transactions of the Royal Society and eight to the Cambridge Philosophical Society. In spite of this promising start and some early work in the theory of light—he was the first to draw attention to the defect of vision known as astigmatism—he developed into the excessively practical type of scientist who is obsessed with elaborate numerical computations and has little use for general scientific ideas. In 1831, the Copley Medal of the Royal Society was awarded to him for this research. Then he investigated the mass of Jupiter and discovered a large discrepancy in the motions of the Earth and Venus. The investigation of this inequality was the most laborious work in planetary theory that had been made up to Airy's time, and represented the first specific improvement in the solar tables effected in England since the establishment of the theory of gravitation. In recognition of this work, the Gold Medal of the Royal Astronomical Society was awarded to him in 1833 (he won it again in 1846). In 1835, Airy was appointed Astronomer Royal in succession to John Pond and began his long career at the national observatory that constitutes his chief claim to fame. Airy received a testimonial from the Royal Astronomical Society in 1848. In 1851, Airy established a new Prime Meridian at Greenwich. This line, the fourth *Greenwich Meridian*, became the definitive, internationally recognized line in 1884. In 1862, Airy presented a new technique to determine the strain and stress field within a beam. This technique, sometimes called the Airy stress function method, is often used to find solutions to many two-dimensional problems in solid mechanics. At the age of 71, Airy conceived the idea of treating the lunar theory in a new way. Some of this work was published in 1886, when he was 85 years old. In his life he published more than 518 research articles. The differential equation

$y'' - xy = 0, \ -\infty < x < \infty$, which bears his name is of particular interest due to the fact that for x negative the solutions are oscillatory, similar to trigonometric functions, and for x positive they are monotonic, similar to hyperbolic functions. Airy and his wife are buried at St. Mary's Church in Playford, Suffolk (England). A cottage owned by Airy still stands, adjacent to the church and now in private hands.

Niels Henrik Abel (1802–1829)

Niels Henrik Abel (1802–1829) was a brilliant Norwegian mathematician whose tragic death at age 26 of tuberculosis was an inestimable loss for mathematics. Along with Gauss and Cauchy, Abel was one of the pioneers in the development of modern mathematics. He was one of six children in the family of a poor country minister. At 16, influenced by a perceptive teacher, he read the works of Newton, Euler, and Lagrange. As a comment on this experience, he inserted the following marginal remark in one of his later mathematical notebooks: "It appears to me that if one wants to make progress in mathematics, one should study the masters and not the pupils." When Abel was only 18 his father died and left the family destitute. They subsisted with the aid of friends and neighbors, and somehow Abel, helped by contributions from several professors, managed to enter the University of Oslo in 1821. His earliest research was published in 1823 and included his solution of the classic tautochrone problem by means of an integral equation. He devoted much of his early life to the study of algebraic solutions of equations and gave the first rigorous proof that the general fifth order equation cannot be solved by purely algebraic operations. This disposed of a problem that had baffled mathematicians for almost 300 years. He published his proof in a small pamphlet at his own expense. Later, Abel spent some time in Berlin and wrote his classic study of the binomial series, in which he founded the general theory of convergence and gave the first satisfactory proof of the validity of this series expansion. He also inspired August Leopold Crelle (1780–1855) to launch the prestigious *Journal für die Reine und Angewandte Mathematik*, which was the world's first periodical devoted wholly to mathematical research. The first three volumes of this journal contained 22 of Abel's articles. He had sent his pamphlet on the fifth degree equation to Gauss, hoping that it would serve as a kind of scientific passport. However, for some reason Gauss put it aside without looking at it, for it was found uncut among his papers after his death 30 years later. Abel felt that he had been snubbed and without visiting Gauss went from Berlin to Paris. Soon after his arrival in Paris, he finished his great *Mémoire sur une Propriété Générale d'une Classe Trés Étendue des Fonctions Transcendantes*, which he regarded as his masterpiece. This work contains the discovery about integrals of algebraic functions now known as Abel's theorem and is the foundation for the later theory of Abelian integrals, Abelian functions, and much of algebraic geometry. Abel submitted his masterpiece to the French Academy with the hope that it would bring him to the notice of the French mathematicians. He waited for a reply in vain until he could no longer afford to stay, and he was forced to return to Berlin. His manuscript was given to Cauchy and Legendre for examination; Cauchy took it home, mislaid it, and forgot all about it. It was not published until 1841, when again the manuscript was lost before the proof sheets were read. The original finally turned up in Florence in 1952. In Berlin, Abel finished his first revolutionary article on elliptic functions, a subject he had been working on for several years, and then went back to Norway deeply in debt. He had expected to be appointed to a professorship at the university on his return, but once again his hopes were dashed. He made a living by tutoring, and for a brief time he held a substitute teaching position. During this period he worked incessantly, mainly on the theory of elliptic functions, which he had discovered to be the inverses of elliptic integrals. This theory quickly took its place as one of the major fields of nineteenth century analysis, with many applications to number theory, mathematical physics, and algebraic geometry. Meanwhile, Abel's fame had spread to all of the mathematical centers of Europe, and he stood among the elite of the world's mathematicians but in his isolation was unaware of it.

As late as 1828, in a letter to his formal teacher, Bernt Michael Holmboe (1795–1850), Abel expressed the following views on divergent series: "The divergent series are the invention of the devil, and it is a shame to base on them any demonstration whatsoever. By using them, one may draw any conclusion he pleases and that is why these series have produced so many fallacies and so many paradoxes.... I have become prodigiously attentive to all this, for with the exception of the geometrical series, there does not exist in all of mathematics a single infinite series the sum of which has been determined rigidly. In other words, the things which are the most important in mathematics are also those which have the least foundation. That most of these things are correct in spite of that is extraordinarily surprising. I am trying to find a reason for this; it is an exceedingly interesting question."

Just a few days before Abel died, he arranged his fiancee's (Crelly Kemp) marriage with his friend Mathias Kielhau. As an ironic postscript, shortly after his death Crelle wrote that his efforts had been successful and that Abel would be appointed to the Chair of Mathematics in Berlin. Later Crelle eulogized Abel in his Journal as follows: "All of Abel's works carry the imprint of an ingenuity and force of thought which is amazing. One may say that he was able to penetrate all obstacles down to the very foundation of the problem, with a force which appeared irresistible... He distinguished himself equally by the purity and nobility of his character and by a rare modesty which made his person cherished to the same unusual degree as was his genius." Mathematicians, however, have their own ways of remembering their great men, and so we speak of Abel's integral equation, Abelian integrals and functions, Abelian groups, Abel's series, Abel's partial summation formula, Abel's limit theorem in the theory of power series, and Abel summability. Few have had their names linked to so many concepts and theorems in modern mathematics, and what he might have accomplished in a normal lifetime is beyond conjecture. According to Hermite, Abel left for mathematicians something to keep them busy for 500 years.

Janos Bolyai (1802–1860)

Janos Bolyai (1802–1860) was born in Kolozsvár, Hungary (now Cluj, Romania), but soon moved to Marosvásárhely where his father, Farkas Wolfgang Bolyai, taught mathematics, physics, and chemistry at the Calvinist College. At the age of four, Janos could distinguish certain geometrical figures, knew about the sine function, and could identify the best known constellations; at the age of five he could read, and at the age of seven he started learning the violin and soon began playing difficult concert pieces. At the age of 13 he mastered calculus and analytical mechanics, and at the age of 15 he graduated from Marosvásárhely College. Janos learned most of the elementary mathematics from his father at home. During 1818–1822, he studied at the Royal Engineering College in Vienna, completing the 7 year course in 4 years with distinctions in many subjects. He also mastered nine foreign languages, including Chinese and Tibetan. Then he spent 11 years in military service and was reputed to be the best swordsman and dancer in the Austro-Hungarian Imperial Army.

In 1820, when he was still studying in Vienna, Janos (like his father) tried to replace Euclid's parallel postulate with another axiom that could be deduced from the others; however, soon he gave up this approach and started developing the basic ideas of non-Euclidean (hyperbolic) geometry. Janos published his findings in 1831, in an Appendix to a mathematical work, *Tentamen*, by his father. Farkas sent a reprint of Tentamen to Gauss who, on reading the Appendix, wrote that "I regard this young geometer Bolyai as a genius of the first order." However, he could not praise Bolyai, since this would mean self-praise, because the ideas of the Appendix had been his own for the past 30 or 35 years. Young Bolyai was deeply disappointed and he never published any more mathematics. In 1848, Bolyai discovered that Lobachevsky had published a similar piece of work in 1829. He died of pneumonia at the age of 57 in Marosvásárhely. Although he never published more than the few pages of the Appendix, he left 20,000 pages of manuscript of mathematical work that includes an attempt to develop all of mathematics based on axiomatic systems, as well as a rigorous geometric concept of complex numbers as ordered pairs of real numbers. His work is now in the Bolyai-Teleki library in Tirgu-Mures. In 1945, a university in Cluj was named after him, but it was closed down under Ceaucescu's government in 1959. Once he said, "I have created a new universe from nothing."

Jacques Charles Francois Sturm (1803–1855)

Jacques Charles Francois Sturm (1803–1855) was a Swiss mathematician of German heritage who spent most of his life in Paris. His parents gave him a good education and at school he showed great promise, particularly in Greek and Latin poetry. In 1818, he started to follow the lectures of the Academy of Geneva. In 1819, Sturm's father died and to support the family he was forced to give lessons to children of rich families. In 1823, he became tutor to the youngest son of Madame de Staël at château. During this period he wrote articles on geometry which were published in Gergonne's Annales de mathématiques pures et appliquées. At the end of 1823, he stayed in Paris for a short time following the family of his student. In Paris the family introduced him to the leading scientists, such as Laplace, Poisson, and Fourier. In December of 1825, Sturm went to Paris to take courses in mathematics and physics. In 1826, with his colleague Jean-Daniel Colladon (1802–1893) he helped make the first experimental determination of the speed of sound in water. In 1829, he discovered a theorem regarding the determination of the number of real roots of a numerical equation included between given limits, which bears his name. In 1830, Sturm was appointed Professor of Mathematics in the Collége Rollin. He became a French citizen in 1833 and was elected to the Académie des Sciences in 1836, filling the seat of André-Marie Ampére (1775–1836). His main work was done in 1836–1837, which is now called the Sturm–Liouville theory of differential equations that has been of steadily increasing importance ever since, in both pure mathematics and mathematical physics. He became répétiteur in 1838 and professor in 1840 at the École Polytechnique. The same year, after the death of Poisson, he was appointed as Mechanics Professor of the Faculté des Sciences of Paris. He made important contributions to infinitesimal geometry, projective geometry, and the differential geometry of curves and surfaces. He also did important work in geometrical optics. From 1851 his health began to fail, and he died in 1855 after a long illness. His works *Cours d'analyse de l'École Polytechnique* (1857–1863) and *Cours de mécanique de l'École Polytechnique* (1861) were first published after his death in Paris. His name is part of the 72 names engraved at the Eiffel Tower.

Giusto Bellavitis (1803–1880)

Giusto Bellavitis (1803–1880) was born in Bassano, Italy. In the beginning, Bellavitis' father taught him at home, and later he studied mathematics on his own. During 1822–1832, he worked without pay for the municipal government of Bassano. In 1834, Bellavitis gave formulas for the areas of polygons that were independently discovered by Karl George Christian von Staudt (1798–1867) in 1842. He believed that algebra had to be founded on geometry and that the number systems could only be defined through geometric concepts. In 1835 and 1837, Bellavitis published two important papers on geometry, and in 1843, he became Professor of Mathematics and Mechanics at Vicenza. In 1845, he was appointed Professor of Geometry at Padua, and in 1867, Bellavitis accepted the Chair of Complementary Algebra and Analytic Geometry. In 1866, he was appointed as a Senator of the Kingdom of Italy. He is often remembered for his significant contributions to algebraic and descriptive geometry, solutions of various mechanical problems, calculus of probabilities, theory of errors, and his place in the history of mathematics.

Pierre Francois Verhulst (1804–1849)

Pierre Francois Verhulst (1804–1849) was a Belgian mathematician who introduced the differential equation $y' = y(r - ay)$ as a model for human population growth in 1838. He referred to it as logistic growth; hence this equation is often called the logistic equation. He was unable to test the accuracy of his model because of inadequate census data, and it did not receive much attention until many years later. Reasonable agreement with experimental data was demonstrated by Raymond Pearl (1879–1940) in 1930 for *Drosophila melanogaster* (fruit fly) populations, and by G.F. Gause (1935) for *Paramecium* and *Tribolium* (flour beetle) populations. As an undergraduate at the University of Ghent, he was awarded two academic prizes for his work on the calculus of variations. Later, he published papers on number theory and physics. His interest in probability theory had been triggered by a new lottery game, but he applied it to political economy and later to demographical studies as well. His main work is *Traité des fonctions elliptiques* (1841), for which he was admitted with a unanimous vote to the Académie de Bruxelles. He was elected President of the Académie in 1848, shortly before the decline of his fragile health, which had troubled him for many years. He died at the young age of 45.

Carl Guslov Jacob Jacobi (1804–1851)

Carl Guslov Jacob Jacobi (1804–1851) was born in Potsdam, Prussia (now Germany) to a Jewish family. He received his early education from his uncle on his mother's side, and at the age of 12 he entered the Gymnasium in Potsdam. Jacobi had remarkable talents, and in 1817, while still in his first year of schooling, he was put into the final year class. However, due to age restrictions, he could not register in the University of Berlin and had to continue in the same class at the Gymnasium until the spring of 1821. He received the highest awards for Latin, Greek, and history. During this period he studied several advanced mathematics texts on his own and attempted to solve quintic equations by using radicals. In 1821, Jacobi entered the University of Berlin and obtained a doctoral degree in 1825 for his thesis *An analytical discussion of the theory of fractions*. Around 1825, Jacobi converted to Christianity, which entitled him to a teaching position in a university. For the academic year 1825–1826 he taught at the University of Berlin. In 1827, he became associate, in 1829 ordinary, and in 1832 full Professor of Mathematics at Königsberg University, where he remained until 1842. In Königsberg he joined Neumann and Bessel. Before moving he had already made major discoveries in number theory and had remarkable new ideas about elliptic functions. He wrote to several mathematicians, including Gauss and Legendre, who were very impressed with his results. In 1829, Jacobi met Legendre, Fourier, Poisson, and Gauss during his travels across Europe. His classic treatise on elliptic functions, published in 1829, along with its later supplements, particularly his development of the theta function, made fundamental contributions to the theory of elliptic functions. Jacobi was the first person to apply elliptic functions to number theory. During 1833–1835, Jacobi became interested in physics. He carried out important research in partial differential equations of the first order and applied them to the differential equations of dynamics. He also worked on determinants and studied the functional determinant now called the Jacobian, which has played an important part in many analytical investigations. However, he was not the first to study the Jacobian; it appeared first in an 1815 paper by Cauchy. In 1834, Jacobi proved that if a univariate single-valued function of one variable is doubly periodic, then the ratio of the periods is imaginary, and such a function cannot have more than two periods. This result was extended further by Liouville and Cauchy. In 1842, Jacobi proved that the spherical image of the normal directions along a closed differentiable curve in space divides the unit sphere into regions of equal area. This is considered to be the prettiest result in the global theory of curves. In 1842, Jacobi was diagnosed with diabetes. He was advised by his doctor to spend some time in Italy, where the climate did indeed help him to recover. On his return he moved to Berlin, then eventually to Gotha, but he still lectured in Berlin. In January 1851 he contracted influenza, then he contracted smallpox before he had regained his strength. He died a few days after contracting smallpox and was buried at a cemetery in the Kreuzberg section of Berlin. He left a vast store of manuscripts, some of which have been published in Crelle's Journal. His most publicized work is the Hamilton–Jacobi theory in rational mechanics.

Jacobi was widely considered to be the most inspiring teacher of his time. He introduced the seminar method for teaching students the latest advances in mathematics. He became an FRS in 1833, and a Fellow of the Royal Society of Edinburgh in 1845. He is best remembered for Jacobi's elliptic functions, Jacobi's formula, Jacobi's integral, Jacobi's polynomials, Jacobian, Jacobi's symbol, Jacobi's identity, Jacobi's method, Jacobi's laws, and the Carathéodory–Jacobi–Lie theorem. The Jacobi crater on the Moon is named after him. The phrase 'Invert, always invert,' is associated with Jacobi, for he believed that it is in the nature of things that many hard problems are best solved when they are addressed backward. It is not God who geometrizes, as Plato said, or arithmetizes, as Jacobi said, but man. Mathematics is no longer an absolute truth; there is no objective reality, no π in the sky. Mathematics is only a very useful tool, and like any other tool, it is man made. According to a story, to drive home the point to a gifted but diffident young man who always put off doing anything until he had learned something more, Jacobi delivered the following parable. "Your father would never have married, and you wouldn't be here now, if he had insisted on knowing all the girls in the world before marring one." He is also remembered for the statement "The real end of science is the honor of the human mind."

Victor Jakowlewitsch Buniakovski (1804–1899)

Victor Jakowlewitsch Buniakovski (1804–1899) was born in Bar, Ukraine. He received a doctorate in Paris in 1825. He carried out additional studies in St. Petersburg and then had a long career there as a professor. He became the Vice President of the Petersburg Academy of Sciences. Buniakovski worked in theoretical mechanics, geometry, hydrodynamics, and number theory, in which his conjecture is well known. Twenty-five years earlier than Schwarz, in the year 1859, he proved the infinite dimensional case of the famous Cauchy–Schwarz inequality. Variations of this inequality occur in many different settings and under various names. Buniakovski died in St. Petersburg.

Peter Gustav Lejeune Dirichlet (1805–1859)

Peter Gustav Lejeune Dirichlet (1805–1859) was born into a French family living near Cologne, Germany, where his father was postmaster. He began to study mathematics at an early age; by age 12, he used his own allowance to purchase mathematics texts. In 1822, he went to the University of Paris to study higher mathematics. Dirichlet held positions at the University of Breslau and the University of Berlin. In 1855, he was chosen to succeed Gauss at the University of Göttingen. Dirichlet was the first person to master Gauss' *Disquisitiones Arithmeticae*, which appeared 20 years earlier. He is said to have kept a copy at his side even when he traveled, and he slept with it under his pillow. Before going to bed he would struggle with some tough paragraph in the hope (which was frequently fulfilled) that he would wake up in the night to find that a rereading made everything clear. In 1829, he gave the first set of conditions sufficient to guarantee the convergence of a Fourier series. The definition of function that is usually used today in elementary calculus is essentially the one given by Dirichlet in 1837. While he is best known for his work in analysis and differential equations, Dirichlet was also one of the leading number theorists of the nineteenth century. He proved Fermat's Last Theorem for $n = 5$ and showed that there are infinitely many primes in the arithmetic progression $an + b$, where a and b are relatively prime. His study of absolutely convergent series also appeared in 1837. Dirichlet's important convergence test for series was not published until after his death. Dirichlet also engaged in studies of mathematical physics. These led, in part, to the important *Dirichlet principle* in potential theory, which establishes the existence of certain extremal harmonic functions. This was important historically because it was the key to finally obtaining a rigorous proof of the Riemann mapping theorem. It is still used today in partial differential equations, the calculus of variations, differential geometry, and mathematical physics. The famous *Dirichlet Pigeonhole Principle* was originally called the Dirichletscher Schubfachschluss, or Dirichlet's drawer-shutting principle. Eisenstein, Kronecker, and Lipschitz were his students. Dirichlet's last few years were unfortunate. In 1858, he went to Montreux, Switzerland, to give a speech in honor of Gauss, but had a heart attack and was forced to remain there for some months. While he was away, his wife died of a stroke, and he himself died in Göttingen in 1859. After Dirichlet's death, his lectures and other results in number theory were collected, edited, and published by Dedekind under the title *Vorlesungen über Zahlentheorie*. Dirichlet was a son-in-law of Jacobi. His brain and Gauss' are preserved in the Department of Physiology at the University of Göttingen.

William Rowan Hamilton (1805–1865)

William Rowan Hamilton (1805–1865) was born in Dublin. His father was a successful lawyer and his mother came from a family noted for their intelligence. He was a classic child prodigy as a linguist and mathematician. He was educated by an eccentric but learned clerical uncle. At the age of three he could read English; at four he began Greek, Latin, and Hebrew; at eight he added Italian and French; at ten he learned Sanskrit and Arabic; and at 13 he is said to have mastered one language for each year he had lived. This forced flowering of his linguistic faculties was broken off at the age of 14, when he turned to mathematics, astronomy, and optics. At 18 he published a paper correcting a mistake in Laplace's *Mécanique Céleste*; and while still an undergraduate at Trinity College in Dublin he was appointed Professor of Astronomy and automatically became Astronomer Royal of Ireland. His first important work was in geometrical optics, particularly the discovery that light refracts as a conical configuration of rays. He became famous at the age of 27. Even more significant was his demonstration that all optical problems can be solved by a single method that includes Fermat's principle of least time as a special case. He then extended this method to problems in mechanics and by the age of 30 had arrived at a single principle known as *Hamilton's principle* that exhibits optics and mechanics as merely two aspects of the calculus of variations. In 1835, he turned his attention to algebra and constructed a rigorous theory of complex numbers based on the idea that a complex number is an ordered pair of real numbers. This work was done independently of Gauss, who had already published the same ideas in 1831, but with emphasis on the interpretation of complex numbers as points in the complex plane. Hamilton subsequently tried to extend the algebraic structure of the complex numbers, which can be thought of as vectors in a plane, to vectors in three-dimensional space. This project failed; however, the year 1843 marked a triumph for Hamilton after several years of searching for a way to multiply *his* quaternions (Hamilton never learned that Gauss had discovered quaternions in 1819 but kept his ideas to himself). A *quaternion* is a kind of hypercomplex number since it represents a force acting in three dimensions rather than in the plane. Quaternions involve the symbols j, k, and i, the imaginary number. Hamilton could add and subtract quaternions, but he could not find any product that did not violate the commutative law, $AB = BA$. Hamilton was strolling along the Royal Canal in Dublin when he suddenly realized that a mathematical system could be consistent without obeying the commutative law. This happened at Brougham Bridge, and he carved on it the famous formula $i^2 = j^2 = k^2 = ijk = -1$. This insight freed algebra and paved the way for multiplication of matrices, which is noncommutative. After Hamilton's death, a 50 year battle began that involved mathematicians and scientists on both sides of the Atlantic. The importance of this discovery to science can be seen in the work done in vector algebras, which were meant to replace quaternions. By 1900, the battle over the various vector algebras began to die down, and quaternions had left the picture.

Hamilton married his third love in 1833, but his marriage worked out poorly because his wife, a semi-invalid, was unable to cope with his household affairs. He suffered from alcoholism and lived reclusively for the last two decades of his life. He died from gout, leaving masses of papers containing unpublished research. Mixed in with these papers were a large number of dinner plates, many containing the desiccated remains of uneaten chops. All his life, Hamilton loved animals and respected them as equals. During his final years the newly founded National Academy of Sciences of the United States elected him as its first foreign associate. As a gesture of rare honor, in 1945 he was lodged for a week in the sacred rooms of Trinity College, where Newton had composed his Principia.

Augustus De Morgan (1806–1871)

Augustus De Morgan (1806–1871) was the son of a member of the East India Company. He was born in Madura (India), but his family moved to England when he was 7 months old. He lost sight in his right eye shortly after birth, which made him shy and solitary and exposed him to jolly schoolboy pranks. He attended private schools during his early teenage years, where he developed a strong interest in mathematics. De Morgan entered Trinity College, Cambridge, in 1823 and graduated in 1827. Although he considered entering medicine or law, he decided on a career in mathematics. He won a position at University College, London, in 1828, but resigned when the college dismissed a fellow professor without giving their reasons. However, he resumed this position in 1836 when his successor died and stayed there until 1866. De Morgan was a noted teacher who stressed principles over techniques. His students included many famous mathematicians, including Ada Augusta, Countess of Lovelace, who collaborated with Babbage in his work on computing machines. He was an extremely prolific writer. He wrote more than 1,000 articles for more than 15 periodicals. De Morgan also authored textbooks on many subjects, including logic, probability, calculus, and algebra. His book *A Budget of Paradoxes* is an entertaining text. In 1838, he presented what was perhaps the first clear explanation of an important proof technique known as *mathematical induction*, a term that he coined. In the 1840s, De Morgan made fundamental contributions to the development of symbolic logic. He invented notation that helped him prove propositional equivalences, such as the laws that are named after him. He regarded mathematics as an abstract study of symbols that are subjected to sets of symbolic operations. In 1842, De Morgan presented what was perhaps the first precise definition of a limit and developed some tests for the convergence of infinite series. De Morgan was also interested in the history of mathematics and wrote biographies of Newton and Halley. In 1837, De Morgan married Sophia Frend, who wrote his biography in 1882. In 1866, he was a cofounder of the London Mathematical Society and became its first President. In the same year De Morgan was elected a Fellow of the Royal Astronomical Society. De Morgan was never an FRS, as he refused to let his name be put forward. He also refused an honorary degree from the University of Edinburgh. De Morgan's research, writing, and teaching left little time for his family or social life. Nevertheless, he was noted for his kindness, humor, fair ability as a flutist, and wide range of knowledge. He once remarked that it is easier to square the circle then to get round a mathematician. According to him *the moving power of mathematical invention is not reasoning but imagination.* He proposed the conundrum: "I was x years old in the year x^2." When was he born? ($x = 43$.) He was an original man even among mathematicians.

Hermann Günter Grassmann (1809–1877)

Hermann Günter Grassmann (1809–1877) was born in Stettin, Prussia (now Szczecin, Poland). His father taught mathematics and physics, wrote several school books, and undertook research on crystallography. He was the third of 12 children. One of Hermann's brothers, Robert, also became a mathematician and the two collaborated on many projects. In the beginning Hermann was a slow learner; however, at the age of 18 he passed his secondary school examinations with the second rank in the school. In 1827, Hermann joined the University of Berlin to study theology, classical languages, philosophy, and literature. In 1830, he returned to Stettin, and started taking interest in mathematics due to his father's influence. In 1844, he published *Die Lineale Ausdehnungslehre, ein neuer Zweig der Mathematik* (Linear Extension Theory, a new branch of mathematics), which he began writing in 1832. In this masterpiece work, which includes Hamilton's quaternions as a special case, he developed the idea of an algebra in which the symbols representing geometric entities such as points, lines, and planes are manipulated using certain rules. He represented subspaces of a space by coordinates leading to point mapping of an algebraic manifold, now called the Grassmannian. In 1844, Grassmann published the first edition of his remarkable *Ausdehnungslehre* (Calculus of Extension), but because of its poor exposition and obscure presentation this book remained practically unknown. For the next 10 years, Grassmann used his new concepts in different fields and published several papers, which were also mostly ignored. A second reformulation of Ausdehnungslehre in 1862 was also unsuccessful. Discouraged, Grassmann gave up mathematics and concentrated on the study of Sanskrit language and literature, a field in which he contributed a number of brilliant papers. He wrote a dictionary for the Rigveda and translated the Rigveda in verse. Grassmann also prepared a treatise on German plants, edited a missionary paper, investigated phonetic laws, harmonized folk songs in three voices, and raised nine of his 11 children. One of his sons, Hermann Ernst Grassmann, also became a mathematician at the University of Giessen and wrote a treatise on projective geometry. Grassmann returned to mathematics in the last couple of years of his life and prepared another edition of the 1844 Ausdehnungslehre, which was published only after his death. Grassmann died in 1877 of heart problems. According to him, pure mathematics is the science of the individual object in as much as it is born in thought.

No doubt, Grassmann was ahead of his time. His mathematical methods were slowly adopted and eventually influenced the works of Elie Cartan, Hankel, Peano, Whitehead, and Klein. In the early twentieth century, his work was applied to the theory of general relativity.

Joseph Liouville (1809–1882)

Joseph Liouville (1809–1882) was a highly respected professor at the Collége de France in Paris, and the founder, and for 39 years the editor, of the *Journal des Mathématiques Pures et Appliquées*, a famous periodical that played an important role in French mathematical life throughout the nineteenth century. For some reason, his remarkable achievements as a creative mathematician have not received the appreciation that they deserve. The fact that his collected works have never been published is an unfortunate and rather surprising oversight on the part of his countrymen.

He was the first to solve a boundary value problem by solving an equivalent integral equation, a method developed by Fredholm and Hilbert in the early 1900s into one of the major fields of modern analysis. His ingenious theory of fractional differentiation answered the long-standing question of whether reasonable meaning can be assigned to the symbol $d^n y/dx^n$ when n is not a positive integer. He discovered the fundamental result in complex analysis now known as *Liouville's theorem* (that a bounded entire function is necessarily a constant) and used it as the basis for his own theory of elliptic functions. There is also a well-known Liouville theorem in Hamiltonian mechanics, which states that volume integrals are time invariant in phase space. His theory of the integrals of elementary functions was perhaps the most original of all his achievements. In it, he proved that integrals like $\int e^{-x^2} dx$, $\int \sin x/x dx$, and $\int dx/\ln x$, as well as the elliptic integrals of the first and second kinds, cannot be expressed in terms of a finite number of elementary functions.

The fascinating and difficult theory of transcendental numbers is another important branch of mathematics that originated in Liouville's work. The irrationality of π and e, that is, the fact these numbers are not roots of any linear equation of the form $ax + b = 0$ whose coefficients are integers, had been proved in the eighteenth century by Lambert and Euler. In 1844, Liouville showed that e is also not a root of any quadratic equation with integral coefficients. This led him to conjecture that

e is *transcendental*, which means that it does not satisfy any polynomial equation $a_n x^n + a_{n-1} x^{n-1} + \cdots + a_1 x + a_0 = 0$ with integral coefficients. His efforts to prove this failed, but his ideas contributed to Hermite's success in 1873 and Lindemann's 1882 proof that π is also transcendental. Lindemann's result showed at last that the age-old problem of squaring the circle using a ruler-and-compass construction is impossible. One of the great mathematical achievements of modern times was Gelfond's proof in 1934 that e^{π} is transcendental, but nothing yet is known about the nature of any of the numbers $\pi + e$, πe, or π^e. Liouville also discovered a sufficient condition for transcendence and used it in 1844 to produce the first examples of real numbers that are provably transcendent. One of these is

$$\sum_{n=1}^{\infty} \frac{1}{10^{n!}} = \frac{1}{10^1} + \frac{1}{10^2} + \frac{1}{10^6} + \cdots = 0.11000100\ldots.$$

His methods regarding transcendence have led to extensive research.

Ernst Eduard Kummer (1810–1893)

Ernst Eduard Kummer (1810–1893) was born in Sorau, Brandenburg (then part of Prussia). His father died when Eduard was only 3 years old. He received private coaching before entering the Gymnasium in Sorau when he was 9 years old. In 1828, Kummer enrolled at the University of Halle. In 1831, he was awarded a prize for a mathematical essay he wrote. In the same year, he was awarded the certificate that entitled him to teach in schools, and he was also awarded a doctoral degree. For the next 10 years Kummer taught mathematics and physics at the Gymnasium in Liegnitz (now Legnica in Poland). In 1836, he published a paper on hypergeometric series in Crelle's Journal. This paper attracted the attention of Jacobi and Dirichlet. In 1839, Kummer was elected to the Berlin Academy on Dirichlet's recommendation. In 1840, Kummer married a cousin of Dirichlet's wife, but she died in 1848, and so he remarried. Overall, he had 13 children. In 1842, he was appointed full professor at the University of Breslau, now Wroclaw in Poland. He was called to Berlin in 1855 as successor to Dirichlet, where he taught until 1883, when he voluntarily stopped doing mathematical work because he felt the onset of a decline in productivity. Kummer's Berlin lectures were always carefully prepared and covered analytic geometry, mechanics, the theory of surfaces, and number theory. His lectures attracted great number of students, as many as 250. He supervised a large number of doctoral students, many of whom went on to hold mathematics chairs at other universities, including Cantor and Schwarz, who later married one of his daughters. Kummer was not only a father to his students but also something of a brother to their parents. He appointed many talented young lecturers including Christoffel and Fuchs, who succeeded him. The three great mathematicians of Berlin, Kummer, Weierstrass, and Kronecker, were close friends for 20 years while they worked closely and effectively together. In mathematics he is remembered for codifying some of the relations between different hypergeometric series known as contiguity relations, the Kummer surface, the Kummer function, the Kummer ring, and the Kummer sum. He is also famous for introducing the notion of *ideal* numbers in the theory of algebraic domains of rationality. This theory was inspired partly by Kummer's attempts to prove Fermat's last theorem and partly by Gauss' theory of biquadratic residues, in which the concept of prime factors had been introduced within the domain of complex numbers. Kummer's ideal factors allowed a unique decomposition of numbers into prime factors in a general domain of rationality. In 1857, the Paris Academy of Sciences awarded Kummer the Grand Prize for his proof of Fermat's last theorem for a considerable class of prime exponents. In 1863, he was elected an FRS. He received numerous other honors in his long career. Kummer died after a short attack of influenza at the age of 83 in Berlin.

Evariste Galöis (1811–1832)

Evariste Galöis (1811–1832) was a French mathematician whose discoveries are considered to be some of the most original mathematical ideas of the nineteenth century. He was born in Bourg-la-Reine to a happy and respected family. His mother was well educated, clever, and intellectually sophisticated, and father was a charming, fun-loving man who delighted in making up clever verses about his friends. Galöis' life was one of disillusionment and disappointment. He was tutored at home by his mother, and at the age of 12 Galöis was admitted to Louis-le-Grand, a boarding school in Paris. As a boy, he had boundless energy and enthusiasm. His love for learning led him to read books of mathematics much like other boys would read mysteries. Because of his superior intelligence, school was extremely boring and unchallenging for Galöis, and he rebelled against his teacher's harsh, domineering treatment, which he found unmotivating. Galöis loved to do mathematics in his head and would not bother to write out the proofs systematically on paper. This proved to be highly detrimental; he failed the entrance test for the ecole Polytechnique as a result. Galöis was extremely disappointed by the unjust test results because he didn't find the questions difficult. When Galöis was 17 years old, the great teacher Professor Louis-Paul-Émile Richard (1795–1849) came to his school. Professor Richard immediately recognized Galöis' genius and encouraged him to compile his discoveries and send them to the French Academy, which was French's finest group of scholars. Galöis promptly took up the challenge and sent his papers to Augustin-Louis Cauchy, a professor at ecole Polytechnique. But, most unfortunately for Galöis, Cauchy never read his work and lost his paper as well. Later, Galöis decided to take the ecole Polytechnique entrance test again. Only two tries were allowed, and this would be his second and final chance. However, the examiners had already heard of his intelligence and had their minds set against him even before the test. They taunted him during the oral test so much that Galöis lost his temper and hurled an eraser at them. This ended all his hopes for admission into the best school of mathematics in Europe. At the age of 19, he produced some very important papers on algebra, in particular an important function of algebra to solve equations. Galöis discovered the types of equations that can be solved and those that cannot be. In a moment of optimism, he submitted his work to the Academy of Science in competition for the Grand Prize in mathematics. The Secretary received his work but died before reading it. To make matters worse, when officials went to retrieve the papers, they had mysteriously disappeared. Disappointments and setbacks finally took their toll on Galöis and he became a very bitter man. He developed a distrust for teachers and institutions and got involved in politics instead. He joined the Republicans, which was a forbidden radical group. Galöis was arrested a couple of times but was acquitted of every charge because they were difficult to prove. He was finally jailed for 6 months on the trumped up, trivial charge of wearing the uniform of the nonexistent National Guard. During his sojourn in jail, Galöis renewed his interest in mathematics. Following his release, he sorted out his papers but was unable to complete them before he was forced into a duel over a girl that he was barely acquainted with and hardly cared for. The duel proved to be his end.

The death of Galöis was indeed a terrible waste. If he had known that his work would be of such great significance to mathematics, life would have been meaningful for him. As it was, he died a very disappointed man. It was not until 1846 that his research was published. Mathematicians began to appreciate the importance of Galöis' work, which centered on solving equations using groups. Galöis found a way to derive a group that corresponds to each equation. This group, called "the group of the equation" contains the important properties of equations. The so-called Galöis groups form an important part of modern abstract algebra.

Ludwig Otto Hesse (1811–1874)

Ludwig Otto Hesse (1811–1874) was born in Königsberg, Prussia. He studied in his hometown of Albertina under Jacobi. Ludwig earned his doctorate in 1840 at the University of Königsberg and earned his habilitation in 1841. He taught in Königsberg, Halle, and Heidelberg, and in 1868 moved to Munich. In 1869, he joined the Bavarian Academy of Sciences. He worked on algebraic invariants and the Hessian matrix, the Hesse normal form, the Hesse configuration, the Hessian group, Hessian pairs, Hesse's theorem, and the Hesse pencil are among his contributions. His doctoral students included Kirchhoff and Schröder. He died in Munich.

Urbain Jean Joseph Le Verrier (1811–1877)

Urbain Jean Joseph Le Verrier (1811–1877) was born in St. Lô, France. He was educated at the Polytechnic school and in 1837 was appointed lecturer of astronomy there. He communicated his first results in astronomy to the Academy in 1839. In this work he calculated, within much narrower limits than Laplace had used, the extent within which the inclinations and eccentricities of the planetary orbits vary. The independent discovery in 1846 by Le Verrier and John Couch Adams of the planet Neptune, by means of the disturbance it produced in the orbit of Uranus, attracted general attention to physical astronomy and strengthened the opinion of the universality of gravity. In 1855, he was elected a foreign member of the Royal Swedish Academy of Sciences. The same year, Le Verrier succeeded Arago as director of the Paris observatory and reorganized it in accordance with the requirements of modern astronomy. His last work consisted of a theoretical investigation of planetary motions and a revision of all tables that involve them. He died in Paris in 1877 and was buried in the Cimetiére Montparnasse. A large stone celestial globe sits over his grave. He will be remembered by the phrase attributed to Arago, "the man who discovered a planet with the point of his pen." Le Verrier's name is one of the 72 engraved on the Eiffel Tower.

Pierre Wantzel (1814–1848)

Pierre Wantzel (1814–1848) was born in Paris. He earned fame for his work on solving equations using radicals. In 1837, Wantzel proved that the problems of doubling the cube and trisecting an angle could not be solved with a straight edge and compass. Drawing on the work of Gauss, Wantzel also showed that if p is an odd prime, then a regular p-gon is constructible only if p has the form $2^n + 1$. He died at the young age of 34 in Paris, apparently due to overwork, as his body was unable to keep pace with the rigors of his mind and was weakened from drinking too much coffee and smoking too much opium.

James Joseph Sylvester (1814–1897)

James Joseph Sylvester (1814–1897) was born in London, England, to a merchant Jewish family. He attended a boarding school in Highgate until 1827, then a school in Islington for 18 months. At the age of 14, Sylvester began attending the University of London, where he was a student of De Morgan; however, when he was only 5 months into his studies he was expelled for stealing a knife from the refectory in order to attack a fellow student. He then attended the Royal Institution in Liverpool and excelled academically, but he was tormented by his fellow students on account of his Jewish origins. Because of the abuse he received, he ran away, taking a boat to Dublin. There he was recognized on the street by Richard Keatinge, whose wife was a cousin of Sylvester. Keatinge arranged for him to return to Liverpool. In 1831, he entered John's College, Cambridge, to study mathematics; however, his studies were interrupted for almost 2 years during 1833–1835 due to a prolonged illness. In 1837, he was ranked second in Cambridge's famous mathematical examinations, the tripos, but he did not obtain a degree because graduates at that time were required to take the oath of the 39 Articles of the Church of England and Sylvester, being a Jew, refused. For the same reason he was not eligible to compete for Robert Smith's Prize. During 1838–1841, he held the Chair of Natural Philosophy at the University of London. Before stepping down, Sylvester had published 15 papers on fluid dynamics and algebraic equations. In 1841, he was awarded a BA and an MA by Trinity College, Dublin, where Jews were allowed to graduate. In the same year, Sylvester moved to the USA to become the Chair of Mathematics at the University of Virginia in Charlottesville; however, just a few months later he fled to New York because he struck a rowdy student with a stick and assumed that he had killed him. In New York he applied for university positions, but was not successful and therefore boarded a ship back to England in 1843. During this period he met a local girl and proposed to her, but she turned him down because of his Jewish religion. On his return to England he studied law, and worked as a Secretary at the Equity Law and Life Assurance Company. He also gave mathematics tuition, and one of his private pupils was Florence Nightingale (1820–1910), the British nurse considered to be the founder of modern medical nursing, who used statistical data and graphs as evidence in her fight to improve sanitation in hospitals. She used the polar-area diagram, her own invention, to dramatically show that British soldiers hospitalized in the Crimean War were more likely to die of infection and disease than their war wounds. Using these graphs, she lobbied the British government and public to make sanitary reforms, which saved many lives. In the early 1850s, Sylvester met Cayley in London, who was practicing law there. Both discussed mathematics, particularly matrix theory, as they walked around the courts, and although very different in temperament, they became life long friends. The term 'minor' in matrix theory is apparently due to Sylvester, who wrote in a paper published in 1850: "Now conceive any one line and any one column be struck out, we get ... a square, one term less in breadth and depth than the original square; and by varying in every possible selection of the line and column excluded, we obtain, supposing the original square to consist of n lines and n columns, n^2 such minor squares, each of which will represent what I term a First Minor Determinant relative to the principal or complete determinant." In 1851, Sylvester discovered the *discriminant* of a cubic equation and in 1852 and 1883 published the important papers *On the principle of the calculus of forms* and *On the theory of syzygetic relations and two rational integer functions*. In particular, he used matrix theory to study higher dimensional geometry. He also contributed to the creation of the theory of elementary divisors of lambda matrices. He did not hold any teaching position until 1855, when he was appointed Professor of Mathematics at the Royal Military Academy, Woolwich, from which he retired in 1869 at the age 55. In 1877, Sylvester accepted the Chair at

Johns Hopkins University, Baltimore, with a salary of $5,000 (quite generous for the time), which he demanded be paid in gold. During his stay at Johns Hopkins (1877–1883), he supervised nine doctorate students. In 1878, he founded the *American Journal of Mathematics*, the first mathematical journal in the USA. At the age of 68, Sylvester was appointed to the Savilian Chair of Geometry at Oxford. By 1892, he was partially blind and suffered from loss of memory, and he returned to London where he spent his last years at the Athenaeum Club.

Sylvester wrote papers on elimination theory, matrix theory, transformation theory, determinants, the theory of invariants (the word invariant describes the properties that are unaffected under transformations), the calculus of forms, partition theory and combinatorics, the theory of equations, multiple algebra, number theory, and probability theory. He is remembered for Chebyshev–Sylvester constant, the coin problem, the greedy algorithm for Egyptian fractions, Sylvester's sequence, Sylvester's formula for evaluating matrix functions, Sylvester's determinant theorem, the Sylvester matrix (resultant matrix), the Sylvester–Gallai (Tibor Gallai, 1912–1992) theorem, and Sylvester's law of inertia. One of Sylvester's lifelong passions was poetry; he read and translated works from the original French, German, Italian, Latin, and Greek, and many of his mathematical papers contain illustrative quotes from classical poetry. He published his *Laws of Verse* in 1870, applying principles of "phonetic syzygy," which was a curious little booklet of which he had a high regard. Sylvester became an FRS in 1839, was elected to Paris Academy of Sciences in 1863, made President of the London Mathematical Society in 1866, and became a Fellow of the Royal Society of Edinburgh in 1874. In 1861, he obtained the Royal Society's Royal Medal, and in 1880, the Royal Society of London awarded him the Copley Medal, its highest award for scientific achievement; in 1901, it instituted the Sylvester Medal in his memory, to encourage mathematical research. Sylvester House, a portion of an undergraduate dormitory at Johns Hopkins, is named in his honor. Often he is called "the Adam of mathematics."

Ada Augusta, Countess of Lovelace (1815–1852)

Ada Augusta, Countess of Lovelace (1815–1852) was the only child of the marriage of the famous poet Lord Byron (1788–1824) and Annabella Millbanke (1792–1860), who separated when Ada was 1 month old. She was raised by her mother, who encouraged her intellectual talents. She was taught by the mathematicians William Frend and Augustus De Morgan. In 1838, she married Lord King, who was later elevated to Earl of Lovelace (this title was created in 1838). Together they had three children. Ada Augusta continued her mathematical studies after marriage, assisting Babbage in his work on an early computing machine called the Analytic Engine. The most complete account of this machine is found in her writing. After 1845 she and Babbage worked toward the development of a system for predicting horse races. Unfortunately, their system did not work well, leaving Ada heavily in debt at the time of her death. The programming language Ada is named in honor of the Countess of Lovelace.

George Boole (1815–1864)

George Boole (1815–1864) was born in Lincoln, England. His father, John Boole (1777–1848), was a cobbler and a remarkable philosopher. George built optical instruments and invited people "to observe the works of God in a spirit of veneration." Once, George Boole's wife (a niece of Sir George Everest) said to her father-in-law, "He seemed to be capable of doing everything but look after his own business." Although Boole's family was well off, he did not get his education easily. He learned Latin and Greek, largely out of borrowed books, and at one time wanted to be a minister. At 16 he worked in an elementary school and by 20 had opened his own school. He occupied himself with mathematics only because books on mathematics were cheaper than those on the Classics. In his preparation for teaching mathematics, unsatisfied with textbooks of his day, Boole decided to read the works of great mathematicians such as Abel, Galöis, and Laplace. While reading papers by Lagrange he made discoveries in the calculus of variations, the branch of analysis that deals with finding curves and surfaces optimizing certain parameters. In 1849, Boole published *The Mathematical Analysis of Logic*, the first of his contributions to formal logic. Although this notion had already been used by Leibniz, it was Boole who first developed a practical logical algebra. His book, which aroused the admiration of de Morgan, freed him from the drudgery of elementary teaching. In 1849, he was appointed Professor of Mathematics at Queen's College in Cork, Ireland. In 1854, he published *An Investigation of the Laws of Thought*, his most famous work. In this work, Boole introduced what is now called 'Boolean Algebra', and showed how it could be applied to the theory of probability, among other fields. This work founded a new scientific discipline, formal logic, which may be considered to be either a part of philosophy or mathematics. Today, boolean algebra is widely used as a tool to aid sound reasoning and the design of electronic computers. (A computer is a device that can take a set of instructions, referred to as a program, and follow them to produce an output). His *Finite Differences* remains a standard work on this subject. In 1864, Boole died in Cork from pneumonia, which he contracted as a result of keeping a lecture engagement even though he was soaking wet from a rainstorm. One of his last wishes was that his children not be allowed to fall into the hands of those who were commonly thought religious. He is considered to be one of the most important pure mathematicians of the 1800s. In fact, Bertrand Russell once said "Pure mathematics was discovered by Boole, in a work which he called The Laws of Thought."

Karl Theodor Wilhelm Weierstrass (1815–1897)

Karl Theodor Wilhelm Weierstrass (1815–1897) was born in Ostenfelde, Germany. His father was a customs officer and had a broad knowledge of the arts and the sciences. In his youth, Weierstrass showed outstanding skill in language and mathematics. At the urging of his dominant father, Weierstrass entered the law and commerce program at the University of Bonn. To the chagrin of his family, the rugged and congenial young man concentrated instead on fencing and beer drinking. Four years later he returned home without a degree. In 1839, Weierstrass entered the Academy of Münster to study for a career in secondary education, and he met and studied under an excellent mathematician named Christof Gudermann (1798–1852). Gudermann's ideas greatly influenced the work of Weierstrass. After receiving his teaching certificate, Weierstrass spent the next 15 years in secondary education teaching German, geography, and mathematics. He also taught handwriting to small children. During this period much of Weierstrass's mathematical work on Abelian functions was ignored because he was a secondary school teacher and not a college professor. Then in 1854 he published a paper of major importance, *Zur Theorie der Abelschen Functionen*, in Crelle's Journal, which created a sensation in the mathematics world and catapulted him to international fame overnight. He was immediately given an honorary Doctorate at the University of Königsberg and began a new career in college, teaching at the University of Berlin in 1856. In 1859, the strain of his mathematical research caused a temporary nervous breakdown and led to spells of dizziness that plagued him for the rest of life. In 1860, he proved the Max–Min Theorem for continuous functions, a theorem that earlier mathematicians had assumed without proof. In 1863, he proved that the complex numbers are the only commutative algebraic extension of the real numbers. Gauss had promised a proof of this in 1831 but failed to give one. In 1870, Sofia Kovalevskaya came to Berlin and Weierstrass taught her privately because she was not allowed admission to the university. It was through Weierstrass' efforts that Kovalevskaya received an honorary doctorate from Göttingen. He also used his influence to help her obtain a post in Stockholm in 1883. Weierstrass and Kovalevskaya corresponded for 20 years, between 1871 and 1890. More than 160 letters were exchanged, but Weierstrass burned Kovalevskaya's letters after her death. She was the closest any woman ever came to claiming the heart of this lifelong bachelor. In 1872, his rigorous work led him to discover a function that, although continuous, had no derivative at any point (Bolzano had already given such a function in 1843). He became an FRS in 1881 and won the Copley Medal in 1895.

Weierstrass was a brilliant teacher and his classes overflowed with a multitude of auditors. According to him, the teacher should let science develop before the eyes of his pupils. As it develops and takes form in the mind of the mature thinker, out of his fundamental ideas, so shall he present it, merely adjusting it to the youthful power of understanding. In spite of his fame, he never lost his early beer-drinking congeniality and was always in the company of students both ordinary and brilliant, such as Cantor, Ferdinand Georg Frobenius (1849–1917, a German mathematician best known for his contributions to the theory of elliptic functions, differential equations, matrix rank, and group theory), Hölder, Adolf Hurwitz (1859–1919), Klein, Adolf Kneser (1862–1930), Lie, Minkowski, Swede Gösta Mittag-Leffler (1846–1927), and Schwarz. Weierstrass has been acknowledged as the "father of modern analysis." He used the power series as the basis of functions, an idea that was key in the development of much of mathematical physics. Weierstrass devised tests for the convergence of series, created the sequential definition of irrational numbers based on convergent series, contributed to the theory of periodic functions, functions of real variables, elliptic functions, Abelian functions, converging infinite products, and the calculus of variations. He also advanced the theory of bilinear and quadratic forms. The present $\epsilon - \delta$ definitions of limit and continuity are due to Weierstrass. When traveling, he kept his unfinished papers and working notes in a large white wooden box. The box was lost while Weierstrass was on a trip in 1880 and never rediscovered. In 1885, Klein remarked that Weierstrass "arithmetized analysis," that is, he freed analysis from geometrical reasoning. Weierstrass is quoted as saying, "It is true that a mathematician who is not also something of a poet will never be a perfect mathematician."

Kummer, Kronecker, and Weierstrass together gave Berlin a reputation as the leading center of mathematics in the world. Although Kronecker was a close friend of Weierstrass for many years, in 1877 Kronecker's opposition to Cantor's work caused a rift between the two men. This became so bad that in 1885 Weierstrass decided to leave Berlin and go to Switzerland. However, he changed his mind and remained in Berlin. Weierstrass spent the last few years of his life confined to a wheelchair and died from pneumonia on February 19, 1897. His last wish was that the priests say nothing in his praise at the funeral, but restrict the services to the customary prayers.

In memory of Weierstrass, Hilbert gave the following address: "The infinite! No other question has ever moved so profoundly the spirit of man; no other idea has so fruitfully stimulated his intellect; yet no other concept stands in greater need of clarification than that of the infinite..."

"When we turn to the question, what is the essence of the infinite, we must first give ourselves an account as to the meaning the infinite has for reality: let us then see what physics teaches us about it."

"The first naive impression of nature and matter is that of continuity. Be it a piece of metal or a fluid volume, we cannot escape the conviction that it is divisible into infinity, and that any of its parts, however small, will have the properties of the whole. But wherever the method of investigation into the physics of matter has been carried sufficiently far, we have invariably struck a limit of divisibility, and this was not due to a lack of experimental refinement but resided in the very nature of the phenomenon. One can indeed regard this emancipation from the infinite as a tendency of modern science and substitute for the old adage *natura non facit saltus* its opposite: Nature does make jumps..."

"It is well known that matter consists of small particles, the atoms, and that the macrocosmic phenomena are but manifestations of combinations and interactions among these atoms. But physics did not stop there: at the end of the last century it discovered atomic electricity of a still stranger behavior. Although up to then it had been held that electricity was a fluid and acted as a kind of continuous eye, it became clear then that electricity, too, is built up of positive and negative electrons."

"Now besides matter and electricity there exists in physics another reality, for which the law of conservation holds; namely energy. But even energy, it was found, does not admit of simple and unlimited divisibility. Max Planck (1858–1947) discovered the *energy—quanta*."

"And the verdict is that nowhere in reality does there exist a homogeneous continuum in which unlimited divisibility is possible, in which the infinitely small can be realized. The infinite divisibility of a continuum is an operation which exists in thought only, is just an idea, an idea which is refuted by our observations of nature, as well as by physical and chemical experiments."

"The second place in which we encounter the problem of the infinite in nature is when we regard the universe as a whole. Let us then examine the extension of this universe to ascertain whether there exists there an infinite great. The opinion that the world was infinite was a dominant idea for a long time. Up to Kant and even afterward, few expressed any doubt in the infinitude of the universe."

"Here too modern science, particularly astronomy, raised the issue anew and endeavored to decide it not by means of inadequate metaphysical speculations, but on grounds which rest on experience and on the application of the laws of nature. There arose weighty objections against the infinitude of the universe. It is Euclidean geometry that leads to infinite space as a necessity... Einstein showed that Euclidean geometry must be given up. He considered this cosmological question too from the standpoint of his gravitational theory and demonstrated the possibility of a finite world; and all the results discovered by the astronomers are consistent with this hypothesis of an elliptic universe."

John Couch Adams (1819–1892)

John Couch Adams (1819–1892) was born in Lidcott, England, to a farming family. When he was very young he could do very difficult problems in his head and became known for his ability for accurate mental numerical calculations. In 1835, while at Landulph, he observed Halley's comet. He determined that an annular eclipse of the Sun would be visible in Lidcot in 1836. In 1839, Adams began his undergraduate mathematics course at St John's College, Cambridge, and graduated as Senior Wrangler in 1843, winning the first prize in Greek testament every year. The same year he became the first Smith's Prizeman and a Fellow of St John's College. During this period he tutored undergraduates in order to earn money, which he sent home to help his brother's education. In 1841, while an undergraduate student, Adams used mathematics to predict the presence of a new planet beyond Uranus. A few months later, the French astronomer Le Verrier also made the same prediction. In September of 1845, Adams gave accurate information on the position of this new planet to James Challis, then the director of the Cambridge Observatory. Now most of the scientists consider Adams and Le Verrier to be the codiscoverers of the planet Neptune in 1846. After a year of unemployment, Adams was elected to a fellowship at Pembroke College, Cambridge in 1853, which he held until his death. In addition, he became Regius Professor of Mathematics at St. Andrews in 1857. In 1859, he succeeded George Peacock as Lowndean (Thomas Lowndes, 1692–1748) Professor of Astronomy and Geometry at Cambridge and held the post for over 32 years. In 1861, he became the director of the Cambridge Observatory. In 1863, he married Elizabeth Bruce, from Dublin, who was a friend of Stokes' wife. In 1864, Adams correctly predicted that the Leonid meteor shower would occur in November of 1866. He discovered that this was due to Tempel's comet, which comes very close to Earth once a year. Adams also studied the motion of the Moon, giving a theory that was more accurate than Laplace's. He also studied terrestrial magnetism, determined the Gaussian magnetic constants at every point on the Earth, and produced maps with contour lines of equal magnetic variation that were published after his death. His procedure for numerical integration of differential equations appeared in 1883 in a book with Francis Bashforth (1819–1912) on capillary action. Adams calculated Euler's constant to 236 decimal places. He wrote about 50 research papers, 11 of which were on pure mathematics.

Adams was happily married, profoundly devout, and enjoyed social visits, music, dancing, parties, and long daily walks. He was awarded honorary degrees from Oxford, Dublin, Edinburgh, and Bologna; however, he refused to be knighted, which was offered to him in 1847. He won the Copley Medal in 1848, FRS in 1849, and Fellow of the Royal Society of Edinburgh in 1849. After a long illness, Adams died in 1892 in Cambridge, England.

George Gabriel Stokes (1819–1903)

George Gabriel Stokes (1819–1903) was born in County Sligo in the northwestern part of Ireland, the youngest of six children. His father was a minister and his mother was a minister's daughter. He had his basic education in schools in Skreen, Dublin, and Bristol, before he matriculated in 1837 at Pembroke College, Cambridge. After 4 years at Cambridge, Stokes graduated at the top of his class, with mathematics as his major field of study. He was then awarded a fellowship that allowed him to begin research in fluid dynamics. He also worked on the physics of light and on problems in geodesy. In 1849, Stokes was appointed Lucasian Professor of Mathematics, and 2 years later, he was elected to the Royal Society. A year after that, he was awarded the Rumsford (Count Rumford, 1753–1814) Medal of the Royal Society. After another 2 years, he became Secretary of the Royal Society, a position he held until becoming President of the Royal Society in 1885. He continued as President until 1890 and received the Society's Copley Medal in 1893. He was also President of the Victoria Institute from 1896 until he died in 1903. His theoretical and experimental investigations covered hydrodynamics, elasticity, light, gravity, sound, heat, meteorology, and solar physics. He is remembered for Stokes' law in fluid dynamics; the Stokes' radius in biochemistry; Stokes' theorem in differential geometry; the Stokes line in Raman scattering; Stokes relations, which relate the phase of light reflected from a non-absorbing boundary; the Stokes shift in fluorescence; the Navier–Stokes equations, the Stokes drift, the Stokes stream function, and the Stokes boundary layer in fluid dynamics; the Stokes phenomenon in asymptotic analysis; the Stokes unit (a unit of viscosity); Stokes parameters and Stokes vectors, which are used to quantify the polarization of electromagnetic waves; and the Campbell–Stokes (John Francis Campbell, 1821–1885) recorder, an instrument for recording sunshine that was improved by Stokes and is still widely used today. It is one of those delightful quirks of history that the theorem we call Stokes' theorem in vector calculus is not his theorem at all. He learned of it from Thomson in 1850 and a few years later included it among the questions on an examination he wrote for the Smith Prize. It has been known as Stokes' theorem ever since. As usual, things have balanced out. Stokes was the original discoverer of the principles of spectrum analysis that we now credit to Bunsen and Kirchhoff. Stokes' mathematical and physical papers were published in a series of five volumes: the first three (Cambridge, 1880, 1883, and 1901) under his own editorship and the last two (Cambridge, 1904 and 1905) under that of Sir Joseph Larmor (1857–1942). The Stokes craters on the Moon and Mars are named in his honor.

George Salmon (1819–1904)

George Salmon (1819–1904) was born in Dublin, but he spent his boyhood in Cork City, Ireland. During his long life he was connected with Trinity College (Dublin). He graduated in 1839 with very high honors in mathematics, and then became an instructor of both mathematics and divinity. He married Frances Anne in 1844, with whom he had six children; however, only two survived. In the late 1840s–1850s Salmon published about 36 papers in algebraic geometry. His main merit lies in his well-known textbooks, which excel in clarity and charm. These books opened the road to analytical geometry and invariant theory to several generations of students in many countries and even now have hardly been surpassed. These are *Conic Sections* (1848), *Higher Plane Curves* (1852), *Modern Higher Algebra* (1859), and the *Analytic Geometry of Three Dimensions* (1862). In 1863, Salmon was elected an FRS, in 1868 he was awarded their Royal Medal for his research in analytical geometry and the theory of surfaces, and in 1889, he received the Copley Medal. Salmon became Provost of Trinity College in 1888 and remained at this post until his death in 1904. His books on theology, *Infallibility of the Church* and *An Historical Introduction to the Study of the Books of the New Testament*, were also widely read. Salmon was a keen chess player and was President of the Dublin Chess Club during 1890–1903.

Heinrich Edward Heine (1821–1881)

Heinrich Edward Heine (1821–1881) was born in Berlin. Heine was a student of Dirichlet and a colleague of Cantor at Halle. He is known for results on special functions and in real analysis. His classical treatise on spherical harmonics and

Legendre functions, *Handbuch der Kugelfunctionen*, is famous. He is known for the Heine–Borel theorem, the Heine–Cantor theorem, the Heine definition of continuity, and Heine's Reciprocal Square Root Identity. He died in Halle.

Pafnuty Lvovich Chebyshev (1821–1894)

Pafnuty Lvovich Chebyshev (1821–1894) was the most eminent Russian mathematician of the nineteenth century. He holds the record for the most variant spellings of his last name. Chebyshev was a contemporary of the famous geometer Lobachevsky, but his work had a much deeper influence throughout Western Europe and he is considered to be the founder of the great school of mathematics that has flourished in Russia for the past century. Chebyshev was born into the gentry in Okatovo, Russia. His father was a retired army officer who fought against Napoleon. In 1832, the family and their nine children moved to Moscow, where Chebyshev completed his high school education at home. As a boy, he was fascinated by mechanical toys and was first attracted to mathematics when he saw the importance of geometry for understanding machines. After his years as a student in Moscow he became Professor of Mathematics at the University of St. Petersburg, a position that he held until his retirement. His father was a member of the Russian nobility, but after the famine of 1840 the family estate was so diminished that for the rest of his life Chebyshev was forced to live very frugally, and he never married. He spent much of his small income on mechanical models and occasional journeys to Western Europe, where he particularly enjoyed seeing windmills, steam engines, and the like.

Chebyshev was a remarkably versatile mathematician with a rare talent for solving difficult problems using elementary methods. Most of his effort went into pure mathematics, but he also valued practical applications of his subject, as the following remark suggests: "To isolate mathematics from the practical demands of the sciences is to invite the sterility of a cow shut away from the bulls." He worked in many fields, but his most important achievements were in probability, the theory of numbers, and the approximation of functions, which he was led to by his interest in mechanisms. In probability, he introduced the concept of mathematical expectations and variance, gave a beautifully simple proof of the law of large numbers based on what is now known as Chebyshev's inequality [also independently developed by the statistician **Jules Bienaymé** (1796–1878)], and worked extensively on the central limit theorem. Complementing Newton's result that there exist three relations among the exponents m, s, and p (rational numbers) so that the integral

$$\int x^m(a + bx^s)^p dx$$

can be expressed in terms of elementary functions, Chebyshev proved that in all other cases the above integral cannot be expressed in terms of elementary functions.

In the late 1840s Chebyshev helped to prepare an edition of some of the works of Euler. It appears that this task caused him to turn his attention to the theory of numbers, particularly to the very difficult problem of the distribution of primes. The primes are distributed among all the positive integers in a rather irregular way, for as we move out they seem to occur less and less frequently, and yet there are many adjoining pairs separated by a single even number. The problem of discovering the law governing their occurrence, and of understanding the reasons for it, is one that has challenged the curiosity of men for hundreds of years. In 1751, Euler expressed his own bafflement in these words: "Mathematicians have tried in vain to this day to discover some order in the sequence of prime numbers, and we have reason to believe that it is a mystery into which the human mind will never penetrate." Many attempts have been made to find simple formulas for the nth prime and for the exact number of primes among the first n positive integers. All such efforts have failed, and real progress was achieved only when mathematicians started to look instead for information about the average distribution of primes among the positive integers. It is customary to denote the number of primes less than or equal to a positive number x by $\pi(x)$. Thus, $\pi(1) = 0$, $\pi(2) = 1$, $\pi(3) = 2$, $\pi(4) = 2$, and so on. In his youth Gauss studied $\pi(x)$ empirically, with the aim of finding a simple function that seems to approximate it with a small relative error for large x. On the basis of his observations he conjectured (perhaps at the age of 14 or 15) that $x/\ln x$ is a good approximation function, in the sense that

$$\lim_{x\to\infty} \frac{\pi(x)}{(x/\ln x)} = 1. \tag{2}$$

This statement is the famous *prime number theorem*; and as far as anyone knows, Gauss was never able to support his guess with even a fragment of proof. Chebyshev, unaware of Gauss' conjecture, was the first mathematician to establish firm conclusions about this question. In 1848 and 1850, he proved that

$$0.9213\ldots < \frac{\pi(x)}{x/\ln x} < 1.1055\ldots$$

for all sufficiently large x, and also that if the limit in (2) exists, then its value must be 1. As a by-product of his work, he also proved Bertrand's postulate of 1845: for every integer $n \geq 1$ there is a prime p such that $n < p \leq 2n$. Chebyshev's efforts did not bring him to a final proof of the prime number theorem (this came in 1896), but they did stimulate many other mathematicians' continued work on the problem. He is regarded as the intellectual father of a long series of well-known Russian scientists who contributed to the mathematical theory of probability, including Markov, Sergei Natanovich Bernstein (1880–1968), Kolmogorov, Aleksandr Yakovlevich Khinchin (1894–1959), and others.

Arthur Cayley (1821–1895)

Arthur Cayley (1821–1895) was born in Richmond, Surrey, England. Cayley spent his first 8 years in St. Petersburg, Russia, because his father was a merchant there. In 1829, they moved to Blackheath, near London. There, Cayley first attended a private school and at the age of 14 went to King's College School. Right from the beginning he showed great skill in numerical calculation and an aptitude for advanced mathematics, which were encouraged by his mathematics teacher. In 1838, instead of entering his father's business, Cayley began his studies at Trinity College, Cambridge, where he graduated in 1842. During this period, although he was an undergraduate student, he published three papers in the newly founded Cambridge Mathematical Journal, edited by Duncan Farquharson Gregory (1813–1844), on subjects that had been suggested by his reading of the Mécanique analytique by Lagrange and some of the works of Laplace. Cayley excelled in Greek, French, German, and Italian, and graduated as Senior Wrangler and won the first Smith's Prize. He continued to reside at Cambridge for the next 4 years on a fellowship, and during this period he published 28 papers in the Cambridge Mathematical Journal. After completing this fellowship, Cayley, like De Morgan, chose law for his profession and at the age of 25 entered Lincoln's Inn, London. He worked as a successful lawyer for 14 years from 1848 to 1862; however, he always considered it to be a means for making money so that he could pursue mathematics. During these 14 years he published about 250 mathematical papers. Sylvester, who was also in the legal profession at that time, became Cayley's friend. They used to walk together round the courts of Lincoln's Inn, discussing the theory of invariants and covariants. In 1863, Cayley became the newly created Sadlerian Professor (Lady Mary Sadleir, died 1706) of Pure Mathematics at Cambridge and gave up his lucrative practice for a modest salary. His duty there was to explain and teach the principles of pure mathematics and to apply himself to the advancement of that science. In 1872, he was made an honorary fellow of Trinity College, and 3 years later an ordinary fellow. In 1876, he published a treatise on *Elliptic Functions*, which was his only book. During the first 5 months of 1882, he gave lectures on Abelian and Theta Functions at Johns Hopkins University, Baltimore, where his friend Sylvester was then Professor of Mathematics. In 1883, Cayley became President of the British Association for the Advancement of Science.

Cayley is ranked third among the most prolific writers of mathematics. He published over 900 research papers as well as works on the theories of matrices and skew surfaces. He touched and enriched almost every area in pure mathematics. In addition to matrix algebra and invariant theory, he made pioneering contributions to analytical geometry, transformation theory, the theory of determinants, higher-dimensional geometry, partition theory, the theory of curves and surfaces, and the theory of Abelian, theta, and elliptic functions. His development of n-dimensional geometry has been applied in physics to the study of the space-time continuum. His work on matrices served as a foundation for quantum mechanics, which was developed by Heisenberg in 1925. Cayley also suggested that Euclidean and non-Euclidean geometry are special types of geometry. He united projective geometry and metric geometry, which is dependent on sizes of angles and lengths of lines. He was the first to define the concept of a group in the modern way: as a set with a binary operation satisfying certain laws. He is often remembered for Cayley's theorem, the Cayley–Hamilton theorem in linear algebra (Cayley proved this theorem for 2×2 matrices and asserted its truth in the general case, whereas Hamilton proved it for 4×4 matrices), Grassmann–Cayley algebra, the Cayley–Menger determinant (Karl Menger, 1902–1985), Cayley–Dickson construction, Cayley algebra, Cayley graph, Cayley table, Cayley–Purser algorithm (Michael Purser), Cayley's formula, Cayley–Klein model of hyperbolic geometry, Cayley transform, and Cayley's sextic. By developing algebras that satisfied structural laws different from those obeyed by the common algebra of arithmetic, Hamilton, Grassmann, and Cayley opened the flood gates of modern abstract algebra. He became an FRS in 1852, a Fellow of the Royal Society of Edinburgh in 1865, and the President of the London Mathematical Society during 1868–1870. He was awarded the Royal Society's Royal Medal in 1859, the Copley Medal in 1882, and the London Mathematical Society's De Morgan Medal in 1884. However, sometimes even brilliant mathematicians have feet of clay. For example, Cayley asserted that if the product of two nonzero square matrices, A and B, is zero, then at least one of the factors must be singular. Cayley was correct, but surprisingly he overlooked an important point, namely, that if $AB = 0$, then A and B must both be singular.

Cayley took special pleasure in paintings and architecture, and he practiced watercolor painting, which he found useful for making mathematical diagrams. During his life he read thousands of novels. He was an amateur mountain climber, making frequent trips to the Continent for long walks and mountain scaling. His massive *Collected Mathematical Papers* were published in 13 large volumes by the Cambridge University Press during 1889–1897. The first seven were edited by Cayley himself, and the remaining six were edited by Andrew Forsyth, his successor in the Sadlerian chair. While editing these volumes, he suffered from a painful internal malady. He died in 1895 at the age of 74. During his funeral a great assemblage met in Trinity Chapel, comprising members of the University, official representatives of Russia and America, and many of the most illustrious philosophers of Britain. According to Hermite, the mathematical talent of Cayley was characterized by clarity and extreme elegance of analytical form; it was reinforced by an incomparable capacity for work that has caused the distinguished scholar to be compared with Cauchy. He has been called "the mathematicians' mathematician."

Phillip Ludwig von Seidel (1821–1896)

Phillip Ludwig von Seidel (1821–1896) was born in Zweibrücken, Germany. He studied in Berlin, Königsberg, and Munich. Among his teachers were Bessel, Jacobi, Dirichlet, and Johann Franz Encke (1791–1865), a German astronomer. He was an assistant to Jacobi and helped him solve problems regarding a system of linear equations arising out of the work on least squares by Gauss. Phillip became professor at the University of Munich in 1855. In 1857, he decomposed the first order monochromatic aberrations into five constituent aberrations which are now known as Seidel aberrations. He is credited for the Gauss–Seidel iteration scheme for a linear system, which he published in 1874 (it is unclear how Gauss' name became associated with it, since Gauss, who was aware of the method much earlier, declared it to be worthless!), that has been used more extensively with the advent of digital computers. He also worked in analysis, astronomy, probability theory, and optics. Unfortunately, Phillip suffered from eye problems and retired at a young age. He died in Munich.

Gregor Johann Mendel (1822–1884)

Gregor Johann Mendel (1822–1884) was born in Hyncice, Austrian Silesia, Austria (now Czech Republic), to a peasant family that managed to send him to school. In 1843, he entered the Augustinian Abbey of St. Thomas in Brunn, present-day Brno, Czech Republic, and by 1847 he had been ordained and had begun teaching there. In 1851, he went to the University of Vienna to study and returned to the Abbey in 1853 as a teacher of mathematics and natural science. Mendel experimented on plants in the abbey garden, especially pea plants, whose distinct traits (characteristic features) he puzzled over. In his findings, which were published in 1866, he showed the existence of paired elementary units of heredity (now called genes) and established the statistical laws governing them. Although Mendel is now called the "father of modern genetics," his work was not recognized at that time because the common belief was that pangenes were responsible for inheritance. After completing his work with peas, he began experimenting with honeybees in order to extend his work to animals. He produced a hybrid strain, but failed to generate a clear picture of their heredity because of the difficulty of controlling the mating behaviors of queen bees. Mendel was elevated as abbot in 1868, which left him no time for further research. He died in Brno, Austria–Hungary (now Czech Republic), from chronic nephritis.

Charles Hermite (1822–1901)

Charles Hermite (1822–1901) was born to a well-to-do middle class family in Dieuze, Lorraine, France, with a deformity of his right leg. Hermite was lame all his life, requiring a cane to get about. As a student, he courted disaster by neglecting his assigned work and instead by studying the classic masters of mathematics; and though he nearly failed his examinations, he became a first-rate creative mathematician while still in his early 20s. In 1870, he was appointed to a professorship at the Sorbonne, where he trained a whole generation of well-known French mathematicians, including Picard, Borel, and Poincaré.

The unusual character of his mind is suggested by the following remark by Poincaré: "Talk with Mr. Hermite. He never evokes a concrete image, yet you soon perceive that the most abstract entities are to him like living creatures." He disliked geometry, but was strongly attracted to number theory and analysis, and his favorite subject was elliptic functions; these two fields touch in many remarkable ways. Although Abel proved many years before that the general polynomial equation of the fifth degree cannot be solved by functions involving only rational operations and root extractions, one of Hermite's most surprising achievements (in 1858) was his discovery that this equation can be solved by elliptic functions. His 1873 proof of the transcendence of e was another high point in his career.

Several of his purely mathematical discoveries had unexpected applications to mathematical physics many years later. For example, the Hermitian forms and matrices that he invented in connection with certain problems of number theory turned out to be crucial for Heisenberg's 1925 formulation of quantum mechanics, and Hermite polynomials and Hermite functions are useful in solving Schrödinger's wave equation.

Hermite was one of the most eminent French mathematicians of the nineteenth century, who is particularly distinguished for the elegance and high artistic quality of his work. He died in Paris in 1901. According to him, we are servants rather than masters in mathematics.

Francis Galton (1822–1911)

Francis Galton (1822–1911) was born in Sparkbrook (near Birmingham), England. He was Darwin's half-cousin, sharing a common grandparent. Galton was a child prodigy. He learned to read at age three, at age five he knew some Greek, Latin, and long division, and by the age of six he had moved on to adult books, including poetry, and Shakespeare for pleasure. After attending several schools, he studied for 2 years at Birmingham General Hospital and King's College London Medical School. During 1840–1844, Galton was interested in mathematics and mechanics, but was an indifferent mathematics student at Trinity College, Cambridge. In 1847, he was awarded a Master of Arts degree without further study. After several years spent as a traveling gentleman of leisure, Galton wrote popular books on his experience, *Narrative of an Explorer in Tropical South Africa*, and *The Art of Travel*, which has gone through many editions. In 1853, he married and settled down in South Kensington. Galton produced over 340 papers and books throughout his lifetime, in many fields of science. In meteorology, he described the anticyclone and pioneered the introduction of weather maps based on charted air pressure data. In statistics, his key work is *Natural Inheritance*, in which he set forth his ideas on regression and correlation, and posed the problem of multiple regression. He showed that the study of heredity could only be placed on a scientific basis by introducing new statistical concepts like regression and correlation. As an investigator of the human mind he founded psychometrics, the science of measuring mental faculties, and differential psychology, sometimes called the "London School" of experimental psychology. In eugenics, he coined the very term itself as well as the phrase "nature versus nurture" and campaigned extensively for the improvement of the human stock. In criminology, he devised a method for classifying fingerprints that proved useful in forensic science. In biology, he studied the nature and mechanism of heredity. Galton is also remembered for the bell curve, the product-moment correlation coefficient, the importance of quickness of response (reaction time), his proof that a normal mixture of normal distributions is itself normal, and for testing differential hearing ability.

In 1853, Galton was awarded the Royal Geographical Society's gold medal for his explorations and mapmaking of southwest Africa, in 1855 he became a member of the Athenaeum Club, in 1860 he was made an FRS, and he was knighted in 1909. He died in 1911 in Grayshott House, Haslemere, Surrey, England. His statistical heir, Pearson, first holder of the Galton Chair of Eugenics at University College London, wrote a three-volume biography of Galton after his death. Despite his colossal achievements, contemporary reputation, and far-reaching influence, Sir Francis Galton is no longer widely known or appreciated except among specialists.

Leopold Kronecker (1823–1891)

Leopold Kronecker (1823–1891) was born in Liegnitz, Prussia (now Legnica, Poland). His father, Isidor Kronecker, was a successful businessman and his mother, Johanna Prausnitzer, came from a wealthy family. They were Jewish, a religion that Kronecker kept until a year before his death, when he converted to Christianity. Because of his early talent for mathematics, his parents hired him a private tutor. Kronecker later entered the Liegnitz Gymnasium, where Kummer was his math teacher. Kummer encouraged him to explore mathematics on his own as well as in the classroom. In 1841, Kronecker became a student at Berlin University and studied under Dirichlet and Steiner. Besides mathematics, he was also attracted to astronomy, meteorology, chemistry, and philosophy. He spent some time at the University of Bonn, and then at the University of Breslau, where Kummer had been appointed to a Chair in 1842. In 1845, at the University of Berlin, Kronecker wrote his dissertation on number theory under the supervision of Dirichlet, giving special formulation to units in certain algebraic number fields. After obtaining his degree, Kronecker managed the estate and business of his uncle. In 1848, he married Fanny Prausnitzer, the daughter of this uncle. He produced nothing mathematical for 8 years. In his 1853 memoir on the algebraic solvability of equations, Kronecker extended Galöis' work on the theory of equations. He studied continuity and irrational numbers, and vigorously opposed Weierstrass' continuous nondifferentiable function. In an 1858 paper, *On the Solution of the General Equation of the Fifth Degree*, Kronecker obtained a solution to the quintic equation similar to Hermite's using group theory. Although he held no official position there, from 1861 to 1883 Kronecker lectured at the University of Berlin. In his lectures he tried to simplify and refine existing theories and to present them from new perspectives. However, he could not attract many students, and only a few continued until the end of the semester. Berlin was so attractive to Kronecker that he declined the Chair of Mathematics in Göttingen in 1868. In 1883, he succeeded Kummer, who filled the vacancy that opened when Dirichlet left for Göttingen in 1855.

Kronecker's primary contributions were in the theory of equations and higher algebra; he also contributed to elliptic functions, the theory of algebraic equations, and the theory of algebraic numbers. The function δ_{ik}, defined as 0 if $i \neq k$ and 1 if $i = k$, is called Kronecker's delta. He is also remembered for the Kronecker product, the Kronecker–Weber theorem, Kronecker's theorem in number theory, and Kronecker's lemma (a lemma is a proposition that is assumed or demonstrated, preliminary to the demonstration of some other proposition). He also supervised Kurt Hensel (1861–1941), Kneser, Mathias Lerch (1860–1922), and Franz Mertens (1840–1927). He believed that mathematics should deal only with whole numbers and with a finite number of operations and is credited with saying, "God made the natural numbers; all else is the work of man." He felt that irrational, imaginary, and all other numbers excluding the positive integers were man's work and therefore unreliable. He doubted the significance of nonconstructive existence proofs, and from the early 1870s opposed the use of irrational numbers, upper and lower limits, the Bolzano–Weierstrass theorem, and believed that transcendental numbers do not exist. Kronecker's views on mathematics weren't similar to others of his time. Heine, Cantor, Lindemann, Dedekind, and Weierstrass were just some of the mathematicians who held beliefs in mathematics different from those of Kronecker. He tried to persuade Heine to withdraw his 1870 paper on the trigonometric series in Crelle's Journal, and in 1877, he tried to prevent the publication of Cantor's work in the same journal. In fact, he always vociferously opposed Cantor's work. Lindemann had proved that π is transcendental in 1882, and in a lecture given in 1886 Kronecker complimented Lindemann on a beautiful proof but, he claimed, one that proved nothing since transcendental numbers do not exist. He created some friction between himself and Weierstrass; in 1888, Weierstrass felt that he could no longer work with Kronecker in Berlin and decided to go to Switzerland, but then, realizing that Kronecker would be in a strong position to influence the choice of his successor, he decided to remain in Berlin. However, Hermann von Helmholtz (1821–1894), who was a professor in Berlin from 1871, managed to stay on good terms with Kronecker. Helmholtz made significant contributions in physiology, physiological optics and acoustics, electricity and magnetism, thermodynamics, theoretical mechanics, hydrodynamics, physical optics, the theory of heat, chemistry, mathematics, meteorology, biology, and psychology.

Despite the disagreements between Kronecker and other mathematicians, Kronecker was still able to gain international fame. He became a codirector of a mathematical seminar in Berlin, which increased his influence there. In 1884, he was honored by being elected a foreign member of the Royal Society of London. He was a very influential figure within German mathematics.

Kronecker was of very small stature and was extremely self-conscious about his height. He died in 1891 and was buried in the St. Matthäus Kirchhoff Cemetery in Schöneberg, Berlin, close to Gustav Kirchhoff. Although he did not make many discoveries, he will always be remembered for breaking down his predecessors' work and adding new insights.

Oskar Xaver Schlömlich (1823–1901)

Oskar Xaver Schlömlich (1823–1901) was a German analyst born in Weimar. He is recognized for solving differential equations using infinite series and for his work with Bessel functions. Schlömlich was the founding editor of the prestigious journal Zeitschrift für Mathematik und Physik (1856). He is often remembered for the Schlömlich remainder, the Schlömlich series, and the Schlömlich equation.

Johann Martin Zacharias Dase (1824–1861)

Johann Martin Zacharias Dase (1824–1861) was born in Hamburg, Germany. He was a calculating prodigy. (Prodigies have extraordinary ability in some area of mental calculation, such as multiplying large numbers, factoring large numbers, or finding roots of large numbers). Johann attended schools in Hamburg, but he made only a little progress there. He used to spend a lot of time developing his calculating skills; people around Johann found him quite dull. He suffered from epilepsy throughout his life, beginning in his early childhood. At the age of 15 he gave exhibitions in Germany, Austria, and England. His extraordinary calculating powers were timed by renowned mathematicians including Gauss. He multiplied 79,532,853 × 93,758,479 in 54 s; two 20-digit numbers in 6 min; two 40-digit numbers in 40 min; and two 100-digit numbers in 8 h 45 min. In 1840, he struck up an acquaintance with L.K. Schulz von Strasznicky (1803–1852), who suggested that he apply his powers to scientific purposes. When he was 20, Strasznicky taught him the use of the formula

$$\frac{\pi}{4} = \tan^{-1}\left(\frac{1}{2}\right) + \tan^{-1}\left(\frac{1}{5}\right) + \tan^{-1}\left(\frac{1}{8}\right)$$

and asked him to calculate π. In 2 months he carried the approximation to 205 decimal places, of which 200 are correct. He then calculated a 7-digit logarithm table of the first 1,005,000 numbers during his off-time from 1844 to 1847, while he was also occupied by work on the Prussian survey. His next contribution was the compilation of a hyperbolic table in his spare time, which was published by the Austrian Government in 1857. Next, he offered to make a table of integer factors of all numbers from 7,000,000 to 10,000,000; on Gauss' recommendation, the

Hamburg Academy of Sciences agreed to compensate him, but Dase died shortly thereafter in Hamburg. He had an uncanny sense of quantity; he could just tell, without counting, how many sheep were in a field or words in a sentence, and so forth, up to quantities of about 30. Several other calculating prodigies have been discovered, such as Jedediah Buxton (1707–1772), Thomas Fuller (1710–1790), André-Marie Ampére, Richard Whately (1787–1863), Zerah Colburn (1804–1840), George Parker Bidder (1806–1878), Truman Henry Safford (1836–1901), Jacques Inaudi (1867–1950), and Shakuntala Devi (1929–2013) of India.

Gustav Robert Kirchhoff (1824–1887)

Gustav Robert Kirchhoff (1824–1887) was born in Königsberg, East Prussia (now Kaliningrad, Russia) to a flourishing intellectual family. After completing his schooling, he entered the Albertus University of Königsberg, where he graduated in 1847. During his studies he regularly attended the mathematico–physical seminar directed by Franz Ernst Neumann (1798–1895) and Jacobi. In 1845, with the guidance of Neumann, Kirchhoff formulated his electric circuit laws: The current at each point in a network may be determined by solving the equations that result from

1. the sum of the currents into (or away from) any point is zero, and
2. around any closed path, the sum of the instantaneous voltage drops in a specified direction is zero.

These laws are familiar to every student of elementary physics. In 1847, he married Clara Richelot, the daughter of his mathematics Professor Friedrich Julius Richelot (1808–1875), and the couple moved to Berlin. After working as a privatdozent in Berlin for some time, he became extraordinary Professor of Physics at Breslau in 1850. Immediately after arriving in Breslau, Kirchhoff employed variational calculus to solve an outstanding problem related to the deformation of elastic plates. In 1854, he was appointed Professor of Physics at Heidelberg, where he collaborated in spectroscopic work with Robert Bunsen. Kirchhoff's three laws of spectroscopy are:

1. A hot solid object produces light with a continuous spectrum.
2. A hot tenuous gas produces light with spectral lines at discrete wavelengths (i.e., specific colors) that depend on the energy levels of the atoms in the gas.
3. A hot solid object surrounded by a cool tenuous gas (i.e., cooler than the hot object) produces light with an almost continuous spectrum that has gaps at discrete wavelengths depending on the energy levels of the atoms in the gas.

In 1861, Kirchhoff and Bunsen discovered cesium and rubidium while studying the chemical composition of the Sun via its spectral signature. In 1862, Kirchhoff coined the term *black body radiation*, which was important in the development of quantum theory. He presented his law of radiation, stating that for a given atom or molecule the emission and absorption frequencies are the same. The same year, he was awarded the Rumford Medal. In 1869, his wife Clara died, with whom he had two sons and two daughters. In 1872, he remarried, to Luise Brömmel. In 1875, Kirchhoff accepted the first Chair specifically dedicated to theoretical physics at Berlin, which allowed him to continue to make contributions to teaching and research. His four-volume masterpiece treatise *Vorlesungen über mathematische Physik*, which had a lasting value, was published during 1876–1894. In this treatise the principles of dynamics as well as various special problems are treated in a novel and original manner.

Kirchhoff's contributions to mathematical physics were numerous and important, his strength lying in his powers of stating a new physical problem in terms of mathematics and then finding solutions. He is also remembered for Kirchhoff equations and the Piola–Kirchhoff stress tensor. He became a Fellow of the Royal Society of Edinburgh in 1868 and an FRS in 1875. Kirchhoff died in 1887 in Berlin.

George Friedrich Bernhard Riemann (1826–1866)

George Friedrich Bernhard Riemann (1826–1866) was born in Breselenz, Kingdom of Hanover (now Germany). Bernard Riemann, as he is commonly known, was the son of a poor Protestant minister. He received his elementary education from his father and showed brilliance in arithmetic at an early age. In fact, he studied the works of Euler and Legendre while he was still in secondary school, and it is said that he mastered Legendre's treatise on the theory of numbers in less than a week. In 1846, he enrolled at Göttingen University to study theology and philology, but he soon transferred to mathematics. He studied physics under Weber and mathematics under Gauss. In 1851, Riemann received his Ph.D. under Gauss, after which he remained at Göttingen to teach. In 1862, 1 month after his marriage, Riemann suffered an attack of pleuritis and for the remainder of his life was a sick and dying man. He finally succumbed to tuberculosis in 1866 at the age of 39 in Selasca, Italy. Riemann published comparatively little, but his work permanently altered the course of mathematics in analysis, geometry, and number theory.

His first published paper was his celebrated dissertation of 1851 on the general theory of functions of a complex variable. Here he described what are now called *Cauchy–Riemann equations*, created the ingenious device of *Riemann surfaces* for clarifying the nature of multiple-valued functions, and was led to the *Riemann mapping theorem*. The notion of the definite integral, as it is presented in most basic calculus courses, is due to him. An interesting story surrounds Riemann's work in geometry. In 1854, for his introductory lecture prior to becoming an associate professor, Riemann submitted three possible topics to Gauss. Gauss surprised Riemann by choosing the topic Riemann liked the least, the foundations of geometry. The lecture "On the hypotheses that underlie geometry" was like a scene from a movie. The old and failing Gauss, a giant in his day, watching intently as his brilliant and youthful protege skillfully pieced together portions of the old man's own work into a complete and beautiful system. Gauss is said to have gasped with delight as the lecture neared its end, and on the way home he marveled at his student's brilliance. Gauss died shortly thereafter. The results presented by Riemann that day opened a new field, now known as *Riemannian geometry* (elliptic geometry), which eventually evolved into a fundamental tool that Einstein used 50 years later to develop relativity theory, and ultimately to the development of nuclear fission. In 1859, Riemann published his only work on the theory of numbers, a brief but exceedingly profound paper of less than ten pages devoted to the prime number theorem. Here he introduced the *Riemann zeta function*,

$$\zeta(s) = 1 + \frac{1}{2^s} + \frac{1}{3^s} + \cdots, \qquad s = \sigma + it.$$

In his paper he proved several important properties of this function, and simply stated a number of others without proof. After his death, many of the finest mathematicians of the world have exerted their strongest efforts and created new branches of analysis in attempts to prove these statements. Since then, with one exception, every statement has been settled in the sense Riemann expected. The exception is the famous *Riemann hypothesis*: that all the zeros of $\zeta(s)$ in the strip $0 \le \sigma \le 1$ lie on the central line $\sigma = 1/2$. It stands today as the most important unsolved problem of mathematics, and perhaps the most difficult problem that the mind of man has ever conceived. On a fragmentary note found among his papers after his death, Riemann wrote that these theorems "follow from an expression for the function $\zeta(s)$ which I have not yet simplified enough to publish." Writing about this fragment in 1944, Hadamard remarked with justified exasperation, "We still have not the slightest idea of what the expression could be." He added a further comment, "In general, Riemann's intuition is highly geometrical, but this is not the case for his memoir on prime numbers, the one in which that intuition is the most powerful and mysterious."

Riemann's early death was a great loss to mathematics, for his mathematical work was brilliant and of fundamental importance. Dedekind described Riemann's death as follows: "On the day before his death he lay beneath a fig tree, filled with joy at the glorious landscape, writing his last work, unfortunately left incomplete. His end came gently, without struggle or death agony; it seemed as though he followed with interest the parting of the soul from the body; his wife had to give him bread and wine, he asked her to convey his love to those at home, saying 'Kiss our child.' She said the Lord's prayer with him, he could no longer speak; at the words 'Forgive us our trespasses' he raised his eyes devoutly, she felt his hand in hers becoming colder." The inscription on his tombstone, erected by his Italian friends, closes with the words *Denen die Gott lieben müsen alle Dinge sum Besten dienen*, meaning 'all things work together for good to them that love the Lord.'

Henry John Stephen Smith (1826–1883)

Henry John Stephen Smith (1826–1883) was born in Dublin and was the fourth child of his parents. When Henry was just 2 years old his father died, and so his mother moved the family to England. He was first privately educated by his mother and tutors, and then in 1841 he entered Rugby boarding school in Warwickshire. In 1844, he won an entrance scholarship to Balliol College, Oxford. He graduated in 1849 with high honors in both mathematics and the classics. During the 1846–1847 academic year, Smith also took classes in mathematics at the Sorbonne in Paris. He was elected a fellow of Balliol in 1850 and Savilian Professor of Geometry in 1861, and in 1874, he was appointed keeper of the university museum. Smith started his mathematical career by publishing a few short papers in geometry and number theory. He is especially remembered for his report on the theory of numbers, which appeared in the British Association volumes from 1859 to 1865. Smith analyzed, with remarkable clarity and order, the works of mathematicians from the preceding century on the theory of congruences and binary quadratic forms. During the preparation of the report, he also published several original contributions to higher arithmetic in the Philosophical Transactions of the Royal Society of London and in the Proceedings of that society. In 1868, he returned to geometrical research and published a memoir, *Sur quelques problémes cubiques et biquadratiques* (Certain Cubic and Biquadratic Problems), for which the Royal Academy of Sciences of Berlin awarded him the Steiner Prize. His final work was a memoir on the theta and omega functions, which he left nearly complete. Smith was elected FRS in 1861. He served on various committees and royal commissions: he was a Mathematical Examiner for the University of London, a member of a Royal Commission to review scientific education practice, a member of the commission to reform University of Oxford governance, chairman of the committee of scientists overseeing the Meteorological Office, and twice president of the London Mathematical Society. Smith died in Oxford.

Elwin Bruno Christoffel (1829–1900)

Elwin Bruno Christoffel (1829–1900) was born in Monschau, Germany. He studied at an elementary school in Monschau and then was tutored for several years at home in mathematics, classics, and languages. Then in 1844 he attended the Jesuit Gymnasium in Cologne, but transferred to the Friedrich Wilhelm Gymnasium (a royal grammar school) in Berlin. He obtained his final school certificate, with distinction, in 1849. Elwin then attended the University of Berlin starting in 1850, where he came in close contact with his illustrious teachers, such as Eisenstein and Dirichlet. Dirichlet had a strong influence on him. Elwin received his Ph.D. from the University of Berlin in 1856 for his dissertation on the motion of electricity in homogeneous bodies. The period 1850–1856 included 1 year of his military service away from the University. In 1859, Elwin became a lecturer at the University of Berlin, and he was appointed in 1862 to a chair at the Swiss Federal Polytechnic School in Zurich that was left vacant by Dedekind. In the 1890s, Einstein attended the same school, where he received in-depth training in quantitative analysis and developed a foundation for his future work in mathematical physics. In 1869, Elwin moved to the Gewerbe Akademie (Trade School) in Berlin, which is currently part of the Technical University of Berlin. In 1872, he was appointed as a professor at the University of Strasbourg (established in 1537), France, one of the country's oldest universities (over 575 years old), where he continued until his retirement in 1894 at the age of 65. He contributed to potential theory, invariant theory, mathematical physics, geodesy, shock waves, tensor analysis, and conformal maps. He is remembered for the Christoffel symbol, the Riemann–Christoffel tensor, the Christoffel–Darboux formula, and Schwarz–Christoffel mapping. He breathed his last on March 15, 1900, in Strasbourg at the age of 70.

Julius Wilhelm Richard Dedekind (1831–1916)

Julius Wilhelm Richard Dedekind (1831–1916) was born in Brunswick, Germany, the birthplace of Gauss. His father, Julius Levin Ulrich Dedekind, was a professor at the Collegium Carolinum in Brunswick. Richard was the youngest of four children, never married, and lived with his unmarried sister. After schooling in Brunswick, in 1848 he entered the Collegium Carolinum, where he received a solid grounding in mathematics. In 1850, he entered the University of Göttingen and became Gauss' last student. Dedekind received his doctorate in 1852 on the thesis *On the Theory of Eulerian integrals*. At that time, the University of Berlin was the leading center for mathematical research in Germany, and therefore Dedekind went to Berlin for 2 years, obtaining his habilitation in 1854, which qualified him to teach at a university. Dedekind returned to Göttingen and began teaching courses on probability and geometry. During 1858–1862, he taught at the Polytechnic in Zürich, and in 1862, when the Collegium Carolinum was upgraded to a Technische Hochschule (Institute of Technology), Dedekind returned to his native Braunschweig, where he spent the rest of his life teaching at the Institute.

Dedekind made a number of highly significant contributions to mathematics. In 1858, while teaching calculus for the first time at the Polytechnic, he came up with the technique now called a *Dedekind cut*, whose history dates back to Eudoxus. He published this in *Stetigkeit und Irrationale Zahlen* (Continuity and Irrational Numbers) in 1872. The central idea of a Dedekind cut is that an irrational number divides the rational numbers into two sets, with all the members of one set (upper) being strictly greater than all the members of the other (lower) set. For example, $\sqrt{2}$ puts all the negative numbers and the numbers whose squares are less than 2 into the lower set, and the positive numbers whose squares are greater than 2 into the upper set. Every point on the real line is either a rational or an irrational number. Therefore, on the real line there are no empty locations, gaps, or discontinuities. This led to the famous Dedekind–Cantor axiom: It is possible to assign to any point on a line a unique real number, and conversely, any real number can be represented in a unique manner by a point on the line. In 1874, Dedekind admired Cantor's work on infinite sets and called two sets *similar* if there exists a one-to-one correspondence between the sets. Using this terminology, he gave the first precise definition of an infinite set: a set is infinite when it is similar to its proper subset. Among Dedekind's other notable contributions to mathematics are his editions of the collected works of Dirichlet, Gauss, and Riemann. Dedekind's study of Dirichlet's work led to his own study of algebraic number fields, as well as to his notion of an ideal, which is fundamental to ring theory (introduced by Hilbert). In 1882, Dedekind and Weber applied the theory of ideals to the theory of Riemann surfaces and established powerful results, such as a purely algebraic proof of the Riemann–Roch (Gustav Roch, 1839–1866) theorem. Around 1900, he wrote the first paper on modular lattices. He is also remembered for the Dedekind domain, the Dedekind eta function, Dedekind-infinite sets, Dedekind sums, and the Dedekind zeta function.

Dedekind received many honors for his outstanding work; he was elected to the Göttingen Academy in 1862 and the Berlin Academy in 1880, as well as the Academy of Rome, the Leopoldino–Carolina Naturae Curiosorum Academia, and the Académie des Sciences in Paris in 1900. He also received honorary doctorates from the universities of Oslo, Zurich, and Braunschweig. Twelve years before his death, Teubner's *Calendar for Mathematicians* listed Dedekind as having died on September 4, 1899, much to Dedekind's amusement. The day, September 4, 'might possibly prove to be correct,' he wrote to the editor, but the year certainly was wrong. "According to my own memorandum I passed this day in perfect health and enjoyed a very stimulating conversation on 'system and theory' with my luncheon guest and honored friend Cantor of Halle." In his commemorative address to the Royal Society of Göttingen in 1917, Edmund Landau (1877–1938) said "Richard Dedekind was not only a great mathematician, but one of the wholly great in the history of mathematics, now and in the past, the last hero of a great epoch, the last pupil of Gauss, for four decades himself a classic, from whose works not only we, but our teachers and the teachers of our teachers, have drawn."

Charles Lutwidge Dodgson (1832–1898)

Charles Lutwidge Dodgson (1832–1898) began using the pen name **Lewis Carroll** to write humorous pieces for magazines. Dodgson, the son of a clergyman, was the third of 11 children, all of whom stammered. He was uncomfortable in the company of adults and is said to have spoken without stuttering only to young girls, many of whom he entertained, corresponded with, and photographed (often in the nude). Although attracted to young girls, he was extremely puritanical and religious. The books *Alice's Adventures in Wonderland* and *Through the Looking-Glass* made Carroll famous. Although the books have a child's point of view, many would argue that the audience best equipped to enjoy them is an adult one. Dodgson, a mathematician, logician, and photographer, used the naiveté of a 7-year-old girl to show what can happen when the rules of logic are taken to absurd extremes. Dodgson shunned attention and denied that he and Carroll were the same person, even though he gave away hundreds of signed copies to children and children's hospitals. Dodgson–Carroll merged again only in *A Tangled Tale*, ten short stories, each with a mathematical problem, and in *The Game of Logic*, which was designed to make learning logic fun. He is credited with inventing cofactor expansion, the method for expressing the determinant of a matrix in terms of determinants of lower order, which he called condensation. He accumulated a library of 5,000 volumes, bought a skeleton so that he could learn anatomy, and rigged up his rooms with thermometers and oil stoves because he had a horror of drafts. Dodgson had the habit of working throughout the night while standing at his tall writing desk. He also worked in bed without any light, using an instrument of his own invention called a nyctograph, which kept his lines straight and his pen from running off the paper. He said: "It may well be doubted whether, in all the range of science, there is any field so fascinating to the explorer—so rich with hidden treasures—so fruitful in delightful surprises—as Pure Mathematics."

Rudolf Otto Sigismund Lipschitz (1832–1903)

Rudolf Otto Sigismund Lipschitz (1832–1903) was born in Bönkein, which is near Königsberg, Germany. He began his studies at the University of Königsberg and then moved to Berlin, where he studied under Dirichlet. Lipschitz completed his doctorate in 1853 and then taught for 4 years at the Gymnasiums in Königsberg and Elbing. In 1857, he joined University of Berlin as a Privatdozent, and in 1862 he became an extraordinary professor at Breslau. In 1864, Lipschitz moved to the University of Bonn, where he spent most of his life. He is remembered chiefly for the 'Lipschitz continuity condition,' which simplified Cauchy's original theory of the existence and uniqueness of solutions of differential equations. He also extended Dirichlet's theorem on the representability of a function by its Fourier series, obtained the formula for the number of ways that a positive integer can be expressed as a sum of four squares as a consequence of his own theory of the factorizations of integral quaternions, and made useful contributions to theoretical mechanics, the calculus of variations, Bessel functions, quadratic differential forms, and the theory of viscous fluids. Lipschitz also wrote several treatises: *Wissenschaft und Staat* in 1874, *Bedeutung der theoretischen Mechanik* in 1876, two volumes of *Lehrbuch der Analysis* in 1877 and 1880, and *Untersuchungen über die Summen von Quadraten* in 1886.

Karl Gottfried Neumann (1832–1925)

Karl Gottfried Neumann (1832–1925) was born in Königsberg, Prussia, and studied there and in Halle. He was a professor at the universities of Halle, Basel, Tübingen, and Leipzig. He made significant contributions to differential equations, integral equations, and complex variables. The Neumann boundary conditions for certain types of ordinary and partial differential equations are named after him. In 1868, along with Clebsch, Neumann founded the mathematical research journal *Mathematische Annalen*, a leading mathematical journal. He died in Leipzig.

Rudolf Friedrich Alfred Clebsch (1833–1872)

Rudolf Friedrich Alfred Clebsch (1833–1872) was born in Königsberg, Germany. He attended the University of Königsberg and was habilitated at Berlin. Clebsch was a professor at Karlsruhe, Giessen, and then Göttingen, where he died at 39 years of age. He published a book on elasticity in 1862, which was translated in 1883 by Adhémar Jean Claude Barré de Saint-Venant (1797–1886) into French and published as *Théorie de l'élasticité des Corps Solides*. Clebsch applied his theory of invariants to projective geometry. His collaboration with Paul Albert Gordan (1837–1912) in Giessen led to *Theorie der Abelschen Funktionen* (1866) in which they introduced Clebsch–Gordan coefficients for spherical harmonics, which are now widely used in quantum mechanics. His lectures on geometry, published by Lindemann, remain a standard text on projective geometry. Together with Karl Neumann at Göttingen, he founded the mathematical research journal Mathematische Annalen.

Immanuel Lazarus Fuchs (1833–1902)

Immanuel Lazarus Fuchs (1833–1902) was born in Moschin, located in Grand Duchy of Poznan. Fuchs first attended the Friedrich Wilhelm Gymnasium in Berlin, and then the University of Berlin where he received his doctorate under the supervision of Weierstrass in 1858. After teaching at several schools, in 1884 he filled Kummer's Chair at the University of Berlin, where he remained for the rest of his life. Fuchs was a gifted analyst whose works form a bridge between the fundamental researches of Cauchy, Riemann, Abel, and Gauss and the modern theory of differential equations discovered by Poincaré, Paul Painlevé (1863–1933), and Picard. His most important research was on singular points of linear differential equations. He recognized the significance of regular singular points, and equations whose only singularities, including the point at infinity, are regular points are known as Fuchsian equations. He is also remembered for Fuchsian functions. During the final 10 years of his life he edited Crelle's Journal and the Journal für die reine und angewandte Mathematik. He died in Berlin, Germany.

Edmond Nicolas Laguerre (1834–1886)

Edmond Nicolas Laguerre (1834–1886) was a professor at the Collége de France in Paris and worked primarily in geometry and the theory of equations. He studied the polynomials named for him in about 1879. He was one of the first to point out that a "reasonable" distance function (metric) can be imposed on the coordinate plane of analytic geometry in more than one way.

John Venn (1834–1923)

John Venn (1834–1923) was born into a London suburban family noted for its work in philosophy. He attended London schools and received a mathematics degree from Caius College, Cambridge, in 1857. He was elected a fellow of the college and held his fellowship there until his death. He took holy orders in 1859, and after a brief stint of religious work he returned to Cambridge, where he developed programs in the moral sciences. Besides his mathematical work, Venn had an interest in history and wrote extensively about his college and family. Venn's book, *Symbolic Logic*, clarifies ideas originally presented by Boole. In it, Venn presents a systematic development of a method that uses geometric figures, now known as Venn diagrams. Today these diagrams are primarily used to analyze logical arguments and to illustrate relationships between sets. In addition to his work in symbolic logic, Venn made contributions to probability theory that are described in his widely used textbook on that subject.

Felice Casorati (1835–1890)

Felice Casorati (1835–1890) was born and studied in Pavia. He received a diploma in engineering and architecture in 1856. In 1858, together with Enrico Betti (1823–1892) and Francesco Brioschi (1824–1897), he visited Göttingen, Berlin, and Paris. This visit is considered to be the point when Italian mathematics joined mainstream European mathematics. In 1859, Casorati was appointed as an extraordinary professor of algebra and analytic geometry at the University of Pavia. In 1862, he was appointed as an ordinary professor at Pavia, and in the following year he succeeded Gaspare Mainardi (1800–1879) to the chair of infinitesimal calculus. In 1864, Casorati again visited Germany and became familiar with the ideas of Riemann and Weierstrass on complex function theory, which he spread through his lectures and contacts. During 1865–1868, he taught geodesy and analysis at Pavia and then moved to Milan where he taught until 1875, although he continued to hold the chair at Pavia. Casorati returned to Pavia in 1875, where he taught analysis until his death in Casteggio. He wrote an important book on complex function theory, *Teoria delle funzioni di variabili complesse* (1868), and 49 research papers, mainly regarding the existence of solutions of differential equations, inversion of Abelian integrals, and the existence of many-valued functions with an infinity of distinct periods. Casorati is best remembered for the Casorati–Weierstrass theorem. He left a large number of papers that are now preserved in Pavia. They contain hundreds of letters to mathematicians, memoranda on conversations, observations on contemporary papers, a diary, some lecture notes, and manuscripts. Casorati received many honors; he was elected to the Göttingen Academy of Sciences (1877), the Turin Mathematical Society (1880), the Bologna Academy of Sciences (1885), and the Berlin Academy of Sciences (1886).

Paul Gustav Heinrich Bachmann (1837–1920)

Paul Gustav Heinrich Bachmann (1837–1920) was the son of a Lutheran pastor and shared his father's pious lifestyle and love of music. His mathematical talent was discovered by one of his teachers, even though he experienced difficulty in some of his early mathematical studies. After recuperating from tuberculosis in Switzerland, Bachmann studied mathematics, first at the University of Berlin and later at Göttingen, where he attended Dirichlet's lectures. He received his doctorate under Kummer in 1862 with a thesis on group theory. Bachman was a professor at Breslau and then at Münster. After he retired from his professorship he continued his mathematical writing, played the piano, and served as a music critic for newspapers. Bachmann's mathematical writing includes a five-volume survey of results and methods in number theory, a two-volume work on elementary number theory, a book on irrational numbers, and a book on the famous conjecture known as Fermat's Last Theorem. He introduced big-O notation in his 1892 book, *Analytische Zahlentheorie*.

Thorvald Nicolai Thiele (1838–1910)

Thorvald Nicolai Thiele (1838–1910) was a Danish astronomer, mathematician, and actuary. He is known for his work in astronomy, statistics, the three-body problem, and an interpolation formula that is named after him. He was the first to propose a mathematical theory of Brownian motion and introduced the likelihood function.

George William Hill (1838–1914)

George William Hill (1838–1914), an American mathematician and astronomer, was born in New York city. At the age of eight, he moved to West Nyack (New York) with his family. After attending high school, Hill entered Rutgers University in 1855 and graduated in 1859. In the same year he won first prize for his essay *On the Confrontation of the Earth* in a competition organized by Runkle's Mathematical Monthly, founded by John Daniel Runkle (1822–1902). From 1861 he worked at the Nautical Almanac Office in Cambridge, Massachusetts. After 2 years in Cambridge he returned to West Nyack where he worked from his home. This suited the reclusive Hill, who preferred being on his own. Except during the 10-year period from 1882 to 1892 when he worked in Washington on the theory and tables of the orbits of Jupiter and Saturn, this was to be the working pattern for the rest of his life. He was essentially the type of scholar and investigator who seems to feel no need for personal contact with others. While the few who knew him speak of the pleasure of his companionship in their frequent tramps over the country surrounding Washington, he was apparently quite happy alone, whether at work or play. George described mathematically the three-body problem and later the four-body problem. His mathematics focused on computing the orbits of the Moon around the Earth and the planets around the Sun. His new idea about how to approach the solution to the three-body problem involved assuming that the three bodies lie in the same plane, that two bodies orbit their common center of mass, and that the third body orbits the other two but is of negligible mass so that it does not affect the orbits of the other two (assumed massive) bodies. He had no interest in modern developments in other areas of mathematics that were not related to solving problems about orbits. He described the *Hill sphere*, which approximates the gravitational sphere of influence of one astronomical body in the face of perturbations from another, heavier body around which it orbits. The term *linear combination* is also due to Hill, who introduced it in a research paper on planetary motion published in 1900. During 1898–1901, George lectured at Columbia University. Hill never married and his income was small, but so were his needs, and several of his brothers lived nearby. A recluse, he was happiest in his large library, where he was free to pursue his research. He died at his home in West Nyack at the age of 76. During 1894–1896, George was the president of the American Mathematical Society. He was elected to the Royal Society of Edinburgh in 1908, as well as the academies of Belgium (1909), Christiania (1910), and Sweden (1913).

Marie Ennemond Camille Jordan (1838–1922)

Marie Ennemond Camille Jordan (1838–1922) was born in Lyon and educated at the École Polytechnique. By profession he was an engineer, but he also taught mathematics at the École Polytechnique and the Collége de France. He made important contributions to analysis, topology, and especially algebra. He is remembered for the Jordan curve theorem, Jordan measure, Jordan–Chevalley (Claude Chevalley, 1909–1984) decomposition, the Jordan totient function, the Jordan–Schönflies (Arthur Moritz Schöenflies, 1853–1928) theorem, and the Jordan–Hölder theorem. The Jordan normal form and the Jordan matrix appeared in his influential book *Traité des substitutions et des équations algébriques*, which was published in 1870. He also refined Beltrami's theory of singular value decompositions of matrices, which was later advanced by Sylvester and Weyl. Jordan's work brought attention to Galöis' theory. He also investigated Mathieu (Émile Léonard Mathieu, 1835–1900) groups. The asteroid 25593 Camillejordan and the Institute of Camille Jordan are named in his honor.

Josiah Willard Gibbs (1839–1903)

Josiah Willard Gibbs (1839–1903) was born in New Haven, Connecticut, USA. His father was a Professor of Sacred Literature at the Yale Divinity School. He started his education at the Hopkins School, and in 1854, he matriculated at Yale College where he graduated in 1858. He continued his studies there and was appointed tutor in 1863, which he continued for 3 years. He then went to Europe, studying in Paris during 1866–1867, in Berlin in 1867, and in Heidelberg in 1868, where he was influenced by Kirchhoff and Helmholtz. In 1869, he returned to New Haven and in 1871 was appointed Professor of Mathematical Physics at Yale College, the first such professorship in the USA, which he held until his death. He did not publish any work until 1876, but within 2 years he wrote a series of papers collectively titled *On the Equilibrium of Heterogeneous Substances* that is now considered to be one of the greatest scientific achievements of the nineteenth century, and is one of the foundations of *physical chemistry*. In these papers, Gibbs applied thermodynamics to interpret physicochemical phenomena, successfully explaining and interrelating what had previously been a mass of isolated facts. During 1880–1884, independently of Oliver Heaviside (1850–1925), Gibbs combined the ideas of Hamilton and Grassmann in order to formulate vector analysis, which advanced mathematical physics. From 1882 to 1889, he wrote a series of research papers on optics and the electromagnetic theory of light. After 1889, Gibbs worked on statistical mechanics, laying a foundation and providing a mathematical framework for quantum theory and Maxwell's theories. In 1901, Gibbs was awarded the Copley medal by the Royal Society of London for being "the first to apply the second law of thermodynamics to the exhaustive discussion of the relation between chemical, electrical, and thermal energy and capacity for external work." This medal, awarded to only one scientist each year, was the highest possible honor granted by the international scientific community of his day. In 1902, he wrote classic textbooks on statistical mechanics, which Yale published. Gibbs also contributed to crystallography and applied his vector methods to the determination of the orbits of planets and comets. Fourier series at points of discontinuity have a special behavior, which is known as the Gibbs phenomenon.

Gibbs' contributions were not fully recognized in his time; however, now he is known for Gibbs free energy, Gibbs entropy, Gibbs inequality, the Gibbs paradox, the Gibbs–Helmholtz equation, the Gibbs algorithm, the Gibbs distribution, the Gibbs state, Gibbs sampling, the Gibbs–Marangoni (Carlo Giuseppe Matteo Marangoni, 1840–1925) effect, the Gibbs–Duhem (Pierre-Maurice-Marie Duhem, 1861–1916) relation, the Gibbs phenomenon, the Gibbs–Donnan (Frederick George Donnan, 1870–1956) effect, and vector analysis. Gibbs' cross product notation, $A \times B$, which he originally referred to as the "skew product," appeared in a published work for the first time in the second edition of the book *Vector Analysis* by Edwin Wilson (1879–1964), a student of Gibbs. In 1910, the Willard Gibbs Medal, founded by William Converse, was established in his honor by the American Chemical Society. Since 1923, the American Mathematical Society has named an annual lecture series in honor of Gibbs. In 1945, Yale University created the Gibbs professorship in Theoretical Chemistry. Gibbs Laboratory at Yale and the Gibbs Assistant Professorship in mathematics at Yale are also named in his honor. In 1950, Gibbs was elected to the Hall of Fame for Great Americans. In 2005, the United States Postal Service issued the American Scientists commemorative postage stamp series, depicting Gibbs, von Neumann, Barbara McClintock (1902–1992), and Richard Feynman (1918–1988). Novelists, including Johann van der Waals (1837–1923) in 1910 in physics, Planck in 1918 in physics, Lars Onsager (1903–1976) in 1968 in chemistry, and Paul Anthony Samuelson (1915–2009) in 1970 in economics, have given full credit to Gibbs' fundamental work.

Gibbs never married, living all his life in his childhood home with a sister and his brother-in-law, who was the Yale librarian. He died at the age of 64 in New Haven, Connecticut. There is a story that Helmholtz, during his visit to Yale in 1893, expressed regret at missing an opportunity to talk with Gibbs; the university officials, perplexed, looked at one another and asked, "Who?"

Julius Peter Christian Petersen (1839–1910)

Julius Peter Christian Petersen (1839–1910) was born in the Danish town of Sorø. His father was a dyer. In 1854, his parents were no longer able to pay for his schooling, so he became an apprentice in his uncle's grocery store. When this uncle died, he left Petersen enough money to return to school. After graduating, he began studying engineering at the Polytechnic School in Copenhagen, but later decided to concentrate on mathematics. He published his first textbook in 1858 on logarithms. When his inheritance ran out, he had to teach to make a living. From 1859 until 1871, Petersen taught at a prestigious private high school in Copenhagen. While teaching high school he continued his studies, entering Copenhagen University in 1862. He obtained a mathematics degree from Copenhagen University in 1866 and his doctorate in 1871. After receiving his doctorate, he taught at a polytechnic and military academy. In 1887, he was appointed to a professorship at the University of Copenhagen. Petersen was well known in Denmark as the author of a large series of textbooks for high schools and universities. One of these books, *Methods and Theories for the Solution of Problems of Geometrical Construction*, was translated into eight languages, with the English language version last reprinted in 1960 and the French version reprinted in 1990. Petersen worked in a wide range of areas, including algebra, analysis, cryptography, geometry, mechanics, mathematical economics, and number theory. His contributions to graph theory, including his results regarding regular graphs, are his best-known work. He was noted for his clarity of exposition, problem solving skills, originality, sense of humor, vigor, and teaching. One interesting fact about Petersen is that he preferred not to read the writing of other mathematicians. This led him to often rediscover results that had been already proved by others, often with embarrassing consequences. However, he was angry when other mathematicians did not read his writings! Petersen's death was front-page news in Copenhagen. A newspaper described him as the Hans Christian Andersen of science; a child of the people who made good in the academic world.

Charles Sanders Peirce (1839–1914)

Charles Sanders Peirce (1839–1914) is considered to be the most original and versatile intellect from the USA. He was born in Cambridge, Massachusetts. His father, Benjamin Peirce, known as the "Father of American Mathematics," was a Professor of Mathematics and Natural Philosophy at Harvard. Peirce attended Harvard (1855–1859) and received a Harvard Master of Arts degree (1862) and an advanced degree in chemistry from the Lawrence Scientific School (1863). His father encouraged him to pursue a career in science, but instead he chose to study logic and scientific methodology. In 1861, Peirce became an aide in the United States Coast Survey, with the goal of better understanding scientific methodology. His service for the Survey exempted him from military service during the Civil War. While working for the Survey, Peirce carried out astronomical and geodesic work. He made fundamental contributions to the design of pendulums and map projections, applying new mathematical developments in the theory of elliptic functions. He was the first person to use the wavelength of light as a unit of measurement. Peirce rose to the position of Assistant for the Survey, where he remained until he was forced to resign in 1891 because he disagreed with the direction taken by the Survey's new administration. Although he made his living from work in the physical sciences, Peirce developed a hierarchy of sciences, with mathematics at the top rung, in which the methods of one science could be adapted for use by those sciences beneath it. He was the founder of the American philosophical theory of pragmatism. The only academic position that Peirce held was lecturer in logic at Johns Hopkins University in Baltimore from 1879 to 1884. His mathematical work during this time included contributions to logic, set theory, abstract algebra, and the philosophy of mathematics. His work is still relevant today; some of his work on logic has been recently applied to artificial intelligence. Peirce believed that the study of mathematics could develop the mind's powers of imagination, abstraction, and generalization. His diverse activities after retiring from the Survey included writing for newspapers and journals, contributing to scholarly dictionaries, translating scientific papers, guest lecturing, and textbook writing. Unfortunately, the income from these pursuits was insufficient to protect him and his second wife from abject poverty. He was supported in his later years by a fund created by his many admirers and administered by the philosopher William James (1842–1910), his lifelong friend. Although Peirce wrote and published voluminously in a vast range of subjects, he left more than 100,000 pages of unpublished manuscripts. Because of the difficulty of studying his unpublished writings, scholars have only recently started to understand some of his varied contributions. A group of people is devoted to making his work available over the Internet, in order to bring a better appreciation of Peirce's accomplishments to the world.

Ernst Schröder (1841–1902)

Ernst Schröder (1841–1902) was born in Mannheim, Baden, Germany. He learned mathematics at Heidelberg, Königsberg, and Zürich; Hesse and Kirchhoff were his teachers. After serving as a school teacher for a few years, in 1874 he moved to the Technische Hochschule Darmstadt. In 1876, Ernst took the position of chair in mathematics at the Polytechnische Schule in Karlsruhe, where he spent the rest of his life. Ernst summarized and extended the work of Boole, De Morgan, Peirce, and Hugh MacColl (1837–1909), a Scottish mathematician and logician. Ernest also made original contributions to algebra, set theory, lattice theory, ordered sets, and ordinal numbers. Along with Cantor, he codiscovered the Cantor–Bernstein–Schröder theorem, although Schröder's proof (1898) was flawed. Felix Bernstein (1878–1956) corrected this proof as part of his doctoral dissertation. Felix later became the director of the Institute for Mathematical Statistics in Göttingen. Ernst remained unmarried throughout his life. He passed away in 1902 in Karlsruhe.

François Édouard Anatole Lucas (1842–1891)

François Édouard Anatole Lucas (1842–1891) was a French mathematician. He studied primary numbers, figurative arithmetic, and tricircular and tetraspheric geometries. Lucas also popularized the Fibonacci sequence. In 1875, Lucas challenged the readers of the *Nouvelles Annales de Mathématiques* to prove that: A square pyramid of cannonballs contains a square number of cannonballs only when it has 24 cannonballs along its base. This problem remained unsolved until 1918.

François Marius Sophus Lie (1842–1899)

François Marius Sophus Lie (1842–1899) was a Norwegian mathematician, the youngest of six children. He was educated at Christiania and obtained a traveling scholarship. In the course of his journeys, Lie became acquainted with Klein, Darboux, and Jordan. In 1870, he discovered the transformation by which a sphere can be made to correspond to a straight line, and by application of this result he showed that theorems on the aggregates of lines can be translated into theorems on aggregates of spheres. This was followed by a Ph.D. thesis on the theory of tangential transformations of space. In 1872, he became professor at Christiania. His earliest research there regarded the relationship between differential equations and infinitesimal transformations. This naturally led him to the general theory of finite continuous groups of substitutions, which are now called Lie groups; the results of his investigations on this subject are embodied in his *Theorie der Transformationsgruppen*, published in Leipzig in three volumes during 1888–1893. Lie groups play an important role in quantum mechanics. He then proceeded to consider the theory of infinite continuous groups, and his conclusions, edited by George Scheffers (1866–1945), were published in 1893. About 1879, Lie turned his attention to differential geometry; a systematic exposition of this was documented in *Geometrie der Berührungstransformationen*. In 1886, he moved to Leipzig, and in 1898 back to Christiania. He became an Honorary Member of the London Mathematical Society in 1878, Member of the French Academy of Sciences in 1892, FRS in 1895, and a foreign associate of the National Academy of Sciences of the United States of America in 1895. Lie died at the age of 57, due to pernicious anemia, a disease caused by impaired absorption of vitamin B12. According to him, among all the mathematical disciplines the theory of differential equations is the most important. It furnishes the explanation of all those elementary manifestations of nature that involve time. To promote the theory of Lie groups and their applications, in 1991 the 'Seminar Sophus Lie' was arranged at the University of Leipzig, which continues on regular basis once a semester and attracts participation from all over Europe.

Wilhelm Jordan (1842–1899)

Wilhelm Jordan (1842–1899) was born in southern Germany. He attended college in Stuttgart and in 1868 became full Professor of Geodesy at the technical college in Karlsruhe, Germany. From 1881 until his death, he was Professor of Geodesy and practical geometry at the technical university in Hanover. He participated in surveying several regions of Germany and Africa. Jordan was a prolific writer, whose major work *Handbuch der Vermessungskunde* (Handbook of Geodesy) was translated into French, Italian, and Russian. He was regarded as a superb writer and an excellent teacher. Unfortunately, the Gauss–Jordan reduction method has been widely attributed to Camille Jordan. Moreover, it seems that the method was also discovered independently at the same time by B.I. Clasen, a priest who lived in Luxembourg.

Jean Gaston Darboux (1842–1917)

Jean Gaston Darboux (1842–1917) was born in Nimes, France. He first attended the Lycée at Nimes and then the Lycée at Montpellier. In 1861, Darboux entered the École Polytechnique, and then the École Normale Supérieure, one of the most prestigious French higher education establishments outside the public university system, where he received his Ph.D. in 1866. His thesis, written under the direction of Michel Chasles, was titled *Sur les surfaces orthogonales*. He taught at the Collége de France for the academic year 1866–1867, at the Lycée Louis le Grand during 1867–1872, the École Normale Supérieure during 1872–1881, and from 1873 to 1878 he was suppléant to Liouville in the Chair of Rational Mechanics at the Sorbonne. In 1878, Darboux became suppléant to Chasles in the Chair of Higher Geometry, and in 1880, he succeeded him to the Chair of Higher Geometry, where he remained until his death. He was also Dean of the Faculty of Science during 1889–1903.

Darboux made important contributions to differential geometry, analysis, algebra, kinematics, and dynamics. His results in calculus, such as the intermediate value property and upper and lower sums for Riemann integral, are taught even today. During 1887–1896, he produced four volumes on infinitesimal geometry, *Lecons sur la théorie général des surfaces et les applications géométriques du calcul infinitésimal*, which included most of his earlier work. He also edited Joseph Fourier's *Oeuvres* during 1888–1890. He is remembered for the Darboux equation, the Darboux integral, the Darboux function, Darboux net invariants, the Darboux problem, Darboux's theorem in symplectic geometry, Darboux's theorem in real analysis, the Christoffel–Darboux identity, the Christoffel–Darboux formula, Darboux's formula, the Darboux vector, the Euler–Darboux equation, the Euler–Poisson–Darboux equation, the Darboux cubic, the Darboux–Goursat (Édouard Jean-Baptiste Goursat, 1858–1936) problem, the Darboux transformation, and the Darboux Envelope theorem. Throughout his life, Darboux was elected as a member of over 100 scientific societies; the list includes Honorary Member of the London Mathematical Society in 1878, Académie des Sciences in 1884, becoming its permanent Secretary in 1900, and FRS and Fellow of the Royal Society of Edinburgh in 1902. In 1916, he won the Royal Society's Sylvester Medal. Darboux was renowned as an exceptional teacher, writer, and administrator. Émile Borel, Élie Cartan, Gheorghe Tzitzéica (1873–1939), and Stanislaw Zaremba are his notable students. He occupied a position in France somewhat similar to that of Klein in Germany. He died in 1917 in Paris.

Joseph Valentin Boussinesq (1842–1929)

Joseph Valentin Boussinesq (1842–1929) was born in Saint-André de Sangonis, Hérault, France. Much of his early education came through one of his uncles, who taught him Greek, Latin, and how to study on his own. He obtained a bachelor's degree in mathematics in 1861 at the age of 19. In 1865, he began working for a Ph.D. on a mechanical theory of light supervised by Émile Verdet (1824–1866), a French physicist. However, Émile died in 1866, before the completion of Joseph's dissertation. He then worked under a different supervisor on a new topic to suit his new supervisor's interest. Joseph wrote a dissertation titled *Study of Propagation of Heat in Homogenous Media*. Lamé played a major role in advising Joseph. Another mathematician who influenced his work was Saint-Venant. Joseph defended his doctorate dissertation in Paris on May 13, 1867. He was a professor at the Faculty of Sciences of Lille during 1872–1886, teaching differential and integral calculus at Institut industriel du Nord (École centrale de Lille). During 1896–1918, he was professor of mechanics at the Faculty of Sciences of Paris. Joseph is known for the Boussinesq approximation in buoyancy, water waves, and turbulence. His awards include the Poncelet Prize from the French Academy of Sciences in 1871. In 1894, his first wife Jeanne passed away. They did not have children. In the following year, he married Claire Onfroy de Véretz (died 1905). This marriage lasted for 10 years, until Claire's death. Joseph married for the third time in the year following his wife's death, to Jeanne Le Bouteiller. This marriage lasted for only 3 years; they separated in 1909. In 1918, at the age of 76, Joseph retired from his university positions. He left his mortal body at the age of 87 in Paris.

Karl Hermann Amandus Schwarz (1843–1921)

Karl Hermann Amandus Schwarz (1843–1921) was born in Jerzmanowa (Poland) but was educated in Germany. He was a protégé of Weierstrass and Kummer, whose daughter, Marie Kummer, he married. They had six children. He worked in Halle, Zürich, Göttingen, and then became a member of the Berlin Academy of Science and a professor at the University of Berlin. His main contributions to mathematics are in the geometric aspects of analysis, such as conformal mappings and minimal surfaces. In connection with the latter, he sought certain numbers associated with differential equations, numbers that have since come to be called eigenvalues. The famous Cauchy–Schwarz inequality was used in the search for these numbers. He is also remembered for the Additive Schwarz method, the Schwarzian derivative, the Schwarz lemma, Schwarz–Christoffel mapping, the Schwarz–Ahlfors–Pick (Lars Valerian Ahlfors, 1907–1996; George Alexander Pick, 1859–1942) theorem, the Schwarz reflection principle, the Schwarz triangle, and the Schwarz integral formula. He died in Berlin. Schwarz was noted for his preciseness and would start an oral examination as follows:

> Schwarz: Tell me the general equation of fifth degree.
> Student: $x^5 + bx^4 + cx^3 + dx^2 + ex + f = 0$.
> Schwarz: Wrong!
> Student: ... where e is not the base of the natural logarithm.
> Schwarz: Wrong!
> Student: ... where e is not *necessarily* the base of the natural logarithm.

William Kingdon Clifford (1845–1879)

William Kingdon Clifford (1845–1879), an English mathematician and philosopher, was born in Exeter, Devon, England. He was primarily a geometer, and his name is given to Clifford algebra, the Clifford Theorem, Klein–Clifford space, Clifford–Klein form, the Bessel–Clifford function, and the Clifford parallel. In 1860, at the age of 15, he attended King's College, London, and later Trinity College, Cambridge, where he became Second wrangler in 1867 at the age of 22. In the very next year he became fellow. At the age of 26, he became professor of mathematics at University College, London. In 1874, he was elected an FRS. In the following year, William married Lucy Lane. During 1876–1878, he had repeated breakdowns, perhaps due to overwork. He went to the island of Madeira to recuperate, where he breathed his last due to tuberculosis at the tender age of 34, leaving behind his widow and two children. Eleven days later, a legend called Einstein was born, who went on to develop the geometry of gravity that William had suggested 9 years earlier, in 1870.

Georg Cantor (1845–1918)

Georg Cantor (1845–1918) was born in St. Petersburg, Russia, to a Danish family. His father, a Jew who had converted to Protestantism, was a successful merchant. Cantor developed his interest in mathematics during his teens. He began his university studies in Zurich in 1862, but when his father died he left Zurich. He continued studying at the University of Berlin in 1863, where he studied under the eminent mathematicians Weierstrass, Kummer, and Kronecker. He received his doctoral degree in 1867 with a dissertation on number theory. Cantor assumed a position at the University of Halle in 1869, where he continued working until his death. Cantor married in 1874 and had five children. His melancholy temperament was balanced by his wife's happy disposition. He is regarded as the founder of set theory and is believed to be the one who introduced this theory to the mathematical world in about 1875. His two major books on set theory, *Foundations of General Theory of Aggregates* and *Contributions to the Founding of the Theory of Transfinite*

Numbers, were published in 1883 and 1895, respectively. For him, mathematics was a sort of auxiliary science to metaphysics, and his set theory was actually a part of metaphysics. Cantor defined an infinite set as one that can be placed in one-to-one correspondence with a part of itself. He compared infinite sets by establishing one-to-one correspondence between their elements. For the set N of natural numbers $1, 2, 3, \ldots$ and the set E of even numbers this can be done as follows:

$$\begin{array}{ccccc} 1 & 2 & 3 & 4 & \cdots \\ \updownarrow & \updownarrow & \updownarrow & \updownarrow & \\ 2 & 4 & 6 & 8 & \cdots \end{array}$$

(see Galileo paradox). Such a correspondence also exists between the set of natural numbers and the set Q of all rational numbers. Cantor called two sets equivalent, or of the same power, if a one-to-one correspondence can be established between their elements. The number, or 'power,' of an infinite set that can be placed in one-to-one correspondence with the positive integers is the number of all the positive integers. It is a cardinal number because it answers the question, 'how many?' Cantor called this a transfinite cardinal and boldly presented it with a name. As the Greeks had named their numbers after the letters in their alphabet, he named his after the first letter of the Hebrew alphabet, aleph. Thus, the sets N, E, and Q are equivalent and have the same power, aleph. The aleph of the positive integers is not the last number. In fact, it is not a positive integer at all. Having defined cardinality for infinite quantities, Cantor proceeded to establish three important facts: (1) the cardinal number of all the positive integers is the smallest transfinite cardinal, which he denoted by $\aleph_0$; (2) for every transfinite cardinal there exists a larger transfinite cardinal, which he denoted by $\aleph_1$; (3) for every transfinite cardinal there exists a next larger transfinite cardinal, which he denoted by $\aleph_2$; and so on. Cantor showed, by mathematical proof, that these alephs include all possible transfinite cardinals. In 1874, he succeeded in proving that the set of real numbers, that is, the points on the real line, is of higher power than the set N. Cantor denoted the number of the continuum by the German character C. This departure from the Hebrew alphabet was extremely significant, for Cantor could not establish where the number of the continuum stood among the alephs, $\aleph_0$, $\aleph_1$, ... Sets that have a greater power than C are known, for example, the cardinality of all correspondences that can be established between two continua (functional manifold) is denoted by f. A set having a power less than f but greater than C is not known, and beyond f there are still greater cardinal numbers.

Although Cantor's work on infinite sets is now considered to be a masterpiece, it generated heated controversy when it was originally published. Cantor's claim that the infinite set was unbounded offended the religious view of the time that God had created a complete universe, which could not be wholly comprehended by man. Like all revolutionary ideas, his work was opposed with emotion, much of it blind and bitter. Even the great Gauss, who thought far ahead of his time and disembarked on many mathematical shores long before their official discoverers, could not accept the idea of a conceived infinite. Cantor's former mentor, Kronecker, ridiculed

Cantor's theories and prevented him from gaining a position at the University of Berlin (see Kronecker). However, Cantor stood firm: "I was logically forced, almost against my will, because in opposition to traditions which had become valued by me, in the course of scientific researches, extending over many years, to the thought of considering the infinitely great, not merely in the form of the unlimitedly increasing... but also to fix it mathematically by numbers in the definite form of a 'completed infinite.' I do not believe, then, that any reasons can be urged against it which I am unable to combat." But, by 1908 several paradoxes had arisen in Cantor's set theory. Here, the word "paradox" is synonymous with "contradiction." Cantor created set theory 'naively,' meaning non-axiomatically. Consequently, he formed sets with such abandon that he himself, Cesare Burali–Forti (1861–1931), Russell, Julius König (1849–1914), and others found several paradoxes within his theory. This resulted in numerous attempts at finding a solution and gave rise to the three main philosophies or schools of thought, namely, the so-called logistic (see Frege), formalist (see Hilbert), and intuitionist (see Brouwer) schools. The logicists considered these paradoxes to be common errors, caused by errant mathematicians and not by a faulty mathematics. The intuitionists, on the other hand, considered these paradoxes to be clear indications that classical mathematics itself is far from perfect. They felt that mathematics had to be rebuilt from the bottom up. Eventually Cantor was given due recognition, but by then the criticism had taken its toll on his health. He had several nervous breakdowns and was hospitalized in 1884, 1902, 1904, 1907, 1911, and spent his last days in a mental hospital. His statement "the essence of mathematics resides in its freedom" is often quoted. In a letter to Hermite, Cantor writes about his failure to get a decent job: "I thank God, the all-wise and all-good, that He always denied me the fulfillment of this wish (for a good position), for He thereby constrained me, through a deeper penetration into theology, to serve Him and His Holy Roman Catholic Church better than I have been able with my exclusive preoccupation with mathematics." One of Cantor's last compositions was a love poem to his wife. After 40 years of marriage, Cantor talks about

> The love you gave me my good wife,
> You cared for me so well.

It has been recently discovered that set theory had existed long before Cantor at the Jain School of mathematics in India, which dealt with cosmological, philosophical, karmic, and number sets. The words employed for 'set,' for example, are rasi, samuha, samudaya, and many other equivalents in the Prakrit language, which is a language derived from Sanskrit. The foundation of measure rests on the number of elements or units contained in any set. The largest set is an omniscient set. The conceptual set containing no element is the null set. Finite, infinite, and transfinite sets are considered. For example, a set that cannot become a null set by being emptied for an unending amount of time is called a transfinite set. Variable sets, for example, are Karmic sets, which change every instant. The set of instants in the past is increasing, the present instant set is in a state of flux, and the future set of instants is a decreasing set. The concept of a union of sets, the method of one-to-one correspondence for comparing transfinite sets, the method of reductio ad absurdum, and the theory of logarithms to all types of bases are employed as analytical tools of set operations. All types of comparability are treated in three different ways: in one's own place, in others' place, and in general. In order to locate the order of comparability of all sets, 14 types of monotone sequences are considered. Such ordering implies the use of a well ordering principle and the stronger axiom of choice. Various structures of sequences contain only those sets that are admissible in the production process, characterizing a particular sequence. The all sequence contains all the sets, starting from unity and ending at the omniscient set. The first term of the dyadic-square-sequence, for example, is the square of two, and each of the succeeding terms is the square of the preceding term. If to each of these terms unity is added, the terms form Fermat's numbers.

Nicolai Yegorovich Zhukovski (1847–1921)

Nicolai Yegorovich Zhukovski (1847–1921) was born in Orekhovo, Vladimir Governorate, Russian Empire, and passed away in Moscow. In 1868, at the age of 21, he graduated from Moscow University where he was supervised by August Davidov (1823–1885). He became a professor in 1872, at the age of 25, at the Imperial Technical School. In 1904, he established the world's first Aerodynamic Institute in Kachino. From 1918 he was the head of Central Aero-Hydro-Dynamics Institute. He was a founder of modern aerodynamics and hydrodynamics and undertook the study of airflow around a wing, developing wing theory. This work occasionally involved integral equations. The conformal mapping of a circle into the shapes of wing sections is due to him. He is well known for his mathematical development of aerodynamic lift through his circulation hypothesis. The first wind tunnel in Russia was built by him. He developed the "water hammer" equation, usually known as the Zhukovsky (also Joukowsky) equation, which is employed by civil engineers. A crater on the Moon and a city near Moscow were named after him. The Russian Air Force academy has been named Zhukovsky Air Force Engineering Academy in his honor.

Friedrich Ludwig Gottlob Frege (1848–1925)

Friedrich Ludwig Gottlob Frege (1848–1925) was born in Wismar, in the state of Mecklenburg–Schwerin (the modern German federal state of Mecklenburg-Vorpommern). Frege studied at a Gymnasium in Wismar and graduated at the age of 15. In 1869, he entered the University of Jena to study mathematics and physics. In 1871, Frege moved to the University of Göttingen and was deeply influenced by the "ingenious philosopher" Rudolf Hermann Lotze (1817–1881). In 1873, Frege obtained his doctorate with Ernst Christian Julius Schering (1824–1897) on the dissertation *On a Geometrical Representation of Imaginary Forms in a Plane*, in which he tried to solve fundamental problems in geometry. In 1874, he earned a habilitation in mathematics from the University of Jena and continued working there in various positions until his retirement in 1917. His 1879 *Begriffsschrift* (Concept Script) marked a turning point in the history of logic. He developed: (1) a formal system that formed the basis of modern logic, (2) an elegant analysis of complex sentences and quantifier phrases that showed an underlying unity to certain classes of inferences, (3) a deep understanding of proof and definition, (4) an insightful analysis of statements about number (i.e., answers to the question 'How many?'), (5) a theory of extensions that, though seriously flawed, offered an intriguing picture of the foundations of mathematics, (6) definitions and proofs of some of the basic axioms of number theory from a limited set of logically primitive concepts and axioms, and (7) a conception of logic as a discipline, which has some compelling features. Frege's purpose was to defend the view that classical mathematics is a branch of logic, a view known as *logicism*. In his two-volume series *Grundgesetze der Arithmetik* (Fundamental Laws of Arithmetic), published in 1893 and 1903, he attempted to derive all of the laws of arithmetic from axioms he asserted as logical. Although it was his lifelong project, logicism was not successful. However, his ideas spread through the writings of his student Rudolph Carnap (1891–1970) and other admirers, particularly George Boole, Peano, Russell, Whitehead, and Ludwig Josef Wittegenstein (1889–1951). Frege died in 1925, in Bad Kleinen (now in Mecklenburg-Vorpommern). He said, "Every good mathematician is at least half a philosopher, and every good philosopher is at least half a mathematician."

Alfred Bray Kempe (1849–1922)

Alfred Bray Kempe (1849–1922) was born in Kensington, London. He was a barrister and a leading authority on ecclesiastical law. However, having studied mathematics at Cambridge University he retained his interest in it, and later in life he devoted considerable time to mathematical research. Kempe made contributions to mathematical logic and kinematics, the branch of mathematics dealing with motion. Kempe is best remembered for his fallacious proof of the *Four Color Theorem*: Given any separation of a plane into contiguous regions, producing a figure called a map, no more than four colors are required to color the regions of the map so that no two adjacent regions have the same color. Two regions are called adjacent if and only if they share a border segment, not just a point. Kempe was elected a fellow of the Royal Society in 1881. He was the President of the London Mathematical Society during 1892–1894. He was also a mountain climber, mostly in Switzerland. He died in London.

Felix Christian Klein (1849–1925)

Felix Christian Klein (1849–1925) was born in Düsseldorf. After graduating from the Gymnasium in Düsseldorf, he entered the University of Bonn to study mathematics and physics. Klein received his doctorate, supervised by Plücker, in 1868. In his thesis, Klein classified second degree line complexes using Weierstrass' theory of elementary divisors. Later that year Plücker died, leaving his book on the foundations of line geometry unfinished, which Klein later completed. He was appointed as a lecturer at Göttingen in early 1871 and then professor at Erlangen in 1872, at the age of 23. In 1875, he took a chair at the Technische Hochschule at Münich. There he taught advanced courses to Hurwitz, Walther Franz Anton von Dyck (1856–1934), Karl Rohn (1855–1920), Runge, Planck, Luigi Bianchi (1856–1928), and Ricci-Curbastro. After 5 years in Münich, Klein was appointed to a chair of geometry at Leipzig. In 1886, he accepted a chair at the University of Göttingen, where he remained until his retirement in 1913. He reestablished Göttingen as the world's leading mathematics research center. Due to his efforts, Göttingen began admitting women in 1893. He supervised the first Ph.D. thesis in mathematics written at Göttingen by a woman, Grace Chisholm Young (1868–1944), who later married William Henry Young. Under Klein's editorship, the journal Mathematische Annalen became one of the best in the world. In 1870, with Sophus Lie, he discovered the fundamental properties of the asymptotic lines on the Kummer surface. They also investigated W-curves, which are curves invariant under a group of projective transformations. In 1871 and 1873, Klein published two works on non-Euclidean geometry in which he showed that non-Euclidean geometry is consistent if and only if Euclidean geometry is consistent. This put non-Euclidean geometry on an even footing with Euclidean geometry. His 1872 Erlanger Program, which regarded geometry as the study of the properties of a space that are invariant under a given group of transformations, gave a unified approach to geometry and was a hugely influential synthesis of much of the mathematics of the day. Klein developed a theory of automorphic functions, connecting algebraic and geometric results in his important 1884 book on the icosahedron. With Robert Fricke (1861–1930), who came to Leipzig in 1884, Klein wrote a major four volume classic on automorphic and elliptic modular functions that was produced over the following 20 years. Other major contributions that are named after him are the Klein bottle, the Klein four-group, the Klein quartic, the Kleinian group, Klein geometry, the Klein quadric, and the Klein model. He became an Honorary Member of the London Mathematical Society in 1875 and was awarded the De Morgan Medal in 1893. In 1885, he became an FRS and was awarded its Copley Medal in 1912. He retired in the following year due to ill health, but he continued to teach mathematics at his home for several more years. He died in Göttingen in 1925. Klein's favorite maxim was, "Never be dull."

Sofia Vasilyevna Kovalevskaya (1850–1891)

Sofia Vasilyevna Kovalevskaya (1850–1891) was born in Russia. In 1869, Sofia traveled to Heidelberg to study mathematics and the natural sciences, only to discover that women could not matriculate at the university. Eventually, she persuaded the university authorities to allow her to attend lectures unofficially, provided that she obtained the permission of each of her lecturers. In 1871, Kovalevskaya moved to Berlin to study with Weierstrass. Despite the efforts of Weierstrass and his colleagues the senate refused to permit her to attend courses at the university. Ironically, this actually helped her, since over the next 4 years Weierstrass tutored her privately. By the spring of 1874, Kovalevskaya had completed three papers. Weierstrass deemed each of these worthy of a doctorate. The three papers were on partial differential equations, Abelian integrals, and Saturn's rings. In 1874, Kovalevskaya was finally granted her doctorate, summa cum laude, from Göttingen University. In 1886, she was awarded the Prix Bordin for her paper *Mémoire sur un cas particulier du probléme de le rotation d'un corps pesant autour d'un point fixe, oú l'intégration s'effectue á l'aide des fonctions ultralliptiques du temps*. In recognition of the brilliance of this work the prize money was raised from 3,000 to 5,000 francs. Kovalevskaya's further research on this subject won her a prize from the Swedish Academy of Sciences in 1889, and Stockholm appointed her a full professor of mathematics. In the same year, on the initiative of Chebyshev, Kovalevskaya was elected a corresponding member of the St. Petersburg Academy of Sciences. She died of influenza, which at the time was epidemic, in 1891 at age 41 after returning from a pleasure trip to Genoa. She is buried in Solna, Sweden. Her motto was: "Say what you know, do what you must, come what may." On her death, Mittag-Leffler wrote: "She came to us from the center of modern science full of faith and enthusiasm for the ideas of her great master in Berlin, the venerable old man who has outlived his favorite pupil. Her works, all of which belong to the same order of ideas, have shown by new discoveries the power of Weierstrass' system."

Jörgen Pederson Gram (1850–1916)

Jörgen Pederson Gram (1850–1916) was a Danish actuary, and **Erhard Schmidt** (1876–1959) was a German mathematician. Their names are as closely linked in the world of mathematics as the names of William Schwenck Gilbert (1836–1911) and Arthur Sullivan (1842–1900) in the world of musical theater. However, Gram and Schmidt probably never met, and the Gram–Schmidt process that bears their names was not discovered by either one of them. Gram's name is linked to the process as a result of his Ph.D. thesis, in which he used it to solve least squares problems; Schmidt's name is linked to it as a result of his studies of certain kinds of vector spaces. Gram loved the interplay between theoretical mathematics and applications, and he wrote four mathematical treatises on forest management, whereas Schmidt was primarily a theoretician. He taught at several leading German universities and made important contributions to the study of integral equations and partial differential equations. In 1908, he wrote a paper on infinitely many linear equations of infinitely many unknowns, in which he founded the theory of Hilbert spaces. Schmidt's work significantly influenced the direction of mathematics in the twentieth century. During World War II (1939–1945), Schmidt held positions of authority at the University of Berlin and had to carry out various Nazi resolutions against the Jews, a job that he apparently did not do well, since he was criticized at one point for not understanding the "Jewish question." At the celebration of Schmidt's 75th birthday in 1951, a prominent Jewish mathematician who had survived the Nazi years spoke of the difficulties that Schmidt faced during that period, without criticism.

Carl Gustav Axel Harnack (1851–1888)

Carl Gustav Axel Harnack (1851–1888) was born to the theologian Theodosius Harnack (1817–1889) in Tartu, the second largest city of Estonia, which is often considered to be the intellectual center of the country, especially since it is the home of Estonia's oldest and most renowned university. He was the twin brother of theologian Adolf von Harnack (1851–1930), who long outlived him. Carl contributed to potential theory. He is well known for Harnack's inequality applied to harmonic functions. In complex analysis, Harnack's principle or Harnack's theorem is one of several closely related theorems about the convergence of sequences of harmonic functions, which follows from Harnack's inequality. He also worked on the real algebraic geometry of plane curves, proving Harnack's curve theorem for real plane algebraic curves. Carl passed away in Dresden, the capital city of the Free State of Saxony in Germany, at the age of only 36.

Carl Louis Ferdinand von Lindemann (1852–1939)

Carl Louis Ferdinand von Lindemann (1852–1939) was born in Hanover (Germany) and studied mathematics at Göttingen, Erlangen, and Munich. He obtained his doctorate at Erlangen on non-Euclidean geometry, under the supervision of Felix Klein. He taught at Würzburg and Freiburg. In 1882, while Lindemann was in Freiburg, he proved that π is transcendental. His result showed at last that the age-old problem of squaring the circle with a ruler-and-compass construction is impossible. Lindemann then spent several years working on his proof of Fermat's Last Theorem, which is unfortunately wrong. He also worked on projective geometry, Abelian functions, and developed a method of solving equations of any degree using transcendental functions. From Freiburg he moved to the University of Königsberg, where he supervised the doctoral theses of Hilbert, Minkowski, and Arnold Sommerfeld (1868–1951).

Jules Henri Poincaré (1854–1912)

Jules Henri Poincaré (1854–1912) was born in Nancy, where his father was Professor of Medicine at the University. He was ambidextrous and nearsighted. During his childhood he had poor muscular coordination and was seriously ill for a long time with diphtheria. In 1862, Henri entered the Lycée in Nancy, which was renamed the Lycée Henri Poincaré in his honor. During his 7 years' stay at the Lycée, he proved to be the top student in every topic he studied. His mathematics teacher described him as a "monster of mathematics." Henri entered the École Polytechnique in 1873 and graduated in 1875. Henri received his doctorate in mathematics from the University of Paris in 1879. He began his academic career at Caen in 1879, but only 2 years later he was appointed to a Chair in the Faculty of Science in Paris. In 1886, Poincaré was nominated for the Chair of Mathematical Physics and Probability at the Sorbonne. He remained there for the rest of his life, lecturing on a different subject each year. In his lectures, which were edited and published by his students, he treated virtually all known fields of pure and applied mathematics (including many that were not known until he discovered them) with great originality and mastery of technique. Altogether, he produced more than 30 technical books on mathematical physics and celestial mechanics, and almost 500 research papers on mathematics. He also wrote half a dozen popular books about science, including *The Value of Science* and *Science and Hypothesis*. In *Foundation of Science* he talks about mathematical creativity and problem solving. He was a quick, powerful, and restless thinker, not given to lingering details, and was described by one of his contemporaries as "a conquerer, not a colonist." According to him, mathematicians do not deal in objects but in relations between objects; therefore, they are free to replace some objects by others so long as the relations remain unchanged. To them, content is irrelevant: they are only interested in form. Poincaré also had the advantage of a prodigious memory, any book that he read (at incredible speed) became a permanent possession, and he could always state the page and line where a particular thing occurred. Poincaré habitually did his mathematics in his head as he paced back and forth in his study, writing it down only after it was complete in his mind. Poincaré tried for 15 days to prove that a certain type of function could not exist, but without success. On one sleepless night: "Ideas came in crowds. I felt them collide until pairs interlocked, making a stable combination." By morning he was able to prove the existence of one class of the functions in question, which took only a few hours. Around this time, Poincaré went on a geology field trip and his mathematical work was forgotten. One day when boarding a bus he experienced a flash of insight: "At the moment when I put my foot on the step the idea came to me, without anything in my former thoughts seeming to have paved the way for it." This idea was the connection between the functions in question and non-Euclidean geometry. Poincaré felt at the time a "perfect certainty" about his idea. Once back home, he verified it at his leisure. He was elected to the Academy of Science at the very early age of 32. The academician who proposed him for membership said that "his work is above ordinary praise, and reminds us inevitably of what Jacobi wrote of Abel—that he had settled questions which, before him, were unimagined."

Poincaré's first great achievement in mathematics, before the age of 30, was in analysis. He generalized the idea of the periodicity of a function by creating his theory of automorphic functions. He applied them to solve linear differential equations with algebraic coefficients, and also showed how they can be used to uniformize algebraic curves, that is, to express the coordinates of any point on such a curve by means of single-valued functions $x(t)$ and $y(t)$ of a single parameter t. In the 1880s and 1890s, automorphic functions developed into an extensive branch of mathematics, involving (in addition to analysis) group theory, number theory, algebraic geometry, and non-Euclidean geometry. Poincaré is also considered to be the originator of the theory of analytic functions of several complex variables. In 1883, he used the Dirichlet principle to prove that a meromorphic function of two complex variables is a quotient of two entire functions. Another focal point of his thought was his research in celestial mechanics; he produced his *Les Méthodes Nouvelle de la Mécanique Céleste* in three volumes during 1892–1899. In the course of this work he developed his theory of asymptotic expansions, studied the stability of orbits, and initiated the qualitative theory of nonlinear differential equations. His celebrated investigation into the evolution of celestial bodies led him to study the equilibrium shapes of a rotating mass of fluid held together by gravitational attraction. Many of the problems he encountered in this period were the seeds of new ways of thinking, which have grown and flourished in twentieth century mathematics. He was the first person to think of chaos, in connection with his work in astronomy. His attempts to master the qualitative nature of curves and surfaces in higher dimensional spaces resulted in his famous memoir *Analysis situs* (1895), which marks the beginning of the modern era in algebraic topology. In his study of periodic orbits he founded the subject of topological dynamics. Another remarkable discovery in this field, now known as the Poincaré recurrence theorem, relates to the long-range behavior of conservative dynamic systems. This result seemed to demonstrate the futility of contemporary efforts to deduce the second law of thermodynamics from classical mechanics, and the ensuing controversy was the historical source of modern ergodic theory. One of the most striking of Poincaré's many contributions to mathematical physics was his famous paper of 1906 on the dynamics of the electron. He had been thinking about the foundations of physics for many years and had obtained many of the results of the special theory of relativity, independently of Einstein. The main difference was that Einstein's treatment was based on elemental ideas relating to light signals, while Poincaré's was founded on the theory of electromagnetism and was therefore limited in its applicability to phenomena associated with this theory. Poincaré had a high regard for Einstein's abilities and in 1911 recommended him for his first academic position.

All his life, Poincaré was very fond of animals. He frequently forgot his meals and almost never remembered whether or not he had breakfast. Poincaré is often heralded as the last universalist and the greatest mathematician of his generation. He said, "The scientist does not study nature because it is useful; he studies it because he delights in it, and he delights in it because it is beautiful. If nature were not beautiful, it would not be worth knowing, and if nature were not worth knowing, life would not be worth living. Of course, I do not here speak of that beauty that strikes the senses, the beauty of quantities and appearances; nor that I undervalue such beauty, far from it, but it has nothing to do with science; I mean that profounder beauty which comes from the harmonious order of the parts, and which a pure intelligence can grasp." The English logician and follower of Russell, Philip Edward Bertrand Jourdain (1879–1919) wrote in his obituary, "One of the many reasons for which he will live is that he made it possible for us to understand him well as to admire him." Jourdain had an interesting calling card. On one side was written "The statement on the other side of this card is true." On the other side was written "The statement on the other side of this card is false." This is the liar's paradox, which was first formulated by Eubulides (about 400 BC), a Greek philosopher who asked, "Does the liar speak the truth when he says he is lying"? According to Painlevé, Poincaré was "the living brain of the rational sciences." The following quotations of Poincaré will always be remembered: "Mathematicians are born, not made" and "Mathematics is the art of giving the same name to different things."

Andrei Andreyevich Markov (1856–1922)

Andrei Andreyevich Markov (1856–1922) was born in Ryazan, Russia, and died in St. Petersburg, Russia. At the age of 21, he was awarded the gold medal for his outstanding solution to the problem "About Integration of Differential Equations by Continuous Fractions with an Application to the Equation." He defended his theses for his master's and doctorate degrees at St. Petersburg University in 1880 and 1885, respectively. After serving the same university in various capacities, he became a Professor of Mathematics in 1894. At that time he became involved in liberal movements and expressed his opposition to the tsarist regime. His early contributions are in number theory and analysis; however, he is best known for his work on theory of stochastic processes. He later developed the field that is now known as Markov chains. He used them to analyze the alternation of vowels and consonants in the poem *Eugene Onegin* by Alexander Sergeyevich Pushkin (1799–1837), a great Russian poet of the Romantic era and a founder of modern Russian literature. Markov believed that the only applications of his chains were to the analysis of literary works. Today, Markov chains are widely used in many applications, such as modern physics, the fluctuations of stock prices, and genetics. He and his younger brother, **Vladimir Andreevich Markov** (1871–1897), proved the famous Markov's inequality. His son, **Andrey Andreevich Markov**

(1903–1979), was also a leading mathematician, who made significant contributions in constructive mathematics and recursive function theory.

Carl David Tolmé Runge (1856–1927)

Carl David Tolmé Runge (1856–1927) was born in Bremen, Germany. At the age of 19, Runge enrolled at the University of Munich to study literature; however, after just a few weeks he changed to mathematics and physics. Runge attended courses with Planck and they became close friends. In 1877, both went to Berlin, but influenced by Kronecker and Weierstrass, Runge turned to pure mathematics and obtained his doctoral degree in 1880. He obtained a Chair at Hanover in 1886 and remained there for 18 years. In 1904, Runge moved to Göttingen and continued there until his retirement in 1925. He is known for his work in complex variable theory. Runge devised a numerical method to solve ordinary differential equations around 1895. In mathematics, Runge also did notable work in the field of Diophantine equations. In physics, Runge is remembered for his work on the Zeeman effect (the spectral lines in the emission spectrum of an element are affected by the presence of a magnetic field). Runge died in 1927, in Göttingen, of a heart attack.

Charles Emile Picard (1856–1941)

Charles Emile Picard (1856–1941) was born in Paris, France. His mother was the daughter of a medical doctor and his father was the manager of a silk factory, who died during the siege of Paris in 1870. After his secondary education at Lycée Napoléon, Picard took the entrance examinations for École Polytechnique and École Normale Supérieure; he placed second and first, respectively, in these examinations. In 1874, he entered the École Normale, where he received his doctorate in 1877, and he remained there for a year as an assistant. He was appointed lecturer at the University of Paris in 1878 and then professor at Toulouse in 1879. In 1881, he was nominated for membership of the mathematics section of the Académie des Sciences. During this year he married Hermite's daughter. Picard and his wife had three children, a daughter and two sons, who were all killed in World War I (1914–1918). His grandsons were wounded and captured in World War II. In 1885, Picard was appointed to the Chair of Differential Calculus at the Sorbonne in Paris. In 1897, he left this position for the Chair of Analysis and Higher Algebra so that he could train research students. He was elected to the Académie des Sciences in 1889 and was its permanent Secretary from 1917 until his death in 1941. Picard was awarded the Poncelet Prize in 1886 and the Grand Prix des Sciences Mathématiques in 1888. He was awarded honorary doctorates from five universities and honorary membership of 37 learned societies, including the London Mathematical Society in 1898 and FRS in 1909. He was made the President of the International Congress of Mathematicians (first held in 1897) at Strasbourg in 1920. Picard received the Grande Croix de la Légion d'Honneur in 1932 and the Mittag-Leffler Gold Medal in 1937.

Picard made two outstanding contributions to analysis: his method of successive approximations, which enabled him to perfect the theory of differential equations that Cauchy had initiated in the 1820s; and his famous theorem (called Picard's Great Theorem) about the values assumed by a complex analytic function near an essential singularity, which has stimulated much important research up to the present day. He also did very significant work in the field of algebraic geometry. His three-volume masterpiece *Traité d'analyse* was published between 1891 and 1896. He is also remembered for the Picard functor, the Picard group, Picard's theorem, Picard variety, the Picard–Lefschetz (Solomon Lefschetz, 1884–1972) formula, the Picard–Lindelöf (Ernst Leonhard Lindelöf, 1870–1946) theorem, and Painlevé transcendents. During 1894–1937, he trained over 10,000 engineers at the École Centrale des Arts et Manufactures. Like a true Frenchman, he was a connoisseur of fine food and was particularly fond of bouillabaisse. He died in 1941 in Paris, France.

Alexander Mikhailovich Liapunov (1857–1918)

Alexander Mikhailovich Liapunov (1857–1918) was educated at home before attending the Gymnasium at Nizhny Novgorod. He graduated in 1876 and entered the faculty of physics and mathematics at St. Petersburg University, where he was taught by Chebyshev. Liapunov obtained his degree in 1880 and published two papers on hydrostatics in 1881. In 1885, he defended his master's thesis, *On the stability of ellipsoidal forms of equilibrium of a rotating liquid.* Immediately afterward, Liapunov joined Kharkov University to teach mechanics and continue research for his doctoral thesis. He presented his doctoral thesis *The General Problem of the Stability of Motion* to the University of Moscow and was awarded his doctorate in 1892. This classic work, which has influenced the development of the subject over a long period of time, was originally published in 1892 in Russian but is now available in its English translation, *Stability of Motion*, Academic Press, New York, 1966. At Kharkov University he played a major role in the Kharkov Mathematical Society, serving as its Vice President from 1891 to 1898 and then President from 1899 until he left Kharkov in 1902. He also edited the Communications of the Kharkov Mathematical Society. In 1901, Liapunov was elected to the Russian Academy of Sciences in St. Petersburg, and in 1902, he became an academician in applied mathematics of the Academy. During 1900–1901, Liapunov published two papers on probability theory, proving the central limit theorem using a technique based on characteristic functions. In 1908, he participated at the fourth Mathematical Congress in Rome. At this time he took part in the publication of Euler's selected works and edited two volumes. In 1917, he moved to Odessa because of his wife's frail health. In the spring of 1918 his wife's health began to deteriorate rapidly, and she died on October 31, 1918. Liapunov shot himself later that day and died 3 days later in a hospital. He was honored for his outstanding contributions by the Accademia dei Lincei in 1909 and the French Academy of Sciences in 1916.

Karl (Carl) Pearson (1857–1936)

Karl (Carl) Pearson (1857–1936) was an English mathematician and mathematical statistician, born in Islington, England. In 1879, while enrolling at the University of Heidelberg, Germany, "Carl Pearson" inadvertently became "Karl Pearson" due to a spelling error. He used both variants of his name until 1884, when he permanently adopted Karl. He was an influential English mathematician and is credited for establishing the discipline of mathematical statistics. Carl was educated privately at University College School, an independent school in Hampstead, northwest London. He studied mathematics at King's College, Cambridge, in 1876, graduating in 1879 as Third Wrangler in the Mathematical Tripos. He then went to Germany to study physics at the University of Heidelberg under Georg Hermann Quincke (1834–1924), a physicist, and metaphysics under Ernst Kuno Berthold Fischer (1824–1907), a German philosopher, historian of philosophy, and critic. He next went to the University of Berlin, where he attended lectures on Darwinism by Emil du Bois-Reymond (1818–1896), a German physician and renowned physiologist who discovered nerve action potential. In Berlin, he also studied Roman Law, medieval and sixteenth century German literature, and socialism, becoming quite familiar with German literature. On his return to England in 1880, Pearson went to Cambridge. His first book, *The New Werther*, was published at this time. In it he gives a clear indication of why he studied so many diverse subjects, varying so widely in nature: "I rush from science to philosophy, and from philosophy to our old friends the poets; and then, over-wearied by too much idealism, I fancy I become practical in returning to science. Have you ever attempted to conceive all there is in the world worth knowing—that not one subject in the universe is unworthy of study? The giants of literature, the mysteries of many-dimensional space, the attempts of Ludwig Eduard Boltzmann (1844–1906) and Crookes to penetrate Nature's very laboratory, the Kantian theory of the universe, and the latest discoveries in embryology, with their wonderful tales of the development of life—what an immensity beyond our grasp!... Mankind seems on the verge of a new and glorious discovery. What Newton did to simplify the planetary motions must now be done to unite in one whole the various isolated theories of mathematical physics." During 1881–1882, Carl decided to study law so that he might, like his father, be called to the bar. However, he never practiced law. During 1882–1884, he lectured around London on a wide variety of topics, such as German social life, the influence of Martin Luther, and history. He also wrote several essays, articles, and reviews, and substituted for professors of mathematics at King's College and University College London. Karl was appointed to the Goldsmid Chair of Applied Mathematics and Mechanics at University College London in 1884. He became the editor of

Common Sense and the Exact Sciences in 1885 and the Professor of Geometry at Gresham College in 1891, where he came in contact with Professor Walter Frank Raphael Weldon (1860–1906), an FRS, English zoologist, evolutionary biologist, a founder of biometry, and later the joint founding editor of *Biometrika* along with Galton and Pearson. Weldon was working on some interesting problems requiring quantitative solutions; his collaboration with Pearson, in biometry and evolutionary theory, was fruitful and lasted until Walter passed away in 1906. Walter introduced Pearson to Darwin's cousin Galton, who was interested in aspects of evolution like heredity and eugenics. Galton left what remained of his estate to the University of London for the creation of a Chair in Eugenics at the time of his death. According to his wishes, Karl was appointed to the Galton Chair of Eugenics, which was later named the Galton Chair of Genetics. Karl created the department of Applied Statistics, which included the Biometric and Galton laboratories. He remained with the department until his retirement in 1933, working until his death in Coldharbour, Surrey. Among other awards received by Karl, the Darwin Medal (1898), the Rudolf Virchow medal from the Berliner Anthropologische Gesellschaft (1932), FRS (1896), and a knighthood (1935) (offered and refused) are notable.

Charlotte Angas Scott (1858–1931)

Charlotte Angas Scott (1858–1931) was born in England. Her father was President of Lancashire College, a Congregational minister, and a social reformer. He encouraged his daughter to pursue a university education, which was unusual for women in those days. She joined Hitchin College, which was renamed to Girton College, part of Cambridge University. All the girls in the class faced severe restrictions on their participation and activities. She was unofficially permitted to take the traditional oral exam at the end of Cambridge's program. She secured eighth position in the overall ranking, including all of the male students. However, at the award ceremony the women's names were not included in the ranking that was read. Charlotte continued to graduate studies at the University of London while serving as a lecturer at Girton College. She was the first woman to receive a doctorate from the University of London. In 1885, she moved to the USA to join the first faculty of the newly founded Bryn Mawr College in Pennsylvania, the first women's college to offer graduate degrees. She published 20 research articles and three books: *An Introductory Account of Certain Modern Ideas and Methods in Plane Analytical Geometry*, first edition in 1894, second edition in 1924, and a third edition in 1961, published as *Projective Methods in Plane Geometry*; *A Proof of Noether's Fundamental Theorem* in 1899; and *Cartesian Plane Geometry, Part I: Analytical Conics* in 1907. In 1909, Charlotte was given the first Endowed Chair at Bryn Mawr, in recognition of her achievements. Charlotte was a member of the council that transformed the New York Mathematical Society into the American Mathematical Society in 1895, and she served as the society's Vice President in 1905. She was coeditor of the American Journal of Mathematics in 1899 and continued editing for that journal until her retirement. Charlotte never married, and she often visited her relatives in England. She died in England in 1931.

Guiseppe Peano (1858–1932)

Guiseppe Peano (1858–1932) was born to a poor farming family in Spinetta, Piedmont, Italy. He attended primary school in Cuneo, and in 1870, his uncle took him to Turin for secondary schooling and preparation for university studies. Peano graduated in 1876 and entered the University of Turin in the same year to study engineering, but he later changed to mathematics. Peano obtained his doctoral degree in mathematics in 1880 with high honors, and the University immediately employed him to assist Enrico D'Ovidio (1842–1933) and then Angelo Genocchi (1817–1889), the Chair of Infinitesimal Calculus. He published his first mathematical paper in 1880 and three subsequent papers in 1881. Peano took over some of Genocchi's classes and revised and edited his lectures in the book *Calcolo Differenziale e Principii di Calcolo Integrale*, which was published in 1884. In 1886, Peano proved that if $f(x, y)$ is continuous then the initial value problem $y' = f(x, y)$, $y(x_0) = y_0$ has a solution. Unfortunately, his proof was inadequate, and a satisfactory proof was not found until many years later. The existence of solutions with a stronger hypothesis on f had been given earlier by Cauchy and then Lipschitz. In 1886, he began teaching concurrently at the Royal Military Academy. In 1887, Peano married Carola Crosio, but had no children. He published his first book in the same year, *Applicazioni Geometriche del Calcolo Infinitesimale*, which began with mathematical logic and introduced modern symbols for the union and intersection of sets for the first time. This book also contains axioms for a vector space, and the concept of dimension for general vector spaces. His results in the book were not appreciated by many of his contemporaries. In 1889, Peano published the following axioms, which define the natural numbers in terms of sets:

1. 1 is a natural number.
2. For each natural number n there is a unique successor $n + 1$.
3. 1 is not the successor of any natural number.
4. Two natural numbers are equal if their successors are equal.
5. Any set of natural numbers which contains 1 and the successor of every natural number p whenever it contains p is the set $\{1, 2, 3, \dots\}$ of natural numbers.

These landmark axioms appeared in a pamphlet *Arithmetices principia, nova methodo exposita*; however, his axioms have led generations of students to wonder whether all of modern algebra is some kind of conspiracy to render the obvious obscure. In 1889, Peano was promoted to Professor First Class at the Royal Military Academy. In 1890, he astounded the mathematical world with his remarkable construction of a continuous curve in parametric form $x = f(t)$, $y = g(t)$ that passes through every point of a square at least once as the parameter t varies from 0 to 1. This curve, now known as the *Peano curve*, destroyed the notion that curves are one-dimensional objects and squares are two-dimensional objects. Peano's discovery motivated mathematicians to reexamine the entire concept of dimension and work toward putting it on a precise mathematical footing. This was an early example of what came to be known as a fractal. In 1890, he was appointed Extraordinary Professor of Infinitesimal Calculus (Genocchi's chair), and in 1891, Peano founded *Rivista di matematica*, a journal devoted mainly to logic and the foundations of mathematics. The first paper in the Journal is by Peano, summarizing his work on mathematical logic up to that time. The same year he was elected a member of the Academy of Sciences in Turin. In 1892, he embarked on an extremely ambitious project, *Formulaire de mathématiques*. It was to be an "encyclopedia of mathematics," containing all known formulas and theorems of mathematical science using a standard notation that was invented by him. The project was finished in 1908, and the fifth and the final version of the book contained over 4,200 symbolized formulas and theorems, with proofs, in only 516 pages. The Formulaire, though impressive in its accomplishment, received little interest in the mathematical community. This was because of confusion over the symbols used in Peano's universal language *Latino sine flexione*, which nobody really understood. In 1908, Peano took over the Chair of Higher Analysis at Turin for 2 years. In 1925, Peano unofficially switched chairs from Infinitesimal Calculus to Complementary Mathematics, a field that better suited his style of mathematics. This move became official in 1931. Peano continued teaching at Turin University until the day before he died, when he suffered a fatal heart attack.

Peano's work strongly influenced Hilbert's axiomatic treatment of plane geometry and the work of Whitehead and Russell on mathematical logic. He was also active in organizations for primary and secondary education. In 1901 and 1905, Peano was honored by the Italian government with several knighthoods. In 1905, he was elected a corresponding member of the Accademia dei Lincei in Rome, the highest honor for Italian scientists. In 1917, Peano was made an officer of the Crown of Italy, and in 1921, he was promoted to Commendatore of the Crown of Italy. He published over 200 books and research papers.

Ernesto Cesáro (1859–1906)

Ernesto Cesáro (1859–1906) was born in Naples, Italy, to a farmer's family. He studied in Naples, Nola, and at the École des Mines in Liége. During his studies at Liége, Cesáro visited Paris and attended lectures by Hermite, Darboux, Serret Briot, Jean-Claude Bouquet (1819–1885), and Chasles at the Sorbonne. In 1884, Luigi Cremona (1830–1903), Giuseppe Battaglini (1826–1894), and Ulisse Dini (1845–1918) supported him in his application for a scholarship to do research at the University of Rome. In just 2 years, Cesáro wrote eighty works on infinite arithmetics, isobaric problems, holomorphic functions, the theory of probability, and intrinsic geometry. He was awarded his doctorate in 1887. In the same year, Cesáro accepted the Chair of Mathematics at Palermo, where he remained until 1891, when he moved to Naples where he held the Chair of Mathematical Analysis until his death.

Cesáro's main contribution is in differential geometry and the theory of numbers. He studied the distribution of primes, trying to improve the results of Chebyshev. He is also known for the Stolz-Cesáro (Otto Stolz, 1842–1905) theorem, Cesáro's theorem, and for his 'averaging' method for the summation of divergent series, known as the Cesáro mean. He wrote very successful texts on calculus and mathematical physics. Two further works on the mathematical theory of heat and hydrodynamics were being prepared at the time of his death. Cesáro died due to injuries that he sustained while trying to save his drowning son.

Otto Ludwig Hölder (1859–1937)

Otto Ludwig Hölder (1859–1937) was a German group theorist born in Stuttgart. He was a student of Kronecker, Weierstrass, and Kummer. He is well known for his work with the summability of series, Hölder's inequality, and the Jordan–Hölder theorem.

Alicia Boole Stott (1860–1940)

Alicia Boole Stott (1860–1940) was the third daughter of the mathematician George Boole, born in Cork, Ireland. George Boole died when Alicia was only 4 years old, so she was brought up by her grandmother and great-uncle. When she was 12 years old she went to London to live with her mother and sisters. She is best known for coining the term "polytope" to refer to a convex solid in four dimensions, and having an impressive grasp of four-dimensional geometry from a very early age. She found that there were exactly six regular polytopes on four dimensions and that they are bounded by 5, 16, or 600 tetrahedra, 8 cubes, 24 octahedra, or 120 dodecahedra. She then produced three-dimensional central cross sections of all the six regular polytopes by purely Euclidean constructions and synthetic methods for the simple reason that she had never learned any analytic geometry. She made beautiful cardboard models of all these sections. She met and married Walter Stott in 1890. Stott learned of Pieter Schoute's (1846–1923) work on central sections of the regular polytopes in 1895, and Alicia Stott sent him photographs of her cardboard models. Schoute came to England and worked with Alicia Stott, persuading her to publish her results, which she did, in two papers published in Amsterdam in 1900 and 1910. The University of Groningen honored her by inviting her to attend the tercentenary celebrations of the university and awarding her an honorary doctorate in 1914. In 1930, she was introduced to Harold Scott MacDonald Coxeter (1907–2003) and they worked together on various problems. Alicia Stott made two further important discoveries concerning constructions of polyhedra related to the golden section. Coxeter described his time doing joint work with her saying: "The strength and simplicity of her character combined with the diversity of her interests to make her an inspiring friend."

Vito Volterra (1860–1940)

Vito Volterra (1860–1940) was an eminent Italian mathematician whose interest in mathematics began at age 11 when he studied Legendre's Geometry. At the age of 13, he had already considered the three-body problem and made some contributions by partitioning the time into small intervals over which the force could be considered to be constant. Volterra's family was extremely impoverished (his father died when he was just 2 years old), but after attending lectures at Florence he was able to proceed to Pisa in 1878. There he studied under Enrico Betti, graduating as a doctor of physics in 1882. His thesis on hydrodynamics included some results of Stokes, which Volterra discovered later, independently. Volterra became Professor of Mechanics at Pisa in 1883. After Betti's death, he occupied the Chair of Mathematical Physics. He was later appointed to the Chair of Mechanics at Turin and then to the Chair of Mathematical Physics in Rome in 1900. His early work on integral equations (along with work by Fredholm and Hilbert) began the full-scale development of linear analysis that dominated so much of mathematics during the first half of the twentieth century. One of the major classes of integral equations is named for him. During World War I, Volterra joined the Air Force. He made many journeys to France and England in order to promote scientific collaboration. After the war, he returned to the University of Rome; at that time his interest shifted to mathematical biology. He enriched both mathematics and biology. Fascism overtook Italy in 1922, and Volterra fought valiantly against it in the Italian Parliament. In 1930, however, the Parliament was abolished; when Volterra refused to take an oath of allegiance to the Fascist Government in 1931 he was forced to resign from the University of Rome. He spent the rest of his life living abroad, mostly in Paris. In 1938, Volterra was offered an honorary degree by the University of St. Andrews, but his doctor would not allow him to travel abroad to receive it. It is noteworthy that Volterra gave a total of four plenary lectures to various International Congresses of mathematics, more than any other scholar in history.

Ivar Otto Bendixson (1861–1935)

Ivar Otto Bendixson (1861–1935) was a Swedish mathematician. He obtained his doctorate in 1890 from Uppsala University and was appointed as a docent at Stockholm University. He then worked as an assistant to the Professor of Mathematical Analysis from 1891 until 1892. During 1892–1899, he taught at the Royal Technological Institute in Stockholm and at Stockholm University. In

1899, Bendixson substituted for the Professor of Pure Mathematics at the Royal Technological Institute, where he was promoted to professor in 1900. In 1905, he became Professor of Higher Mathematical Analysis at Stockholm University, and from 1911 until 1927 he was the rector of the University. His first research work was on set theory and the foundations of mathematics, in which he closely followed the ideas of Cantor. He contributed important results in point set topology; one of these states that "every uncountable closed set can be partitioned into a perfect set and a countable set." The perfect set in this partition is now called the Bendixson derivative of the original set. Hence the derived set is either of Baire class 1 or Baire class 2. He also gave an example of a perfect set that is totally disconnected. Bendixson also made fundamental contributions to the classical problem of the algebraic solution of equations. He extended Abel's methods to describe precisely which equations could be solved by radicals. The Poincaré–Bendixson theorem, which says an integral curve that does not end in a singular point has a limit cycle, was first proved by Poincaré but a more rigorous proof with weaker hypotheses was given by Bendixson in his memoir in 1901. This result has entered into the textbooks. As a teacher he was always well prepared. For his outstanding contributions, Bendixson received many honors.

Alfred North Whitehead (1861–1947)

Alfred North Whitehead (1861–1947) was born in Ramsgate, Kent (England). Alfred was taught at home until the age of 14. He graduated in 1884 from Trinity College, Cambridge, and became a lecturer of mathematics there, a post he held until 1911. During 1911–1914, Whitehead was a lecturer in applied mathematics and mechanics at the University of London and Professor of Mathematics from 1914–1924. In 1924, he moved to Harvard as Professor of Philosophy and remained there for the rest of his life. He died in 1947 at the age of 86. His life is often described as having three distinct phases: mathematician and logician (1884–1910), philosopher of science (1910–1924), and philosopher of metaphysics (1924–1947). In mathematics he extended the range of algebraic procedures. In philosophy, with Russell, he wrote three volumes of *Principia Mathematica* that together comprise more than 2,000 pages. In metaphysics, he developed a comprehensive system that is known as Process Philosophy. Whitehead asserted the essential inter-relationship of matter, space, and time; that objects may be understood as a series of events and processes. His other works include *A Treatise on Universal Algebra with Applications* (1898), *An Introduction to Mathematics* (1911), *The Organization of Thought Educational and Scientific* (1917), *The Concept of Nature* (1920), *The Principle of Relativity with Applications to Physical Science* (1922), *Science and the Modern World* (1925), *An Enquiry Concerning the Principles of Natural Knowledge* (1925), *Religion in the Making* (1926), *Symbolism, Its Meaning and Effect* (1927), *Process and Reality: An Essay in Cosmology* (1929), *The Aims of Education and Other Essays* (1929), *Function of Reason* (1929), *Adventures of Ideas* (1933),

Nature and Life (1934), *Modes of Thought* (1938), and *Essays in Science and Philosophy* (1947). According to him, "the science of pure mathematics, in its modern developments, may claim to be the most original creation of the human spirit"; "By relieving the brain of all unnecessary work, a good notation sets it free to concentrate on more advanced problems, and, in effect, increases the mental powers of the race"; "While most mathematicians are neither philosophers nor deeply engaged in the methodological investigations of philosophy, mathematics has appealed widely to scholars since 1600 as the exemplar for reaching relatively true statements"; "God is neither omnipotent nor omniscient, but is changed by events as they unfold."

David Hilbert (1862–1943)

David Hilbert (1862–1943), the son of a district judge, was born in Königsberg, East Prussia. In 1872, he entered the Friedrichskolleg Gymnasium, but then enrolled at the Wilhelm Gymnasium where he graduated in 1880. Hilbert then moved to the University of Königsberg and obtained his doctorate in 1885 with a dissertation written under Lindemann entitled Über invariante Eigenschaften spezieller binärer Formen, insbesondere der Kugelfunktionen (On the invariant properties of special binary forms, in particular the spherical harmonic functions). Minkowski and Hurwitz joined the University of Königsberg in 1882 and 1884, respectively, and the three of them often had scientific discussions. During 1886–1895, Hilbert worked at the University of Königsberg as a professor. In 1892, Hilbert married Käthe Jerosch and had a child, Franz Hilbert (1893–1969) in 1893, who suffered his entire life from an undiagnosed mental illness. In 1895, Hilbert obtained the position of chairman of mathematics at the University of Göttingen, where he worked until his retirement in 1930. Hilbert is recognized as one of the most influential and universal mathematicians of the nineteenth and early twentieth centuries. He invented or developed a broad range of fundamental ideas, finding results in invariant theory, the axiomatization of geometry, and introducing the notion of Hilbert space, which is one of the foundations of functional analysis. He adopted and warmly defended Cantor's set theory and transfinite numbers. In 1899, he published a 92-page book *Grundlagen der Geometrie* (The Foundations of Geometry), in which he stated a set of axioms that removed the weaknesses from Euclidean geometry. This book attracted attention all over the mathematical world, and until 1972 at least 11 editions had been published. In the same year, 1899, a 19-year-old American student named Robert Lee Moore independently published an equivalent set of axioms. Some of the axioms in both systems were the same, but there was an interesting feature about those axioms that were different: Hilbert's axioms could be proved as theorems from Moore's, and vice versa. In 1900, Hilbert proposed a list of 23 problems as a challenge to the International Congress of Mathematicians, held in Paris under the presidency of Poincaré, that he felt would be important to mathematicians of the twentieth century. This is generally regarded as the most successful and deeply considered compilation of open problems ever to be produced by an individual mathematician. Some of these were solved within a short time, and others have been discussed throughout the twentieth century, but a few are now assumed to be too open ended to come to closure. Some remain a challenge for mathematicians to this day. In any case, such a large body of new mathematics has arisen from Hilbert's problems that we can say that the problems he posed became the backbone for mathematical research in the twentieth century. The solution of one of Hilbert's problems provides instant recognition in the mathematical community. These problems are:

1(a). *Cantor's problem of the cardinal number of the continuum.* (Is there a transfinite number between that of a denumerable set and the numbers of the continuum?) For this problem Gödel in 1938 and Paul Cohen in 1963 showed that the answer depends on the particular version of set theory assumed. Consistent set theories may either include it or deny it.

1(b). *Can the continuum of numbers be considered a well ordered set*? This has been shown to be impossible to prove or disprove within the Zermelo–Frankel set theory, with or without the Axiom of Choice.

2. *The compatibility of the arithmetical axioms.* (Can it be proven that the axioms of logic are consistent?) There is no consensus on whether the results of Gödel and Gerhard Gentzen (1909–1945) give a solution to the problem as stated by Hilbert. Gödel's second incompleteness theorem, proved in 1931, shows that no proof of its consistency can be carried out within arithmetic itself. Gentzen proved in 1936 that the consistency of arithmetic follows from the well-foundedness of the ordinal ϵ_0.
3. *The equality of the volumes of two tetrahedra of equal bases and equal altitudes.* (Give two tetrahedra that cannot be decomposed into congruent tetrahedra directly or by adjoining congruent tetrahedra.) Max Dehn (1878–1952) in 1902 showed this could not be done by inventing the theory of Dehn invariants, and Venyamin Fedovović Kagan (1869–1953) obtained the same result in 1903, independently.
4. *Problem of the straight line as the shortest distance between two points.* (Find geometries whose axioms are closest to those of Euclidean geometry if the ordering and incidence axioms are retained, the congruence axioms weakened, and the equivalent of the parallel postulate omitted.) This problem was shown by Georg Karl Wilhelm Hamel (1877–1954) to be too vague as stated.
5. *Lie's concept of a continuous group of transformations without the assumption of the differentiability of the function defining the group.* This problem was solved by von Neumann in 1930 for bicompact groups and the Abelian case, and the solvable case was addressed by Deane Montgomery (1909–1992) and Leo Zippin (1905–1995) in 1952, followed by Hidehiko Yamabe (1923–1960) in 1953. Andrew Gleason (1921–2008) showed in 1952 that the answer is also "yes" for all locally bicompact groups.

6. *Mathematical treatment of the axioms of physics.* This problem is partially solved. In particular, von Neumann and others have axiomatized quantum mechanics.
7. *Irrationality and transcendence of certain numbers.* (Is a^b transcendental, for algebraic $a \neq 0, 1$, and irrational algebraic b?) The affirmative answer was given by Gelfond in 1934 and Theodor Schneider (1911–1988) in 1935, now known as the Gelfond–Schneider theorem. Here irrationality of b is necessary.
8. *The Riemann hypothesis.* This conjecture has neither been proved nor disproved.
9. Proof of the most general law of reciprocity in any number field. Solved by Emil Artin (1898–1962) in 1927 for Abelian extensions of the rational numbers, but the non-Abelian case remains open. (After a lecture by Artin someone asked "Your talk was most interesting. But of what use is such work anyway?" Artin's reply was, "I make my living at it!")
10. *Determination of the solvability of a diophantine equation.* (Find an algorithm to determine whether a given polynomial Diophantine equation with integer coefficients has an integer solution.) In 1970, the impossibility of this was shown by Yuri Vladimirovich Matiyasevich (born 1947). It was later independently solved by Gregory Volfovich Chudnovsky (born 1952) at the age of 17. In appreciation, Matiyasevich described Chudnovsky's method as the right approach to solve this problem.
11. Quadratic forms with any algebraic numerical coefficients. In 1924, Helmut Hasse devised a theory of quadratic forms over the rational numbers and later extended this solution to any algebraic number field.
12. *Extension of Kronecker's theorem on Abelian fields to any algebraic realm of rationality.* This problem is unresolved. It calls for the construction of holomorphic functions in several variables that have properties analogous to the exponential function and elliptic modular functions.
13. *Impossibility of solution of the general equation of the seventh degree by means of functions of only two arguments.* This problem has been resolved by Vladimir Igorevich Arnold (1937–2010).
14. *Proof of the finiteness of certain complete systems of functions.* This problem has been resolved with the answer no, by a counterexample, by Masayoshi Nagata (1927–2008) in 1959.
15. *Rigorous foundations of Schubert's enumerative calculus.* This problem has been partially solved, but no general theory has been developed.
16. *Problem of the topology of algebraic curves and surfaces.* The Shimura–Taniyama Conjecture postulates just this connection. This problem is unsolved.
17. *Expressions of definite forms by squares.* Artin showed in 1927 that a positive definite rational function is a sum of squares.
18. *Building up of space from congruent polyhedra.* This problem has been resolved by Karl August Reinhardt (1895–1941) and Thomas Callister Hales (born 1958).

19. *Are the solutions of regular problems in the calculus of variations always necessarily analytic*? This problem has been solved in the affirmative by Ennio de Giorgi (1928–1996) and, independently and using different methods, by John Forbes Nash (born 1928).
20. *The general problem of boundary values*. This problem has been resolved. This was a significant area of research throughout the twentieth century, which culminated in solutions for the nonlinear case.
21. *Proof of the existence of linear differential equations having a prescribed monodromic group*. Several special cases proving their existence have been solved, but a negative solution was found in 1989 by B. Bolibruch.
22. *Uniformization of analytic relations by means of automorphic functions*. This problem has been solved.
23. *Further development of the methods of the calculus of variations*. This problem has been solved.

Hilbert originally included 24 problems in his list, but decided against including one of them in the published list. The "24th problem" (in proof theory, on a criterion for simplicity and general methods) was rediscovered in Hilbert's original manuscript notes by German historian Rüdiger Thiele in 2000.

More important than the specific problems was Hilbert's proclamation of faith that in mathematics there can be no *ignorabimus*. His argument was that it is in the nature of mathematics to pose and solve problems, and there was no possibility of never knowing. The tools of pure thought employed by creative mathematicians should be sufficient to solve any specific mathematical problem. Based on these ideas, in 1910 Hilbert developed a program to axiomatize all of mathematics, and to achieve this goal he began what is known as the "formalist school" of mathematics. According to the formalist, mathematics is a game devoid of meaning in which one plays with symbols according to formal rules that are agreed upon in advance. Although there were already formalists in the nineteenth century, the modern concept of formalism, which includes finitary reasoning, must be credited to Hilbert. It is interesting to observe that both logicists and formalists formalized the various branches of mathematics, but for entirely different reasons. The logicists wanted to use such a formalization to show that the branch of mathematics in question belongs to logic; the formalists wanted to use it to prove mathematically that every branch is free of contradictions. However, like the logicists, Gödel showed in 1931 that formalism cannot be carried out successfully as stated.

Among his 69 Ph.D. students in Göttingen there were many who later became famous mathematicians, including Wilhelm Ackermann (1896–1962), Felix Bernstein, Ludwig Otto Blumenthal (1876–1944), Courant, Max Dehn, Carl Gustav Hempel (1905–1997), Erich Hecke (1887–1947), Hellmuth Kneser (1898–1973), Robert König (1885–1979), Erhard Schmidt, Hugo Steinhaus, Teiji Takagi (1875–1960), Weyl, Zermelo, and the chess champion Emanuel Lasker (1868–1941). Von Neumann was his assistant. During 1900–1914, many mathematicians from the USA, who later played an important role in the development of mathematics, came to Göttingen to study under him. Hilbert was also surrounded by a social circle of some of the most important mathematicians of the twentieth century, such as Emmy Noether and Alonzo Church. He is remembered for the Cayley–Klein–Hilbert metric, Einstein–Hilbert action, the Hilbert class field, the Hilbert cube, the Hilbert curve, the Hilbert function, Hilbert inequality, the Hilbert matrix, the Hilbert polynomial, the Hilbert series, Hilbert space, the Hilbert spectrum, the Hilbert symbol, the Hilbert transform, Hilbert's arithmetic of ends, Hilbert's axioms, Hilbert's basis theorem, Hilbert's constants, Hilbert's irreducibility theorem, Hilbert's Nullstellensatz, Hilbert's paradox of the grand hotel (introduced by physicist George Gamow in 1947), Hilbert's problems, Hilbert's program, Hilbert's theorem (differential geometry), Hilbert's syzygy theorem, Hilbert-style deduction, the Hilbert-Pólya conjecture, the Hilbert–Schmidt operator, the Hilbert–Smith conjecture, the Hilbert–Speiser (Andreas Speiser, 1885–1970) theorem, his principles of theoretical logic, and the relativity priority dispute. In 1904, while developing the theory of integral equations, he introduced the term "eigenvalue" (see Schwarz), which was applied to matrices some time later.

Hilbert praised Cantor's ideas and proclaimed that "no one shall expel us from the paradise which Cantor created." His paradox of the Grand Hotel, a meditation on strange properties of the infinite, is often used in popular accounts of infinite cardinal numbers. He analogized the concept of infinity to a hotel with an infinite number of rooms numbered $1, 2, 3, \ldots$. If a bus full of an infinite number of travelers arrives at the hotel when the hotel is already full, the manager can move the guest already in room 1 to room 2, room 2 to room 4, ..., room n to $2n$. This frees up every odd-numbered room, and thus each traveler on the bus can have a room. Even if an infinite number of buses each with an infinite number of guests arrives, each person can be accommodated. He remarked, "The infinite! No other question has ever moved so profoundly the spirit of man".

Hilbert became an honorary member of the London Mathematical Society in 1901, and an FRS in 1928. During 1902–1939, he was the editor of the Mathematische Annalen, the leading mathematical journal of the time. Hilbert died in 1943. His funeral was attended by fewer than a dozen people, only two of whom were fellow academics. On his tombstone at Göttingen his epitaph reads: "Wir müssen wissen. Wir werden wissen." (We must know. We will know.) According to him, the art of doing mathematics consists of finding the special case that contains all the germs of generality.

The following anecdotes involving Hilbert are often quoted:

There was a party in Hilbert's house, and Mrs. Hilbert noticed that Hilbert had not put on a fresh shirt. She said, "Please go upstairs and put on a fresh shirt." Hilbert went upstairs and took off his coat, his tie, and his shirt. He forgot that he had to go back to the party and in accordance with his usual habit, he went to bed. When his wife came after some time to chide him for his delay, he was fast asleep.

A colleague came to meet Hilbert, took off his hat and put it on the floor, and started talking and did not stop. Hilbert was annoyed that his work had been interrupted, so he put on his colleague's hat and told his wife, "Perhaps we have detained our colleague too long." Then he walked out of the house.

Somebody asked Hilbert, "If you were to be born again after 500 years, what would be the first question you would ask?" His reply was prompt. "I will ask whether anybody has proved Riemann's hypothesis."

William Henry Young (1863–1942)

William Henry Young (1863–1942) was born in London. He was educated at the City of London School and at Peterhouse, Cambridge. Young was Fourth wrangler in 1884, and during 1886–1892 he was a Fellow of Peterhouse. After leaving Cambridge he traveled widely, visiting universities in Europe, America, Asia, and Africa. In 1906, with his wife Grace Chisholm Young he published a very influential book, *The Theory of Sets and Points*. In 1913, he accepted two part-time chairs, Hardinge Professorship of Pure Mathematics at Calcutta University (1913–1917) and Professorship of Philosophy and the History of Mathematics at the University of Liverpool (1913–1919). In 1919, Young was appointed to the Chair of Pure Mathematics at the University College of Wales in Aberystwyth, Wales. He held this post until 1923. He researched the theory of integration and contributed to the theories of Fourier series, orthogonal series, and functions of several variables. He is remembered for Young's inequality and the Hausdorff–Young inequality. Young received many honors for his mathematical achievements: he was elected an FRS in 1907, he received the Sylvester medal from that Society in 1928, he was president of the London Mathematical Society during 1922–1924, he was awarded the Society's De Morgan Medal in 1917, and he was president of the International Mathematical Union (IMU, founded in Strasbourg in 1920) during 1929–1932. He received honorary degrees from the universities of Calcutta, Geneva, and Strasbourg. Young died in Lausanne, Switzerland.

Stanislaw Zaremba (1863–1942)

Stanislaw Zaremba (1863–1942) was born in Romanówka (Ukraine). He was first educated as an engineer, and after obtaining his diploma in 1886 he left his home to study mathematics in Paris. In 1889, he received his doctoral degree at the Sorbonne with the thesis: *Sur un probléme concernant l'état calorifique d'un corps homogéne indéfini*. He stayed in France until 1900, when he moved to Poland and joined the faculty at the Jagiellonian University in Kraków. His time in France enabled him to establish a strong bridge between the Polish and French mathematicians. Zaremba's research in partial differential equations, applied mathematics, and classical analysis (particularly harmonic analysis) was widely recognized. He also authored a number of university textbooks and monographs. He was a member of the Cracow Academy of Learning (since 1903), cofounder and president of the Polish Mathematical Society (1919), and editor of the *Annals of the Polish Mathematical Society* for many years. The Soviet Academy appointed Zaremba a member in 1925, and the universities of Caen, Kraków, and Poznań conferred on him the degree of *doctor honoris causa*. He died in Kraków during the German occupation of Poland.

Hermann Minkowski (1864–1909)

Hermann Minkowski (1864–1909) was a Russian-born mathematician, but his parents were German. His family returned to Germany and settled in Königsberg when Hermann was 8 years old. He read the works of Dedekind, Dirichlet, and Gauss at a very early age. Hermann entered the University of Königsberg in 1880 and became Hilbert's close friend. During 1880–1884, he spent three semesters at the University of Berlin. In 1883, the Paris Academy of Sciences granted the Grand Prize in Mathematics to the young Minkowski and the elderly Henry Smith (at the beginning and end of their careers, respectively) for their solution to the problem of the number of representations of an integer as the sum of five squares. Minkowski's doctoral thesis, submitted in 1885, was a continuation of this prize winning work. After his doctorate, he continued doing research at Königsberg. When Hurwitz joined the staff at the University of Königsberg in 1884, Minkowski became his close friend. Minkowski began teaching at the University of Bonn in 1887, and he was promoted to assistant professor in 1892. He moved back to Königsberg 2 years later, where he taught for 2 years. He then joined the Eidgenössische Polytechnikum Zürich, where he became a colleague of his friend, Hurwitz, who was appointed there in 1892. Einstein was a student in several of his courses there. Minkowski accepted a chair at the University of Göttingen in 1902, where he remained for the rest of his life. At Göttingen, Minkowski worked in mathematical physics, gaining interest from Hilbert and his associates. By 1907, Minkowski realized that the work of Hendrik Antoon Lorentz (1853–1928) and Einstein could be best understood in a non-Euclidean space. In fact, he is credited with being the first person to recognize that space and time, which had previously been viewed as independent quantities, could be coupled together to form a four-dimensional space-time continuum. This space-time continuum provided a framework for all later mathematical work in relativity. In 1909, Minkowski made a four-dimensional treatment of electrodynamics. His work on quadratic forms, continued fractions, and the geometry of numbers is often cited in the literature. He is also known for the Cantor–Minkowski cover and the Minkowski sausage. He married Auguste Adler (1863–1923) in Strasburg in 1897; they had two daughters: Lily, born in 1898, and Ruth, born in 1902. At the young age of 44, Minkowski suddenly died from a ruptured appendix.

William Fogg Osgood (1864–1943)

William Fogg Osgood (1864–1943) was born in Boston, Massachusetts. He studied at the Boston Latin School, and then he studied the classics at Harvard College during 1882–1883. William graduated with an A.B. (Bachelor of Arts) degree in 1886, securing second rank out of 286 students. He then started graduate work at Harvard, where he obtained a master's degree in 1887. He studied at the universities of Göttingen (1887–1889) and Erlangen (1890) in Germany, where he obtained a doctoral degree in 1890. He was an instructor during 1890–1893, an assistant professor during 1893–1903, and a full professor of mathematics during 1903–1933 at Harvard. He became Perkin Professor of Mathematics in 1913 and professor emeritus in 1933. He was the chairman of the department of mathematics at Harvard during 1918–1922. During 1899–1902, he served as an editor of the Annals of Mathematics, and he was the president of the American Mathematical Society during 1904–1905, whose Transactions he edited in 1909–1910. In 1904, he was elected to the National Academy of Sciences. William worked on complex analysis, in particular, conformal mapping and the uniformization of analytic functions. His main work was on the convergence of sequences of continuous functions, solutions of differential equations, the calculus of variations, and space filling curves. Klein invited him to write an article on complex analysis in the *Enzyklopädie der mathematischen Wissenschaften*, which was later expanded in the book *Lehrbuch der Funktionentheorie*. Besides his research in analysis, William worked in mathematical physics and wrote on the theory of the gyroscope. William married and then divorced Teresa. At the age of 68 he married Celeste Phelps Morse, who was 40 years old and the former wife of Harold Calvin Marston Morse (1892–1977), who was also a professor of mathematics at Harvard, best known for his work on the calculus of variations in the large, a subject in which he introduced the technique of differential topology that is now known as Morse theory. This was not only a great shock to Harold, but a disgraceful event as well, which led William to retire from Harvard in 1933. After his retirement, William taught for 2 years at the National University of Peking. During this time he published two books, *Functions of Real Variables* and *Functions of a Complex Variable*. William was said to be "A kindly but reserved man, he liked to travel by car, playing tennis and golf, and smoking cigars ... he smoked until little of the cigar was left, then inserted the small blade of a penknife in the stub so as to have a convenient way to continue." After returning from China, William lived in Belmont, Massachusetts, where he passed away.

Charles Proteus Steinmetz (1865–1923)

Charles Proteus Steinmetz (1865–1923) was born in Breslau, Prussia, 5 days before Abraham Lincoln (1809–1865) was assassinated. Steinmetz suffered from dwarfism (he was 4 ft, 3 in. tall), was a hunchback, and had hip dysplasia, as did his father and grandfather. He first attended Johannes Gymnasium and then Wroclaw University, where he received his undergraduate degree in 1883. He then moved to the University of Breslau, where he was about to finish his doctorate in 1888 when he was investigated by the German police for his socialist political activities. He fled to Switzerland and then emigrated to the USA in 1889 with his friend, Oscar Asmussen. He worked for Rudolf Eickemeyer in Yonkers, New York, and then the General Electric Company. In 1977, Steinmetz was inducted into the United States Patent Office's National Inventor's Hall of Fame. The three accomplishments that gained him this honor were the Law of Hysteresis (formulated in Yonkers), a formula for alternating current (he pioneered the use of complex numbers in the study of electrical circuits), and the theory of electrical transients. In 1983, the United States Post office issued a postage stamp in his honor. In 1901, Harvard University conferred on him an Honorary Degree, and in 1903, Union College awarded him the degree of Doctor of Philosophy. He never married, fearful that his children would be deformed, like himself. At the time of his death, Steinmetz held over 200 US patents. Steinmetz once remarked, "I want to say that absolutely all the success I have had has been due to my thorough study of mathematics."

Jacques Salomon Hadamard (1865–1963)

Jacques Salomon Hadamard (1865–1963), an outstanding French mathematician and an FRS, was born to Amédée Hadamard (a Jewish teacher) and Claire Marie Jeanne Picard (a pianist) in Versailles, France. Jacques studied in Paris at the Lycée Charlemagne, where his father was a teacher. During his early years, Jacques excelled in all subjects except mathematics. His mathematical knowledge improved remarkably after his admission to the Lycée Louis-le-Grand in 1876. He was awarded prizes in algebra and mechanics in 1883. In 1884, Jacques secured first rank in the entrance examinations for the ecole Polytechnique and the ecole Normale Superieure. In 1888, at the age of 23, Jacques graduated from the ecole Normale. He completed his Ph.D. in 1892 and received the Grand Prix of Mathematical Sciences award for his innovative essay on the Riemann zeta function. Jacques was married in 1892 at the age of 27 to Louise-Anna Trénel, who was a Jewish girl. He fathered five children—three sons and two daughters. In 1893, he was appointed as a lecturer at the University of Bordeaux, where he proved his famous inequality regarding determinants. This resulted in his discovery of Hadamard matrices (for equality). In 1896, he proved the prime number theorem using complex function theory, and he received the prestigious Bordin Prize of the French Academy of Sciences for his original contributions regarding geodesics in the differential geometry of surfaces and dynamic systems. He became Professor of Astronomy and Rational Mechanics in Bordeaux in the same year. He moved from Bordeaux to Paris in 1897 after resigning his chair at Bordeaux. In Paris, at the request of Darboux, he published the first volume of a book on two-dimensional geometry, *Elementary Geometry Lessons* in 1898, which was followed by a second volume on three-dimensional geometry in 1901. These volumes had a major impact on mathematics education in France. He was awarded the Prix Poncelet in 1898 for his contributions during the preceding 10 years. Jacques was not limited to his teaching and research activities; he was also involved in politics during his time in Bordeaux. Alfred Dreyfus, his wife's relative and a Jewish army official in the War Ministry, came from Alsace. In 1894, Alfred was accused of selling military secrets to the Germans and was sentenced to life imprisonment. Initially many people, including Jacques, felt that Alfred was guilty. After moving to Paris in 1897, Jacques discovered that evidence against Alfred had been forged. Jacques could not keep silent, and he became actively involved in rectifying the injustice done to Alfred. Jacques was elected President of the French Mathematical Society in 1906 and was appointed as the head of mechanics at the Collége de France in 1909. In 1910, he published *Lectures on the Calculus of Variations*, which was a significant contribution to the development of functional analysis. He was appointed as professor of analysis at the École Polytechnique, where he succeeded Jordan in 1912. In the same year he was elected to the Academy of Sciences to succeed Poincaré. The events of World War I brought tragedy to Jacques' life; his two eldest sons, Pierre and Etienne, were killed in action in 1916. Jacques' extraordinary spiritual strength lessened the pain of these tragedies and allowed him to dive deeper into mathematics. In 1920, he was appointed to the Paul Émile Appell (1855–1930) chair of analysis at the École Centrale while retaining his positions in the École Polytechnique and the Collége de France. During 1920–1933, he visited the USA, Spain, Czechoslovakia, Italy, Switzerland, Brazil, Argentina, and Egypt. He continued to produce high quality books and research articles. Perhaps the most famous among them was a text that he published in 1922, *Lectures on Cauchy's Problem in Linear Partial Differential Equations*, which was based on a lecture course that he gave at Yale University. Besides this, he published several papers on new topics such as the theory of probability, specifically Markov chains, and education. At the start of World War II, France was quickly occupied by Nazi Germany in 1940. Jacques escaped to the USA with his family, where he was appointed as a visiting faculty member at Columbia University. In 1944, he received the devastating news that his third and youngest son, Mathieu, had been killed in the war on July 26, 1943. Jacques left the USA soon afterward, spending about a year in England before returning to Paris after the end of the war in 1945. After the devastating World War II, he took an active part in the international peace movement. He was allowed to enter the USA due to the extraordinary support of mathematicians there, and he attended the International Congress held in Cambridge, Massachusetts, in 1950. He was eventually made honorary president of the Congress. In 1962, when he was 96, yet another tragedy struck him; his grandson Étienne was killed in a mountaineering accident. This time his spirit could not withstand the tragedy, and afterward he remained in his house to await his death. The legend known to the world as "Jacques Hadamard" left his mortal body in Paris on October 17, 1963, at the age of about 98, leaving behind the invaluable treasure of his discoveries that enriched mathematics and the mathematical sciences enormously. Specifically, he is remembered for the Hadamard product, his prime number theorem, the Hadamard matrix, the Cartan–Hadamard theorem, the Cauchy–Hadamard theorem, his infinite product expansion for the Riemann zeta function, Hadamard's inequality, the Hadamard manifold, the Hadamard transform, the Ostrowski–Hadamard (Alexander Markowich Ostrowski, 1893–1986) gap theorem, and Hadamard space. The famous quotation, "the shortest path between two truths in the real domain passes through the complex domain," is due to him.

Erik Ivar Fredholm (1866–1927)

Erik Ivar Fredholm (1866–1927) was born in Stockholm, Sweden, to a well-educated, wealthy, merchant family. Fredholm studied at the Beskowska School in Stockholm and was awarded his baccalaureate in 1885. He then studied at the Royal Technological Institute in Stockholm for a year before moving to the University of Uppsala. Fredholm was awarded a master's degree in 1888, a doctorate in 1893 under the supervision of Mittag-Leffler, and the degree of Doctor of Science in 1898. In that same year he was appointed as a lecturer in mathematical physics at the University of Stockholm, where he was assigned to a Chair in Mechanics and Mathematical Physics in 1906. During 1909–1910, he was Pro-Dean and then Dean of the same university. Fredholm also held various other positions, becoming a civil servant in 1899 and then Head of Department for the Swedish State Insurance Company in 1902. He was an actuary in the Skandia Insurance Company during 1904–1907. Fredholm also served on the International Committee for Weights and Measures.

Fredholm's first paper was published in 1890 by the Royal Swedish Academy of Sciences. In it he constructed a function that is analytic on the unit disk and infinitely differentiable on the closed disk, but has no analytic continuation outside the disk. The inspiration for this work came from the heat equation. In his 1898 doctoral dissertation, inspired by an equilibrium problem in elasticity, he studied partial differential equations. This work was published in 1900, and its general case was published in 1908. Fredholm's best work is on integral equations and spectral theory, which he published in 1903. In this work he emphasized the similarities between integral equations and systems of linear algebraic equations. This work provided the foundations for much of the research later carried out by Hilbert and others. He is best remembered for the Analytic Fredholm theorem, the Fredholm alternative, the Fredholm determinant, the Fredholm equation, the Fredholm kernel, the Fredholm operator, and Fredholm theory.

Fredholm received many honors for his mathematical contributions, including the V.A. Wallmarks Prize for the theory of differential equations in 1903, the Poncelet Prize of the French Academy of Sciences in 1908, and an honorary doctorate from the University of Leipzig in 1909. He was also a musician. At the time of his death he was working on the mathematics of the acoustics of the violin, but the unfinished work he left on this topic has been impossible to understand. He died in 1927 in Danderyd, County of Stockholm, Sweden.

Maxime Bôcher (1867–1918)

Maxime Bôcher (1867–1918) was born in Boston, Massachusetts. Because of his family's strong academic background, he received a high quality education from his parents and several public and private schools in Boston and Cambridge. Maxime graduated from the Cambridge Latin School in 1883. He received his first degree from Harvard in 1888. At Harvard he studied a wide range of topics that included Latin, chemistry, philosophy, zoology, geography, geology, meteorology, Roman art, mathematics, and music. Maxime received many awards, which allowed him to travel to Europe, especially to Göttingen, for research. At Göttingen he attended various lectures, mainly by Klein and Schwarz. He was awarded a Ph.D. in 1891 for his dissertation *Development of the Potential Function into Series*, which was supervised by Klein. He was also awarded a prize by the University of Göttingen for this work. He then returned to Harvard where he was appointed as an instructor. In 1894, he was made assistant professor due to his impressive work. He became a full professor of mathematics in 1904. The terms *augmented matrix*, *linearly independent*, and *linearly dependent* appear to have been introduced by Maxime in his book *Introduction to Higher Algebra*, which was published in 1907. He was president of the American Mathematical Society during 1908–1910. He wrote about 100 research articles on differential equations, series, and algebra, as well as elementary textbooks on analytical geometry and trigonometry, which were greatly appreciated by students and are still in demand today. Maxime is known for Bôcher's theorem in complex analysis, his theorem regarding harmonic functions, Bôcher's equation, and the first satisfactory treatment of the Gibbs phenomenon. The Bôcher Memorial Prize is named after him. He breathed his last at his Cambridge home in Massachusetts at the age of 51 after a prolonged illness.

Martin Wilhelm Kutta (1867–1944)

Martin Wilhelm Kutta (1867–1944) was born in Pitschen, Upper Silesia (now Byczyna, Poland). His parents died when he was very young, after which he went to live with his uncle in Breslau. After graduating from the Gymnasium in Breslau, Kutta studied at the University of Breslau from 1885 to 1890, and then he studied at the University of Munich from 1891 to 1894. During 1894–1897, Kutta worked as an assistant to Walther Franz Anton von Dyck at the Technische Hochschule in Munich. He spent 1898–1899 at the University of Cambridge, and in 1900, he was awarded a doctorate from the Technische Hochschule at Munich for his thesis *Beiträge zur näherungsweisen Integration totaler Differentialgleichungen*, advised by Lindemann and Gustav Bauer. This work contains the now famous Runge–Kutta method for solving ordinary differential equations. In 1902, he submitted his habilitation thesis on aerodynamics to the Technische Hochschule in Munich, which contains his most significant contribution, the Zhukovsky–Kutta

(Joukowski–Kutta) theorem for the lift on an aerofoil. In 1907, Kutta was promoted to extraordinary Professor of Applied Mathematics. In 1909, he moved to the University of Jena, and in the following year he was appointed as an ordinary professor at the Technische Hochschule at Aachen. He became an ordinary professor at the Technische Hochschule in Stuttgart in 1911, where he remained until he retired in 1935. At Stuttgart, Kutta focused his teaching on engineers, who benefited greatly from his inspiring presentations. Kutta also made significant research on glaciers and the history of mathematics. He died in 1944, in Fürstenfeldbruck, Germany.

Felix Hausdorff (1868–1942)

Felix Hausdorff (1868–1942) was born in Breslau, Germany; however, his family moved to Leipzig, where he was raised. Hausdorff obtained his Ph.D. in 1891 from the University of Leipzig for his application of mathematics to astronomy. He submitted his habilitation thesis in 1895, also based on his research in astronomy and optics. Hausdorff taught mathematics in Leipzig until 1910, and then became Professor of Mathematics at the University of Bonn. During 1913–1921, he was professor at the University of Greifswald, and then he returned to Bonn. Hausdorff, who was Jewish, lost his position in 1935 when the Nazis came to power. When they could not avoid being sent to a concentration camp in 1942, Hausdorff, his wife, and her sister all committed suicide. They were buried in Bonn, Germany. He is considered to be one of the founders of modern topology; he also contributed significantly to set theory, descriptive set theory, measure theory, function theory, and functional analysis. In mathematics he is mainly remembered for the Hausdorff measure, the Hausdorff dimension, the Hausdorff maximal principle, Hausdorff spaces, and the generalized continuum hypothesis. He introduced the term 'metric space' in 1914. Hausdorff also published several philosophical and literary books and articles under the pseudonym *Paul Mongré*.

Élie Joseph Cartan (1869–1951)

Élie Joseph Cartan (1869–1951) was born in Dolomieu, France. His father was a poor blacksmith. Élie showed remarkable talent in primary school, which impressed the young school inspector, Antonin Dubost (1842–1921), who later became an important politician. Dubost acquired state funding for Élie's tuition, which allowed him to attend the Lycée in Lyons, where he completed his school education with distinction in mathematics. In 1888, the state stipend was extended to allow him to study at the École Normale Supérieure in Paris. Cartan obtained his doctorate in 1894 and then moved to the University at Montpellier. During 1896–1903, he was a lecturer at the University of Lyon, and he was then appointed professor at the

University of Nancy. In 1909, he became a lecturer at the Sorbonne, and 3 years later he was appointed to the Chair of Differential and Integral Calculus in Paris. Cartan served as Professor of Rational Mechanics beginning in 1920 and then Professor of Higher Geometry from 1924 to 1940, when he retired. In mathematics, he is noted for his work on differential equations, differentiable manifolds, his discovery of spinor groups as exponentials of the orthogonal algebras, and his development of a system of calculus called exterior differential forms. In 1945, he published the book *Les systémes différentiels extérieurs et leurs applications géométriques*. He married Marie-Louise Bianconi (1880–1950) in 1903 and they had four children—three sons (Henri, Jean, and Louis) and one daughter. Unfortunately, Jean died of tuberculosis at the age of 25, and Louis was beheaded by the Nazis in December of 1943. Cartan received honorary degrees from the University of Liege (1934), Harvard University (1936), the Free University of Berlin (1947), the University of Bucharest (1947), the Catholic University of Louvain (1947), and the University of Pisa (1948). He was elected an FRS in 1947, and he also became a fellow of Accademia dei Lincei and the Norwegian Academy. He was elected to the French Academy of Sciences in 1931, was made its Vice President in 1945, and then became President in 1946. Cartan died in 1951 in Paris.

Alfred Marie Liénard (1869–1958)

Alfred Marie Liénard (1869–1958) was a French scientist who spent most of his career teaching applied physics at the School of Mines in Paris, of which he became director in 1929. His physical research was mainly in the areas of electricity and magnetism, elasticity, and hydrodynamics. From time to time he worked on mathematical problems arising from his other scientific investigations, and in 1933, he was elected President of the French Mathematical Society. He was an unassuming bachelor whose life was devoted entirely to his work and his students. He is best remembered for the differential equation $y'' + f(y)y' + g(y) = 0$, which bears his name.

Ernst Friedrich Ferdinand Zermelo (1871–1953)

Ernst Friedrich Ferdinand Zermelo (1871–1953) was born in Berlin. After graduating from the Luisenstädtisches Gymnasium in Berlin in 1889, he studied mathematics, physics, and philosophy at the universities of Berlin, Halle, and Freiburg. In 1894, the University of Berlin awarded him a doctorate for his dissertation *Untersuchungen zur Variationsrechnung*, which followed the Weierstrass approach to the calculus of variations. He then worked as an assistant to Planck. In 1897, he moved to Göttingen, where he completed his habilitation thesis in 1899 and was appointed professor in 1905. In 1910, Zermelo was appointed to the Chair of Mathematics at Zürich University, which he resigned in 1916. In 1926, he was appointed to an honorary chair at Freiburg im Breisgau, but he resigned in 1935 because he disapproved of Hitler's (Adolf Hitler, 1889–1945) regime. At the end of World War II, Zermelo was reinstated to his honorary position at Freiburg. In 1902, he published his first work concerning the addition of transfinite cardinals and discovered the so-called Russell paradox. In 1904, he took the first decisive step toward the continuum hypothesis when he proved the well-ordering theorem, which states that every set can be well ordered. This proof was based on the powerset axiom and the axiom of choice: For any family A of sets there is a function that assigns to each set S of the family A a member of S. The antagonists of the axiom of choice were then known as realists and included Baire, Borel, and Lebesgue. They rejected the axiom of choice because it asserts the existence of a set, which is called the selector of a given family of non-void disjoint sets, without providing any effective possibility of defining that set. Moreover, they considered it to not be self-evident, and they found that it has paradoxical consequences that are incompatible with intuition. The proponents of the axiom of choice, the so-called idealists, including Hadamard, Frankel, and Hausdorff, pointed out that the axiom appears in many proofs of very intuitive theorems; some of the idealists considered the axiom of choice to be authorized in the same degree as the other axioms of set theory, which nobody challenged. In 1908, Zermelo gave an improved proof that made use of Dedekind's notion of the "chain" of a set, which became more widely accepted; this was mainly because in the same year he also offered an axiomatization of set theory. In 1922, Frankel and Albert Thoralf Skolem (1887–1963) independently improved Zermelo's axiomatic system. The resulting 10-axiom system, now called Zermelo–Frankel set theory, is now the most commonly used system for axiomatic set theory. He died in Freiburg in Breisgau.

Félix Édouard Justin Émil Borel (1871–1956)

Félix Édouard Justin Émil Borel (1871–1956) was born in Saint-Affrique, France. He was educated at the École Normale Supérieure in Paris. In 1893, at the age of 22, Borel was appointed to the Chair of Mathematics at Lille University, and in 1896, he was appointed to the École Normale Supérieure. In 1909, at Sorbonne, a new Chair of Theory of Functions was created for him, where he remained until 1941. In 1921, Borel was elected to the Académie des Sciences, and in 1934, he became its President. Along with two other French mathematicians, Rene Baire (1874–1932) and Henri Lebesgue, he founded measure theory, which marked the beginning of the modern theory of functions of a real variable. Borel also contributed greatly to the study of divergent series, probability, and game theory. In addition, he bridged the gap between hyperbolic geometry and special relativity. In mathematics, he is best remembered for Borel algebra, Borel's lemma, the Borel measure, Borel's paradox, Borel space, the Borel–Cantelli (Francesco Paolo Cantelli, 1875–1966) lemma, the Borel–Carathéodory theorem, the Heine–Borel theorem, and Borel summation.

After 1924 he became active in politics, serving in the French Chamber of Deputies during 1924–1936 and as Minister of the Navy during 1925–1940. Borel was awarded the Resistance Medal in 1945 and the Grand Croix Légion d'Honneur in 1950. For his scientific work he received the first gold medal of the Centre National de la Recherche Scientifique in 1955. He died in Paris in 1956.

Forest Ray Moulton (1872–1952)

Forest Ray Moulton (1872–1952) was an American astronomer and administrator of science. While calculating ballistic trajectories during World War I, he made substantial improvements to the Adams formula for numerical integration of ordinary differential equations. The crater Moulton on the Moon is named after him.

Bertrand Arthur William Russell (1872–1970)

Bertrand Arthur William Russell (1872–1970) was born in Wales into an aristocratic British family that was active in the progressive movement and had a strong commitment to liberty. He became an orphan at an early age, so his childhood was lonely and he often contemplated suicide. Russell was placed in the care of his father's parents, who had him educated at home. His brother Frank introduced him to the work of Euclid, which had a profound impact on Russell's life. In these formative years, he was also deeply influenced by the work of Percy Bysshe Shelley (1792–1822), a major English romantic poet. At the age of 18, he was awarded a scholarship at Trinity College, Cambridge, to read for the Mathematical Tripos. He came under the influence of George Edward Moore (1873–1958), a distinguished English philosopher, and Whitehead, an English mathematician and philosopher who recommended him to the Cambridge Apostles, an intellectual secret society at the University of Cambridge. In 1893, Russell graduated as Seventh Wrangler, distinguishing himself as a mathematician. But later remarked, "When I finished my Tripos, I sold all my mathematical books and made a vow that I would never look at a mathematical book again." He became a fellow of the College in 1895. Russell taught German social democracy at the famous London School of Economics in 1896. In 1901, he intensively studied the foundations of mathematics at Trinity, where he discovered Russell's paradox (also called Russell's antinomy), which challenged the foundation of set theory. It demonstrated that the naive set theory of Dedekind and Frege leads to contradictions. An interesting version of his paradox is as follows: (1) The men in a village are of two types: men who do not shave themselves and men who do. (2) The village barber shaves all men who do not shave themselves, and he shaves only those men. But who shaves the barber? The barber cannot shave himself. If he did, he would fall into the category of men who shave themselves. However, (2) states that the barber does not shave such men. So barber does not shave himself, and he falls into the category of men who do not shave themselves. According to (2), the barber shaves all of these men; hence, the barber shaves himself, too. We find that the barber cannot shave himself, yet the barber must shave himself. In 1903, Russell published his first important book on mathematical logic, *The Principles of Mathematics*, which started with clear-cut, fundamental assumptions and proceeded to principles of strict logic. He showed that mathematics can be derived from a very small number of principles. Russell published a significant and influential philosophical essay, *On Denoting*, in the journal Mind in 1905. He became an FRS in 1908. In 1910, along with Whitehead, he published the first of three volumes of *Principia Mathematica*. This publication, along with his earlier book (The Principles of Mathematics) made Russell's name known worldwide. In 1910, Trinity College appointed him to a lectureship in logic and the philosophy of mathematics. He became the Ph.D. advisor of an Austrian engineering student, Wittegenstein, whom he considered a genius and a potential successor who could continue his work on logic. Although it was often a drain on his energy, he patiently spent long hours addressing Wittegenstein's apprehensions and queries. Russell contributed significantly to Wittegenstein's academic development, which culminated in the publication in 1922 of Wittegenstein's only philosophical work, *Tractatus Logico-Philosophicus*. This work dealt with the identification of relationships between language and reality and a definition of the limits of science. Russell passed through both World Wars in Europe. During World War II, Russell was one among the few intellectuals engaged in pacifist activities, and in 1916, he lost his job at Trinity College due to his conviction and consequent 6-month imprisonment under the Defense of the Realm Act. Russell went to Russia on an official visit and met Lenin in 1920, but was disappointed when he perceived "cruelty" in him. Consequently, he no longer supported the Russian revolution. He then visited Beijing for a year and delivered talks on philosophy. In China, Russell suffered from severe pneumonia, and the Japanese press published an incorrect report of his death. His second wife Dora, who was with him at the time, reported to the world that "Mr Bertrand Russell, having died according to the Japanese press, is unable to give interviews to Japanese journalists." The press did not like this sarcasm. Russell divorced Dora, the mother of his two children, who left him for the American journalist Griffin Barry (1894–1986). He then married his third wife, Patricia Spence, an Oxford undergraduate and his children's governess since 1930.

Russell taught at the University of Chicago before World War II. He was appointed professor at the City College of New York in 1940. Due to the views he had presented in his book *Marriage and Morals*, published 11 years earlier in 1929, in which he questioned the Victorian notion of morality regarding sex and marriage, there was a public outcry and his appointment was annulled by a court order. He was judged "morally unfit" to teach at the college. This protest was initiated by the mother of a student who would not have been eligible for his graduate-level course in mathematical logic. However, many intellectuals, led by John Dewey (1859–1952), an American philosopher, psychologist, and educational reformer, protested against this treatment. During this time, Einstein's often-quoted aphorism, viz.,

"Great spirits have always encountered violent opposition from mediocre minds," originated in his open letter in support of Russell. Russell returned to Britain in 1944 and joined the faculty of Trinity College for the second time. On June 9, 1949, when Russell was given the Order of Merit award, King George VI (1895–1952) was friendly but slightly embarrassed by decorating a former prisoner, saying that "You have sometimes behaved in a manner that would not do if generally adopted." Russell only smiled, but later said "That's right, just like your brother." In 1950, at the age of 78, Russell was awarded the Nobel Prize in Literature. Spence divorced Russell in 1952, and Russell married his fourth wife, Edith Finch (1900–1978), on December 15, 1952. Russell and Finch had known each other since 1925. Finch remained with him until his death in 1970, and their marriage was happy. In 1961, at the age of 89, he was imprisoned for the second time for his protests in support of nuclear disarmament.

Russell, the philosopher, logician, mathematician, historian, socialist, pacifist, and social critic, spent most of his relatively long life in England. He made brief appearances in France, Russia, China, Japan, and the USA for various lecture tours, visits, and academic appointments. He passed away in Wales due to influenza on February 2, 1970, at the age of 97. According to him, mathematics is the science concerned with the logical deduction of consequences from the general premises of reasoning. Mathematics, rightly viewed, possesses not only truth but also supreme beauty—a beauty cold and austere, like that of sculpture, without appeal to any part of our weaker nature, without the gorgeous trappings of painting or music, yet sublimely pure and capable of a stern perfection such as only the greatest art can show. Russell's famous epigram is: "Pure mathematics is the subject in which we do not know what we are talking about, nor whether what we are saying is true." He once remarked that to create a healthy philosophy, you must renounce metaphysics but be a good mathematician.

There are in all eternity as many years as there are days, which Russell calls the *Tristram Shandy* paradox. Shandy spent 2 years recounting the events of the first 2 days of his life and bemoaned the fact that at this rate he would fall farther and farther behind in his autobiography. This is quite true for a mortal Shandy, but an immortal Shandy, with all eternity at his disposal, would recount the first day's events in the first year, the second day's in the second year, and so on; eventually he would complete his narrative for any given day.

Russell has written: "Hardy once told me that if he could find a proof that I was going to die within a few minutes, he would be sorry to lose his friend, but his sorrow would be far outweighed by the pleasure in finding the proof. I entirely sympathized with him and was not at all annoyed."

Constantin Carathéodory (1873–1950)

Constantin Carathéodory (1873–1950) was born in Berlin to Greek parents and grew up in Brussels. Constantin attended a private school in Vanderstock in 1881 and after 2 years went back to his father in Berlin. He spent two successive winters, during 1883–1885, on the Italian Riviera, and then he returned to Brussels in 1885 to attend grammar school for a year, where he developed an interest in mathematics. He entered the high school Athénée Royal d'Ixelles in 1886 and studied there until his graduation in 1891. Constantin won the prize of best mathematics student in Belgium twice during his high school days. During 1891–1895, he attended the École Militaire de Belgique and was trained as a military engineer. He also studied at the École d'Application during 1893–1896. During 1897–1900, he was offered a job in the British colonial service and was assigned to Egypt, where he worked on the construction of the Assiut Barrage, a dam. Constantin studied at the University of Berlin in 1900, where he attended lectures by Frobenius and a fortnightly colloquium run by Schwarz. He also became close friends with Fejér, a Hungarian mathematician. He received his doctorate in 1904 from Göttingen University for his thesis *Über die diskontinuierlichen Lösungen in der Variationsrechnung*, which he submitted to Minkowski. He remained in Göttingen to write his habilitation thesis, *Über die starken Maxima und Minima bei einfachen Integralen*, which he submitted in 1905. He then lectured as a Privatdozent at Göttingen until 1908. Constantin then worked at Hanover Technical High School and Breslau Technical High School. In 1913, he became professor at the University of Göttingen, and in 1919, he became professor at the University of Berlin. During 1920–1938, he worked at the Ionian University of Smyrna, the University of Athens, Athens Polytechnic, and the University of Munich. After his retirement in 1938, Constantin continued working for the Bavarian Academy of Science. He contributed significantly to the calculus of variations, measure theory, and the theory of functions of a real variable, including the theory of boundary correspondence and conformal representations. In 1909, Constantin pioneered the axiomatic formulation of thermodynamics through a purely geometrical approach. He is remembered for the Borel–Carathéodory theorem, Carathéodory's theorem, the Carathéodory metric, Carathéodory's lemma, the Carathéodory kernel theorem, the Carathéodory–Jacobi–Lie theorem, and the Carnot–Carathéodory metric. He breathed his last at the age of 76 in Munich.

Leonard Eugene Dickson (1874–1954)

Leonard Eugene Dickson (1874–1954) was born in Independence, Iowa, USA. Leonard attended both primary and secondary schools in Cleburne, Texas, USA. He then studied at the University of Texas at Austin and quickly came under the influence of George Bruce Halsted (1853–1922), who introduced Leonard to Euclidean and non-Euclidean geometries. He obtained his bachelor's and master's degrees in 1893 and 1894, respectively, under George's guidance. Leonard, at the age of only 22, was the first to receive his doctorate in mathematics from the University of Chicago for a dissertation titled *The Analytic Representation of Substitutions on a Power of a Prime Number of Letters with a Discussion of the Linear Group*, under the guidance of Eliakim Hastings Moore (1862–1932), another American mathematician. Leonard then went to Leipzig and became a student of Sophus Lie. He also studied in Paris under Jordan. Leonard returned to the USA, and in 1899, at the age of only 25, became associate professor at the University of Texas. The University of Chicago countered by offering him a permanent assistant professorship in 1900. Leonard readily accepted the offer and spent the rest of his career there, becoming an associate professor in 1907, a full professor in 1910, and professor emeritus in 1939, when he returned to Texas where he breathed his last in Harlingen. At the University of Chicago, he supervised 53 Ph.D. theses; his most accomplished student was probably Abraham Adrian Albert (1905–1972). He was also a visiting professor at the University of California in 1914, 1918, and 1922. Leonard had a major impact on American mathematics. He published 18 books and more than 250 research papers. The use of the term 'adjoint' for the transpose of the matrix of cofactors appears to have been introduced by him in a research paper published in 1902. His 3-volume *History of the Theory of Numbers* (1919–1923), which contains more than 1,600 pages, is still referenced. He was a member of several academies and societies, such as the National Academy of Sciences, the American Philosophical Society, the American Academy of Arts and Sciences, the London Mathematical Society, the French Academy of Sciences, and the Union of Czech Mathematicians and Physicists. He was the first recipient of a prize created in 1924 by The American Association for the Advancement of Science for his work on the arithmetic of algebras. Harvard (1936) and Princeton (1941) awarded him honorary doctorates. Leonard presided over the American Mathematical Society in 1917–1918. He criticized American mathematicians for falling short of those in Britain, France, and Germany in his 1918 presidential address, entitled *Mathematics in War Perspective*. He stressed: "Let it not again become possible that thousands of young men shall be so seriously handicapped in their Army and Navy work by lack of adequate preparation in mathematics." He was also the first recipient of the Cole Prize (named after the algebraist Frank Nelson Cole 1861–1926) for algebra in 1928, for his book *Algebren und ihre Zahlentheorie*.

André-Louis Cholesky (1875–1918)

André-Louis Cholesky (1875–1918) was born in Montguyon, France. He was a French military officer who, in addition to being a battery commander, was involved with geodesy and surveying in Crete and North Africa before World War I. Cholesky was killed in action during the war. His work on matrices, known as the *Cholesky decomposition*, was published posthumously on his behalf by a fellow officer.

Giuseppe Vitali (1875–1932)

Giuseppe Vitali (1875–1932) was born in Ravenna, Italy. After completing his elementary school education in Ravenna in 1886, Giuseppe spent 3 years at the Ginnasio Comunale in Ravenna where his performance in the final examinations of 1889 was not outstanding. He continued his secondary education at the Dante Alighieri High School in Ravenna. Here his mathematics teacher, Giuseppe Nonni, realized the great potential of his young student and advised Giuseppe's father to allow his son to study mathematics. After graduating from the Dante Alighieri High School, Giuseppe studied during 1895–1897 at the University of Bologna. His teachers were impressed by their young, talented pupil, and they supported his application for a scholarship at the Scuola Normale Superiore in Pisa, where he studied during 1897–1899. During 1899–1901, he assisted Ulisse Dini, who taught him infinitesimal calculus, while he studied for a teaching diploma. After receiving his teaching diploma, Giuseppe left university level mathematics and became a secondary school teacher, possibly due to financial problems. During 1904–1923, he taught at the Liceo C Colombo in Genoa, where he was involved in politics and became a Socialist councillor. When the Fascists came to power in 1922 they dissolved the Socialist Party, so his political career came to an end and Giuseppe devoted most of his time to mathematics. In December of 1925, Giuseppe was appointed to the chair of mathematical analysis at the University of Padua. Despite serious health problems, he contributed a lot to Padua during the 5 years that he worked there. He became the first director of the Seminario Matematico of the University of Padua, which was founded by him. In 1930, he started a journal, *The Rendiconti del Seminario Matematico dell'Universitá di Padova*, and published his paper *Determinazione della superficie di area minima nello spazio hilbertiano* in its first volume. In mathematics he is mainly remembered for the Vitali convergence theorem, the Vitali covering theorem, the Vitali–Hahn–Saks (Hans Hahn, 1879–1934; Stanislaw Saks, 1897–1942) theorem, and the Vitali set. In 1930, Giuseppe moved to the chair of mathematics at the University of Bologna, Italy. In spite of his ill health (with a paralyzed arm that did not permit him to write), nearly half of his research papers were written during the last 4 years of his life. In his last few years he worked on a new absolute differential calculus and a geometry of Hilbert spaces, both of which, however, were not pursued by later mathematicians. He breathed his last in Bologna at the age of 56. Giuseppe was honored with elections to the Academy of Sciences of Turin in 1928, the Accademia Nazionale dei Lincei in 1930, and the Academy of Bologna in 1931.

Henri Léon Lebesgue (1875–1941)

Henri Léon Lebesgue (1875–1941) was born in Beauvais, Oise, Picardie, France. His father was a typesetter who died of tuberculosis when Henri was very young, so his mother had to work tirelessly to support him. Henri suffered from poor health throughout his life. Beginning with primary school, Henri was a brilliant student. He later studied at the École Normale Supérieure. Lebesgue is most famous for his theory of integration, which was a generalization of the seventeenth century concept of integration—summing the area between an axis and the curve of a function defined for that axis. His theory was originally published in his dissertation *Intégrale, longueur, aire* (Integral, length, area) at the University of Nancy in 1902. This is considered to be one of the finest mathematical dissertations ever written. He also studied trigonometric functions and made significant contributions in topology and potential theory. Besides integration, he is also remembered for the Lebesgue dominated convergence theorem, the Lebesgue covering dimension, the Lebesgue point, Lebesgue's number lemma, the Lebesgue spine, and the Lebesgue constant (interpolation). His integral paved the way for several generalizations, such as the Denjoy integral and the Haar integral, proposed by the Frenchman Arnaud Denjoy (1884–1974) and the Hungarian Alfréd Haar (1885–1933), respectively. Another well-known integral is the Lebesgue–Stieltjes integral, a combination of the ideas of Lebesgue and the Dutch analyst Thomas Joannes Stieltjes (1856–1895). He was elected to the Academy of Sciences (1922), the Royal Society, the Royal Academy of Science and Letters of Belgium (1931), the Academy of Bologna, the Accademia dei Lincei, the Royal Danish Academy of Sciences, the Romanian Academy, and the Kraków Academy of Science and Letters. Lebesgue was also awarded honorary doctorates from many universities. He was a recipient of a number of prizes, including the Prix Houllevigue (1912), the Prix Poncelet (1914), the Prix Saintour (1917), and the Prix Petit d'Ormoy (1919). Lebesgue once wrote, "reduced to general theories, mathematics would be a beautiful form without content." He died in Paris.

Issai Schur (1875–1941)

Issai Schur (1875–1941) was born in Mogilev province, Russian Empire (now Belarus) on January 10 and died on the same date in 1941 in Tel Aviv, Palestine (now Israel). He spoke German without a trace of an accent, and nobody even guessed that it was not his first language. At the age of 13, Schur went to Latvia, and in 1894, he entered the University of Berlin to read mathematics and physics. He obtained his doctorate in 1901, became lecturer in 1903, and during 1911–1916 held a professorship in mathematics at the University of Bonn. He returned to Berlin in 1916 and was promoted to full professor in 1919. Schur was a brilliant mathematician and a popular lecturer who attracted many students and researchers to the University of Berlin. His lectures sometimes attracted so many students that opera glasses were needed to see him from the back row. Schur's life became increasingly difficult under the Nazi government, and in April of 1933 he was forced to "retire" from the university under a law that prohibited non-Aryans from holding "civil service" positions. There was an outcry from many of his students and colleagues, who respected and liked him, but this did not stave off his complete dismissal in 1935. Schur, who thought of himself as German rather than a Jew, never understood the persecution and humiliation he received at the hands of the Nazis. He left Germany for Palestine in 1939, a broken man. Lacking financial resources, he had to sell his beloved mathematics books and lived in poverty until his death. He is best remembered today for his result regarding the existence of the Schur decomposition and for his work on group representations (Schur's lemma). He became a foreign member of the Russian Academy of Sciences in 1929.

Ludwig Prandtl (1875–1953)

Ludwig Prandtl (1875–1953) was born in Freising, a town about 20 miles northeast of Munich. Ludwig was admitted to the Technische Hochschule Munich in 1894, where he received a Ph.D. for his work on beam buckling in 1899 under the supervision of August Otto Föppl (1854–1924), a professor of Technical Mechanics and Graphical Statics, who is credited for introducing the Föppl–Klammer theory and the Föppl–von Kármán (Theodore von Kármán, 1881–1963) equations. In 1901, Ludwig became a professor of fluid mechanics at the Technical School in Hanover (now the Technical University, Hanover), where he developed many of his most important theories. In 1904, he produced a vitally important paper on fluid flow in very little friction, in which he described the boundary layer and its relation to drag and streamlining. The paper also discussed flow separation due to the boundary layer, unambiguously explaining the concept of stall for the first time. Although closed-form solutions could not be obtained, the approximation presented in his original paper is still widely used. This single paper was so impressive and its usage was so great that he was appointed professor of technical physics from 1904 to 1947 and professor of applied mechanics in 1907 at the University of Göttingen. Here he established a school of aerodynamics and hydrodynamics that achieved world renown. In 1912, Ludwig became the founding member and Chief Executive of the Scientific Aerospace Society in Berlin. In 1922, he founded and became Chief Executive of the Society for Applied Mathematics and Mechanics (GAMM). He remained the Chief Executive of GAMM until 1950. In 1925, Ludwig was made director of the Kaiser Wilhelm (later the Max Planck) Institute for Fluid Mechanics, where he worked until 1947. He became Emeritus Professor at the University of Göttingen in 1947 and continued until his death in 1953.

Ludwig and his student Theodor Meyer (1882–1972), a German mathematician and a founder of the scientific discipline now known as compressible flow or gas dynamics, developed the first theories of supersonic shock waves and flow in 1908. Following earlier work done by Frederick William Lanchester (1868–1946), an English polymath and engineer who made important contributions to automotive engineering, aerodynamics, and coinvented the topic of operations research, Ludwig worked on developing a useful mathematical tool for examining lift from "real-world" wings. This ground breaking work on wing theory was published in 1918–1919 and is known as Lanchester–Prandtl wing theory. Ludwig introduced the concept of swept wings and created a method for designing a supersonic nozzle in 1929. In Göttingen, he not only developed low-speed wind tunnels but also supervised the construction of the first supersonic wind tunnel in Germany. He also devised a soap-film analogy for analyzing the torsion forces of structures with noncircular cross sections. He is also remembered for the Prandtl number, the Prandtl–Glauert singularity, the Prandtl–Glauert method, and the Prandtl–Meyer function. Ludwig supervised 85 dissertations and was awarded honorary doctorates from the Universities of Danzig, Zurich, Prague, Trondheim, Bucharest, and Istanbul. He was decorated for outstanding services and was awarded the Grosses Verdienstkreuz (highest order of merit) of the Federal Republic of Germany. He is widely known as the father of modern aerodynamics. The crater Prandtl on the far side of the Moon is named in his honor. The Ludwig–Prandtl Ring is awarded by the Deutsche Gesellschaft für Luft-und Raumfahrt in his honor for outstanding contributions to the field of aerospace engineering.

Waldemar Hermann Gerhard Kowalewski (1876–1950)

Waldemar Hermann Gerhard Kowalewski (1876–1950) was born in Pomerania, then part of the German Empire. He studied at the University of Königsberg and University of Greifswald and received his doctorate at the University of Leipzig under the supervision of Sophus Lie. During 1901–1909, he worked at the University of Greifswald and the University of Bonn, and then he moved to Prague to join a German-based school. In 1912, Kowalewski was hired at the German University of Prague. In 1935, he became the Rector of the Dresden University of Technology. In 1945, he moved to the Munich University of Technology. Kowalewski is known for the introduction of matrix notation in 1909. He published more than 100 works in the fields of determinants, transformation groups, natural and differential geometry, approximation, and interpolation. The terms 'natural equation' and 'natural geometry' were introduced by him. He also published over 24 mathematical textbooks. Kowalewski encouraged women in mathematics, and the first female scientific assistant in a mathematical department in Germany worked under him. Kowalewski had 13 students, including William Threlfall (1888–1949), Herbert Seifert (1907–1996), Hilmar Wendt, and Alfred Kneschke (1902–1979), as well as 495 descendants.

Gilbert Ames Bliss (1876–1951)

Gilbert Ames Bliss (1876–1951) was born in Chicago. He entered the University of Chicago in 1893 and received his B.S. in 1897. He then began his graduate studies at Chicago in mathematical astronomy and his first publication was in that field. However, mathematics was his real love, and in 1898, he began his doctoral studies with work on the calculus of variations. Gilbert received his doctorate in 1900 for the dissertation *The Geodesic Lines on the Anchor Ring*, published in the Annals of Mathematics in 1902. After 2 years as an instructor at the University of Minnesota, Gilbert spent the 1902–1903 academic year at the University of Göttingen, interacting with renowned mathematicians such as Klein, Hilbert, Minkowski, Zermelo, Schmidt, and Carathéodory. He then taught 1 year each at the University of Chicago and the University of Missouri. During 1905–1908, Gilbert worked at Princeton University with a strong group of young mathematicians, including Luther Pfahler Eisenhart (1876–1965), Robert Lee Moore, and Oswald Veblen (1880–1960), an American mathematician, geometer, and topologist, whose work found application in atomic physics and the theory of relativity. During this period he was also an associate editor of the Annals of Mathematics. In 1908, he joined the University of Chicago, where he remained until his retirement at the age of 65. During 1908–1916, he served as editor of the Transactions of the American Mathematical Society, and during 1927–1941 he chaired the Department of Mathematics at the University of Chicago. Gilbert was elected to the National Academy of Sciences in 1916. He was the American Mathematical Society's Colloquium Lecturer in 1909, Vice President in 1911, and President during 1921–1922. He was awarded the Mathematical Association of America's first Chauvenet (William Chauvenet, 1870–1920) Prize in 1925 for his article *Algebraic functions and their divisors*, which resulted in his book *Algebraic Functions* in 1933. His 1946 monograph, *Lectures on the Calculus of Variations*, was the outcome of his work on the calculus of variations. This monograph depicted its topic not as an adjunct of mechanics but as an end in itself. He greatly simplified the transformation theories of Clebsch and Weierstrass and strengthened the necessary conditions given by Euler, Weierstrass, Legendre, and Jacobi. Gilbert set out the canonical formulation and solution of the problem given by Oskar Bolza (1857–1942), a student of Klein, with side conditions and variable end points. He breathed his last at the age of 75 in Harvey, Illinois.

Edmund Georg Hermann (Yehezkel) Landau (1877–1938)

Edmund Georg Hermann (Yehezkel) Landau (1877–1938) was born in Berlin, Germany, to a wealthy Jewish family. His father was a gynecologist and his mother was from a well-known German banking family. Landau was a child prodigy; legend has it that at the age of three, when his mother forgot her umbrella in a carriage, he replied "It was number 354" and her umbrella was quickly found. Landau attended high school and university in Berlin and received his doctorate in 1899 for a thesis on number theory under the direction of Frobenius. He was always interested in mathematical puzzles, and even before he received his doctorate he had published two books on mathematical problems in chess. In 1900, he wrote a letter to Hilbert giving an outline of his ideas for proving the *prime ideal theorem* for algebraic number fields. Landau completed his habilitation (postdoctorate) in 1901 with work on the Dirichlet series, a topic in analytic number theory that examines number-theoretic propositions. From 1899 to 1909, he taught at the University of Berlin. During this period his publication list grew rapidly; in fact, by 1909 he already had nearly 70 research papers in print. In 1903, Landau gave a simpler proof of the prime number theorem, compared to those given in 1896 by Vallée-Poussin and Hadamard. His masterpiece was published in 1909, entitled *Handbuch der Lehre von der Verteilung der Primzahlen*, a two-volume work that gave the first systematic presentation of analytic number theory. He also made important contributions to the theory of analytic functions of a single variable. In calculus he is often remembered by the small o notation. In 1909, Landau moved to Göttingen, where he succeeded Minkowski. Hilbert and Klein were his colleagues there until Klein retired in 1913. Despite his outstanding talent as a teacher, researcher, and mathematician (with an encyclopedic knowledge of the literature in his areas of expertise), Landau annoyed many of his colleagues at Göttingen. He criticized their results, and he used to tell people who would ask for his address in Göttingen, "You'll find it easily; it is the most splendid house in the city." In 1933, the Nazis forced him to stop teaching, and thereafter he lectured only outside of Germany. Landau wrote over 250 papers on number theory, which had a major influence on the development of the subject. Hardy once wrote that no one was more passionately devoted to mathematics than Landau. In 1938, he died from a heart attack.

Godfrey Harold Hardy (1877–1947)

Godfrey Harold Hardy (1877–1947) was born in Cranleigh, Surrey, England. He was the older of the two children of Isaac Hardy and Sophia Hall Hardy. His father was the geography and drawing master at the Cranleigh School, where he also gave singing lessons and played soccer. His mother gave piano lessons and helped run a boardinghouse for young students. Hardy's parents were devoted to their children's education. Hardy demonstrated his numerical ability at the early age of two when he began writing down numbers into the millions. He had a private mathematics tutor rather than regular classes at the Cranleigh School. He moved to Winchester College, a private high school, when he was 13 and was awarded a scholarship. He excelled in his studies and demonstrated a strong interest in mathematics. He entered Trinity College, Cambridge, in 1896 on a scholarship, graduating in 1899. Hardy held the position of lecturer in mathematics at Trinity College at Cambridge University from 1906 to 1919, when he was appointed to the Sullivan Chair of Geometry at Oxford. He had become unhappy with Cambridge for the dismissal of the famous philosopher and mathematician Bertrand Russell from Trinity for antiwar activities, and he did not like a heavy load of administrative duties. In 1931, he returned to Cambridge as the Sadleirian Professor of Pure Mathematics, where he remained until his retirement in 1942. He was a pure mathematician and held an elitist view of mathematics, hoping his research would never be applied. Ironically, he is perhaps best known as one of the developers of the Hardy–Weinberg (Wilhelm Weinberg, 1862–1937) law, which predicts patterns of inheritance. Hardy worked primarily in number theory and function theory, on such topics as the Riemann zeta function, Fourier series, and the distribution of primes. Hardy once said that the theory of numbers, more than any other branch of mathematics, began by being an experimental science. Its most famous theorems had all been conjectured, sometimes a 100 years or more before they were proved, and they have been suggested by the evidence of a mass of computations. Hardy is also remembered for his collaboration with Littlewood, his colleague at Cambridge, with whom he wrote more than 100 papers, and the famous Indian mathematical prodigy Ramanujan. Hardy had the wisdom of recognizing Ramanujan's genius from the unconventional but extremely creative writing Ramanujan sent him. However, he felt that "Oriental mathematics may be an interesting curiosity, but Greek mathematics is the real thing. As Littlewood said to me once, 'The Greeks are not clever schoolboys or scholarship candidates, but fellows at another college.'" Hardy received numerous medals and honorary degrees for his accomplishments. Hardy's book, *A Course of Pure Mathematics*, which gave the first rigorous English exposition of functions and limits for the college undergraduate, had a great impact on university teaching. Hardy also wrote *A Mathematician's Apology*, in which he gives his answer to the question of whether it is worthwhile to devote one's life to the study of mathematics. It presents Hardy's view of what mathematics is and what a mathematician does. He said: "I believe that mathematical reality lies outside us, and that our function is to discover or observe it, and that the theorems which we prove, and which we describe grandiloquently as our 'creations' are simply our notes of our observations." Hardy had a rebellious spirit; he once listed his most ardent wishes: (1) to prove the Riemann hypothesis (a famous unsolved mathematical problem), (2) to make a brilliant play in a crucial cricket match, (3) to prove the nonexistence of God, (4) be the first man at the top of Mt. Everest, (5) be proclaimed the first president of the USSR, Great Britain, and Germany, and (6) to murder Mussolini. According to Hardy, a mathematician, like a painter or poet, is a maker of patterns. If his patterns are more permanent than a painter's, it is because they are made with ideas. The mathematician's patterns, like the painter's or the poet's, must be beautiful; his ideas, like the colors or the words, must fit together in a harmonious way. Beauty is the first test: there is no permanent place in the world for ugly mathematics.

The following anecdotes regarding Hardy are often quoted:

Hardy was an agnostic and was also afraid of travel. Once when he left by air, he left a note on his table: "I have found a proof of Fermat's last theorem". All Cambridge was agog with excitement to know the proof. When Hardy returned, he simply stated: "I left this note to ensure that God will bring me back alive to show that I am an imposter."

Hardy once took a risky boat and wrote a post card to Bohr, "I have proved Riemann's hypothesis." His argument was that if the boat sank and he drowned, everybody would believe that Hardy had proved the hypothesis, but since God would not afford Hardy such a great honor, he could not allow the boat to sink.

There was a bad weather for a number of days. Hardy said to a friend who was leaving: "Please when the train starts, you stick your head through the window, look up to the sky and say in loud voice, 'I am Hardy.' When God thinks Hardy has gone, he would make good weather just to annoy me."

Hardy and Bohr were working together and used to make an agenda for work every day. Hardy insisted on making the first item of work for every day "Prove Riemann's hypothesis." He knew it would not be done, but he wanted to express faith in the human will power to meet challenges.

While lecturing at Cambridge, Hardy wrote a formula on the blackboard and said: "This is obvious, of course." Then on seeing the blank faces of his listeners, he remarked: "At least I think this is obvious," and went to his room. After half an hour, he returned from his room with sheets of written paper and said: "Yes, it is obvious."

Polya had once given a good idea, but had not followed it up. Once, when Hardy went with Polya to a zoo, he saw a bear coming to the lock, sniffing at it, hitting it with his paw, and going back. Hardy remarked: "This bear is like Polya. He has excellent ideas, but does not carry them to the end."

Louis Pierre Joseph Fatou (1878–1929)

Louis Pierre Joseph Fatou (1878–1929) was born in Lorient (France). In 1898, he entered the École Normale Supérieure in Paris to study mathematics. Fatou graduated in 1901, and he was immediately appointed at the Paris Observatory. However, he continued his work on integration and complex function theory, resulting in a doctoral degree in 1907. Fatou wrote many papers, developing a fundamental theory of complex iteration in 1917. His results were very similar to those of Julia. Their work is now commonly referred to as the generalized Fatou–Julia theorem. Fatou also introduced Mandelbrot sets. He died in Pornichet (France).

Jan Lukasiewicz (1878–1956)

Jan Lukasiewicz (1878–1956) was born into a Polish-speaking family in Lvov. Lvov was then part of Austria, but it is now in the Ukraine. His father was a captain in the Austrian army. Lukasiewicz became interested in mathematics while in high school. He studied mathematics and philosophy at the University of Lvov, at both the undergraduate and graduate levels. After completing his doctoral work he became a lecturer there, and in 1911, he was appointed to a professorship. When the University of Warsaw was reopened as a Polish university in 1915, Lukasiewicz accepted an invitation to join the faculty. In 1919, he served as the Polish Minister of Education. He returned to the position of professor at Warsaw University, where he remained from 1920 to 1939, serving as rector of the university twice. Lukasiewicz was one of the cofounders of the famous Warsaw School of Logic. He published his famous text, *Elements of Mathematical Logic*, in 1928.

With his influence, mathematical logic was made a required course for mathematics and science undergraduates in Poland. His lectures were considered to be excellent and even attracted students of the humanities. Lukasiewicz and his wife experienced great suffering during World War II, which he documented in a posthumously published autobiography. After the war they lived in exile in Belgium. Fortunately, in 1949 he was offered a position at the Royal Irish Academy in Dublin. Lukasiewicz worked on mathematical logic throughout his career. His work on a three-valued logic was an important contribution to the subject. Nevertheless, he is best known in the mathematical and computer science communities for his introduction of parenthesis-free notation, now called Polish notation.

Maurice René Fréchet (1878–1973)

Maurice René Fréchet (1878–1973) was born in Maligny, France. He studied in Lycée Buffon secondary school in Paris, where he received the special attention of Hadamard. Fréchet then enrolled at the École Normale Supérieure to study mathematics. During 1907–1908, he served as a Professor of Mathematics at the Lycée in Besançon, in 1908 he moved to the Lycée in Nantes to stay there for a year, and then he worked at the University of Poitiers between 1910 and 1919. Then for two and a half years he worked at the front for the British Army as an interpreter. Fréchet then served as a professor of higher analysis and Director of the Mathematics Institute at the University of Strasbourg. Later, Fréchet held various positions in Paris until his retirement in 1948. He made major contributions to the topology of point sets and introduced the concept of metric spaces. He also made several important contributions to the field of statistics and probability, as well as calculus. His dissertation opened the entire field of functionals on metric spaces and introduced the notion of compactness. Independently of Friedrich Riesz (1880–1956), he discovered the representation theorem in the space of Lebesgue square integrable functions. However, despite his major achievements in mathematics, his work was not appreciated in France. In fact, he was nominated numerous times for membership of the Academy of Sciences, but he was not accepted until the age of 78. He is remembered for Fréchet space, Fréchet–Urysohn space, the Fréchet derivative, the Fréchet distribution, the Fréchet mean, the Fréchet distance, and the Fréchet filter. He died at the age of 94 in Paris.

Albert Einstein (1879–1955)

Albert Einstein (1879–1955) was a superb German-American theoretical physicist, philosopher, and passionate humanitarian. He was born in Ulm (Germany) on March 14, 1879. His father was Hermann Einstein (1847–1902), a salesman and electrical engineer. His mother was Pauline Einstein (1858–1920). Einstein attended a Catholic elementary school during 1884–1889. He ranked first in the school although he had childhood speech difficulties. Einstein's father once showed him a pocket compass; Albert realized that there must be something causing the needle to move, despite the apparent empty space. As he grew, he constructed models and mechanical devices and began to show a talent for mathematics. In 1889, Max Talmud (Talmey), a poor Polish medical student who took meals with the Einsteins on Thursdays for 6 years, introduced the 10-year old Einstein to key texts on science, mathematics, and philosophy. During this time, Talmud wholeheartedly guided Einstein through many secular educational interests. In 1894, Einstein's family moved to Milan and then to Pavia, Italy, while Albert stayed in Munich, Germany, to complete his studies at the Luitpold Gymnasium. His father wanted him to study electrical engineering, but Einstein clashed with authorities and resented the school's regimen and teaching method. In 1895, he withdrew to join his family in Pavia, convincing the school to let him go by using a doctor's note. He later wrote that the spirit of learning and creative thought were lost in strict rote learning. During this time, Einstein wrote his first scientific work, "The Investigation of the State of Ether in Magnetic Fields." Einstein applied to the Swiss Federal Institute of Technology (Eidgenössische Polytechnische Schule, also called Polytechnic), Zürich, Switzerland, for admission. He appeared for an entrance examination that he could not pass, although he secured excellent scores in physics and mathematics. The Einsteins sent Albert to Aarau (capital of the northern Swiss canton) to complete secondary schooling. Einstein studied Maxwell's electromagnetic theory there. In 1896, at the age of 17, he graduated and then enrolled in the 4-year teaching diploma program in mathematics and physics at the Polytechnic in Zürich. At the Polytechnic he met Mileva Maric (1875–1948), his future wife and the only woman among six students in the mathematics and physics group, who enrolled in the same year. Einstein and Maric's friendship grew deeper. They read books together on physics, in which Einstein became increasingly engrossed. Einstein received his teaching diploma from Zürich Polytechnic in 1900, but Maric failed the examination for the "Theory of Functions" course in mathematics. Einstein published a paper in 1901 on the capillary forces of a straw, entitled *Capillarity*, in the top journal *Annalen der Physik*, 4, pp. 513–523. He worked with Alfred Kleiner (1849–1916), the Professor of Experimental Physics and his Ph.D. advisor, and completed his dissertation entitled *A New Determination of Molecular dimensions* on April 30, 1905, for which he received his Ph.D. from the University of Zürich. In his earlier research articles, Einstein attempted to establish the physical reality of atoms and molecules. In 1905, at the age of 26, he published four ground-breaking research articles on Brownian motion (Robert Brown, 1773–1858), the special theory of relativity, photoelectricity, and the equivalence of matter and energy, which later brought him to the attention of the world scientific community. During the early twentieth century, each zigzag motion of a particle suspended in a liquid was believed to have been caused by collision with a single molecule of the liquid. Einstein showed in his article on Brownian motion how this effect could be caused by an uneven bombardment of many tiny molecules. He could also estimate the size of the molecules, which was later verified to be correct by the French physicist

Jean Perrin (1870–1942). Einstein's explanation led to the accurate estimation of the Avogadro number, a constant denoting the number of atoms per mole. In 1907, Einstein established that mass is related to energy by the simple equation $E = mc^2$. In 1908, he became a lecturer at the University of Bern, and in 1909, he relinquished the post of lecturer and took the position of physics docent (faculty with mid-level seniority) at the University of Zürich. In 1911, he became a full professor at Karl-Ferdinand University in Prague, Czech Republic, and during the same year he extended his special theory of relativity and predicted that light from another star would be bent by the gravitational pull of the Sun, which was confirmed through observations made by a British expedition headed by Sir Arthur Stanley Eddington (1882–1944) during the solar eclipse of May 29, 1919. He refined these ideas in his general theory of relativity in 1915. According to this theory, masses distort the structure of space-time, which explained an anomaly in the orbit of the planet Mercury that could not be accounted for by Newtonian mechanics. His famous statement, "God does not play dice with the universe," was a rejection of the probabilistic way that quantum mechanics works. Unfortunately for Einstein, the laws of quantum mechanics work incredibly well, and one can't successfully explain quantum effects with traditional classical physics.

Newton's "System of the World" was based on Euclidean geometry. He viewed space and time as absolute and infinite. Space, to him, was a three-dimensional entity; time, a separate one-dimensional entity. This way of looking at the world seemed to be the only possible way. Men were so certain of these assumptions that they never questioned them or regarded them as assumptions. They were truths, incontestable, absolute, and self-evident. Einstein's system differs from Newton's in three major respects: Space is non-Euclidean, space is not absolute, and space is not a separate entity from time. The geometry of space is not Euclidean, according to Einstein, nor is it even a uniform non-Euclidean system. The presence of matter alters the properties of space in such a way that no one geometry can describe it. That is, the geometry on Earth is not the same as that, for instance, on the Sun. If a man on Earth were to measure a certain distance or period of time on Mars, he would obtain different measurements from those made by a man on Mars that measured that same distance or time. This discrepancy would occur not because of errors or optical illusions, but because each man's measurements are based on his own local length and local time. Newton's three dimensions have been replaced in Einstein's system by four, the fourth being time. One of the most important aspects of this new view of the universe is that it dispenses with the approximation of gravity that Newton introduced to explain the course of planets, stars, and falling objects. The new theory asserts that all objects move in straight lines, but non-Euclidean straight lines. Thus when planets move in elliptical orbits around the Sun they are simply moving in the "straight lines" of their particular geometry.

In 1914, Einstein was appointed director of the Kaiser Wilhelm Institute for Physics and professor at the Humboldt University of Berlin. He continued at the Institute until 1932. He became a member of the Prussian Academy of Sciences and was the President of the German Physical Society during 1916–1918. Einstein was awarded the Nobel Prize in Physics in 1921 for his discovery in the field of photoelectricity. Based on the theory that light energy is capable of behaving as particles, called photons, he observed that light particles striking the surface of certain metals produce the photoelectric effect: an electron is emitted by the metal that carries energy equal to the energy of the photon that sets the electron free. He was one of the first physicists to prove the wave/particle nature of light. In 1922, Einstein traveled extensively for 6 months in Asia on a lecture tour. He visited Singapore, Ceylon (Sri Lanka), Japan, and Palestine, and was given an exceptionally warm welcome everywhere. In 1925, he was awarded the Copley Medal by the Royal Society of London.

In 1933, Einstein was compelled to emigrate from Germany to the USA due to the rise of Nazism under the new chancellor, Hitler, who imposed unacceptable restrictions on Jews (14 Nobel Laureates and 26 of the 60 professors of physics were forced to flee Germany and the neighboring countries). Einstein did not return to Germany, but accepted a position at the Princeton Institute of Advanced Study where he spent the rest of his life. A few months before the start of World War II in Europe, Einstein, Edward Teller (1908–2003), Eugene Wigner (1902–1995), the Hungarian physicist Leo Szilard (1898–1964), and others felt that it was their responsibility to alert Americans to the possibility that German scientists might build the atom bomb first and that Hitler would be willing to use such a weapon. Initially, the President of the USA, Franklin Delano Roosevelt (1882–1945), did not assign this problem much importance, but he changed his mind rather quickly. The USA entered the arms race, and the secret Manhattan Project that drew immense material, financial, and scientific resources was born. The outcome of this project was that the USA became the only country to successfully develop the atom bomb during World War II.

Einstein lived in the USA for only about 15 years after he became a US citizen in 1940. On April 17, 1955, he experienced internal bleeding due to an abdominal aortic aneurysm, which had been reinforced surgically by Dr. Rudolph Nissen (1896–1981) in 1948. At the time, he was preparing a speech for a television appearance commemorating Israel's seventh anniversary. He took the draft of his speech along with him to the hospital, but he did not live long enough to complete it. He refused further surgery, saying "... It is tasteless to prolong life artificially. I have done my share, it is time to go. I will do it elegantly." Early in the morning on April 18, at the age of 76, he breathed his last at Princeton Hospital. The following three quotations are due to him: "So far as the theories of mathematics are about reality, they are not certain; so far as they are certain, they are not about reality"; "Do not worry about your difficulties in mathematics, I assure you that mine are greater"; and "Teaching should be such that what is offered is perceived as a valuable gift and not as a hard duty."

The following anecdotes about Einstein, who was named the person of the century by Time magazine in 1999, have been recounted for decades and are both interesting and revealing:

One day during a lecture tour, Einstein's driver, who often sat at the back of the lecture hall during his talks, remarked that he could possibly give the talk himself, having heard it so many times. At the next stop on the tour, Einstein and the driver switched places, with Einstein sitting at the back in his driver's uniform. Having delivered a flawless talk, the driver was asked a difficult question by a member of the audience. "Well, the answer to the question is quite simple," he casually replied. "I bet my driver, sitting up at the back there, could answer it!"

Einstein's wife often suggested that he dress more professionally when he headed off to work. "Why should I?" he would invariably argue. "Everyone knows me there." When the time came for Einstein to attend his first major conference, she begged him to dress up a bit. "Why should I?" said Einstein. "No one knows me there!"

Einstein was often asked to explain the general theory of relativity. "Put your hand on a hot stove for a minute, and it seems like an hour," he once declared. "Sit with a pretty girl for an hour, and it seems like a minute. That's relativity!"

During an address at the Sorbonne in Paris, Einstein said, "If my theory of relativity is proven successful, Germany will claim me as a German and France will declare that I am a citizen of the world. Should my theory prove untrue, France will say that I am a German and Germany will declare that I am a Jew."

When Einstein was working at Princeton, 1 day he was going back home and he forgot his address. The driver of the cab did not recognize him. Einstein asked the driver if he knew Einstein's home. The driver said "Who does not know Einstein's address? Everyone in Princeton knows. Do you want to meet him?" Einstein replied "I am Einstein. I forgot my home address, can you take me there?" The driver took him to his home and did not even collect the fare.

Einstein was once traveling from Princeton on a train when the conductor came down the aisle, punching the tickets of every passenger. When he came to Einstein, Einstein reached in his vest pocket. He couldn't find his ticket, so he reached in his trouser pockets. It wasn't there, so he looked in his briefcase but couldn't find it. Then he looked in the seat beside him. He still couldn't find it. The conductor said, "Dr. Einstein, I know who you are. We all know who you are. I'm sure you bought a ticket. Don't worry about it." Einstein nodded appreciatively. The conductor continued down the aisle punching tickets. As he was ready to move to the next car, he turned around and saw the great physicist down on his hands and knees looking under his seat for his ticket. The conductor rushed back and said, "Dr. Einstein, Dr. Einstein, don't worry, I know who you are. No problem. You don't need a ticket. I'm sure you bought one." Einstein looked at him and said, "Young man, I too, know who I am. What I don't know is where I'm going."

Robert Daniel Carmichael (1879–1967)

Robert Daniel Carmichael (1879–1967) was born in Goodwater, Alabama. He received his undergraduate degree from Lineville College in 1898 and his Ph.D. in 1911 from Princeton. Carmichael held positions at Indiana University from 1911 until 1915 and at the University of Illinois from 1915 until 1947. He was an active researcher in a wide variety of areas, including number theory, real analysis, differential equations, mathematical physics, and group theory. Carmichael is best known for Carmichael numbers, Carmichael's theorem, and the Carmichael function, which all play a significant role in number theory. His Ph.D. thesis, written under the direction of George Birkhoff, is considered to be the first significant American contribution to the subject of differential equations.

Alfred James Lotka (1880–1949)

Alfred James Lotka (1880–1949), an American biophysicist, mathematician, physical chemist, and statistician, was born in what is now the Ukraine and was educated mainly in Europe. He is remembered chiefly for his formulation of the Lotka–Volterra equations in population dynamics. He was also, the author, in 1924, of the first book on mathematical biology.

Lipót Fejér (1880–1959)

Lipót Fejér (1880–1959) was born as Leopold Weiss in Pécs (Hungary) and changed to the Hungarian name Fejér around 1900. The word "Weiss" means white in German, while the word "Fejér" means white in Hungarian. Leopold's schooling was in Pécs but initially things did not move smoothly. His father took him out of school for a while and taught him himself. However, Leopold developed a tremendous interest in mathematics due to one of his high school teachers. He graduated from the high school in Pécs in 1897. He was also awarded the second prize in the Eötvös Mathematics Competition. Leopold studied mathematics and physics in Budapest and Berlin, where he was taught by Schwarz. Leopold taught at the University of Pázmány Péter during 1902–1905 and at Franz Joseph University in Hungary during 1905–1911. Leopold was appointed in 1911 to the chair of mathematics at the University of Budapest. He held this position until his death. It was Leopold's discussions with Schwarz that led him to look at the convergence of Fourier series and prove the highly important Fejér's theorem. During his period in the chair at the University of Budapest, Leopold headed a highly successful Hungarian school of analysis. He was the doctoral advisor of several renowned mathematicians, such as von Neumann, Erdös, Pólya, and Paul Turán (1910–1976), who worked primarily in number theory. Leopold became a corresponding member in 1908 and a member in 1930 of the Hungarian Academy of Sciences. He passed away in Budapest at the age of 79 and is buried in Kerepesi Cemetery in Budapest. In his tribute Pólya said: "Why did Hungary produce so many mathematicians of our time? Many people have asked this question which, I think, nobody can fully answer. There were, however, two factors whose influence on Hungarian mathematics is manifest and undeniable, and one of these was Leopold Fejér, his work, his personality. The other factor was the combination of a competitive examination in mathematics with a periodical."

The following anecdotes on Fejér are entertaining:

Polya's wife stopped Leopold Fejér and some other professors of a university in Germany where Polya worked in a temporary capacity. She took one picture and was going to take another when Fejér spoke up: "What a good wife! She put up all the professors on the track of the street car so that they may be all run over and her husband may get a permanent job."

Once when Fejér was angry at a Hungarian topologist, he said, "What he says is a topological map of the truth."

Leonard Bairstow (1880–1963)

Leonard Bairstow (1880–1963) was born in 1880 in Halifax, West Yorkshire. His father, Uriah Bairstaw, was a keen mathematician. He studied in Queens Road and Moorside Council Schools, followed by Heath Grammar and Council Secondary Schools. He then attended the Royal College of Science and conducted research on the explosion of gases under a Whitworth scholarship. Subsequently he joined the National Physical Laboratory, the largest applied physics organization in the UK, based at Bushy Park in Teddington, London, England. Here he rose to the position of head of aircraft research work. During 1920–1949, he held the prestigious Zaharoff Chair of Aviation at Imperial College and became Sir Leonard Bairstow. Leonard is best known for his work in aerospace engineering and for the Bairstow method in numerical analysis to compute a zero of a polynomial. He was elected a fellow of the Royal Society of London and the Royal Aeronautical Society.

Oskar Perron (1880–1975)

Oskar Perron (1880–1975) was a German mathematician. During 1914–1922, he taught at the University of Heidelberg, and from 1922 to 1951 at the University of Munich. He made numerous contributions to ordinary and partial differential equations. He wrote an encyclopedic book on continued fractions *Die Lehre von den Kettenbrüchen*. Positive matrices were introduced by Perron in connection with the three-dimensional continued fractions of Jacobi. These matrices play a significant role in many areas of pure and applied mathematics. They are of particular importance in connection with computing, and much of the fundamental work was done by Garrett Birkhoff. He is remembered for the Perron–Frobenius theorem, Perron's formula, the Perron method, and Perron's paradox.

Luitzen Egbertus Jan Brouwer (1881–1966)

Luitzen Egbertus Jan Brouwer (1881–1966) was born in Overschie (since 1941 part of Rotterdam), the Netherlands. In 1897, he entered the University of Amsterdam to study mathematics and physics, where he graduated in 1904. In 1907, he received his doctorate with the dissertation *Over de Grondslagen der Wiskunde* (On the Foundations of Mathematics), under the supervision of Diederik Johannes Korteweg (1848–1941). In 1909, Brouwer became a privaat-docent (unpaid lecturer) at the University of Amsterdam. The same year, he met Hilbert, whom he admired; however, 20 years later, Brouwer's relationship with Hilbert turned sour. During 1909–1913, he proved a number of breakthrough theorems in the emerging field of topology, which include the topological invariance of dimension, the fixed-point theorem, the mapping degree, and the definition of dimension. In 1912, he was elected to be a member of the Royal Dutch Academy of Sciences and was appointed full Professor Extraordinarius in the field of set theory, function theory, and axiomatics. In 1913, Brouwer became full Professor Ordinarius, succeeding Korteweg, who had generously offered to vacate his chair. In 1914, he joined the editorial board of the Journal Mathematische Annalen and continued there until 1928. In 1919, Brouwer declined offers for professorships in Göttingen and in Berlin.

Intuitionism views mathematics as a free activity of the mind, independent of any language or Platonic realm of objects, and therefore bases mathematics on a philosophy of mind. The implications are twofold. First, it leads to a form of constructive mathematics, in which large parts of classical mathematics are rejected. Second, the reliance on a philosophy of mind introduces features that are absent from classical mathematics as well as from other forms of constructive mathematics: unlike those, intuitionistic mathematics is not a proper part of classical mathematics. It is important to realize that, like logicism, intuitionism is rooted in philosophy. There are theorems which hold in intuitionism but are false in classical mathematics. Brouwer's little book of 1905, *Life, Art and Mysticism*, contains his basic ideas on mind, language, ontology, and epistemology (philosophical questions related to knowledge). He applied these ideas to mathematics in his doctorate dissertation in 1907 that marked the beginning of his intuitionistic reconstruction of mathematics. In 1911, the terms 'intuitionism' and 'formalism' appeared in Brouwer's writings for the first time. In 1918, Brouwer began a systematic intuitionistic reconstruction of mathematics. Thus Brouwer founded the mathematical philosophy of intuitionism as an opponent to the then-prevailing formalism of Hilbert and his collaborators, Paul Isaac Bernays (1888–1977), Wilhelm Ackermann, von Neumann, and others. Intuitionism, like logicism and formalism, although acceptable to the majority of mathematicians, have failed to give mathematics an adequate foundation. In fact, Gödel's paper of 1931 left us without a firm foundation for mathematics, and mathematicians on the whole threw up their hands in frustration and turned away from the philosophy of mathematics.

Hilbert (the formalist) and Brouwer (the intuitionist) were involved in a very public and eventually demeaning controversy in the late 1920s. It reached the point that in 1929, Hilbert expelled Brouwer from the board of the Mathematische Annalen in an unlawful way, in spite of objections from Einstein. Brouwer vehemently protested, but in the end, the whole board was dissolved and immediately reassembled without Brouwer (Einstein and Carathéodory declined). In 1929, his briefcase was stolen and he despaired, doubting that he would ever be able to reconstruct its contents. The conflict and the loss of his briefcase left Brouwer mentally broken and isolated. In 1934, Brouwer became the founding editor of Compositio Mathematica. However, because of his damaged reputation, he had to step down after only a few years. In 1942, he published some brief notes on intuitionistic foundations. During 1947–1951, he gave lectures in Cambridge, which were published posthumously. He retired from the University of Amsterdam in 1951. During 1952–1953, he gave lectures at several places, including London, Cape Town, Helsinki, MIT, Princeton, the University of Wisconsin-Madison, Berkeley, Chicago, and Ontario. He published his last paper in 1955.

Brouwer received honorary doctorates from the University of Oslo in 1929 and Cambridge in 1954, and was made Knight in the Order of the Dutch Lion in 1932. He was also a member of the Royal Dutch Academy of Sciences, the Royal Society in London, the Akademie der Wissenschaften in Berlin, and the Akademie der Wissenschaften in Göttingen. Brouwer died in 1966 at the age of 85 after being struck by a vehicle while crossing the street in front of his house. His funeral speech was given by Max Euwe (1901–1981), who later became world chess champion. Brouwer's archive is kept at the Department of Philosophy, Utrecht University, the Netherlands.

Amalie Emmy Noether (1882–1935)

Amalie Emmy Noether (1882–1935) was a German mathematician whose work was of great importance to the development of modern algebra, which includes such topics as groups, rings, and fields. Born in the year 1882, Emmy had always been different from other girls of her generation. No doubt she attended the usual school for girls, where she learned to play the piano, to manage a house, and to act polite at dances, but her heart was never in it. Her passion was the study of mathematics. Perhaps it was her father's influence; he would always come home excited after teaching his mathematics classes at the University of Erlangen. He loved to gather Emmy and her three younger brothers around him and explain complicated ideas in a way that his children could understand. Especially Emmy was fascinated with her father's ideas about algebra, and she caught on very quickly.

Though she was an extremely bright girl, nineteenth century Germany was not ready for a young woman who wanted to study mathematics. Women were not allowed to enroll for a course in mathematics at the University of Erlangen. Fortunately for Emmy, she had very supportive parents who arranged for a tutor to teach her mathematics. By 1902, when the university decided to admit women, Emmy was more than ready. She was the only woman in her class but she worked hard at producing her best work. In no time at all, Emmy managed to distinguish herself as a very creative scholar with the ability to see larger concepts underlying mathematical processes. In 1907, Emmy obtained her doctoral degree in mathematics with highest honors after working under Paul Gordan.

Getting her education had been a struggle, but Emmy's problems were far from over. While the universities began to enroll women as students, they were definitely not ready to employ them as university professors. Fortunately, Emmy's family came to her aid again. She lived at home while working in the university without any pay. She worked there for 8 years, during which time Emmy would sometimes teach in place of her father, who had grown increasingly weak from the polio that he contracted as a child. Emmy's reputation grew quickly as her publications appeared. In 1908, she was elected to the *Circolo Matematico di Palermo*, in 1909 she was invited to become a member of the *Deutsche Mathematiker-Vereinigung*, and in the same year she was invited to address the annual meeting of the Society in Salzburg. In 1913, she lectured in Vienna.

After Emmy's father retired and her mother had passed away, she moved to the city of Göttingen in 1915. At the university there, Hilbert and Klein were working on Einstein's theory of relativity. Emmy was invited to join them in their research, as they were very enthusiastic about what she could contribute. They even appealed to the university to hire her but to no avail. Gradually, through hard work and brilliant results, she managed to win their respect and was finally given a modest salary. Emmy's first piece of work is a result in theoretical physics, sometimes referred to as Noether's Theorem, which establishes a relationship between symmetries and conservation principles in physics. It was her work in the theory of invariants that led to formulations for several concepts of Einstein's general theory of relativity. Emmy published many papers, changing the way that mathematicians understand algebra. Emmy had a unique ability to work with abstract concepts, visualize complex connections, and help others see them too. In 1924, B.L. van der Waerden (1903–1996) came to Göttingen and spent a year studying with Noether. After returning to Amsterdam, van der Waerden wrote his book *Moderne Algebra* in two volumes. The greater part of the second volume consists of Noether's work. From 1927 on, Noether collaborated with Helmut Hasse and Richard Dagobert Brauer (1901–1977) in work on noncommutative algebras. In addition to teaching and research, Noether helped edit *Mathematische Annalen*. Much of her work appears in papers written by colleagues and students, rather than under her own name. Further recognition of her outstanding mathematical contributions came with invitations to address the International Mathematical Congress at Bologna in 1928 and again at Zurich in 1932. (The next woman had to wait 58 years, until the Kyoto 1990 IM congress). In 1932, she received, jointly with Artin, the Alfred Ackermann-Teubner (1857–1940) Memorial Prize for the Advancement of Mathematical Knowledge. At Göttingen, Emmy led a quiet lifestyle. She was well loved by her friends. Students enjoyed her company and her courses were never boring. Emmy expected her students to work hard and she taught them the importance of understanding the structures underlying the algebraic systems.

By 1933, Emmy had achieved many of her goals in life. She was respected as a professor and privileged to work with the top mathematicians in Europe. She was able to explore mathematical ideas to her heart's content. All of this changed when Hitler came to power. In 1933, Hitler placed Emmy and many of her colleagues "on leave until further notice" in an effort to maintain absolute control, even over ideas. Emmy remained calm and courageous despite the disadvantages that she faced, namely, she was an intellectual woman and a Jew, and she was politically liberal. It became obvious that Emmy would have to leave her country.

Emmy was offered a "visiting professorship" position at Bryn Mawr College near Philadelphia in the USA. She enjoyed her days at Bryn Mawr even though she missed her home country. For the first time, Emmy received a decent salary. More importantly, she valued her relationship with her students. They would often accompany her on her Saturday afternoon jaunts where she would become totally absorbed in talking about mathematics.

While Emmy was at Bryn Mawr, she also worked at the Institute for Advanced Study at Princeton. Einstein and Weyl were there at the same time, and the three of them attracted many common admirers. Through her studies in algebra, she showed mathematicians how to formulate general theorems that would apply to many problems.

In 1935, Emmy Noether died unexpectedly from complications of a routine surgery. Her death was a terrible loss to the field of mathematical research as she was at the peak of her career. Many came to pay their respects and to offer tributes to her contribution to mathematics. To quote Einstein, "In the judgment of the most competent living mathematicians, "Fraulein Noether" was the most significant creative mathematical genius thus far produced since the higher education of women began…" In his memorial address, Weyl said, "Her significance for algebra cannot be read entirely from her own papers, she had great stimulating power and many of her suggestions took shape only in the works of her pupils and coworkers." Further, van der Waerden writes, "For Emmy Noether, relationships among numbers, functions, and operations became transparent, amenable to generalization, and productive only after they have been dissociated from any particular objects and have been reduced to general conceptual relationships."

Certainly, Emmy Noether's contribution to mathematics cannot be challenged. But she had also made another contribution to human history. She showed that following one's dreams and standing up to society's opposition can lead to great achievement and satisfaction. After her the sex barrier in mathematics was broken, and universities became open to the acceptance of women on their faculties.

Harry Bateman (1882–1946)

Harry Bateman (1882–1946), the youngest son of Samuel and Marnie Elizabeth, was born in Manchester, England, and was an outstanding mathematical physicist. He attended Manchester Grammar School, and in the final year won a scholarship from Trinity College, Cambridge, where in 1903 he became a Senior Wrangler, and in 1905, he won Smith's Prize. Before moving to the USA in 1910 he studied in Göttingen and Paris, was a lecturer at the University of Liverpool, and a reader at the University of Manchester. In the USA, he taught at Bryn Mawr College followed by Johns Hopkins University, where he worked with Frank Morley (1860–1937) in geometry and received his Ph.D. in 1913. By then he was already an eminent mathematician, with 60 publications to his name. He then joined Throop Polytechnic Institute, now known as California Institute of Technology (CalTech). He was elected a fellow of the Royal Society of London in 1928, a fellow of the National Academy of Sciences in 1930, and Vice President of American Mathematical Society in 1935. While on his way to New York to receive an award from the Institute of Aeronautical Science, he passed away in Pasadena, California, on January 21, 1946. His interests were in geometry, differential equations, and electromagnetism. He is known for his excellent teaching, his ability to inspire his students, his textbooks, and for the conformal group of space-time. The term 'singular value' is apparently due to him; Harry used it in a research paper published in 1908.

Joseph Henry Maclagan Wedderburn (1882–1948)

Joseph Henry Maclagan Wedderburn (1882–1948) was born to Alexander Wedderburn, a physician, and Anne Ogilvie as their tenth of 14 children, in Forfar, Angus, Scotland. He attended Forfar Academy from 1887 to 1895. During 1895–1898, he studied in George Watson's College in Edinburgh. He entered the University of Edinburgh in 1898 on a scholarship. During 1902–1903, he worked as an assistant in the Physical Laboratory of the University. He began mathematical research while still an undergraduate student, and published his first paper, *On the isoclinal lines of a differential equation of the first order*, in the Proceedings of The Royal Society of Edinburgh in 1903. His two other papers, published in 1903 in the same journal, were on the scalar functions of a vector and an application of quaternions to differential equations. He received his M.A. in mathematics with First Class Honors from the University of Edinburgh in 1903. Joseph then pursued postgraduate studies in Germany, spending the academic session 1903–1904 at the University of Leipzig and the summer semester of 1904 at the University of Berlin, where he met Frobenius and Schur. A Carnegie Scholarship allowed Joseph to spend the 1904–1905 academic year at the University of Chicago, where he worked with Veblen, Hastings Moore, and most importantly, Dickson. Joseph came back to Scotland in 1905 and worked for 4 years at the University of Edinburgh as an assistant to George Chrystal (1851–1911). George supervised his D.Sc. degree, which was awarded in 1908 for a thesis titled *On Hypercomplex Numbers*. During 1906–1908, Joseph edited the Proceedings of the Edinburgh Mathematical Society. He returned to the USA in 1909 and became a preceptor (teacher) in mathematics at Princeton University. His colleagues included the renowned mathematicians Eisenhart, Veblen, Bliss, and George Birkhoff. At the outbreak of the First World War, Joseph was the first person at Princeton to enlist in the British Army as a private. As a Captain in the Fourth Field Survey Battalion in France, he devised sound-ranging equipment to locate enemy artillery. He returned to Princeton after the war as an associate professor in 1921, edited the Annals of Mathematics until 1928, and supervised three doctoral students. He taught at Princeton University for most of his career. As a significant algebraist, he proved that a finite division algebra is a field and proved part of the Artin–Wedderburn theorem on simple algebras. He also worked on group theory and matrix algebra. Joseph became increasingly lonely and possibly developed depression. His isolation after his 1945 retirement at the age of 63 was such that his death from a heart attack was not noticed for several days. His *Nachlass* (the collection of manuscripts, notes, correspondence, and so on left behind when a scholar dies) was destroyed according to his instructions. Joseph was awarded the MacDougall–Brisbane Gold Medal and Prize from the Royal Society of Edinburgh in 1921 and was elected a Fellow of the Royal Society of London in 1933. As to why Wedderburn never married: "It seems that an old Scottish tradition required that a man, before marrying, accumulate savings equal to a certain percentage of his annual income. In Wedderburn's case his income had gone up so rapidly that he had never been able to accomplish this."

Waclaw Franciszek Sierpiński (1882–1969)

Waclaw Franciszek Sierpiński (1882–1969) was born in Warsaw, the son of a well-known physician, Konstanty Sierpiński. He received his secondary education at the then well-known Fifth Grammar School in Warsaw. In 1899, Sierpiński entered the University of Warsaw, and in 1903, he was awarded a gold medal for his essay on lattice points, which he withheld until 1907, when he published it in the Polish magazine *Prace Matematyczno-Fizyczne* (The Works of Mathematics and Physics), instead of immediately publishing it in Russian. He graduated in 1904 with the degree of Candidate of Science and went on to work in a secondary school. In 1905, when the school was closed due to strike connected with the revolution, he moved to the Jagiellonian University at Cracow and received his doctoral degree in 1906. Then Sierpiński returned to Warsaw and continued teaching at secondary schools. He also did research work and published his results in the most important journals in Warsaw and Cracow. In 1908, he was appointed a regular member of the reactivated Warsaw Scientific Society. In the same year he passed the qualifying examination for docent at Jan Kazimierz University at Lwów and 2 years later became an associate professor of the same University. He gave lectures on set theory beginning in 1909, and from these lectures he developed material for his first systematically organized book, *Outline of set theory*, in 1912. During World War I, Sierpiński was in Moscow working with the famous Russian mathematician Nikolai Luzin. Together they studied analytic and projective sets and the theory of real functions. In 1916, Sierpiński gave the first example of an absolutely normal number. In February 1918, he returned to Lwów and in the autumn was elected to a chair at the reborn Polish university in Warsaw. In 1919 he was promoted to professor. He spent the rest of his life in Warsaw. In 1920, after Janiszewski's death, he assumed the editorship of *Fundamenta Mathematicae* along with Mazurkiewicz and directed it for several decades. He worked as editor-in-chief of Acta Arithmetica, and as an editorial-board member of Rendiconti del Circolo Matematico di Palermo, Composito Matematica, and Zentralblatt für Mathematik. Sierpiński authored 724 research articles in set theory (on the axiom of choice and the continuum hypothesis), number theory, and the theory of functions and topology, as well as 50 books. Three well-known fractals are named after him: the Sierpiński triangle, the Sierpiński carpet, and the Sierpiński curve. He is also remembered for Sierpiński numbers and the associated Sierpiński problem. Sierpiński received honorary degrees from several institutions, including Lwów (1929), St. Marks of Lima (1930), Amsterdam (1931), Tarta (1931), Sofia (1939), Prague (1947), Wroclaw (1947), Lucknow (1949), and Moscow (1967). He was also elected as a member of numerous societies and academies. His contributions to science were met with the Polish authorities' highest appreciation. Sierpiński died on October 21, 1969, in Warsaw. To honor his memory, the Mathematical Institute of the Polish Academy of Sciences, one of Warsaw's streets, and one of the Moon's craters have all been named after him. As he requested in his will, his grave carries the inscription 'investigator of infinity.'

Robert Lee Moore (1882–1974)

Robert Lee Moore (1882–1974), an American mathematician who is known as an outstanding topologist and an unconventional teacher of great repute, was born in Dallas, Texas, USA. He became a student of the University of Texas at Austin in 1898 when he was just 16, and by that time he had already learned calculus on his own through self-study. It took Robert just 3 years, instead of the usual four, to earn his B.A. and M.A. degrees, both in 1901. During 1901–1902, he was a teaching Fellow at Texas, and for 1 year in 1902–1903 he taught high school students in Marshall, Texas. He studied at the University of Chicago as a graduate student during 1903–1905, where he received his doctoral degree in 1905 under the supervision of Hastings Moore and Veblen for his dissertation *Sets of Metrical Hypotheses for Geometry*. Robert is mainly remembered for his work on the foundations of topology, a topic he first encountered in his doctoral dissertation. He then taught at the University of Tennessee, Princeton University, and Northwestern

University for 1, 2, and 3 years, respectively. He married Margaret MacLelland Key in 1910. They remained childless. In 1911, he taught at the University of Pennsylvania. In 1920, Robert returned to the University of Texas at Austin as an associate professor and then became a professor after 3 years. He is also known for Moore road space, the Moore plane, and the Moore space conjecture. He left his mortal body at the age of about 92 in Austin, Texas.

Nikolai Nikolaevich Luzin (1883–1950)

Nikolai Nikolaevich Luzin (1883–1950) was born, the only son of his parents, in Irkutsk, Russia, one of the largest cities in Siberia. He attended Moscow University and studied mathematics in 1901. His advisor was Dimitri Fyodorovich Egorov (1869–1931), a Russian/Soviet mathematician known for his contributions to differential geometry and mathematical analysis. Young Nikolai was deeply moved by the misery of the people during 1905–1906, which caused him great personal upheaval. He wrote to Pavel Alexandrovich Florensky (1882–1937), a Russian orthodox theologian, priest, philosopher, mathematician, physicist, electrical engineer, inventor, and neomartyr: "You found me a mere child at the University, knowing nothing. I don't know how it happened, but I cannot be satisfied any more with analytic functions and Taylor series ... it happened about a year ago. ... To see the misery of people, to see the torment of life, to wend my way home from a mathematical meeting ... where, shivering in the cold, some women stand waiting in vain for dinner purchased with horror—this is an unbearable sight. It is unbearable, having seen this, to calmly study (in fact to enjoy) science. After that I could not study only mathematics, and I wanted to transfer to the medical school. ... I have been here about 5 months, but have only recently begun to study." During 1910–1914, Nikolai studied at Göttingen, where he was influenced by Hermann Landau. Nikolai's first significant result was, in 1912, a construction of an almost everywhere divergent trigonometric series with monotonic convergence to zero coefficients. This example disproved Pierre Fatou's conjecture and was a surprise to most mathematicians at the time. At about the same time, he also proved what is now called Lusin's theorem in real analysis. He then returned to Moscow and received his doctorate in 1915, at the age of 32, for his dissertation entitled *Integral and Trigonometric Series*. During the Russian Civil War (1918–1920), Nikolai left Moscow for the Ivanovo–Vosnesensk Polytechnic Institute. He returned to Moscow in 1920. He became a corresponding member of the USSR Academy of Sciences and then a full member of the Academy first in the Department of Philosophy and then at the Department of Pure Mathematics in 1929. His doctoral students included several renowned Soviet mathematicians such as Khinchin, Kolmogorov, and Pavel Samuilovich Urysohn (1898–1924). Nikolai breathed his last in Moscow.

Richard Edler von Mises (1883–1953)

Richard Edler von Mises (1883–1953) and **Hilda Geiringer von Mises** (1893–1973) are known for applying probability theory to biology and genetics. Richard was born in Lemberg (now Lviv, Ukraine), into a Jewish family. He attended the Akademisches Gymnasium in Vienna, and then joined the Vienna University of Technology, where he studied as an undergraduate and then a doctorate student. He was awarded a doctoral degree in 1908 with the dissertation *The Determination of Flywheel Masses in Crank Drives* and received his habilitation from Brünn (now Brno, Czech Republic) with *Theory of the Waterwheels*. In 1909, Richard became Professor of Applied Mathematics in Strasbourg, France. After World War I, he held the new Chair of Hydrodynamics and Aerodynamics at the Dresden Technische Hochschule. In 1919, he was appointed director of the new Institute of Applied Mathematics created in Berlin. In 1921, he founded the journal *Zeitschrift für Angewandte Mathematik und Mechanik* and became its editor. In the same year, Geiringer became Richard's assistant. In 1933, along with Geiringer, he moved to Turkey to hold the newly created Chair of Pure and Applied Mathematics at the University of Istanbul. Richard and Geiringer fled Germany in 1933 to the USA, and married in 1943. Richard was appointed Gordon-McKay Professor of Aerodynamics and Applied Mathematics at Harvard University in 1944, while Geiringer chaired the math department at Wheaton College. Richard is known as one of the most significant applied mathematicians of the century. In aerodynamics he made notable advances in boundary-layer-flow theory and airfoil design; in solid mechanics he made an important contribution to the theory of plasticity by formulating what has become known as the Von Mises yield criterion; he developed the distortion energy theory of stress, which is one of the most important concepts used by engineers in material strength calculations; he is also credited for the Principle of Maximum Plastic Dissipation. In probability theory, Richard proposed that all probability is grounded in real phenomena that only after a large number of observations can one make a statement about the probability of an event. He proposed the "birthday problem": How many people must be in a room before the probability that some share a birthday, ignoring the year and leap days, becomes at least 50 %? He also defined the impossibility of a gambling system. He is remembered for his contributions to the philosophy of science as a neopositivist. After Richard's death, his wife Geiringer took up his earlier work, reworking and expanding his controversial views. On her own, she was one of the pioneers of disciplines such as molecular genetics, human genetics, plant genetics, heredity in man, genomics, bioinformatics, biotechnology, biomedical engineering, and genetic engineering; however, she was not given proper credit for her work because it was published in obscure Turkish journals.

Henry Maurice Sheffer (1883–1964)

Henry Maurice Sheffer (1883–1964) was born to Jewish parents in the western Ukraine and then emigrated to the USA in 1892 with his parents and six siblings. He studied at the Boston Latin School before entering Harvard, where he completed his undergraduate degree in 1905, his master's in 1907, and his Ph.D. in philosophy in 1908. After holding a postdoctoral position at Harvard, Henry traveled to Europe on a fellowship. Upon returning to the USA, he became an academic nomad, spending 1 year each at the University of Washington, Cornell, the University of Minnesota, the University of Missouri, and City College in New York. In 1916, he returned to Harvard as a faculty member in the philosophy department. He remained at Harvard until his retirement in 1952. Sheffer introduced what is now known as the Sheffer stroke in 1913, but it became known only after its use in the 1925 edition of Whitehead and Russell's *Principia Mathematica*. In this same edition, Russell wrote that Sheffer had invented a powerful method that could be used to simplify the *Principia*. Because of this comment, Sheffer was something of a mystery man to logicians, especially because Sheffer, who published little in his career, never published the details of this method, and only described it in mimeographed notes and in a brief published abstract. Sheffer was a dedicated teacher of mathematical logic. He liked his classes to be small and did not like auditors. When strangers appeared in his classroom, Sheffer would order them (even his colleagues or distinguished guests visiting Harvard) to leave. Sheffer was barely 5 ft tall. He was noted for his wit and vigor, as well as for his nervousness and irritability. Although widely liked, he was quite lonely. He is noted for a quip at his retirement: "Old professors never die, they just become emeriti." Sheffer is also credited with coining the term "Boolean algebra." Sheffer was briefly married and lived most of his later life in small rooms at a hotel packed with his logic books and vast files of slips of papers he used to jot down his ideas. Unfortunately, Sheffer suffered from severe depression during the last two decades of his life.

George David Birkhoff (1884–1944)

George David Birkhoff (1884–1944), the first American mathematician of undisputed world fame, widely known for the development of ergodic theory and Birkhoff curve shortening flow, was born to David Birkhoff, a medical doctor, and Jane Gertrude Droppers in Overisel, Michigan. George received instruction in Chicago at the Lewis Institute from 1896 to 1902. After graduating from the Lewis Institute in 1902, George began his university education and obtained his B.A. and M.A. from Harvard University in 1905 and 1906, respectively. He completed his doctorate in 1907, on *Asymptotic Properties of Certain Ordinary Differential Equations with Applications to Boundary Value and Expansion Problems*, at the University of Chicago. While Hastings Moore was his supervisor, he was most influenced by the writings of Poincaré. George taught at the University of Wisconsin at Madison as an instructor during 1907–1909. In 1908, he married Margaret Elizabeth Grafius, and they had three children. He then went to Princeton University as a preceptor in mathematics and became a professor there in 1911. He moved to Harvard University in 1912 as an assistant professor, where he was promoted to full professor in 1919. In the latter part of the 1920s he started a mathematical treatment of aesthetics. He continued working at Harvard until his death in 1944 in Cambridge, Massachusetts, at the age of 60. He was elected Vice President of the American Mathematical Society in 1919 and became its President during 1925–1926. He was also elected to the National Academy of Sciences, the American Philosophical Society, the American Academy of Arts and Sciences, the Académie des Sciences in Paris, the Pontifical Academy of Sciences, and the London and Edinburgh Mathematical Societies. His son, Garrett Birkhoff (1911–1996), was also a great mathematician. Garett Birkhoff was a tenured faculty member of the Harvard Mathematics Department but he did not possess a Ph.D. When asked why not, supposedly he responded, "Where could I find someone to examine me?"

Tobias Dantzig (1884–1956)

Tobias Dantzig (1884–1956) was born in Latvia in the Baltic region of North Europe. He studied mathematics with Poincaré in Paris. He married a fellow Sorbonne University student, Anja Ourisson, and together they emigrated to the USA in 1910. Working for a time as a lumberjack in Oregon, he received his doctorate in mathematics from Indiana University in 1917. Tobias taught at Johns Hopkins, Columbia University, and the University of Maryland. He is the author of *Number: The Language of Science*, published in 1930, for which Einstein said, "This is beyond doubt the most interesting book on the evolution of mathematics which has ever fallen into my hands." In this book, he tells the following interesting story: "A fifteenth century German merchant wanted his son to have an advanced education on commerce. He sought advice from a well-known university professor as to where he should send his son. The response was that if the mathematical curriculum of his son was to be confined in addition and subtraction, then he could probably be instructed in a German university but for the art of multiplication and division he should go to Italy which, according to him, was the only country where such advanced instruction could be received." The only other book authored by him was *Aspects of Science*, published in 1937. Tobias passed away in Los Angeles. He was the father of George Bermard Dantzig (1914–2006), who was a key figure in the development of the simplex algorithm for linear programming.

Bharati Krishna Tirthaji Maharaja (1884–1960)

Bharati Krishna Tirthaji Maharaja (1884–1960) was born in Tinnevely, Madras, to a highly illustrious family. He was known as Venkatraman Shastri before he became a saint. He received his basic education at National College, Church Missionary Society College, and then Hindu College, all in Tirunelveli. Venkatraman passed his matriculation examination from Madras University in January of 1899. As a student, Venkatraman was recognized for his splendid brilliance, superb retentive memory, and insatiable curiosity. In July of 1899, at the age of 16, he was awarded the title of *Saraswati* for his all round proficiency and talent in Sanskrit by the Sanskrit Association of Madras. At about that time, Venkatraman was profoundly influenced by his Sanskrit guru Sri Vedam Venkatrai Shastri, whom he regarded with the deepest love, reverence, and gratitude. He obtained his bachelor's degree in 1902 with the highest rank. In 1903, he received his master's degree from the American College of Sciences in Rochester, New York (Bombay Center), in six subjects: Sanskrit, philosophy, English, mathematics, history, and science, simultaneously scoring the highest honors in all, which was perhaps a world record at the time. He became a Lecturer of Mathematics and Science at Baroda College, and then he was made Principal of the National College in Rajamundary, Andhra Pradesh. In 1905, he participated in the freedom movement along with his fellow nationalists Shri Aurobindo Ghosh (1872–1950) and Gopal Krishna Gokhale (1866–1915). He wrote to numerous newspapers in support of the freedom movement. In 1908, he was appointed *Warden of the Sons of India* by Dr. Annie Besant (1847–1933). In 1908, he joined the Shringeri Matha in Mysore to study under Shri Shankaracharya Shri Sachchidananda Shiva Abhinava Narasimha Saraswati. However, his spiritual education was interrupted when he was pressured by nationalist leaders to head the newly started National College at Rajamahendri. He taught at the college for 3 years, and during 1911–1918 he practiced deep meditation and studied metaphysics and the Vedas, and he began to lead the arduous life of a Sadhu (saint). He practiced vigorous meditation, Brahma-sadhana and Yoga-sadhana, during those years in the nearby forests, living on roots and fruits. It is believed that he attained spiritual self-realization during these years in the Sringeri Math. In 1919, at the age of 35, Venkatraman was initiated into Sanyas by Shri Trivikrama Tirthaji Maharaja, and in 1921 he became the Shankaracharya of Sharadapith in Benares (Varanasi). Afterward, he was called by his new name, Shri Bharati Krishna Tirthaji Maharaja. In 1925, he became the Head of the Govardhan Matha Monastery in Puri, Orissa, and he dedicated the rest of his life to traveling and preaching philosophy and science all over the world. He was the pontiff until 1960, the year of his *Maha Samadhi*, the departure of a self-realized saint from his mortal existence.

While a Sadhu, he wrote a large number of treatises and books on Sanatana Dharma (the Eternal Religion of the Soul), sciences, mathematics, world peace, and social issues. In 1953, at Nagpur, he founded an organization called *Sri Vishwa Punarnirmana Sangha* (World Reconstruction Association). Maharaja wrote 16 volumes on the detailed application of his 16 Sanksrit sutras in mathematics, one volume for each sutra. With these sutras, he claimed to have reconstructed the missing appendix of Atharvaveda. He was firmly convinced of the wide scope of their application: "The sutras apply to and cover each and every branch of mathematics (including arithmetic, algebra, geometry—plane and solid, trigonometry—plane and spherical, conics—geometrical and analytical, astronomy, calculus—differential and integral, etc.). In fact, there is no part of mathematics, pure or applied, which is beyond their jurisdiction." The manuscripts of these 16 volumes were deposited in the house of one of his disciples in Nagpur for safe keeping prior to publication. However, they vanished before they could be published. This tremendous loss of scientific knowledge remains an unsolved mystery. When the loss of his work was finally reported to Maharaja, he remained unperturbed and said that everything was still in his mind, and that he could rewrite the entire 16 volumes all over again from memory. In 1957, at the age of 73, the Shankaracharya finally started rewriting these volumes. He was only able to complete the first manuscript, an introductory volume, which was supposed to be published in the USA but was sent back to India in 1960 after his death. This manuscript was published in 1965 at Varanasi by Motilal Banarsidass, under the title *Vedic Mathematics*, and it was reprinted four times in the 1970s. This manuscript planted the seeds of a fertile field of research for future scientists and mathematicians to explore and expand. As a final comment, Maharaja has asserted that the names for "Arabic numerals," "Pythagoras' Theorem," and "Cartesian" coordinates are historical misnomers.

Hermann Klaus Hugo Weyl (1885–1955)

Hermann Klaus Hugo Weyl (1885–1955) was born to Anna Dieck and Ludwig Weyl, the director of a bank, in Elmshorn (near Hamburg), Germany. During 1904–1908, he studied at the universities of Munich and then Göttingen and completed his doctoral work under the guidance of Hilbert in 1908. His thesis involved an in-depth investigation of singular integral equations with a critical evaluation of Fourier integral theorems. He was appointed as a lecturer at the University of Göttingen, where he taught until 1913. Hermann gave a series of lectures on Riemann surfaces in 1911–1912, and from these lectures came the material for his first book, which he published in 1913. This book had a greater influence on the development of geometric function theory than any other publication since Riemann's dissertation; it led to three later editions and was reprinted as recently as 1997, which demonstrates more fully the significance of this work to the development of mathematics. Hermann married Helene Joseph in 1913 at the age of 28, a Jewish philosopher and a Spanish-language translator. Both of them had an interest in philosophy as well as a talent for language. During 1913–1930, he held the chair of mathematics at the Technische Hochschule in Zürich, Switzerland. During his first year there, Hermann came in close contact with his esteemed colleague, Einstein, who was at that time deeply engrossed in his work on the theory of general relativity. This contact influenced Hermann immensely; he became fascinated by the mathematical principles underlying Einstein's theory. Hermann conceived the idea of unifying general relativity with the laws of electromagnetism. In 1917, he delivered a series of lectures on a novel approach to relativity through differential geometry. This series formed the basis of his second book, *Space–Time–Matter*, published in 1918 with three subsequent editions in 1919, 1920, and 1923 that represented the gradual development of his theory. These made use of the Weyl metric, which resulted in a gauge field theory. Hermann's approach was not completely accepted by several prominent physicists. During this period, Hermann also made contributions on the uniform distribution of numbers modulo 1, which is fundamental in analytic number theory. During 1930–1933, Hermann was appointed head of mathematics at the University of Göttingen, filling the vacancy created by Hilbert's retirement. He probably would have remained in Göttingen for the rest of his career, but the rise of the Nazis and his wife's Jewish heritage forced him to emigrate to the USA in 1933, where he accepted a faculty position at the Institute for Advanced Study in Princeton. Hermann retired from IAS in 1952. On the eve of his retirement he delivered a series of lectures on symmetry. Hermann passed away in Zürich on December 8, 1955, at the age of 70. He is remembered for his contributions in both mathematics and theoretical physics through Weyl algebra, the Weyl inequality, the Weyl group, the Weyl conjecture, the Weyl dimension formula, the Weyl tensor, the Weyl lemma on the Laplace equation, the Weyl field, the Weyl transformation, the Weyl law, and the asymptotic distribution of eigenvalues.

Herbert Westren Turnbull (1885–1961)

Herbert Westren Turnbull (1885–1961) was born in Tettenhall, Wolverhampton. His father was interested in mathematics and transmitted his enthusiasm for the subject to his son. Herbert was educated at Sheffield Grammar School and then at Trinity College, Cambridge. He was Second Wrangler in the Mathematical Tripos, meaning that he was ranked second among the students who were awarded a First Class degree, and in 1909, he won Smith's Prize. After graduating, Turnbull taught at St Catharine's College, Cambridge, the University of Liverpool, and Hong Kong University. After he returned from the Far East in 1915, he taught for a few years at a public school before he became Regius Professor of Mathematics at United College, St. Andrews University. He held this post from 1921 to 1950 and then was made Professor Emeritus. Turnbull's major written works include *The Theory of Determinants, Matrices, and Invariants* (1928), *The Great Mathematicians* (1929), *Theory of Equations* (1939), *The Mathematical Discoveries of Newton* (1945), and *An Introduction to the Theory of Canonical Matrices* (1945), which he coauthored with Alexander Craig Aitken (1895–1967). During the last years of his life he edited the Royal Society's magnificent *The Correspondence of Isaac Newton*, which has been produced in three volumes and is an exciting and beautifully executed piece of recent scholarship. Turnbull received many honors for his work, the most notable being his election as an FRS in 1932. He was also elected to the Royal Society of Edinburgh, and he received their Keith Medal and the Gunning Victoria Jubilee Prize. He is remembered as a gifted simplifier of mathematical ideas.

Niels Henrik David Bohr (1885–1962)

Niels Henrik David Bohr (1885–1962) was born in Copenhagen, Denmark. His father, Christian Bohr (1855–1911), was a professor of physiology at the University of Copenhagen. His mother, Ellen Adler Bohr (1860–1930), belonged to a wealthy and prominent Jewish family. His brother, Harald August Bohr (1887–1951), was a mathematician and an Olympic soccer (football) player (the first Olympic games were held in 776 BC). Niels Bohr himself was also a soccer player. He matriculated at the Gammelholm Grammar School in 1903 at the age of 18, and he joined Copenhagen University and received his master's degree in physics in 1909 and then his Ph.D. in 1911. As a postdoctoral student, Bohr carried out experiments under Joseph John Thomson (1856–1940), a British physicist and a Nobel Laureate, at Trinity College, Cambridge. In 1912, Bohr married Margrethe Nørlund. In the same year, he worked at Manchester University with Ernest Rutherford (1871–1937), a British–New Zealand chemist and physicist and a Nobel Laureate (in chemistry). Bohr adapted Rutherford's nuclear structure to Planck's quantum theory. This resulted in a theory of atomic structure that remains valid today. This theory was greatly influenced by the work of Werner Heisenberg, a German theoretical physicist and a Nobel Laureate. Bohr published his model of atomic structure, using Rutherford's theory, in 1913. He introduced the concept of electrons traveling in orbits around the atom's nucleus. The chemical properties of each element were mainly determined by the number of electrons in the outer orbits of that element's atoms. Bohr discovered that an electron could drop from a higher energy orbit to a lower energy orbit, emitting a photon (light quantum) of discrete energy. This formed the very basis of quantum theory. Bohr's work with Rutherford during 1913–1917 was a golden period of his life. In 1918, Bohr returned to Denmark and became the director of the newly established Institute of Theoretical Physics. Bohr was awarded Nobel Prize in Physics for his fundamental contribution to atomic structure, known as Bohr's atomic model, and his work on quantum mechanics in 1922. He was also the recipient of the Benjamin Franklin medal (first awarded in 1915) in 1929. Bohr was awarded Order of the Elephant in 1947, a prestigious Danish distinction usually reserved for royalty or distinguished generals. He mentored many of the top physicists of the twentieth century at his institute in Copenhagen. He was one of the members of the team of physicists working on the Manhattan project that was responsible for producing an atom bomb during World War II. He worked on this project in the top secret Los Alamos Laboratory in New Mexico, USA. After World War II, Bohr came back to Copenhagen advocating the peaceful use of nuclear energy for the benefit of humanity. He is best known for the Copenhagen interpretation, the Bohr atomic model, complementarity, the Sommerfeld–Bohr theory, the Bohr–Einstein debate, the Bohr magnetron, and the

Bohr–Kramers–Slater (Hendrik Anthony Kramers, 1894–1952; John Clark Slater, 1900–1976) (BKS) theory, which was perhaps the final attempt to understand the interaction of matter and electromagnetic radiation.

Bohr had six sons. The eldest one died in a tragic boat accident and another died as a child of meningitis (inflammation of the membranes of the brain). His other children led successful lives. His son Aage Bohr (1922–2009) was also a scientist; he was awarded the Nobel Prize in Physics in 1975. Bohr breathed his last in Copenhagen on November 18, 1962 at the age of 77 after a heart attack. During the last few years of his life, Bohr was keenly interested in new developments in molecular biology. The latest formulation of his thoughts on the problem of life appeared in his final article: "Light and Life revisited," ICSU Rev., 5 (1963), 194, published after his death. Bohr was not a good public speaker. For the first 5 min of his talk, it was difficult to know whether he was talking in Danish, German, or English. The saying was that he stuttered in all languages.

Srinivasa Ramanujan (1887–1920)

Srinivasa Ramanujan (1887–1920) was a famous mathematical prodigy, born and raised in the southern Indian Brahmin family of Ayyangars near the city of Madras. His father, Srinivasa, was a clerk in a cloth shop. His mother, Komalatammal, contributed to the family income by singing at a local temple. In his elementary school class, the teacher was explaining the concept of division (sharing) through an example: if three bananas are divided among three children, then each child will get one banana; if four bananas are divided among four children, then each child will get one banana; if five bananas are divided among five children, and so on. Then the teacher generalized the idea of sharing x bananas among x children. Ramanujan claimed that when x is equal to zero, each child will get one banana. He explained to his classmates that zero divided by zero could be anything, since the zero of the denominator may be any number times the zero of the numerator. Ramanujan studied at the local English language school, displaying his talent and interest for mathematics. At 13 he mastered a textbook used by college students. When he was 15, a university student lent him a copy of George Shoobridge Carr's (1837–1914) *Synopsis of Pure Mathematics*. Ramanujan decided to work out over 6,000 results in this book, stated without proof or explanation, writing on papers that were later collected to form notebooks. He graduated from high school in 1904, winning a scholarship to the University of Madras. He enrolled in a fine arts curriculum, neglected every subject but mathematics, and lost his scholarship. He failed the university examination four times from 1904 to 1907, doing well only in mathematics. During this time, he filled his notebooks with original writings, sometimes rediscovering work that had already been published, and at other times making new discoveries. Without a university degree, it was difficult for Ramanujan to find a decent job. To survive, he had to depend on the goodwill of his friends. He tutored students in mathematics, but his unconventional ways of thinking and failure to stick to the syllabus caused problems. He was married to a 9-year old bride, Janaki Ammal, as per his parents' wishes, but she did not come to live with him until 1912. Needing to support himself and his wife, he moved to Madras and sought a job. He showed his notebooks of mathematical writings to his potential employers, but the books bewildered them. However, a professor at the Presidency College recognized his genius and supported him for a while, and in 1912, he found work as an accounts clerk, earning a small salary. Ramanujan continued his mathematical work during this time and published his first paper on Bernoulli numbers in 1911 in the *Journal of the Indian Mathematical Society*. He realized that his work was beyond that of the other Indian mathematicians, and so he began writing to leading English mathematicians. The first mathematicians he contacted, Ernest William Hobson (1856–1933) and Henry Frederick Baker (1866–1956), turned down his requests for help. But in January 1913 he wrote to Hardy, who was inclined to turn Ramanujan down, but the mathematical statements in his letter, although stated without proof, puzzled Hardy. He decided to examine them closely with the help of his colleague and collaborator, Littlewood. They decided, after careful study, that Ramanujan was probably a genius, since his statements "could only be written down by a mathematician of the highest class, they must be true, because if they were not true, no one would have the imagination to invent them." Hardy arranged a scholarship for Ramanujan, bringing him to England in 1914. At first Ramanujan refused to travel, but in the end he decided to go, which was attributed to the divine intervention of the family goddess Namagiri. Hardy personally tutored him in mathematical analysis, and they collaborated for 5 years, proving significant theorems about the number of partitions of integers. During this time, Ramanujan made important contributions to number theory and also worked on continued fractions, infinite series, and elliptic functions. Unfortunately, in 1917, Ramanujan became extremely ill. At the time, it was thought that he had trouble with the English climate and had contracted tuberculosis. It is now thought that he suffered from a vitamin deficiency, brought on by Ramanujan's strict vegetarianism and the food shortage in wartime England. His illness was certainly exacerbated by Hardy's high expectations and his own strict work ethic. During this period he became severely depressed. He threw himself on the track of an approaching train in London and badly injured his knee. Miraculously, his life was spared. Observing Ramanujan's failing health, Hardy pursued relentlessly to have him elected an FRS. In 1918, Ramanujan became one of the youngest to achieve this honor. He returned to India in 1919 and continued to do mathematics even while he was confined to his bed by debilitating pain. On January 12, 1920, he wrote to Hardy about his discovery of *mock theta functions*, which, according to George Neville Watson (1886–1965), was a very significant achievement. On April 26, 1920, at the age of 32, Ramanujan fell unconsciousness and died, and the world lost one of its mathematical legends.

According to Hardy, "the limitations of Ramanujan's knowledge were as startling as its profundity." "Here was a man who could work out modular equations and theorems of complex multiplication, to orders unheard of, whose mastery of continued fractions was beyond that of any mathematician in the world, who had found for himself the functional equation of the zeta-function, and the dominant terms of the many of the most famous problems in the analytic theory of numbers; and he had never heard of a doubly periodic function or of Cauchy's theorem, and had indeed but the vaguest idea of what a function of a complex variable was."

Ramanujan was a simple, religious man; he always attributed his mathematical talent to his family deity, Namagiri. He considered mathematics and religion to be linked. He said, "an equation for me has no meaning unless it expresses a thought of God." He belongs to the school of hyper-pure mathematicians who regard mathematics as a game that involves the manipulation of symbols according to given rules. He intuitively stated many complicated results and attributed these to his goddess. He was endowed with an astounding memory and remembered the idiosyncrasies of the first 10,000 integers to such an extent that each number became like a personal friend to him. Once, Hardy went to see Ramanujan when he was in a nursing home and remarked that he had traveled in a taxi with a rather dull number, 1729, at which Ramanujan exclaimed, "No, Hardy, 1729 is a very interesting number. It is the smallest number that can be expressed as the sum of two cubes ($1729 = 1^3 + 12^3 = 9^3 + 10^3$), and the next such number is very large." His life can be summed up in his own words, "I really love my subject."

In 1976, George Andrews (born 1938), Evan Pugh Professor of Mathematics at Pennsylvania State University, came across 130 pages of scrap paper in a library at Cambridge University that were filled with notes representing Ramanujan's work during the last year of his life in Madras. These pages are now known as Ramanujan's *Lost Notebook*, and they contain about 4,000 formulas and other work. According to Richard Askey (born 1933), who collaborated with Andrews, "The work of that 1 year, while he was dying, was the equivalent of the lifetime work of a very great mathematician. What he accomplished was unbelievable. If it were in a novel, nobody would believe it." Ramanujan's formulas keep mathematicians busy to this day, waiting to be deciphered and proved. Many of these formulas were given in their final form without any intermediate steps. This is quite consistent with the ancient Indian tradition, in which great mathematicians merely used to state their results, leaving their students and followers to provide commentaries. His 1914 paper on "Modulus functions and approximation to π" was used recently to compute π to a level of accuracy never before attained. As Robert Kanigel (born 1946) wrote in his book *The Man Who Knew Infinity: A Life of the Genius Ramanujan* (Charles Scribner's Sons, New York, 1991) remarks, "What makes Ramanujan's work so seductive is not the prospect of its use in the solution of real-world problems, but its riches, beauty and mystery—its sheer mathematical loveliness". There are many institutions and a temple in India where different portraits and busts of Ramanujan can be found. During 1985–1997, Bruce Carl Berndt (born 1939) meticulously edited Ramanujan's original notebooks in five volumes, published by Springer-Verlag.

Erwin Schrödinger (1887–1961)

Erwin Schrödinger (1887–1961) was born in Vienna. He was educated at home by a private tutor until the age of ten. In 1898, Schrödinger entered the Akademisches Gymnasium, where he graduated in 1906. He wrote later about his time at the Gymnasium, "I was a good student in all subjects, loved mathematics and physics, but also the strict logic of the ancient grammars, hated only memorizing incidental dates and facts. Of the German poets, I loved especially the dramatists, but hated the pedantic dissection of their works." After he left the Gymnasium in 1906, he entered the University of Vienna, where he studied a variety of subjects in theoretical physics and mathematics. In 1910, he was awarded his doctorate for the dissertation *On the Conduction of Electricity on the Surface of Insulators in Moist Air*. Afterward, Schrödinger performed voluntary military service in the fortress artillery. He then joined the University of Vienna as an assistant to Franz Exner (1849–1926) and his work in experimental physics. In 1914, he completed the work for his habilitation and was awarded his degree. He published his first paper that year, in which he expanded Boltzmann's ideas. During World War I he served as an artillery officer, but he continued his theoretical work. In 1917, Schrödinger was sent back to Vienna and assigned to teach a course in meteorology. During 1918–1920, he made substantial contributions to the theory of color vision. In 1920, he accepted an assistantship in Jena, and after only a short time there he moved to a chair in Stuttgart. He then moved to a chair at Breslau, his third move in 18 months. In late 1921 he moved once again to Zürich, accepting a chair of theoretical physics, where he remained for 6 years. This was a fruitful period for Schrödinger; he was actively engaged in a variety of work in theoretical physics. His papers dealt with the specific heats of solids, problems in thermodynamics, and atomic spectra. He published his revolutionary work relating to wave mechanics and the general theory of relativity in a series of six papers in 1926. In 1927, Schrödinger moved to Berlin as Planck's successor. In 1933, Schrödinger shared the Nobel Prize with Dirac and he moved to Oxford on a fellowship. In 1934, he was offered a permanent position at Princeton University, which he declined. After staying for short periods in the Universities of Graz, Oxford, and Ghent, in 1939 he joined the newly created Institute for Advanced Studies in Dublin, where he became Director of the School for Theoretical Physics. He remained in Dublin until his retirement in 1955. After his retirement he returned to Vienna, where he died in 1961 after a long illness. His scientific work can be appreciated only by experts, but he was a man of broad cultural interests and was a brilliant and lucid writer in the tradition of Poincaré. He liked to write short books on big themes: *What Is Life?*, *Science and Humanism*, and *Nature and the Greeks*, Cambridge University Press, New York, 1944, 1952, and 1954, respectively. The huge crater Schrödinger on the far side of the Moon is named after him. The Erwin Schrödinger International Institute for Mathematical Physics has been established in Vienna in 1993.

Hugo Dyonizy Steinhaus (1887–1972)

Hugo Dyonizy Steinhaus (1887–1972) was born in Jaslo to a family of Jewish intellectuals. His paternal uncle, Ignacy, was a well-known political activist and a deputy to the Austrian parliament. Steinhaus studied for 1 year at the University of Lwów, and then he moved to Göttingen, which was famous for its excellent mathematicians, including Hilbert, Klein, Minkowski, Landau, and Carathéodory. He received his doctoral degree there, in 1911 and with distinction, under Hilbert's supervision. However, after his doctorate, he was influenced by Lebesgue's work and decided to follow his line of research. Steinhaus spent the years immediately preceding World War I in his native Jaslo. After the outbreak of war, he served in the Polish Legion until 1916 and then he took up an assistantship at the University of Lwów. From then until 1941, Steinhaus was inseparably connected with Lwów. He became docent there in 1917, associate professor in 1920, and full professor in 1923. In 1926, he was elected a member of the Scientific Society in Lwów. In 1929, with Banach, he founded the prestigious journal *Studia Mathematica*. During the first part of World War II, Steinhaus continued lecturing at the University of Lwów. But after the departure of the Soviet authorities from Lwów, and after the closing of the university by the German occupants, he was compelled to go into hiding. As soon as Poland was liberated in 1945, he settled in Wroclaw; he was the first Dean of the Department of Mathematics, Physics, and Chemistry (jointly for the University and the Technical University), the first chairman of the Wroclaw Division of the Polish Mathematical Society, and the first general secretary (and later, president) for many years of the newly founded Wroclaw Scientific Society. He was also a professor at the University of Notre Dame (Indiana, USA, 1961–1962) and the University of Sussex (1966), and he served many international scientific societies and science academies in various capacities. Steinhaus authored over 170 works on trigonometric series, the theory of real functions, functional analysis, orthogonal series, probability theory, game theory, topology, and set theory. He applied mathematics to various fields, including medicine, electricity supply rates, dendrometry, patrimony, the estimation of loads by means of test borings, and anthropology. His joint monograph with Stefan Kaczmarz (1895–1939), *Theorie der Orthogonalreiben* "Monografie Matematyczne" 6 (1935), and the Banach–Steinhaus theorem are often cited in the mathematical literature. He described mathematics as a "science of nonexistent things." Steinhaus passed away on February 25, 1972, in Wroclaw.

George Polya (1887–1985)

George Polya (1887–1985) was born in Budapest (Hungary) to a Jewish family; however, he was baptized by the Roman Catholic Church shortly after his birth. Early in his education he did not do well in mathematics, and he blamed two of his three mathematics instructors at the gymnasium, calling them "despicable teachers." He enrolled at the University of Budapest in 1905 to study law, but he found it so boring that he gave it up after only one semester, shifting his focus to languages and literature. After 2 years he began to take physics and mathematics courses, which were taught by Fejér and Lóránd Baron von Eötvös (1848–1919); Polya was greatly influenced by Fejér. Polya spent the academic year 1910–1911 studying mathematics and physics at the University of Vienna. He returned to Budapest during the following year, and he was awarded a doctorate in mathematics for his work on geometric probability. During 1912–1913, he visited Göttingen and met Klein, Carathéodory, Hilbert, Runge, Edmund Landau, Weyl, Hecke, Courant, and Otto Toeplitz (1881–1940). In 1914, he was appointed a Privatdozent at Eidgenössische Technische Hochschule Zürich, where Hurwitz, Karl Friedrich Geiser (1843–1934), Bernays, Zermelo, and Weyl were his colleagues. In 1918, he married a Swiss girl, Stella Vera Weber, who was a daughter of the Professor of Physics at the University of Neuchâtel. Polya became extraordinary professor at the Hochschule Zürich in 1920. He was awarded the Rockefeller (John Rockefeller, 1839–1937) Fellowship in 1924 to study with Hardy and Littlewood in England. There, the three of them began writing the famous book *Inequalities*, which was published in 1934. During 1926–1928, Polya wrote 31 research articles on series, number theory, combinatorics, voting systems, astronomy, probability, and integral functions. He was made Ordinary Professor at the Hochschule Zürich in 1928. In 1933, Polya was again awarded the Rockefeller Fellowship, this time to study at Princeton. He returned to Zürich, but because of the political situation in Europe, he had to flee to the USA in 1940. He worked at Brown University for 2 years, Smith College for a short while, and then he took up an appointment at Stanford. In 1953, Polya retired from Stanford, but he continued his association with Stanford as Professor Emeritus, devoting all his time to mathematics education. On December 13, 1977, a dinner was given at Stanford to celebrate his 90th birthday, and many of his friends and colleagues gave him glowing tributes. In 1978 he taught a course on combinatorics in the Stanford Computer Science Department. He died in 1985 in Palo Alto, California.

Polya wrote more than 250 research papers in many languages, as well as many books besides Inequalities, including *Aufgaben und Lehrsätze aus der Analysis* with Gábor Szegö (1895–1985) in 1925; *How to Solve It* (1945); *Isoperimetric Inequalities in Mathematical Physics* with Szegö (1951); *Mathematics and Plausible Reasoning* (1954); and *Mathematical Discovery* in two volumes (1962 and 1965). Polya remarked that "A great discovery solves a great problem, but there is a grain of discovery in the solution of any problem." He proposed an excellent general outline for solving applied problems in his classic book How to Solve It: (1) Understand the problem. (2) Devise a plan. (3) Carry out the plan. (4) Look back. This book sold over one million copies and has been translated into 17 languages. In mathematics, he is remembered for the multivariate Polya distribution, the Polya conjecture, the Polya enumeration theorem, the Polya–Vinogradov inequality, and the central limit theorem. Once he remarked that teaching is not a science; it is an art. If teaching were a science there would be a best way of teaching and everyone would have to teach like that. He was elected an honorary member of the Hungarian Academy, the London Mathematical Society, the Mathematical Association of Great Britain, and the Swiss Mathematical Society. He was elected to the National Academy of Sciences of the United States, the American Academy of Arts and Sciences, the Académie Internationale de Philosophie des Sciences de Bruxelles, and the California Mathematics Council. He was also a corresponding member of the Académie des Sciences in Paris. Polya has become known as the father of problem solving. According to Polya, "Mathematical thinking is not purely 'formal'; it is not concerned only with axioms, definitions, and strict proofs, but many other things belong to it; generalizing from observed cases, inductive arguments, arguments from analog, recognizing a mathematical concept in, or extracting it from, a concrete situation."

The following anecdotes about Polya are often quoted:

Polya was asked why so many good mathematicians came out of Hungary at the turn of the (twentieth) century. He theorized that it was because mathematics is the cheapest science, requiring no expensive equipment, only a pencil and paper.

Polya left Göttingen due to rather unfortunate circumstances. He explained the incident in a letter to Ludwig Georg Elias Moses Bieberbach (1886–1982) in 1921: "On Christmas 1913 I traveled by train from Zürich to Frankfurt and at that time I had a verbal exchange—about my basket that had fallen down—with a young man who sat across from me in the train compartment. I was in an overexcited state of mind and I provoked him. When he did not respond to my provocation, I boxed his ear. Later on it turned out that the young man was the son of a certain Geheimrat; he was a student in Göttingen. After some misunderstandings I was told to leave by the Senate of the University."

An imaginary doctor invented by Polya comforted his patient with the remark, "you have a very serious disease. Of ten persons who get this disease only one survives. But do not worry. It is lucky you came to me, for I have recently had nine patients with this disease and they all dead of it."

While in Switzerland, Polya loved to take afternoon walks in the local garden. One day he met a young couple that was also walking, and so he chose another path. He continued to do this, yet he met the same couple six more times as he strolled in the garden. He asked his wife how it could be possible to meet them so many times while he walked randomly through the garden.

Zygmunt Janiszewski (1888–1920)

Zygmunt Janiszewski (1888–1920) was born in Warsaw and attended secondary schools in Lwów until 1907. He then studied in Zürich, Göttingen, and Paris, where he obtained his doctoral degree in 1911 based on an excellent thesis in topology, *Sur les continus irréductibles entre deux points*. The board that passed his thesis consisted of three eminent mathematicians: Poincaré, Lebesgue, and Fréchet. Poincaré, fascinated by this new branch of mathematics, expressed his attraction to the subject in his *Notice sue les travaux scientifiques*: "All the roads which I entered in turn, always led me to topology." After he received his doctoral degree, Janiszewski returned to Poland, where he taught at the University of Lwów and then joined the University of Warsaw as a professor. At the outbreak of World War I he became a soldier in the Polish Legions of Józef Pilsudski, believing that he was fighting for Polish independence. In 1916, Janiszewski refused to take an oath of allegiance to the Austrian government, as it was not compatible with his patriotic conscience. He was already a professor at the reborn University of Warsaw, where he was very active, lecturing for scientific courses and taking an interest in problems of the organization of mathematics and its methodology and philosophy. Janiszewski, Sierpiński, and Mazurkiewicz formed the driving force behind the mathematical journal *Fundamenta Mathematicae*. Janiszewski proposed the name of the journal in 1919, though the first edition was not published until 1920. Sadly, his life was cut short during the influenza pandemic of 1918–1919, which took his life in Lvov on January 3, 1920, while he was just 31 years old. Besides his doctoral thesis, he presented a sensational report entitled *Über die Begriffe: Linie und Fläche* (On the notations: line and surface) at the International Mathematical Congress in Cambridge (1912). His fundamental results in the field of plane topology, entitled *On cutting plane by continua* published in *Prace Matematyczno-Fizyczne* (1913), are of immense value. Janiszewski donated the inherited family property left by his father to charity and public education. He also donated all prize money he received from mathematical awards and competitions to the education and development of young Polish students. Further, he willed that his body be used for medical research, and his cranium for craniological study, desiring to be "useful after his death."

Stefan Mazurkiewicz (1888–1945)

Stefan Mazurkiewicz (1888–1945) was born in Warsaw. After graduating from secondary school, he left Warsaw to study in Cracow, Münich, Göttingen, and he was finally awarded his doctoral degree in Lwów in 1913 with Sierpiński. His thesis concerned curves filling in the square. In 1915, Mazurkiewicz became the Chair of Mathematics at the newly created University of Warsaw; he also applied for the title of docent at Cracow on the basis of a paper, *The theory of G_δ sets*, which was later published in the journal *Wektor*. During the Polish–Soviet War (1919–1921) in 1919, Mazurkiewicz broke the most common Russian cipher for the Polish General Staff's cryptological agency. He was one of the chief editors of *Fundamenta Mathematicae*, and cofounder of *Monografie Matematyczne*, President of the Polish Mathematical Society during 1933–1935, Dean of the University of Warsaw for 9 years, its Vice Rector after 1937, and general secretary of the Warsaw Scientific Society after 1935. In mathematics, he is remembered for the Hahn–Mazurkiewicz theorem, named after Hans Hahn (1879–1934), a basic result on curves prompted by the phenomenon of space-filling curves, and his 1935 paper *Sur l'existence des continus indécomposables*, which is generally considered to be the most elegant piece of work in point-set topology. Mazurkiewicz published 141 articles on topology, analysis, and probability theory. Karol Borsuk (1905–1982), Bronislaw Knaster (1893–1980), Kazimierz Kuratowski, Stanislaw Saks, and Antoni Zygmund (1900–1992) were his students. Mazurkiewicz was expelled from Warsaw and had no financial means for the proper treatment of his illness. He died in a hospital in the vicinity of Warsaw as another of the countless victims of the war.

Chandrasekhara Venkata Raman (1888–1970)

Chandrasekhara Venkata Raman (1888–1970) was born at Thiruvanaikaval near Tiruchirappalli (Trichy), India. His father, R. Chandrasekhara Iyer (born 1866), was a teacher of mathematics and physics; his mother, Parvati Ammal, was a lady of exceptional perseverance. He was the second of his parent's eight children (they had five sons and three daughters). He grew in an academic atmosphere: an atmosphere of science, Sanskrit literature, and music. At an early age, Raman moved to Visakhapatnam, a port city in the state of Andhra Pradesh, Southeastern India, and there he studied at the St. Aloysius Anglo-Indian High School. At the age of 11, he completed his secondary education. Two years later he moved to the Presidency College in Madras (Chennai) in Tamil Nadu. At the age of 15, he received his bachelor's degree with honors in physics and English, winning the first place and the gold medal in physics. It was then suggested that Raman go to England for further study; however, he was not allowed because the Civil Surgeon of Madras did not find him physically fit for the journey. Raman did not regret it. In fact, he later said that he felt grateful to the Civil Surgeon for being responsible for keeping him in India. At the age of 18, he received his master's degree with highest distinctions. During Raman's time, there were few job opportunities for scientists in India, and so he was compelled to take the Civil Services Competitive Examination for the Finance Department, for which he received the highest score. He accepted the government position of Assistant Accountant General in Calcutta (Kolkata). During this time, Raman came across a 13-year-old girl playing Thyagaraja Keerthana on the Veena (a plucked stringed musical instrument that dates back to the Vedic period) and he immediately developed an intense fondness for her. Against the accepted custom, he arranged his marriage with her. Her name was Lokasundari. In 1907, along with his wife, he departed for Kolkata to begin his new job. If only he knew what a glorious period of his career he would find in Kolkata! one day, during the very first week of his arrival in Kolkata, he spotted a sign board while he was on his way to work that read "The Indian Association for Cultivation of Sciences" (IACS). This was to play a major role in his life, as well as in the history of scientific culture in India. Raman joined this institute and put in several hours every day before and after his other work at the IACS accounts department. He published a research article, "The Small Motion at the Nodes of a Vibrating String," in the journal Nature in 1909. Raman's scientific work was interrupted by an official transfer to Yangon (Rangoon), Myanmar (Burma) for about a year in 1909, and a subsequent transfer to Nagpur (India) in 1910. In both cities, besides performing his official duties, he carried out his experiments at his residence with very few tools. Fortunately, he was transferred back to Kolkata in 1911, and he was able to resume his work at the IACS. Raman started a Bulletin of the IACS, which grew in 1926 to a full-scale scientific journal, the Indian Journal of Physics. Raman performed some research in the field of sound and music, and he discovered the overtones in the sounds of the mridangam, a South Indian drum, and the tabla, another Indian percussion instrument that is mainly used in Hindustani classical music. Raman analyzed them and demonstrated that the richness of these instruments was far superior to the normal stretched membrane of the western drums. Raman published an interesting article on the acoustic knowledge of the ancient Hindus. By that time he had established himself as an international expert on sound and musical instruments.

In 1917, the Vice Chancellor of Calcutta University, Asutosh Mukherjee (1864–1924), advised him to quit his job at the Accounts Department and accept the prestigious Palit Professorship (a chair named after Tarak Nath Palit) at the University. During those days there was a regulation that a candidate for a Palit Professorship must be trained in a foreign country! Raman simply refused to comply, which was very consistent with his character and his pride as an Indian. The highly respected Vice Chancellor Asutosh Mukherjee, known in Bengal as the "Tiger of Bengal", changed the rules in order to appoint Raman. He resigned his government job in 1917 and became Palit Professor of Physics at the University of Calcutta, where he spent 17 of his career's most valuable years, from 1917 to 1933. He continued his research at IACS, where he became Honorary Secretary in 1919. Many meritorious students with a zeal for science gathered around him at both the University of Calcutta and the IACS. Raman used to refer this period in Kolkata as the golden era of his career. In the midst of serious research in physics, he depicted his unbounded affection to his students. Often he lovingly asked them to drink rasam, a south Indian drink (which is not widely known and is seldom drunk in Bengal) that is consumed by south Indians almost daily, in order to benefit their intellectual activities. Rasam has enormous health benefits and is used to cleanse the body of toxins and infections.

In 1921, Raman went to Oxford, England, as a delegate to the Universities' Congress. He visited London's Saint Paul's Cathedral, famous for its whispering gallery. The whispering gallery induced exceptional excitement and curiosity in Raman. He performed a few experiments and analyzed the results. His findings were significant and were published in two research articles in two of the foremost scientific journals, Nature and the Proceedings of the Royal Society. On his way to Europe by ship, Raman was struck by the blueness of the sea. Rayleigh himself observed: "...he much admired blueness of the sea... is simply the blue of the sky seen by reflection." Rayleigh was the first scientist to explain why the sky is blue. While he was on the ship, Raman did a simple experiment and found that the blueness of the sea was not merely due to the reflected color of the sky, as Rayleigh had observed. On the contrary, the blue color of the sea is due to the scattering of the sun's light by molecules of water. It must have been amazing to observe this Indian scientist, who was always seen with a turban, conducting his experiments with a simple nicol prism (a polarizer, an optical device used to generate a beam of polarized light) on the ship's deck with samples of sea water! After his return to India, Raman began further research on (i) the scattering of light by liquid, (ii) the viscosity of liquids, and (iii) the scattering of x-rays by liquid. His work on the scattering of light by liquid, which came to be known as the Raman effect, earned him and India their first Nobel Prize in Physics in 1930. During January of 1928, one of Raman's associates noticed that the scattered light was greenish, instead of the usual blue, in pure glycerin. Furthermore, the reflected light was strongly polarized. During the last week of January in 1928, Raman asked Kariamanickam Srinivasa Krishnan (1898–1961), who was monitored by Raman, to repeat these scattering experiments under more controlled conditions. Krishnan confirmed Raman's findings, and Raman personally verified all the observations. He was excited about these findings because he understood what this discovered phenomenon was. On February 16, 1928, Raman sent a note to Nature mentioning that the modified radiation observed in these experiments could be due to certain molecular fluctuations; this phenomenon was not fully understood yet. On February 27, Raman set up an experiment in which he very keenly observed the behavior that was thought to be due to fluorescence. He used a direct vision spectroscope for this experiment. However, his experiment could not be performed; by the time the experiment had been set up, the winter sun in Kolkata had already set. On the morning of February 28, when Raman finally carried out his experiment, he observed that the emitted spectrum contained not only the incident color but also one more, separated by a dark region. This was the very first evidence of what is known as the Raman effect. On the night of February 28, 1928, he personally went to the office of The Statesman (a daily newspaper renowned for publishing highly reliable news) and asked them to publish the news of his discovery the very next day. At the time, the Statesman press was busy collecting the news that was to be published in the early morning of February 29, 1928 (a leap year). The concerned editor gave this news enormous importance, and it made the front page of the February 29 issue of the newspaper! Several physicists in different parts of the world were working in this area, hence the urgency. In addition, Raman sent a note to Nature on March 8, 1928, announcing his discovery along with an explanation of his experiment. Several scientists then repeated the experiment with the same specified measurements, and they confirmed the findings.

Raman and Suri Bhagavantam (1909–1989), who was the director of the Indian Institute of Science (IISc), Bangalore, during 1957–1962, discovered quantum photon spin in 1932, which further confirmed the quantum nature of light. In 1933, after bidding farewell to Kolkata, Raman joined IISc, Bangalore, as director (1933–1937). He retired from IISc in 1948 at the age of 60. In 1949, he established the Raman Research Institute (RRI) in Bangalore, very close to IISc. He served as its director and remained active in its work until his death in 1970.

One of the authors of this monograph (Sen) was introduced to Raman around 1968 by his friend, Professor Sisir Kumar Chatterjee. This was a pleasant surprise for the author. He observed that Raman spent a significant amount of time talking to Chatterjee about the colors of flowers with enormous enthusiasm. Raman's mind was perhaps completely engrossed in finding the reason that such beautiful colors are created in nature. A few days later the author and Chatterjee happened to be on the RRI campus, which was about 15 min' walk from the IISc main gate. The first thing that attracted the attention of any visitor was the exotic flower garden on the campus and its large variety of flowers, which includes several rare species. Once again, Raman talked to Chatterjee zealously, explaining the various properties of the colors and their possible origin while showing him flowers of different colors in the garden. It appeared that during the last days of his life, Raman's mind was fully occupied by the colors of flowers. It was as if he was in a state of meditation and that only a single thought, the thought of beautiful flowers instead of random ideas, pervaded his whole being! It was a higher state of mind and probably a state not very far from the state of silence, the ultimate state.

Behind this great physicist was Lady Lokasundari Raman, a wonderful wife who showered her husband's students with unstinted motherly affection and shouldered all the responsibilities and unavoidable worries of their day-to-day existence. If not for her, Raman would hardly have been able to devote himself so wholeheartedly to scientific research. On November 21, 1970, Raman passed away in Bangalore at the age of 82, and India lost one of the greatest physicists of the twentieth century. Some of the notable awards bestowed on Raman were FRS in 1924, the Matteucci Medal (established in 1870 to honor Carlo Matteucci, 1811–1868) in 1928, Knight Bachelor in 1929, the Nobel Prize in Physics in 1930, the Bharat Ratna in 1954 (meaning Jewel of India—Republic of India's highest civilian award), and the Lenin Peace Prize (established in 1949 to honor Vladimir Ilyich Lenin, 1870–1924) in 1957.

Richard Courant (1888–1972)

Richard Courant (1888–1972), a German American mathematician and mathematical physicist, was born in Lublinitz in Silesia province in the Prussian Kingdom of the German Empire, which is now part of Poland. His father was an unsuccessful businessman. Richard studied at the University of Göttingen and also in Zürich, Switzerland. In 1908, Hilbert chose him to be his assistant, and in 1910, he received his Ph.D. under the supervision of Hilbert from the University of Göttingen. His thesis was entitled *On the application of Dirichlet's principle to the problems of conformal mappings*. He was drafted to fight in World War I, but was allowed to leave the military after he was seriously injured. In 1912, Courant married Nelly Neumann, a fellow student from Breslau; their marriage lasted only 4 years. He was married in 1919 to Nerina Runge, a daughter of Carl Runge. They had much in common (they shared a passionate love of music), but in many respects they were very different. Richard and Nerina had four children, two sons and two daughters, and all of them were highly successful in their careers. Richard continued his academic activities in Göttingen, and he spent a 2-year period as a professor in the University of Münster. In Göttingen, he established the Mathematical Institute and remained as its director during 1928–1933. After spending 1 year in Cambridge, Richard joined New York University, USA in 1934, where he became a professor in 1936. Here he established an institute for graduate studies in applied mathematics, which was renamed the Courant Institute of Mathematical Sciences in 1964. This has become a revered research institute for applied mathematics. Courant encouraged the publication of high quality mathematical texts and monographs. He and Hilbert authored the renowned text entitled *Methods of Mathematical Physics*, which has been used worldwide as a classic textbook for over eight decades and is still used extensively. Richard gave the solid mathematical basis for the finite element methods used in solving partial differential equations with widely varied boundary conditions. He is remembered for the Courant number, the Courant–Friedrichs–Lewy (Kurt Otto Friedrichs, 1901–1982; Hans Lewy, 1904–1988) condition, and the Courant minimax principle. He is probably best known for his leadership talent. Richard passed away on January 27, 1972, at the age of 84 in New Rochelle, New York. The following anecdote is told about him: Richard Courant encountered a very formidable five-fold multiple integral in his work. He tried everything he knew to evaluate it. After a superhuman struggle stretching over many months he managed to reduce it to a three-fold integral, and then all further progress ceased. In desperation he brought the three-fold integral to Gustav Herglotz (1881–1953). Herglotz looked at the integral for a few minutes and said "We should not evaluate the integral as it stands. However, there is a certain not too obvious transformation which will convert the three-fold integral into a five-fold integral which we can evaluate quite easily."

Balthasar van der Pol (1889–1959)

Balthasar van der Pol (1889–1959) was born in Utrecht and died in Wassenaar. His main interests were radio wave propagation, theory of electrical circuits, and mathematical physics. He initiated the study of a differential equation in the 1920s that bears his name. This led Liénard and others to investigate the mathematical theory of self-sustained oscillations in nonlinear mechanics. He was awarded the Institute of Radio Engineers (now the IEEE) Medal of Honor in 1935. The asteroid 10443 van der Pol is named after him.

Sir Ronald Aylmer Fisher (1890–1962)

Sir Ronald Aylmer Fisher (1890–1962), an English statistician, eugenicist, evolutionary biologist, geneticist, and an FRS, was born in East Finchley in London to George and Katie Fisher. Ronald had a very poor eyesight but was an extraordinarily mature student with an amazing mind. He was tutored in mathematics without the aid of a pen and paper because of his poor eyesight. Consequently, he developed an exceptional ability to visualize problems in terms of geometry, and not so much in terms of derivation or proof. In 1906, at the age of 16, he was awarded the Neeld Medal of an essay competition in mathematics at Harrow School, an English independent school for boys situated in the town of Harrow in north-west London. His ability to conjecture mathematical statements without justifying intermediate steps amazed his peers. He developed a strong interest in biology, and especially evolution. He set the foundation of what was to become known as population genetics. His work had a major influence on the evolutionary biologist William Donald Hamilton (1936–2000) and the development of his later theories on the genetic basis for kin selection. Ronald married Ruth Eileen while they were 27 and 16 years old, respectively. They had two sons and seven daughters, one of whom died in infancy. In 1937, Ronald visited the Indian Statistical Institute (ISI, Calcutta, India), which consisted of one part-time employee, Mahalanobis. He revisited ISI often, encouraging its development. He was the guest of honor at the 25th anniversary of ISI (established in 1931) in 1957, when it had grown to 2,000 employees. Ronald was opposed to the conclusions of Sir William Richard Shaboe Doll (1912–2004), FRS, a British physiologist and the foremost epidemiologist of the twentieth century, and Sir Austin Bradford Hill (1897–1991), FRS, an English epidemiologist and statistician who found evidence that smoking causes lung cancer. He compared the correlations in their papers to a correlation between the import of apples and the rise of divorce to show that correlation does not necessarily imply causation. Ronald published nearly 400 articles and several books. He is widely known in literature for many of his outstanding contributions, such as the Fisher–Kolmogorov equation, Fisher's exact test, Fisher's inequality, Fisher's geometric model, Fisher's method for combining independent tests of significance, Fisher's permutation test, Fisher's z-distribution, the Fisher–Bingham (Christopher Bingham) distribution, the Fisher–Tippett (Leonard Henry Caleb Tippett, 1902–1985) distribution, the von Mises–Fisher distribution, Fisher's theory of the evolution of the sex ratio, and the Fisher kernel. Fisher was honored as a Knight Bachelor by Queen Elizabeth II in 1952, and he was awarded the Linnean Society of London's prestigious Darwin–Wallace (Alfred Russel Wallace, 1823–1913) Medal in 1958. After retiring from Cambridge University in 1957, he spent some time as a senior research fellow at the Commonwealth Scientific and Industrial Research Organization (CSIRO), Australia's national science agency and one of the largest and most diverse research agencies in the world, in Adelaide, Australia. He died of colon cancer there on July 29, 1962.

Ivan Matveevich Vinogradov (1891–1983)

Ivan Matveevich Vinogradov (1891–1983) was born in the Velikiye Luki district, Pskov Oblast. His father was a priest of the Russian Orthodox Church. He graduated from the University of St. Petersburg, and in 1920, he became a professor there. After 1934 he was a Director of the Steklov Institute of Mathematics, a position that he held for the rest of his life, except during the period 1941–1946. In 1937,

Vinogradov used his main problem-solving technique (applied to central questions involving the estimation of exponential sums) to make progress toward proving Goldbach's Conjecture by showing that every odd integer greater than, say, $10^{10^{10}}$ (or some similar bound) is a sum of three prime numbers. Vinogradov was very strong, and he was able to lift a chair with a person sitting on it by holding only the leg of the chair. He was twice awarded the order of Hero of the Socialist Labour. The house where he was born was converted into his memorial—a unique honor among Russian mathematicians. In 1941, he was awarded the Stalin Prize (established in 1941). In mathematics he is also remembered for the Polya–Vinogradov inequality and the Bombieri–Vinogradov theorem.

Stefan Banach (1892–1945)

Stefan Banach (1892–1945) was born in Cracow. In 1902, Banach enrolled in Kraków's Henryk Sienkiewicz Gymnasium, where he became known as a prodigy. At the age of 14, he started studying higher mathematics, and he earned his living by giving private lessons. For a short time he attended Jagiellonian University, and then he entered the Lwów Technical University, but his studies were interrupted by the outbreak of World War I. Banach returned to Cracow, where he continued reading mathematics and corresponded with the mathematicians Otto Marcin Nikodym (1887–1974) and Witold Wilkosz (1891–1941). Steinhaus described his first meeting with Banach as follows: "One summer evening in 1916 while walking in the park that surrounds the old center of Cracow; I overheard a conversation, or rather a couple of words: the words 'Lebesgue's integral' were so unexpected that I went over to the park bench and made the speaker's acquaintance: it was Banach and Nikodym talking about mathematics." Thus was Banach "discovered," and Steinhaus considered this his greatest mathematical discovery. This resulted in a long-lasting collaboration and friendship. In fact, Banach's first paper was published jointly with Steinhaus in the "Bulletin of the Cracow Academy." Antoni Marian Lomnicki (1881–1941) hired Banach as his assistant at Lwów Technical University in 1920, and in the same year he submitted his doctoral dissertation to Jan Kazimierz University in Lwów, which was published in 1922 in the third volume of 'Fundamenta Mathematicae' under the title *Sur les opérations dans les ensembles abstraits et leur application aux équations intégrales*. In 1922, Banach passed his qualifying examination for the title of docent, and in the same year he became professor of the University in Lwów; 2 years later he became a corresponding member of the Academy of Learning. In Lwów, Banach and Steinhaus gathered a large group of young talented mathematicians, and, in their direction there, developed a new, *Lwów School of Mathematics*, which as early as 1929 started its own mathematical journal devoted to functional analysis, *Studia Mathematica*. In 1932, Banach's famous work, *Théories des opérations linéaires*, appeared in print as the first volume of a new established publication *Monografie Matematyczne* (Mathematical Monographs), of which Banach was one of the founders. This monograph was translated into French next year, and it gained wider recognition in European academic circles. He was invited to give the plenary lecture at the International Mathematical Congress in Oslo in 1936. In 1939, he was elected President of the Polish Mathematical Society. For 2 years 1940–1941, he was Dean of the University. Following the German takeover of Lwów in 1941 in Operation Barbarossa (red beard), all universities were closed, and Banach to save his life acted as a lice feeder at Professor Weigel's Institute, which produced antityphoid vaccines (much of which was secretly delivered to the Home Army—the Polish Underground Army). Following the liberation of Lwów by the Soviet Army, Banach returned to his work at the University and maintained lively contacts with Soviet mathematicians. In January 1945, he was diagnosed with a fatal illness, lung cancer. He died on August 31, 1945, aged 53. Notable mathematical concepts named after Banach are Amenable Banach algebra, Banach–Alaoglu (Leonidas Alaoglu, 1914–1981) theorem, Banach algebra, Banach bundle, Banach-fixed point theorem, Banach function algebra, Banach integral, Banach limit, Banach manifold, Banach's matchbox problem, Banach measure, Banach space, Banach–Mazur (Stanislaw Manzur, 1905–1981) game, Banach–Mazur theorem, Banach–Schauder (Juliusz Pawel Schauder, 1899–1943) theorem, Banach–Steinhaus theorem, Banach–Stone (Marshall Harvey Stone, 1903–1989) theorem, Banach–Tarski paradox (a ball could be decomposed into sets, which could then be reassembled to form two identical balls, each equal to the first, i.e., volume is a meaningless concept), and Hahn–Banach theorem. His complete list of publications comprises 58 items, of which six are posthumous.

His many results originated in discussions with his students and collaborators, often while having coffee at a restaurant or at the Scottish Coffee House: Banach was known to spend a considerable part of each day in a café. Surrounded by his collaborators and young adapts, he could discuss and analyze for hours new problems which mostly he himself had posed. A café table became a place of mathematical inspiration.

Gaston Maurice Julia (1893–1978)

Gaston Maurice Julia (1893–1978) was born in Sidi bel Abbés (Algeria), at the time governed by the French. He started studying mathematics and music; however, in 1914, when France became involved with World War I, he was conscripted to serve with the army. During an attack he suffered a severe injury, losing his nose, and for the rest of his life he resigned himself to wearing a leather strap. During hospitalization he buried himself into a mathematical problem: it was the behavioral pattern of the formula behind the so-called Julia fractals (sets). In 1918, he published a 199-page article *Mémoire sur l'itération des fonctions rationnelles* in the Journal de Mathématiques Pures et Appliquées, which earned him the Grand Prix from the Académie des Sciences and immense popularity and respect among mathematicians. However, despite his fame, his work was mostly forgotten until Mandelbrot mentioned it in his works. Julia died in Paris at the age of 85.

Norbert Wiener (1894–1964)

Norbert Wiener (1894–1964) was born in Columbia (Missouri). He entered high school at the age of 9 and graduated 2 years later; he obtained his doctoral degree at the age of 18 from Harvard University. Then he studied philosophy, logic, and mathematics especially Brownian motion, Fourier integral, Dirichlet's problem, harmonic analysis, and the Tauberian (Alfred Tauber, 1866–1942) theorems, in Cambridge (England) and Göttingen under Russell, Hardy, Hilbert, and Landau, among others. In 1919, he joined faculty of mathematics at MIT as an instructor and was promoted to assistant professor in 1929 and to full professor in 1931. In 1933 he was elected to the National Academy of Sciences (USA), from which he resigned in 1941. Wiener's work in mathematical physics helped to establish MIT as one of the premier mathematical research centers in the world. In 1940 he turned to the theory of communications and the development of a field he called *cybernetics* (taken from a Greek word meaning steersman or governor), a term used by nineteenth-century engineers, but which Wiener applied to a new field of mathematics. Cybernetics is concerned with three important concepts: information, control, and communication. He is referred to as the Father of Automation. Wiener wrote many books *The Fourier Integral and Certain of its Applications* (1933), *Extrapolation, Interpolation and Smoothing of Stationary Time Series* (1942), *Cybernetics: or Control and Communication in the Animal and the Machine* (1948), *The Human Use of Human Beings* (1950), *Ex-Prodigy: My Childhood and Youth* (1953), *I am a Mathematician* (1956), *Nonlinear Problems in Random Theory* (1958), *Generalized Harmonic Analysis and Tauberian Theorems* (1966), and about 300 articles. He died in Stockholm of a second heart attack, the first having occurred just over 10 years before. Prizes and medals in memory of Wiener have been created. One of them, the *Norbert Wiener Memorial Gold Medal*, has been presented several times since his death.

The following anecdotes on Wiener are often quoted:

Once he taught a course on differential equations to undergraduates. A student wanted a differential equation to be solved. Wiener thought for sometime and wrote the answer on the blackboard and announced that the equation was solved. The student was obviously not satisfied. Wiener thought again and wrote the same answer once more. The student was still not satisfied. Wiener was annoyed and told him. "Why are you not satisfied when I have solved the equation by two different methods and have got the same answer both times." Of course, Wiener had solved the equation both times in his mind and not on the blackboard.

Once Wiener offered to donate blood to the blood bank. He stood in the queue and underwent all the preliminary tests. He was told that his blood could not be taken as its hemoglobin content was too low. Wiener said: "I knew the fact and I know my blood would not be taken but I spent all this time simply to demonstrate to everyone my willingness to donate blood."

Wiener had been concentrating on a piece of paper in the post office for sometime when a student approached him and said "Good morning, Professor Wiener." Wiener stared at him, slapped his forehead and said "Well, Wiener is the word, which I could not recollect all this time."

Satyendra Nath Bose (1894–1974)

Satyendra Nath Bose (1894–1974) was born in Kolkata. He was the eldest and only son of his parents, Surendra Nath Bose and Amodini Bose. Satyen had six sisters. His father was an official in the engineering department of the East India Railway and one of the founders of the Indian Chemical and Pharmaceutical Works. He raised his children well in spite of his wife's death at an early age. An astrologer made the following prediction about Satyen when he was just 3 years old: In spite of many obstacles throughout his life, he would achieve a great name and fame with his exceedingly sharp intellect. Though Surendra Nath had seven children, he gave special attention to Satyen and saw that nothing came in the way of his education. The father felt that Satyen was not serious about his studies and wondered if the astrologer's prediction would really come true. With time, Satyen changed remarkably. He showed zeal for work, an eagerness to learn, and a thirst for knowledge. Surendra Nath would leave arithmetic problems written on the floor, and young Satyen would sit and solve the problems and proudly show his father when he returned. As a student of the Hindu High School in Kolkata he established a new record, scoring 110 marks for a maximum of 100 in mathematics. Because he had solved some of the problems using more than one method, his teacher gave him more marks than the maximum. Seeing problems in more than one way and finding innovative solutions was typical of Satyen's thinking. He secured the first rank in the degree programs for Intermediate Science (I.Sc.), in B.S.(Hons.) Mathematics, and in M.S. Mathematics from Calcutta University in 1911, 1913, and 1915, respectively. At the time, the ranking system was based on total marks and not on a grade-point average, which is currently in vogue. In 1914, at the age of 20, Satyen married Ushabala Ghosh, the daughter of a prominent Calcutta physician. The next year, several young men including Meghnad Saha (1893–1956) (the discoverer of the path-breaking theory of ionization and its application to stellar atmosphere), Jnan Chandra Ghosh (1894–1959) (known for his Theory of Strong Electrolysis in photochemistry), Bose, and other doctoral candidates requested postgraduate courses in physics and mathematics at Calcutta University. In 1916, the University began M.S. classes in Modern Mathematics and Modern Physics. Saha, Ghosh, and Bose were all appointed as lecturers. Bose served Calcutta University for 5 years during 1916–1921, where he was involved with the Calcutta Mathematical Society. (This society exists to this day and houses the S.N. Bose School of Mathematics and Mathematical Sciences.) Here, Bose worked with Nobel Laureate C.V. Raman. Besides having a lifelong friendship with Saha, he was also close to Prasanta Chandra Mahalanobis (1893–1972), the founder of the

Indian Statistical Institute. Together, these men strived to make Calcutta University a world-class institution, in spite of its inadequate facilities and lack of funds. He joined Dacca University (Dhaka, Bangladesh), in 1921 as a Reader of Physics. At the beginning of the twentieth century, physicists were debating on whether light (energy) flowed in a wave or behaved as particles. The wave theory of light, based on classical mechanics, was predominant.

In 1900, the renowned German scientist Planck, in an article to the German Physical Society, claimed that energy was emitted not in a continuous flow but in discrete bursts that he called quanta. From this observation, he was able to derive his well-known Planck's constant. Although the wave theory opened the way to regarding radiation energy as quanta, Planck was not able to avoid the wave theory of radiation in his explanation. In 1905, Einstein published an article called "Light Quantum Hypothesis." In this famous paper, he introduced the photoelectric effect and argued that light was made up of light quanta, which later became known as "photons." In 1917, Einstein, using theories presented by Bohr in 1913 on the emission of radiation, gave a derivation of Planck's Law that was later experimentally confirmed. However, a complete mathematical derivation of Planck's law that relied only on the particle nature of radiation continued to elude Einstein. Bose then sent a four-page paper in English on Planck's Law and the Light Quantum Hypothesis to Einstein. In an accompanying letter, Bose wrote that he had derived "Planck's law independent of classical electrodynamics." Einstein liked it so much that he himself translated it into German and had it published in the prestigious German journal *Zeitschrift für Physik* in 1924.

On his European tour, Bose stayed in Paris in 1924 for 1 year, and he carried out research at the Madame Curie (1867–1934) Laboratory, which had special facilities. From Paris he went to Berlin in 1925 to work with Einstein. Here he came into close contact with renowned physicists such as Schrödinger and Heisenberg. He took part in all the meetings and discussions held there. While Bose was in Berlin, Dacca University made a professorial position available, for which a doctoral degree was required. Jnan Chandra Ghosh and other friends persuaded Bose to apply for the post. Bose had not yet obtained his Ph.D, so it was difficult for him to secure the professorship. However, a recommendation from Einstein for the post would have made things easy for him. Bose approached Einstein with great hesitation, and Einstein was surprised. He told Bose, "You are so proficient in your subject; is there need for any other certificate or recommendation?" He wrote a letter to the authority of Dacca University in which he said, "Can you find another scientist as proficient as Satyendranath? He is quite fit for the post." This had the desired effect. Bose became professor and head of the department of physics. Bose's contributions added to the foundation of twentieth century physics, and his name was given to physical concepts such as Bose statistics, bosons (subatomic particles that obey Bose–Einstein statistics), and the theory of the Bose–Einstein condensate. The famous Dutch-born American physicist and scientific historian Abraham Pais (1918–2000), a biographer of Einstein, wrote in *Subtle is the Lord*: "The paper by Bose is the fourth and last of the revolutionary papers of the old quantum theory (the other three being by, respectively, Planck, Einstein, and Bohr)."

Around 1963, one of the authors (Sen) was a student in the Electronics Division of the Indian Statistical Institute, Kolkata. His account follows: The division had two first generation digital computers, an HEC 2M (Hollerith Electronic Computer 2M, a British computing machine) and a URAL (a big-bodied Russian Computer). These were the first imported computers in India during the last half of the 1950s. Their computing capability compared to their physical size was almost zero when compared to today's (2014) tiny laptop computers. At the age of 69, Bose frequently visited this division and used, just like a 21-year-old student, these computers to solve his numerical problems.

On February 4, 1974, at the age of 80, Satyen breathed his last after an unexpected heart attack. His death was a great loss to India and the world of science. He was a man with a passion for science, and he possessed a refined taste for music and the art of conversation (known as adda in Bengali). He was a genius and a kind and gentle soul.

Warren Weaver (1894–1978)

Warren Weaver (1894–1978) was born in Reedsburg, Wisconsin, USA. He graduated in 1919 at the University of Wisconsin-Madison with degrees in civil engineering and mathematics. Weaver became an Assistant Professor of Mathematics at Throop College (which became the California Institute of Technology) for a year and then he returned to Wisconsin to complete his doctoral studies. In 1921, he received his doctoral degree and joined the mathematics department at Wisconsin-Madison, serving as the department's chairman during 1928–1932. At the Rockefeller Foundation he was the director of the Division of Natural Sciences (1932–1955), a science consultant (1947–1951), and a trustee (1954). In 1958, he joined the Sloan-Kettering Institute for Cancer Research as its vice president. He was also the President of the American Association for the Advancement of Science in 1954 and chairman of its Board in 1955. Weaver's chief interests included problems of communication and the mathematical theories of probability and statistics. In 1929, he coauthored a monograph, *The Electromagnetic Field*, which was published by the University of Chicago Press (and reprinted in 1952). In 1949, he coauthored a landmark work on communication, *The Mathematical Theory of Communication*, which was published by the University of Illinois Press. In this work he suggested that there are three types of problems in communication, namely, (1) How accurately can the symbols of communication be transmitted? (2) How precisely do the transmitted symbols convey the desired meaning? (3) How effectively does the received meaning affect conduct in the desired way? In 1962 he wrote *Lady Luck*, a popular account of the theory of probability. Weaver was fascinated by Lewis Carroll's Alice's Adventures in Wonderland, and in 1964, he wrote a book about its translation history. He had a deep personal commitment to improving the public understanding of science. In 1965, he was awarded UNESCO's Kalinga Prize and the first Arches of Science Medal for his outstanding contributions to the public understanding of science. He died in 1978 in Upper Milford, Connecticut, USA.

Tadeusz Ważewski (1896–1972)

Tadeusz Ważewski (1896–1972) was born in Galicia, Poland. He attended secondary school in Tarnów (which belonged to Austria) and then studied mathematics at Jagiellonian University under the direction of Zaremba. Ważewski spent 1921–1923 in Paris. There he received his doctorate on the basis of a thesis that contained interesting topological results concerned with dendrites, which was approved by a board that consisted of Borel, Denjoy, and Paul Montel (1876–1975). He returned to Cracow and became docent in 1927 after producing a paper on rectifiable continua. Afterward, his interests shifted to analysis. In 1933, he was appointed associate professor at Jagiellonian University. During the war he suffered along with his colleagues. He was held for several years in the concentration camp at Oranienburg. Following the liberation of the camp in 1945, Ważewski was appointed full professor at Jagiellonian University. He made important contributions to the theories of ordinary and partial differential equations, control theory, and the theory of analytic spaces. He is most famous for his application of the topological concept of retract, which was later introduced by Borsuk to the study of the solutions of differential equations. He presented the results of his research to the International Congress in Amsterdam (1954) in an invited address. Jagiellonian University distinguished him with an honorary doctorate, and he was the president and an honorary member of the Polish Mathematical Society. The state and local authorities also conferred on him a number of highly distinguished awards and scientific prizes. Ważewski died on September 5, 1972.

Kazimierz Kuratowski (1896–1980)

Kazimierz Kuratowski (1896–1980) was the son of a famous Warsaw lawyer. He attended secondary school in Warsaw. During 1913–1914, he studied in Glasgow, Scotland, and at the end of his first year he was awarded the Class Prize in mathematics. He then studied chemistry at the Technical College during the summer and returned to Poland for the holidays. However, he could not return to Glasgow after the outbreak of World War I. In 1915, the Russians withdrew from Poland, and the University of Warsaw reopened. Kuratowski was one of the first students to study mathematics when the university reopened, and he was active in the Polish patriotic student movement. Kuratowski wrote his first paper in 1917, *On the definitions in mathematics*, and he graduated in 1919. He received his Ph.D. in 1921 under Janiszewski and Mazurkiewicz. He was an active member of the group known as the Warsaw School of Mathematics, which worked on the foundations of set theory and topology. He was made an associate professor at the Lwów Polytechnic University, where he stayed for 7 years, collaborating with the important Polish mathematicians Banach and Stanislaw Marin Ulam (1909–1984). With Banach he settled some fundamental problems on measure theory. In 1930, while at Lwów, Kuratowski completed his fundamental work characterizing planar graphs. In 1934, he returned to Warsaw University as a full professor, where he spent the rest of his career. In 1936, he spent a month at Princeton and wrote a joint paper with von Neumann. Until the start of World War II, he was active in research and teaching. During the war, because of the persecution of educated Poles, Kuratowski went into hiding under an assumed name and taught at the clandestine Warsaw University. At the end of World War II the whole educational system in Poland was destroyed. Kuratowski led the revival of Polish mathematics through the Polish Mathematical Society, of which he was the President for 8 years. He also served as the director of the Polish National Mathematics Institute for 19 years, beginning in 1949. He also served as the Vice President of the Polish Academy of Sciences. He joined the editorial board of the journal Fundamenta Mathematicae in 1928, became editor-in-chief in 1952, and remained there for the rest of his life. He was also one of the founders and an editor of the important Mathematical Monographs series. He contributed the third volume in this series, a monograph on *Topology*, which was the crowning achievement of the Warsaw School in point set topology. Kuratowski performed a remarkable job as an ambassador for Polish mathematics and made many foreign visits and lecture tours. He lectured in London (1946), Geneva (1948), many universities in the USA during 1948–1949, Prague, Berlin, Budapest, Amsterdam, Rome, Peking (1955), Canton (1955), Shanghai (1955), and then Agra, Lucknow, and Bombay in 1956. He published three widely used textbooks and over 180 papers, mainly in the areas of topology and set theory, the topology of the continuum, the theory of connectivity, dimension theory, measure theory, compactness, and metric spaces. He used the notion of a limit point to give closure axioms for the definition of a topological space, and he used Boolean algebra to characterize the topology of an abstract space independently of the notion of points. His work in set theory regarded a function as a set of ordered pairs. Kuratowski was elected to membership of the Academy of Sciences of the USSR, the Hungarian Academy, the Austrian Academy, the Academy of the German Democratic Republic, the Academy of Sciences of Argentina, the Accademia Nazionale dei Lincei, the Academy of Arts and Letters of Palermo, and the Royal Society of Edinburgh. He received honorary degrees from many universities including Glasgow, Sorbonne, Prague, and Wroclaw.

Helmut Hasse (1898–1979)

Helmut Hasse (1898–1979) was born in Kassel. His father was a judge. He attended various secondary schools near Kassel until 1913 and then studied for 2 years in Berlin. He served in the German navy after high school. He began his university studies at Göttingen University in 1918, where Edmund Landau, Hilbert, Emmy Noether, and Hecke were his teachers. In 1920, he moved to Marburg University to study under the number theorist Kurt Hensel. During this time, Hasse made fundamental contributions to algebraic number theory. In 1922, he joined the University of Kiel as a lecturer, and he was made a professor after 3 years at the University of Halle. During his time at Kiel, he extended Heinrich Weber's (1842–1913) work on class field theory, wrote several important papers, and began work on his famous report on class field theory, which included the contributions of Kronecker, Heinrich Weber, Hilbert, Philipp Furtwängler (1869–1940), and Takagi. At Halle he obtained fundamental results on the structure of central simple algebras over local fields. In 1930, Hasse became Hensel's successor at Marburg, and in 1934, he was Weyl's replacement at Göttingen. While in Marburg he collaborated with many mathematicians, including Emmy Noether and Richard Brauer on simple algebras and Harold Davenport (1907–1969) on Gauss sums (which are now known as Hasse–Davenport relations). During 1939–1945, Hasse was on war leave from Göttingen and he returned to naval duty, working in Berlin on problems in ballistics. He returned to Göttingen in 1945, but he was dismissed from his post by the British occupation forces. His right to teach was terminated, and he refused their offer of a pure research position. He moved to Berlin in 1946 and took up a research post at Berlin Academy. He resumed teaching in 1948 in Berlin and attracted a large audience. In 1949, Hasse was appointed professor at Humboldt University in East Berlin. At this time, his work on determining the arithmetical properties of Abelian number fields and the textbook *Zahlentheorie* appeared. This book provided a systematic introduction to algebraic number theory based on the local method for the first time. In 1950, Hasse was appointed at Hamburg, where he continued to teach until he retired in 1966. Hasse served for 50 years as an editor of Crelle's Journal, a famous German mathematical periodical. He was honored by many organizations, such as Deutsche Akademie der Naturforscher Leopoldina zu Halle, Akademie der Wissenschaften zu Göttingen, Finnische Akademie der Wissenschaften in Helsinki, Deutsche Akademie der Wissenschaften zu Berlin, Akademie der Wissenschaften und der Literatur in Mainz, Deutscher Nationalpreis Klasse für Wissenschaft und Technik, Spanische Akademie der Wissenschaften zu Madrid, Doctor honoris causa Universität Kiel, Cothenius Medaille der Deutschen Akademie der Naturforscher Leopoldina. Hasse is controversial for his connections with the Nazi party. However, investigations have shown that he was a strong German nationalist. He died in Ahrensburg in 1979.

Werner Karl Heisenberg (1901–1976)

Werner Karl Heisenberg (1901–1976) was born in Würzburg, Germany. His father, Kaspar Ernst August Heisenberg, was a professor ordinarius of medieval and modern Greek studies at the University of Munich. His mother was Annie Wecklein, a daughter of the headmaster of the Maximilian Gymnasium in Munich. Heisenberg attended school in Munich until the age of 19, and then he studied physics and mathematics from 1920 to 1923 at the Ludwig–Maximilians-Universität München and the Georg-August-Universität Göttingen. He received his doctorate in 1923 at Munich under Sommerfeld. The sad account of Heisenberg's doctoral examination is evidence that human emotions can sometimes lead us astray. On July 23, 1923, at the age of 22, Heisenberg appeared before a committee of four professors, including Sommerfeld and Wilhelm Jan Wien (1864–1928), the professor of experimental physics in Munich. Heisenberg had no problem answering Sommerfeld's questions and any related to mathematics, but he stumbled over the questions in astronomy and experimental physics. In his laboratory experiments, Heisenberg had used an interferometer to observe the interference of light waves. This tool was studied by the class extensively, but Heisenberg had no idea how to derive the resolving power of the interferometer. Also, to Wien's surprise, he could not even derive the resolution (the ability to distinguish objects) of common instruments such as the telescope and the microscope. Wien angrily asked him how a storage battery works, and Heisenberg could not respond. Wien was against passing him, regardless of his strength in other branches of physics. Sommerfeld and Wien began arguing about the relative importance of theoretical and experimental physics. Consequently Heisenberg, a future Nobel Laureate, received a III (equivalent to C grade) for his physics examination and his doctorate. Both of these grades were probably the average of Sommerfeld's A grade (excellent) and Wien's F (fail). This shocked Sommerfeld and humiliated Heisenberg, who was accustomed to being the best in his class. He found this mediocre grade hard to accept for his doctorate. That very evening, Heisenberg took the midnight train and left Munich for Göttingen to meet Max Born (1882–1970), who hired him as his teaching assistant the next morning. After informing Born of his disastrous failure in his oral exam, Heisenberg asked, "I wonder if you still want to have me." Born did not react until he carefully read the questions that Heisenberg could not answer. Born felt that the questions were "rather tricky," and told Heisenberg that his offer of the teaching assistantship remained unaltered. Heisenberg's worried father requested that the renowned Göttingen experimental physicist James Franck (1882–1964) teach his son experimental physics. Frank was not able to overcome Heisenberg's total lack of interest, and it became clear that Heisenberg would only survive as a theoretical physicist. In 1924, Heisenberg gained the vania legendi (the process of supplying a person with the means to develop maximum independence in the activities of their daily life through training or treatment) at the University of Göttingen. He also completed his habilitation in 1924 at Göttingen under Born. During 1924–1925, Heisenberg worked with Bohr at the University of Copenhagen with a Rockefeller Grant. In 1926, he became a lecturer of Theoretical Physics at the University of Copenhagen under Bohr. In 1927, at the age of only 25, he became the youngest professor of Theoretical Physics at the University of Leipzig (Germany), one of the oldest universities in Europe. In 1929, he spent several months in the USA, India, and Japan on a lecture tour. World War II began in 1939. Europe became the epicenter of the war, as Hitler's army conquered one European country after another. In 1941, Heisenberg was appointed professor of physics at the University of Berlin and director, a prestigious position, of the Kaiser Wilhelm Institute for Physics, also in Berlin. Einstein was once director of this famous institute, during 1917–1933.

In 1935, the Munich University faculty short-listed three candidates to replace Sommerfeld's position as ordinarius professor of theoretical physics and head of the Institute for Theoretical Physics: Werner Heisenberg (1932 Physics Nobel Laureate), their first choice; Peter Debye (1884–1966, 1936 Chemistry Nobel Laureate); and Richard Becker (1887–1955, an accomplished writer). All had been students of Sommerfeld. But supporters of the Nazi "German Physics" (or Aryan Physics) had their own list of candidates, and the consequent battle continued for 4 years. During this period, these supporters vehemently attacked Heisenberg and published an article in Das Schwarze Korps, the Nazi newspaper of Heinrich Himmler (1900–1945), a military commander and a leading member of the Nazi party. Heisenberg was called a "White Jew" (an Aryan who acts like a Jew) who should be vanished. Heisenberg took this attack seriously, and he responded strongly with an editorial and a letter to Himmler in an attempt to reestablish his honor. Heisenberg's maternal grandfather and Himmler's father knew each other and their respective families. Through this connection, Heisenberg's mother met Himmler's mother. To settle the matter, Himmler sent two letters on July 21, 1938, one to Heisenberg, and the other to Reinhard Tristan Eugen Heydrich (1904–1942), a high-ranking German Nazi General. To Heydrich, Himmler wrote that Germany could not afford to lose or silence Heisenberg since he would be useful in teaching a generation of scientists. In his letter to Heisenberg, Himmler asked Heisenberg to make a distinction between professional physics research and the personal/political views of the involved scientists. This letter was signed under the closing "Mit freundlichem Gruss und, Heil Hitler!" (With friendly greetings, Heil Hitler!). Although the Heisenberg affair seemed to be a victory for academic standards and professionalism, the appointment of Wilhelm Müller, who was not a theoretical physicist and had not published in a physics journal and who was not even a member of the German Physical Society (then the world's largest organization of physicists), to replace Sommerfeld was indeed a political victory over academic standards. During the Nazi investigation of Heisenberg, his three investigators were physicists. Heisenberg had participated in the doctoral examination of one of them at the University of Leipzig. During their investigation, they became supporters of Heisenberg and his position against the ideological policies of the German Physical Society movement. After the end of World War II in 1945, US troops arrested Heisenberg and other German physicists and sent them to England as prisoners. In 1946, Heisenberg was released, and he returned to Göttingen, Germany, and reorganized the Institute for Physics along with his colleagues. This institute was renamed, in 1948, the Max Planck Institute for Physics. Heisenberg remained its director until the institute was moved to Munich in 1958.

Heisenberg is widely known for discoveries such as (i) his *Uncertainty Principle*, which states that the determination of the position and momentum of a mobile particle necessarily contains errors the product of which cannot be less than the Planck's (quantum) constant h, and that, although these errors are negligible on the human scale, they cannot be ignored while studying the atom, (ii) the *Heisenberg*

Microscope, which existed only as a thought experiment and was criticized by his mentor Bohr, and subsequently served as the nucleus of some commonly held ideas and misunderstandings about Quantum Theory, and (iii) his *Matrix Mechanics*, which interpreted the physical properties of particles as matrices that evolve in time and was a formulation (the first and the correct definition that extended Bohr's atomic model by describing how quantum jumps occur) of quantum theory created by Heisenberg along with Max Born and Pascual Jordan in 1925 when Heisenberg was just 23. For this theory and its applications, which resulted (among other things) in the discovery of allotropic forms of hydrogen, Heisenberg was awarded the Nobel Prize for Physics in 1932.

Besides the Nobel Prize for Physics in 1932, Heisenberg also received the Max Planck Medal in 1933, along with many other honors over years. From 1957, Heisenberg took a special interest in plasma physics and nuclear fusion, which is the process by which two or more nuclei of atoms join together to form a single, heavier nucleus, and which is usually accompanied by the release or absorption of large amounts of energy. He also collaborated with the International Institute of Atomic Physics in Geneva. He was a member of the Institute's Scientific Policy Committee, and for several years he was the Committee's chairman. In 1973, Heisenberg gave a lecture at Harvard University on the historical development of the concepts of quantum physics. One of his hobbies was classical music, and he was a distinguished pianist. In 1937, Heisenberg married Elisabeth Schumacher, the daughter of a well-known Berlin economics professor. They had seven children and lived in Munich. Heisenberg passed away at his home on February 1, 1976, at the age of 74 due to cancer of the kidneys and gall bladder. The next evening, his colleagues and friends walked in remembrance from the Institute of Physics to his home and each put a candle near the front door.

Alfred Tarski (1902–1983)

Alfred Tarski (1902–1983) was born in Warsaw as Alfred Teitelbaum to a Jewish family. Alfred was educated in the Schola Mazowiecka in Warsaw, where he studied Russian, German, French, Greek, Latin, and mathematics. He entered the University of Warsaw in 1918, intending to study biology, but he soon decided to study mathematics instead. He was taught by the leading Polish mathematicians Stanislaw Lesniewski (1886–1939), Lukasiewicz, Sierpiński, and Mazurkiewicz, and the philosopher Tadeusz Kotarbinski (1886–1981). His first paper on set theory was published in 1921 when he was only 19 years old. In 1923, Alfred Teitelbaum changed his name to Alfred Tarski and converted to Catholicism. He submitted his doctoral thesis for examination, supervised by Lesniewski, during the same year. In 1924, he received his doctorate in mathematics from the University of Warsaw, becoming the youngest person ever to be awarded this degree at that university. In 1924, he published major results on set theory and a joint paper with Banach on what is now called the Banach–Tarski paradox. During 1925–1939, he served as a professor of mathematics at Zeromski's Lycée in Warsaw. He married Maria Josephine Wilowski in 1929, and they had two children. Tarski met Gödel in 1930, and he moved to the USA in 1939, narrowly escaping internment and probable death at the hands of the Nazis. He was a research associate at Harvard University from 1939 to 1941 and a visiting professor at the College of the City of New York in 1941. During 1941–1942, he was a member of the Institute for Advanced Study, Princeton, and in 1942, he joined the faculty of the University of California, Berkeley, where he was made professor in 1949. He remained at Berkeley for the rest of his career, becoming professor emeritus in 1968. Tarski continued to teach and supervise research students until 1973. Tarski was a visiting professor at the Catholic University of Chile during 1974–1975.

Tarski worked on set theory and algebra and is noted as one of the first to study formalized logical systems as purely algebraic structures. He emphasized the difference between the metalanguage, used to talk about these structures, and the formal language, whose syntax formed the system being studied. His famous paper *The Concept of Truth in Formalized Languages* (1935) is a cornerstone of model theory and has had a profound influence both in logic and the philosophy of language. He published over 100 research papers and books, including *Introduction to Logic* (1941), *Undecidable Theories* (1953), *Ordinal Algebras* (1956), *The Theory of Modules* (editor, 1965), and *Cylindric Algebras* (1971). He received the Alfred Jurzykowski Foundation Award in 1966 and honorary doctorates from the Catholic University of Chile, the University of Marseille, and the University of Calgary. He was a member of the National Academy of Sciences and a president of the Association of Symbolic Logic. In 1971, the University of California sponsored an international conference to discuss Tarski's influence and his ideas regarding math, logic, and philosophy, which Tarski himself attended. He died in Berkeley.

Paul Adrien Maurice Dirac (1902–1984)

Paul Adrien Maurice Dirac (1902–1984) was born in Bristol to a Swiss father and an English mother. Paul had an elder brother, Félix Dirac, who committed suicide in 1925, and a younger sister, Béatrice. He began his education at the Bishop Road Primary School, and then attended Merchant Venturer's Secondary School, where his father was a strict French teacher. Dirac obtained his B.S. in electrical engineering from the University of Bristol in 1921. He then studied mathematics for 2 years at the University of Bristol, and then he joined St. John's College, Cambridge, as a research student of mathematics. He received his Ph.D. in 1926. He became a Fellow of St. John's College in the following year, and during 1932–1969 he served as Lucasian Professor of Mathematics at Cambridge. Dirac traveled extensively and visited several universities, including Copenhagen, Göttingen, Leyden, Wisconsin, Michigan, and Princeton. In 1937, he married Margit Wigner (1904–2002) of Budapest. Following his retirement from Cambridge, Dirac moved to the USA and held a research professorship at Florida State University for the last 14 years of his life. His most celebrated result is the relativistic equation for the electron, published in 1928. From this equation he predicted the existence of an "antielectron," or positron, which was first observed in 1932. He was elected an FRS in 1930 and was awarded the Society's Royal Medal in 1939 and both the Copley Medal and the Max Planck Medal in 1952. He was awarded the Nobel Prize in 1933, with Schrödinger, for the discovery of new productive forms of atomic theory. He became an Honorary Fellow of the American Physical Society in 1948 and an Honorary Fellow of the Institute of Physics, London, in 1971. He was elected a member of the Pontifical Academy of Sciences in 1961. Dirac was awarded the Order of Merit, an outstanding recognition by the land of his birth, in 1973. His publications include the books *Quantum Theory of the Electron* (1928) and *The Principles of Quantum Mechanics* (1930; 3rd ed. 1947). Dirac is remembered for the Dirac comb, the Dirac delta function, Fermi–Dirac statistics, the Dirac sea, the Dirac spinor, the Dirac measure, the Dirac adjoint, the Dirac large numbers hypothesis, the Dirac fermion, the Dirac string, Dirac algebra, the Dirac operator, the Abraham–Lorentz–Dirac (Max Abraham, 1875–1922) force, the Dirac bracket, the Fermi–Dirac integral, negative probability, the Dirac picture, the Uncertainty principle, and the Dirac–Coulomb–Breit (Charles-Augustin de Coulomb, 1736–1806; Gregory Breit, 1899–1981) equation. His doctoral students include Homi Bhabha, Harish Chandra Mehrotra, Dennis William Siahou Sciama (1926–1999), and Behram Kursunoglu (1922–2003). His childhood home in Bristol is commemorated with a blue plaque, and the nearby Dirac Road is named for him. There is a Paul Dirac Drive in Tallahassee, Florida, and the Paul A.M. Dirac Science Library on the Florida State University campus. The BBC named its video codec Dirac in his honor. The Institute of Physics, the United Kingdom's professional body for physicists, awards the Paul Dirac Medal and Prize for "outstanding contributions to theoretical (including mathematical and computational) physics." The first recipient of this award was Stephen Hawking (born 1942) in 1987. The Abdus Salam (1926–1996) International Center for Theoretical physics (ICTP) awards the Dirac Medal of the ICTP each year on Dirac's birthday (August 8).

Dirac was known among his colleagues for his precise and taciturn nature. His colleagues in Cambridge jokingly defined a 'dirac' unit, which represented one word per hour. John Charlton Polkinghorne (born 1930) recalls that Dirac "was once asked what was his fundamental belief. He strode to a blackboard and wrote that the laws of nature should be expressed in beautiful equations." When Bohr complained that he did not know how to finish a sentence in a scientific article he was writing, Dirac replied, "I was taught at school never to start a sentence without knowing the end of it." He criticized the physicist Julius Robert Oppenheimer (1904–1967) for his interest in poetry: "The aim of science is to make difficult things understandable in a simpler way; the aim of poetry is to state simple things in an incomprehensible way. The two are incompatible."

Frank Plumpton Ramsey (1903–1930)

Frank Plumpton Ramsey (1903–1930) was born in Cambridge, England. His father, a mathematician, was the President of Magdalene College, Cambridge. He entered Winchester College in 1915 and completed his secondary school education in 1920. He then entered Trinity College, Cambridge, to study mathematics. Ramsey became a senior scholar in 1921 and graduated as a Wrangler in the Mathematical Tripos of 1923. The same year, he was elected a fellow of King's College, Cambridge. In 1926, he became a university lecturer in mathematics and later the Director of Studies in mathematics at King's College. Ramsey suffered from chronic liver problems, contracted jaundice after an abdominal operation, and died in January of 1930 at Guy's Hospital in London, at the age of 26. Although he had a short career, he established himself as an outstanding lecturer on the foundations of mathematics. Ramsey did an extraordinary amount of pioneering work in economics, mathematics, logic, and philosophy. His works include *Universals* in 1925, *Facts and propositions* in 1927, *Universals of law and of fact* in 1928, *Knowledge* in 1929, *Theories* in 1929, and *General propositions and causality* in 1929. He is remembered for the Ramsey–Lewis (David Kellogg Lewis, 1941–2001) method, the Ramsey–Dvoretzky–Milman (Aryeh Dvoretzky, 1916–2008; Vitali Davidovich Milman, born 1939) phenomenon, and the Ramsey growth model. In one of his papers of 1930, a result which is known as *Ramsey's theorem* was published. This result has stimulated an enormous study of graph theory and has opened an entire branch of mathematics, known as Ramsey theory. The Decision Analysis Society annually awards the Ramsey Medal to recognize his substantial contributions to decision theory and their application to important classes of real decision problems.

John Louis von Neumann (1903–1957)

John Louis von Neumann (1903–1957) was an extraordinary child prodigy, born into a Jewish banking family in Budapest, Hungary. As a child he was taught languages by his German and French governesses. He had a photographic memory; at the age of six he could divide eight-digit numbers mentally, and he could exchange jokes with his father in classic Greek. During 1911–1921, von Neumann studied at the Lutheran Gymnasium in Budapest under Michael Fekete (1886–1957), with whom he published his first mathematics paper at the age of 18. In 1921 he was admitted to the mathematics program at the University of Budapest; however, he did not attend his lectures and instead entered the University of Berlin to study chemistry. In 1923, von Neumann moved to Zürich, and he received his diploma in chemical engineering from the Technische Hochschule in 1925. He also achieved a superb score in the mathematics examinations at the University of Budapest. While von Neumann was in Zürich he continued his interest in mathematics and interacted with Weyl and Pólya. He even took over one of Weyl's courses when he was absent from Zurich for a time. He received his doctorate in mathematics from the University of Budapest at the age of 23 with a thesis on set theory. During 1926–1930, he was a private lecturer in Berlin and he was granted a Rockefeller Fellowship, which enabled him to undertake postdoctoral studies with Hilbert for some time at the University of Göttingen. In 1930, von Neumann was invited to Princeton University, and when the Institute for Advanced Studies was founded there in 1933, he was appointed as one of the original six professors of mathematics (which consisted of James Wadell Alexander (1888–1962), Einstein, Morse, Veblen, von Neumann, and Weyl), a position that he retained for the remainder of his life. In 1933, von Neumann became a coeditor of the Annals of Mathematics, and 2 years later he became coeditor of Compositio Mathematica. He held both of these editorships until his death. From 1940 he was a member of the Scientific Advisory Committee at the Ballistic Research Laboratories at Aberdeen Proving Ground in Maryland. He was a member of the Navy Bureau of Ordnance during 1941–1955 and a consultant to the Los Alamos Scientific Laboratory during 1943–1955. From 1950 to 1955 he was a member of the Armed Forces Special Weapons Project in Washington, DC. In 1955, President Eisenhower appointed him to the Atomic Energy Commission. During 1951–1953, von Neumann was also the President of the American Mathematical Society.

Von Neumann began his work in mathematical logic, axiomatic set theory, and the substance of set theory. He published a definition of ordinal numbers (an ordinal number indicates not only how many, but also answers the question: In what order?) when he was 20, and this definition is used even today. In his doctoral thesis, he completed the axiomatization of set theory. From the substance of set theory he obtained interesting results in measure theory and the theory of real variables. By his mid-twenties, von Neumann's fame had spread worldwide among the mathematical community. At academic conferences, he would find himself pointed out as a young genius. By the age of 25 he had published ten major papers. During 1927–1929, he focused his attention on the axiomatization of quantum mechanics. His text *Mathematische Grundlagen der Quantenmechanik*, published in 1932, built a solid framework for the new quantum mechanics. In a complementary work of 1936, von Neumann proved (along with Garrett Birkhoff) that quantum mechanics also requires a logic that is substantially different from the classical logic. In economics, von Neumann proposed the application of concepts from game theory and general equilibrium theory. Game theory involves making recommendations or formulating techniques to enable a competitor to win a game. His most significant contribution to economics was the minimax theorem of 1928. In 1937, von Neumann used a generalization of Brouwer's fixed-point theorem to prove the existence of an equilibrium, later known as Shizuo Kakutani's (1911–2004) Fixed-Point Theorem. His classic *Theory of Games and Economic Behavior* with Oskar Morgenstern (1902–1977) in 1944 not only extends this minimax principle but also provides the method of proof, which is used in game theory and is known as backward induction. This book also introduced several other important elements used in other fields of economics, such as the axiomatization of utility theory and the axiomatization of choice under uncertainty, that is, the formalization of the expected utility hypothesis. Self-adjoint algebras of bounded linear operators on a Hilbert space, closed in the weak operator topology, were introduced in 1929 by von Neumann in a paper in Mathematische Annalen. In the mid-1930s, he was fascinated by the problem of hydrodynamical turbulence. In the second half of 1930s and early 1940s, von Neumann worked with Francis Joseph Murray (1911–1996) on a fundamental series of papers that aimed to lay the foundations for the study of von Neumann algebras. After 1940, he turned to applications of mathematics, working with Edward Teller and Stanislaw Ulam to find key steps in the nuclear physics involved in thermonuclear reactions and the hydrogen bomb. He also made significant contributions to computer science and linear programming. He advanced the theory of cellular automata, advocated the adoption of the bit as a measurement of computer memory, and solved problems involving obtaining reliable answers from unreliable computer components. He helped to build one of the first modern electronic brains, MANIAC (Mathematical Analyzer, Numerical Integrator and Computer). His later work on parallel processes and networks has earned him the title of "father of the modern computer." His interest in artificial intelligence is expressed in his lecture notes, published as *The Computer and the Brain*. As a member of the team working on the Manhattan Project at Los Alamos, he helped to design nuclear bombs and missiles. He wrote 150 published papers in his life, which include 60 in pure mathematics, 20 in physics, and 60 in applied mathematics. He is best remembered for von Neumann algebra, the von Neumann conjecture, von Neumann entropy, the Stone–von Neumann theorem, von Neumann–Bernays–Gödel set theory, the von Neumann universe, the von Neumann bicommutant theorem, the von Neumann regular ring, the von Neumann architecture, the von Neumann universal constructor, the von Neumann bottleneck, the minimax theorem, and self-replicating spacecraft.

It would be almost impossible to give a complete list of the honors and awards that von Neumann received. In 1938, he was awarded the Bôcher Memorial Prize for his memoir, *Almost periodic functions and groups*, which was published in the Transactions of the American Mathematical Society in two parts, the first in 1934 and the second during the following year. He received a D.Sc. (Hon) from Princeton University in 1947; the Medal for Merit (Presidential Award) in 1947; the Distinguished Civilian Service Award in 1947; a D.Sc. (Hon) from the University of Pennsylvania in 1950; a D.Sc. (Hon) from Harvard University in 1950; a D.Sc. (Hon) from the University of Istanbul in 1952; a D.Sc. (Hon) from the Case Institute of Technology in 1952; a D.Sc. (Hon) from the University of Maryland in 1952; a D.Sc. (Hon) from the Institute of Polytechnics, Munich in 1953; a Medal of Freedom (Presidential Award) in 1956; the Einstein Commemorative Award in 1956; and the Enrico Fermi (1901–1954) Award in 1956. He was elected to many academies, including the American Academy of Arts and Sciences; the American Philosophical Society; the Academiz Nacional de Ciencias Exactas, Lima, Peru; the Acamedia Nazionale dei Lincei, Rome, Italy; the Instituto Lombardo di Scienze e Lettere, Milan, Italy; the National Academy of Sciences; and the Royal Netherlands Academy of Sciences and Letters, Amsterdam, the Netherlands. The John von Neumann Computing Center in Princeton, New Jersey, was named in his honor.

The John von Neumann Lecture is given annually at the Society for Industrial and Applied Mathematics (SIAM) by a researcher who has contributed to applied mathematics, and the selected lecturer is awarded a monetary prize. The IEEE John von Neumann Medal is awarded annually "for outstanding achievements in computer-related science and technology." Von Neumann, a crater on Earth's Moon, is named after him. The professional society of Hungarian computer scientists, Neumann János Számítógéptudományi Társaság, is named after him. The John von Neumann Award of the Rajk László College for Advanced Studies was named in his honor and has been awarded every year since 1995 for outstanding contributions in the field of exact social sciences. In 2005 the United States Postal Service issued the American Scientists commemorative postage stamp series, a set of four 37-cent self-adhesive stamps in several configurations. The scientists depicted were John von Neumann, Barbara McClintock, Josiah Gibbs, and Richard Feynman.

Von Neumann married twice, first to Mariette Kövesi in 1930, whom he divorced in 1937, and then to Klára Dán in 1938. Von Neumann had one child, by his first marriage, a daughter named Marina in 1936. She is a distinguished Professor of International Trade and Public Policy at the University of Michigan. Von Neumann enjoyed throwing extravagant parties, driving recklessly (often while reading a book), and sometimes crashing into a tree or getting arrested. He once reported one of his many car accidents in this way: "I was proceeding down the road. The trees on the right were passing me in orderly fashion at 60 miles per hour. Suddenly one of them stepped in my path." Von Neumann was diagnosed with bone cancer or pancreatic cancer in 1957, which was possibly caused by his exposure to radiation while observing A-bomb tests in the Pacific or in the course of his later work on nuclear weapons at Los Alamos. Von Neumann died within a few months of the initial diagnosis in excruciating pain. The cancer had spread to his brain, inhibiting mental ability. He was developing a theory of the structure of the human brain before he died. He said, "If people do not believe that mathematics is simple, it is only because they do not realize how complicated life is."

Andrey Nikolaevich Kolmogorov (1903–1987)

Andrey Nikolaevich Kolmogorov (1903–1987), an outstanding twentieth-century Soviet mathematician, was born in Tamboy, a city situated 300 miles southeast of Moscow. His unwed mother died during his birth, and he was raised by his aunt (his mother's sister), who embodied high social ideals and independence. She brought him up lovingly and infused in him an independence of opinion and a sense of responsibility. Her example taught him intolerance toward poorly performed tasks and idleness. It was she who taught him not only to memorize but also to understand. Most of what we now see of Andrey was likely due to the character of his aunt and her careful education of the boy. Andrey treated his aunt as his mother until her death in 1950 at Komarovka at the age of 87. It may be noted that Andrey's mother belonged to the highest social class, and that his grandfather (his mother's father) was a district head of the nobles in Uglich, a historic city in Yaroslavl Oblast, Russia. Andrey studied at his aunt's village school. His mathematical and literary articles were published in the school magazine. Moving to Moscow after 1910, he attended a gymnasium (an advanced secondary school), where he graduated in 1920. He then went to Moscow State University. As an undergraduate student, he attended seminars of Russian history and published his first research article on landholding practices in the Novgorod Republic (a large Russian state that stretched from the Baltic Sea to the Ural Mountains during the Middle Ages). During 1921–1922, he proved several results in set theory and in Fourier series. In 1922, he constructed a Fourier series that diverges almost everywhere. In 1925, Andrey graduated from Moscow State University and began to study under the supervision of Luzin. He was awarded a Ph.D. by Moscow State University in 1929. Andrey was appointed as a professor at Moscow University in 1931. Around 1936, he did fundamental research in ecology and generalized the Lotka–Volterra model of predator–prey systems. In 1939, he was elected an academician of the USSR Academy of Sciences. Andrey and the English mathematician and geophysicist Sydney Chapman (1888–1970) independently developed the pivotal set of equations known as the Chapman–Kolmogorov equations in Markov processes. In 1942, Andrey married Anna Dmitriyevna Egorov at the age of 39. He did not have any children. In 1957, with his student Arnold, he solved a particular interpretation of Hilbert's 13th problem. Andrey spent a considerable amount of time in improving mathematics education in secondary schools in Soviet Russia; the subject was often detested by young students if its teaching failed to create interest and inquisitiveness among them. Andrey enriched many areas of mathematics (including topology) and its applications to practical problems in geology, biology, and physics, including turbulence theory, celestial and classical mechanics, and ballistics. He also contributed to probability theory and statistics, operations research including the theory of random processes, theoretical computer science including information theory, automata theory, algorithmic complexity theory (Kolmogorov complexity theory), and intuitionistic logic. He also spent a significant amount of his time providing special schools for those having extraordinary mathematical talent. His effort was remarkably successful. He also attempted to quantify some aspects of Russian poems, notably, those by Pushkin. He was elected a member of several international academies and societies. Among these, the Romanian Academy of Sciences (1956), the German Academy of Sciences Leopoldina (1959), the Royal Netherlands Academy of Sciences (1963), the London Royal Society (1964), the USA National Society (1967), and the Paris Academy of Sciences (1968) are notable. He wrote several papers for the Great Soviet Encyclopedia and spent a significant amount of time in the latter part of his life developing the mathematical and philosophical relationships between abstract probability theory and applied probability. Andrey breathed his last in Moscow on October 20, 1987, at the age of 84.

Alonzo Church (1903–1995)

Alonzo Church (1903–1995) was born in Washington, DC. He received his bachelor's degree from Princeton University in 1924 and his Ph.D. for the dissertation *Alternatives to Zermelo's Assumption* in 1927, under Veblen. Church spent 2 years as a National Research Fellow, 1 year at Harvard University and then a year in Göttingen (with Hilbert) and Amsterdam. During 1929–1967, he was a member of the faculty at Princeton University, and then he moved to the University of California, Los Angeles, as Kent Professor of Philosophy and Mathematics, eventually retiring in 1990. He made many substantial contributions to the theories of mathematical logic and computability, including his solution to the decision problem, his invention of the lambda-calculus (which influenced the design of the LISP programming language and functional programming languages in general), and his statements of what are now known as Church's Theorem (which shows that arithmetic is undecidable) and the Church–Turing thesis (which conjectures that effective computation is equivalent to the notion of a "recursive" function), both published in 1936. Besides mathematical logic, he also wrote some papers on the theory of differential equations and generalizations of Laplace transforms. Church taught 31 doctoral students, including Curtis Anthony Anderson (born 1940), Peter Bruce Andrews (born 1937), Martin David Davis (born 1928), Leon Albert Henkin (1921–2006), John George Kemeny (1926–1992), Stephen Kleene, Gary Mar, Michael Oser Rabin (born 1931), Hartley Rogers, Jr. (born 1926), John Barkley Rosser (1907–1989), Dana Stewart Scott (born 1932), Raymond Smullyan, and Alan Turing. He continued to publish articles past his 90th birthday. He was a founder and editor of the *Journal of Symbolic Logic* (1936–1979) and a member of the National Academy of Sciences. He died in 1995 and was buried in Princeton Cemetery.

Witold Hurewicz (1904–1956)

Witold Hurewicz (1904–1956) was born in Lodz, Russian Poland. His father was an industrialist. Hurewicz received his early education in Russian-controlled Poland and his doctorate in Vienna in 1926 under the supervision of Hahn and Menger. He was a Rockefeller Fellow during 1927–1928 in Amsterdam, an assistant to Brouwer in Amsterdam from 1928 to 1936, and then he moved to the USA. He worked at the Institute for Advanced Study established in 1930 in Princeton, then the University of North Carolina at Chapel Hill's Radiation Laboratory, and then at the Massachusetts Institute of Technology after 1945. Hurewicz's early work was in set theory and topology. He wrote an important text, *Dimension Theory*, with Henry Wallman (1915–1992) in 1941. Hurewicz is best remembered for two remarkable contributions to mathematics: his discovery of the higher homotopy groups in 1935–1936, and his discovery of exact sequences in 1941. He died during an outing at the

International Symposium on Algebraic Topology in Uxmal, Mexico, after tripping and falling off the top of a Mayan ziggurat. His second important text, *Lectures on Ordinary Differential Equations*, appeared after his death in 1958. Regarding the stability of differential equations, the Hurewicz matrix is often used.

Alston Scott Householder (1904–1993)

Alston Scott Householder (1904–1993) was born in Rockford, Illinois, and died in Malibu, California. He received his undergraduate degree from Northwestern University and his Master of Arts from Cornell University, both in philosophy. He received his Ph.D. in mathematics in 1937 from the University of Chicago. His early work in mathematics dealt with the application of mathematics to biology. In 1944, he began to work on problems dealing with World War II. In 1946, he became a member of the mathematics division at Oak Ridge National Laboratory, and he became its director in 1948. At Oak Ridge, his interests shifted from mathematical biology to numerical analysis. He is best known for the Householder transformation, as well as many important contributions in the field of numerical linear algebra. In addition to his research, Householder occupied a number of posts in professional organizations, such as President of the American Mathematical Society, President of SIAM, and President of the Association for Computing Machinery. He also served on a variety of editorial boards, being editor-in-chief of the SIAM Journal on Numerical Analysis; and he organized the important Garlinburg conferences (now known as the Householder Symposia), which continue to this day.

Henri Cartan (1904–2008)

Henri Cartan (1904–2008) was born in Nancy, France. Like his father, Élie Cartan, he was a distinguished and influential French mathematician. His family moved from Nancy to Paris when he was 5 years old. There he attended the Lycée Buffon and the Lycée Hoche in Versailles, and then he studied at the Ëcole Normale Supérieure in Paris. In 1928, Henri Cartan received his Docteur és Sciences mathématiques under the supervision of Montel. After receiving his doctorate, he taught at the Lycée Caen from 1928 to 1929, at the University of Lille from 1929 to 1931, and then he took up a post at the University of Strasbourg. In 1940, Henri Cartan was appointed professor at the Sorbonne in Paris. He taught in Paris until 1969 and then at the Université de Paris-Sud at Orsay from 1970 until his retirement in 1975. His major contributions are in the fields of analytic functions of several complex variables, the theory of sheaves, homological theory, algebraic topology, and potential theory. He was a master of differential geometry and a protege of Poincaré. Henri Cartan was one of the founding members of the Bourbaki group and a member of the Académie des Sciences of Paris, the London Mathematical

Society (1959), the Royal Danish Academy of Sciences (1962), the Royal Society of London (1971), the American Academy (1950), the National Academy of Sciences in Washington (1972), the Royal Academy of Belgium (1978), the Akademie der Wissenschaften Göttingen (1971), the Royal Academy of Science Madrid (1971), Bayerische Akademie der Wissenschaften (1974), the Academy of Japan (1979), the Academy of Finland (1979), the Royal Swedish Academy of Sciences (1981), the Polish Academy of Sciences (1985), and the Russian Academy (1999). He received honorary doctorates from several universities, including ETH Zurich (1955), Münster (1952), Oslo (1961), Sussex (1969), Cambridge (1969), Stockholm (1978), Oxford (1980), Zaragoza (1985), and Athens (1992). He also received the Gold Medal of the National Center for Scientific Research (1976), the Wolf (Johann Rudolf Wolf, 1816–1893) Prize in Mathematics (awarded by the Wolf Foundation in Israel to honor the achievements of a lifetime) in 1980, and he was made Commandeur de la Légion d'honneur in 1989.

Lev Genrikhovich Schnirelmann (1905–1938)

Lev Genrikhovich Schnirelmann (1905–1938) was a Soviet mathematician who graduated from Moscow State University in 1925, and during 1934–1938, he worked at the Steklov Mathematical Institute. Schnirelmann tried to prove Goldbach's conjecture, and in 1931, he showed that any natural number greater than 1 can be written as the sum of not more than 20 primes. He committed suicide in Moscow.

Rózsa Péter (1905–1977)

Rózsa Péter (1905–1977) was born in Budapest, Hungary. She attended the Maria Terezia Girls' School until 1922, and then entered the Loránd Eötvös University in Budapest to study chemistry, but her interest in mathematics was aroused by Fejér's lectures. Another person who had an important influence on Rózsa Péter was László Kalmár (1905–1976), a fellow student at Loránd Eötvös University. Péter graduated in 1927, but could not find a permanent job because of her Jewish heritage; she was forbidden to teach, so she earned a living by tutoring mathematics. Her first research topic was in number theory, but she became discouraged after finding that her results had already been proved by Dickson. For a while Péter wrote poetry, but around 1930 she was encouraged by Kalmár to return to mathematics. He suggested that she study Gödel's work, and in a series of papers beginning in 1934 she developed various deep theorems about primitive recursive functions, most accompanied by explicit algorithmic content. Her first post, at the Budapest Teachers Training College, was obtained in 1945. In 1951, Péter collected all that was known about recursive functions, including her own work, in the book

Rekursive Funtionen, the first book devoted exclusively to this topic. An English translation of this book appeared in 1967. In 1952, Péter became the first Hungarian female mathematician to become an Academic Doctor of mathematics. In the same year, S.C. Kleene described Rózsa Péter in a paper in the *Bulletin of the American Mathematical Society* as "the leading contributor to the special theory of recursive functions." She wrote a charming book in German, *Playing with Infinity: Mathematical Explorations and Excursions* in 1955, which has been translated into at least 14 languages. When the College closed in 1955, she became a professor at Loránd Eötvös University and remained at this post until her retirement in 1975. Beginning in the mid-1950s, Péter applied recursive function theory to computing. Her final book on this topic was *Recursive Functions in Computer Theory* in 1976. Péter received many honors and prizes, including the Kossuth Prize (established in 1948 by Lajos Kossuth, 1802–1894) for her scientific and pedological work from the Hungarian government in 1951, the Mano Beke Prize by the Janos Bolyai Mathematical Society in 1953, and the State Prize Silver Degree in 1970 and the Gold Degree in 1973. In 1973 she was elected as the first female mathematician in the Hungarian Academy of Sciences.

Aleksander Osipovich Gelfond (1906–1968)

Aleksander Osipovich Gelfond (1906–1968) was born in St. Petersburg to the family of a professional physician and amateur philosopher, Osip Isaakovich Gelfond. He joined Moscow State University in 1924 at the age of 18, entered for postgraduation there in 1927, and received his doctorate in 1930 at the age of 24. His major advisor was Khinchin. He proved, in 1934, that all numbers of the form a^b are transcendental if a is algebraic and not equal to 0 or 1 and b is an irrational algebraic number. This is now called the Gelfond–Schneider theorem. This theorem does not help determine whether numbers such as e^e, π^π, or π^e are transcendental, since both the bases and exponents are transcendental numbers and therefore do not satisfy the conditions of the Gelfond–Schneider theorem. Log 2 (base 10) can be shown to be transcendental using the Gelfond–Schneider theorem. The number $2^{\sqrt{2}}$ is known as the Gelfond–Schneider constant, and e^π is known as Gelfond's constant. Besides number theory, Aleksander also obtained important results in several mathematical domains such as analytic functions, integral equations, and the history of mathematics. He passed away in Moscow at the age of 62.

Kurt Friedrich Gödel (1906–1978)

Kurt Friedrich Gödel (1906–1978) was born in Brünn (Austria–Hungary), now Brno (Czech Republic). His father, Rudolf Gödel, was the manager of a textile factory. He became a Czechoslovak citizen at the age 12 when the Austro-Hungarian empire broke up at the end of World War I. He was never able to speak Czech and refused to learn it at school. He became an Austrian citizen by choice at age 23. When Nazi Germany annexed Austria, Gödel became a German citizen at age 32. After World War II, at the age of 42, he became an American citizen. In his family, young Kurt was known as Der Herr Warum ("Mr. Why") because of his insatiable curiosity. He completed his high school (or Gymnasium) education in Brünn with honors in 1923. By that time he had already mastered university-level mathematics. Kurt immediately entered the University of Vienna, taking courses in theoretical physics, mathematics, and philosophy. As an undergraduate, he became acquainted with Bertrand Russell's work on logic and the foundations of mathematics. He completed his doctoral dissertation in 1929 under the supervision of Hahn, *Gödel Completeness Theorem for first-order predicate logic*, which proved the completeness of the first order functional calculus. During 1930–1938, he worked in the school of logical positivism at the University of Vienna. In 1931, Gödel showed that within a rigidly logical system such as that which Russell and Whitehead had developed for arithmetic, propositions can be formulated that are undecidable or undemonstrable within the axioms of the system. That is, within the system there exist certain clear-cut statements that can neither be proved nor disproved. In particular, the consistency of the system's axioms can never be proved. These landmark twentieth-century results, known as *Gödel's Incompleteness Theorems*, settled a 100 years' worth of attempts to establish axioms that would place all of mathematics on an axiomatic basis. Gödel's Incompleteness Theorems (Emil Post (1897–1954) discovered incompleteness before Gödel but published his proofs later) did not destroy the fundamental idea of formalism, but it did demonstrate that any complete, consistent system would have to be more comprehensive than that envisaged by Hilbert. Gödel's Incompleteness Theorems showed that mathematics is not a finished project, as had been believed. It also implies that a computer can never be programmed to answer all mathematical questions.

Gödel's Incompleteness Theorem is so simple, and so sneaky, that it is almost embarrassing to relate. His basic procedure is as follows: Someone introduces Gödel to a UTM, a machine that is supposed to be a Universal Truth Machine, capable of correctly answering any question. Gödel asks for the program and the circuit design of the UTM. The program may be complicated, but it must be finite. Let P(UTM) represent the Universal Truth Machine's program. Smiling a little, Gödel writes out the following sentence: "The machine constructed on the basis of the program P(UTM) will never say that this sentence is true." Call this sentence G for Gödel. Note that G is equivalent to: "UTM will never say G is true." Now Gödel laughs and asks UTM whether G is true or not. If UTM says G is true, then "UTM will never say G is true" is false. If "UTM will never say G is true" is false, then G is false (since G is equal to "UTM will never say G is true"). So if UTM says G is true, then G is in fact false, and UTM has made a false statement. So UTM will never say that G is true, since UTM makes only true statements. We have established that UTM will never say G is true. So "UTM will never say G is true" is in fact a true statement. So G is true (since G = "UTM will never say G is true"). "I know a truth that UTM can never utter," Gödel says. "I know that G is true. UTM is not truly universal." "Think about it—it grows on you... With his great mathematical and logical genius, Gödel was able to find a way [for any given P(UTM)] actually to write down a complicated polynomial equation that has a solution if and only if G is true. So G is not at all some vague or non-mathematical sentence. G is a specific mathematical problem that we know the answer to, even though UTM does not! So UTM does not, and cannot, embody a best and final theory of mathematics." *Rucker, Infinity and the Mind.*

In 1933, Gödel traveled to the USA, where he met and befriended Einstein. During this year, he developed his ideas on computability and recursive functions to the point where he delivered a lecture on general recursive functions and the concept of truth. This work was developed in number theory, using Gödel numbering. In 1934 he gave a series of lectures at the Institute for Advanced Study in Princeton on undecidable propositions in formal mathematical systems. Stephen Kleene, who had just completed his doctorate at Princeton, took notes on these lectures, which were subsequently published. Gödel had his first nervous breakdown when he returned to Europe. He was treated by a psychiatrist and spent several months in a sanatorium recovering from depression. In 1936, Gödel had another breakdown, and after his recovery he was called to a guest professorship in the USA. Despite his health problems, during 1935–1937, he worked on proofs of the consistency of the axiom of choice and the continuum hypothesis; he showed that these hypotheses cannot be disproved within the common system of axioms of set theory. He married Adele Porkert in 1938, whom he had known for over 10 years. His parents did not approve of their relationship because she was a divorced dancer, 6 years older than him. She was not the first girl that Gödel's parents had objected to; while he was at university, he met a girl that was 10 years older than him. Gödel and Porkert had no children. In 1940, Gödel emigrated to the USA via Russia and Japan. He became a US citizen in 1948. He was an ordinary member of the Institute for Advanced Study from 1940 to 1946, and then a permanent member until 1953. He held a Chair at Princeton from 1953 until his death, with no lecturing duties. After settling in the USA, Gödel remained very active and produced work of the greatest importance. His masterpiece, *Consistency of the axiom of choice and of the generalized continuum-hypothesis with the axioms of set theory*, in 1940 is a classic of modern mathematics. During his many years at the Institute, Gödel's interests turned to philosophy and physics. In the late 1940s, Gödel demonstrated the existence of paradoxical solutions to Einstein's field equations in general relativity. These "rotating universes" would allow time travel and caused Einstein to have doubts about his own theory.

Gödel received the Einstein Award in 1951 and the National Medal of Science in 1974. He was a member of the National Academy of Sciences of the United States, an FRS, a member of the Institute of France, a Fellow of the Royal Academy, and an Honorary Member of the London Mathematical Society. The Kurt Gödel Society, founded in 1987, is an international organization for the promotion of research in the areas of logic, philosophy, and the history of mathematics. The Gödel Prize for outstanding papers in the area of theoretical computer science is sponsored jointly by the European Association for Theoretical Computer Science (EATCS) and the Special Interest Group on Algorithms and Computation Theory of the Association for Computing Machinery (ACM-SIGACT).

Gödel was shy, withdrawn, and eccentric, and he believed all his life that he was always right, not only in mathematics but also in medicine, so he was a very difficult patient for his doctors. In 1940s, he suffered from a bleeding ulcer, but his distrust of doctors led him to delay treatment; he risked death and was saved only by an emergency blood transfusion. He would wear warm, winter clothing in the middle of summer. In the middle of winter, he would leave all of the windows open in his home because he believed that conspirators were trying to assassinate him with poison gas. Porkert was a great support to him and she did much to ease the anxieties that troubled him. However, she herself began to experience health problems, suffering two strokes and a major surgical operation. Toward the end of his life Gödel became convinced that he was being poisoned, and he refused to eat to avoid more poison, essentially starving himself to death. At the time of his death he weighed 65 pounds. He is remembered for his quotation "Either mathematics is too big for the human mind or the human mind is more than a machine."

Einstein and Gödel had a legendary friendship, shared in the walks they took together to and from the Institute for Advanced Study. It seems that when Gödel decided to become an American citizen, he studied American history and the American constitution and went to take the usual test. For moral support he brought along Einstein and Ernst Gabor Straus (1922–1983). The judge asked various questions and at first Gödel gave the expected answers. Then the great logical mind took charge and to each of the judge's questions, Gödel gave a masterful analysis ending with four or five logical possibilities. The judge became more and more angry and at last Einstein could stand it no more. So he kicked Gödel under the table and said "Please give the judge the unique answer that he wants. We can discuss the other alternatives at home."

Rear Admiral Grace Murray Hopper (1906–1992)

Rear Admiral Grace Murray Hopper (1906–1992), an American computer scientist and a United States Navy officer, was born Grace Brewster Murray in New York. She was an inquisitive child and remained so throughout her life. She wanted to determine how an alarm clock worked while she was just 7. She dismantled seven alarm clocks before her mother realized what she was doing. She was then limited to one clock. She attended Hartridge School in Plainfield, New Jersey. At the age of 17, she attended Vassar College, a highly selective, residential, liberal arts college located in the heart of the Hudson Valley in New York State. She received her bachelor's degree in mathematics and physics from Vassar in 1928 and then her master's degree from Yale University in 1930. Grace was one of the first programmers of the IBM (International Business Machines) Automatic Sequence Controlled Calculator (ASCC), called the Harvard Mark I computer, which was an electromechanical computer that was designed by Aiken, was built at IBM, and was shipped to Harvard in February of 1944. It began performing computations for the U.S. Navy Bureau of Ships in May and was officially presented to Harvard University on August 7, 1944. Grace conceptualized the idea of machine-independent programming languages, which led to the development of COBOL (Common Business Oriented Language), one of the first modern programming languages that appeared in 1959. She popularized the term "debugging" for removing/fixing programming errors. Grace served as the director of the Navy Programming Languages Group in the Navy's Office of Information Systems Planning during 1967–1977 and was promoted to the rank of Captain in 1973. She developed validation software for COBOL and its compiler as part of a COBOL standardization program for the entire Navy. Due to her extraordinary accomplishments and her naval rank, she is sometimes called "Amazing Grace." The US Navy destroyer USS Hopper and the Cray XE6 "Hopper" supercomputer at NERSC were named for her. Grace retired from the Navy in 1986. To celebrate her retirement, she was awarded the Defense Distinguished Service Medal by the Department of Defense, the highest noncombat award. Grace was then hired as a senior consultant to Digital Equipment Corporation (DEC), a position she retained until her death at the age of 85 at Arlington, Virginia, USA.

Olga Taussky Todd (1906–1995)

Olga Taussky Todd (1906–1995), a Czech–American mathematician, was born in Olomouc, Czech-Republic (then part of the Austro–Hungarian Empire). She felt a powerful call to mathematics early in life. Her first research, at the University of Vienna (where she was awarded a Ph.D. in 1930 under Furtwängler), was on algebraic number theory, and she always regarded that as her primary field. She was awarded a doctorate in 1930 and her thesis was published in Crelle's Journal in 1932. After she received her doctorate, she earned a living by tutoring but also continued to work hard on her mathematics, continuing to develop the ideas from her thesis. During World War II, she worked at the National Physical Laboratory in London, where she was assigned to study flutter in supersonic aircraft. It turned out that key questions about flutter were related to the locations of the eigenvalues of a certain 6×6 complex matrix, so a large staff of young girls was appointed to perform the required calculations on hand-operated machines. Taussky-Todd had heard of a result, called Geršgorin's (Semyon Aranovich Gershgorin, 1901–1933) theorem, which provided a simple way of identifying certain circles containing the eigenvalues of a complex matrix. She quickly realized that this theorem could be used to provide information about flutter that would otherwise require laborious calculations. This observation elevated the theorem of Geršgorin from obscurity to practical importance. In 1955, Taussky-Todd and her husband, John Todd (1911–2007), spent a year's leave at the Courant Institute in New York where she taught a matrix theory course and her husband taught a numerical analysis course. Taussky-Todd enjoyed working at the National Bureau of Standards, but she missed teaching.

In 1957, Olga and Jack (John Todd) both accepted appointments at the California Institute of Technology. In 1977, Olga retired, becoming Professor Emeritus, but she continued to lead a very active mathematical life. Todd received the Ford Prize from the Mathematical Association of America in 1971 for a paper on the sums of squares. She was elected to several academies including the Austrian Academy of Science (1975) and the Bavarian Academy of Sciences (1985). She was also elected a Fellow of the American Association for the Advancement of Science (1991). She was honored by the Austrian government in 1978 with their highest award, the Cross of Honor in Science and Arts, First Class, and in 1980, she received an honorary doctorate from the University of Vienna. She was also awarded an honorary doctorate by the University of Southern California in 1988.

André Weil (1906–1998)

André Weil (1906–1998) was born in Paris to an Alsatian Jewish family. The famous philosopher Simone Weil (1909–1943) was his sister. He fell in love with mathematics and Sanskrit literature; by the age of 16 he had read the Bhagavad Gita in the original Sanskrit. Weil studied at the École Normale in Paris, and then went to Rome, and from there to Göttingen, where he produced his first substantial piece of mathematical research in the theory of algebraic curves. From Göttingen he moved to University of Paris, where he continued his study of the theory of algebraic curves under the supervision of Hadamard. Weil was awarded a doctorate in 1928. He taught at the Aligarh Muslim University in India from 1930 to 1932, then he spent 1 year in Marseilles, and then he taught for 6 years at the University of Strasbourg, France, until World War II began. To avoid military service, Weil fled to Finland to visit Rolf Herman Nevanlinna (1895–1980); however, he was arrested in Finland. Letters written in Russian were found in his room from Pontryagin (describing mathematical research), which made things worse. Before he was executed as a spy, Nevanlinna persuaded the authorities to deport him. Weil returned to France and was charged with failure to report for duty. He was sentenced to 5 years; however, he was able to argue successfully for his release on the condition that he join the army. In January 1941, the first opportunity to escape to the USA came, and he sailed to New York with his family. In the USA he taught at Haverford College and Swarthmore College, Pennsylvania. Beginning in 1945, he spent 2 years at Sao Paulo University, Brazil, where he worked with Oscar Zariski (1899–1986), a Jewish-American mathematician and a well-known algebraic geometer. During 1947–1958, he taught at the University of Chicago and then settled at the Institute for Advanced Study in Princeton. He retired in 1976, becoming Professor Emeritus.

André Weil was a universalist, proficient in almost every field of mathematics. He contributed substantially to various fields, such as algebraic geometry, algebraic number theory, differential geometry, harmonic analysis, functions of several complex variables, and topology. His profound connections between algebraic geometry and number theory that began in the 1940s led to the foundations of abstract algebraic geometry and the modern theory of Abelian varieties. His proof of the Riemann hypothesis for the congruence fo zeta functions of algebraic function fields is another example of his major achievements. He is best remembered for Weil cohomology, the Weil conjecture disambiguation page, the Weil conjectures, the Weil conjecture on Tsuneo Tamagawa (born 1925) numbers, the Weil distribution, the Weil divisor, the Siegel–Weil (Carl Ludwig Siegel, 1891–1981) formula, the Weil group, the Weil–Deligne group scheme, the Weil–Châtelet group, the Chern–Weil (Shiing-Shen Chern 1911–2004) homomorphism, Chern–Weil theory, the Hasse–Weil L-function, the Weil pairing, the Weil reciprocity law, the Weil representation, the Borel–Weil theorem, the de Rham–Weil theorem (Georges de Rham, 1903–1990), and the Louis Joel Mordell (1888–1972)–Weil theorem. He invented the notation Ø for the empty set. He was one of the members of the Bourbaki group and was so involved that he named his daughter Nicolette after Bourbaki's first name. He published several books, including *Foundations of Algebraic Geometry* in 1946, *Basic Number Theory* in 1967, *Elliptic Functions According to Eisenstein and Kronecker* in 1976, *Adeles and Algebraic Groups* in 1982, and *Number Theory: An Approach Through History From Hammurabi to Legendre* in 1984.

Weil received many honors and awards for his outstanding work in mathematics, including honorary membership of the London Mathematical Society in 1959, Fellowship of the Royal Society of London in 1966, the Wolf Prize in 1979, and the American Mathematical Society's Steele Prize (since 1970) in 1980. He was an invited speaker at the International Congress of Mathematicians in 1950, 1954, and 1978. In 1994, he received the Kyoto Prize from the Inamori Foundation of Japan for his outstanding achievement and creativity in the mathematical sciences. He is remembered for his sayings: "First rate people hire other first rate people, Second rate people hire third rate people, Third rate people hire fifth rate people," and "Rigour is to the mathematician what morality is to man." He died in 1998, in Princeton, New Jersey, USA. David Rowe (born 1950) has documented the quarrel between Weil and Sabetai Unguru (born 1931) over the origins of geometric algebra.

Sir Edward Maitland Wright (1906–2005)

Sir Edward Maitland Wright (1906–2005) was born in a village near Leeds to a well-to-do businessman father and a music teacher mother. The family business collapsed when Wright was 3, and his parents separated. He followed his mother in her teaching career in boarding schools. He supported himself from the age of 14 by teaching French. Around this time, he became fascinated with mathematics, which he learned by himself and in evening classes. A few years later, Wright became a research student of Hardy, and wrote his doctoral dissertation under his supervision. He became a Professor of Mathematics at the University of Aberdeen at the unusually early age of 29. Wright held this post until 1962. He then became Principal and Vice Chancellor, a post he held until his retirement in 1976. He was knighted in 1977. His services to the University of Aberdeen are commemorated in the Edward Wright Building there, where the Department of Mathematics was based until 1997. Wright was primarily a number theorist, but later he became interested in graphs. He wrote over 140 research articles, and with Hardy he wrote a highly influential mathematical textbook, *An Introduction to the Theory of Numbers*, in 1938. Wright was widely honored, held numerous honorary degrees, and was a Fellow of the Royal Society of Edinburgh. He was a long-serving member of the London Mathematical Society (elected in 1929) and was an honorary Fellow of Jesus College, Oxford, since 1963. Wright died in Reading, England, when he was only 2 weeks away from his 100th birthday. Professor Edward Paterson, who worked with Sir Edward as a senior maths lecturer in the early 60s, said: "He was a most unusual man who was able to deal with all the administration matters which came with being principal and yet didn't forget his first love of mathematics."

Lev Davidovich Landau (1908–1968)

Lev Davidovich Landau (1908–1968) was born in Baku, Azerbaijan, Russian Empire, into a Jewish family. Landau was child prodigy in mathematics. He was quoted as saying in later life that he scarcely remembered a time when he was not familiar with calculus. He graduated at 13 from the Gymnasium, and at 14 matriculated at Baku University, studying in two departments simultaneously: the physico-mathematical and the chemical. In 1924, he entered the physics department at Leningrad State University, where he graduated in 1927. His first scientific paper, on quantum theory, appeared in the same year. At the age of 21 he received his doctorate from the Leningrad Physico-Technical Institute. In 1929, Landau received the Soviet Government's Traveling Fellowship and a Rockefeller Fellowship to travel abroad for 18 months. He visited Germany, Switzerland, Holland, England, Belgium, and Denmark. In particular, his stay at Bohr's Institute for Theoretical Physics in Copenhagen was most important. Landau always considered himself a pupil of Bohr's and his attitude toward physics was greatly influenced by Bohr's example. During 1932–1937, he headed the department of theoretical physics at the Kharkov Mechanics and Machine Building Institute. In 1937, Landau went to Moscow to become Head of the Theory Division of the Physical Institute of the USSR Academy of Sciences. In 1938, Landau was imprisoned as a suspected German spy. He was released the next year and reinstated as a research fellow at the Institute of Physical Problems, but only after his colleague Pyotr Kapitsa (1894–1984), an experimental low-temperature physicist, and Bohr wrote personal letters to Stalin. In 1962, Landau's car collided with an oncoming lorry, rendering him unconscious for 6 weeks. His injuries included 11 broken bones and a fractured skull. The physics community around him rallied to save his life. His doctors declared him clinically dead several times; however, he regained consciousness and recovered but could not perform creative work. His death in 1968 was a consequence of the injuries from the accident.

Landau worked on atomic collisions, astrophysics, low-temperature physics, atomic and nuclear physics, thermodynamics, quantum electrodynamics, the kinetic theory of gases, quantum field theory, and plasma physics. His accomplishments include the codiscovery of the density matrix method in quantum mechanics, the quantum mechanical theory of diamagnetism, the theory of superfluidity, the theory of second order phase transitions, the Ginzburg–Landau (Vitaly Lazarevich Ginzburg, 1916–2009) theory of superconductivity, the explanation of Landau damping in plasma physics, the Landau pole in quantum electrodynamics, Landau–Hopf (Heinz Hopf, 1894–1971) theory of turbulence, and the two-component theory of neutrinos. Landau wrote a number of outstanding textbooks and research monographs; however, his most important contribution is the ten-volume *Course of Theoretical Physics*, written jointly with his friend and former student Evgeny Mikhailovich Lifshitz (1915–1985). They began work on this in the 1930s and the first part of the book is based on lecture notes. Lifshitz continued to work on the book after Landau's death, and it was not completed until 1979. The work includes many of the results of Landau and Lifshitz's research, including the results of many jointly written research papers. These volumes are still widely used as graduate-level physics texts.

Landau received many international honors for his contributions. He was elected to the Royal Danish Academy of Sciences in 1951, The Netherlands Academy in 1956, the British Physical Society in 1959, and the Royal Society of London in 1960. In 1960, he was elected to the USA National Academy of Sciences and the American Academy of Arts and Sciences. In 1960 he received the Fritz Wolfgang London (1900–1954) Prize and the Max Planck Medal. He received the 1962 Nobel Prize in Physics for his development of a mathematical theory of superfluidity that accounts for the properties of liquid helium II at a temperature below 2.17K (−270.98 °C). Landau was widely considered among Einstein, Paul Dirac, and Richard Feynman to be a brilliant physicist with a superb imaginative intellect.

Lev Semenovich Pontryagin (1908–1988)

Lev Semenovich Pontryagin (1908–1988) was born in Moscow. At the age of 14, he lost his eyesight due to a primus stove explosion. His mother devoted herself for many years to help him succeed, working as his secretary, reading scientific works aloud to him, writing the formulas in his manuscripts, correcting his work, and so on. Pontryagin joined the University of Moscow in 1925. Although he could not take lecture notes like the other students, he was able to remember most of the details of the topics covered in class. He graduated from the University of Moscow in 1929 and was appointed to the Mechanics and Mathematics Faculty. His doctoral advisor was Pavel Sergeyevich Alexandrov (1896–1982), who wrote about three hundred papers, making important contributions to set theory and topology. In 1934, Pontryagin became a member of the Steklov Institute, and in 1935, he became head of the Department of Topology and Functional Analysis at the Institute. He made major discoveries in several fields of mathematics, including algebraic topology and differential topology. Pontryagin worked on the duality theory of homology while only 19 years old. In topology, he posed the basic problem of cobordism theory, which led to Pontryagin classes. He introduced the cohomology operations that are now called Pontryagin squares. He is also remembered for the Andronov–Pontryagin (Aleksandr Aleksandrovich Andronov, 1901–1952) criterion, the Pontryagin–Kuratowski theorem, the Pontryagin duality, Pontryagin's minimum principle in optimal control theory, and his idea of the bang–bang principle. Pontryagin also proved Hilbert's fifth problem for abelian groups using the theory of characters on locally compact abelian groups, which he had introduced. He authored or coauthored several books, including *Topological Groups* (1938), *The Mathematical Theory of Optimal Processes* (1961), and *Ordinary Differential Equations* (1962). He also produced several very influential students. Pontryagin received many honors and awards for his work. He was elected to the Academy of Sciences in 1939, becoming a full member in 1959. In 1941, he was one of the first recipients of the Stalin Prize. In 1962, he was awarded Lenin Prize, and in 1970, he was elected Vice President of the IMU. He died at the age of 79 in Moscow.

Morris Kline (1908–1992)

Morris Kline (1908–1992) grew up in Brooklyn and in Jamaica, Queens. After graduating from Boys High School in Brooklyn, he studied mathematics at New York University, earning a bachelor's degree in 1930, a master's degree in 1932, and a doctorate in 1936. Kline taught at New York University from 1938 to 1975. He made numerous contributions to mathematical thought, wrote extensively on education, especially mathematics education, and lectured as a consultant throughout his very active career. Regarding mathematics education, he said, "Our teachers of mathematics have been teaching carpentry instead of architecture and color mixing instead of painting." He is the author of many popular books, including *Mathematical Thought from Ancient to Modern Times*, *Mathematics and the Search for Knowledge*, *Why Johnny Can't Add: The Failure of the New Mathematics*, and *Mathematics, A Cultural Approach*, in which he claims that the Egyptian and Babylonian contributions to mathematics were "almost insignificant." He further states that "compared with what the Greeks achieved the mathematics of Egyptians and Babylonians is the scrawling of children just learning to write, as opposed to great literature." He continues, "in any case, these civilizations barely recognized mathematics as a distinct discipline so that over a period of 4,000 years hardly any progress was made in the subject." This famous quotation is due to him "God exists since mathematics is consistent, and the devil exists since its consistency cannot be proved." In the same spirit, we have "God exists since computer mathematics is consistent, and the devil does not exist since we can prove its consistency."

Willard Van Orman Quine (1908–2000)

Willard Van Orman Quine (1908–2000) was born in Akron, Ohio, and attended Oberlin College and then Harvard University, where he received his Ph.D. in philosophy in 1932. He became a Junior Fellow at Harvard in 1933 and was appointed to a position on the faculty there in 1936. He remained at Harvard for his entire professional life, except during World War II, when he worked for the USA Navy decrypting messages from German submarines. Quine was always interested in algorithms, but not in hardware. He discovered what is now called the Quine–McCluskey (Edward J. McCluskey, born 1929) method as a device for teaching mathematical logic, rather than as a method for simplifying switching circuits. Quine was one of the most famous philosophers of the twentieth century. He made fundamental contributions to the theory of knowledge, mathematical logic and set theory, and the philosophies of logic and language. His books, including *New Foundations of Mathematical Logic* published in 1937 and *Word and Objects* published in 1960, have had a profound impact on their field. Quine retired from Harvard in 1978 but continued to commute from his home in Beacon Hill to his office there. He used the 1927 Remington typewriter, on which he prepared his doctoral thesis, for his entire life. He even had an operation performed on this machine to add a few special symbols, removing the second period, the second comma, and the question mark. When asked whether he missed the question mark, he replied, "Well, you see, I deal in certainties." There is even a word *quine* that is defined in the *New Hacker's Dictionary* as a program that generates a copy of its own source code as its complete output. Producing the shortest possible quine in a given programming language is a popular puzzle for hackers.

Stephen Cole Kleene (1909–1994)

Stephen Cole Kleene (1909–1994) was born in Hartford, Connecticut, USA. His mother, Alice Lena Cole, was a poet, and his father, Gustav Adolph Kleene, was an economics professor. Kleene attended Amherst College and received his Bachelor of Arts degree in 1930. From 1930 to 1935, he was a graduate student and research assistant at Princeton University, where he received his doctorate in mathematics in 1934 under the supervision of Alonzo Church for a thesis entitled *A Theory of Positive Integers in Formal Logic*. Kleene joined the faculty of the University of Wisconsin in 1935, where he remained until 1979 except for several leaves including stays at the Institute for Advanced Study in Princeton. During World War II he was a navigation instructor at the Naval Reserve's Midshipmen's School and later served as the director of the Naval Research Laboratory. He served as the Acting Director of the Mathematics Research Center and as Dean of the College of Letters and Sciences at the University of Wisconsin. Kleene made significant contributions to the theory of recursive functions, investigated questions of computability and decidability, and proved one of the central results of automata theory. Kleene was a student of natural history. He discovered a previously undescribed variety of butterfly that is named after him. He was an avid hiker and climber. Kleene was also noted as a talented storyteller; he had a powerful voice that could be heard several offices away. He is best remembered for Kleene algebra, the Kleene star, Kleene's recursion theorem, and the Kleene fixed-point theorem.

Subramanyan Chandrasekhar (1910–1995)

Subramanyan Chandrasekhar (1910–1995) was born in Lahore, Punjab, British India (now Pakistan). His father, Chandrasekhara Subrahmanya Ayyar (1885–1960), was the Deputy Auditor General of the North-Western Railways and his mother, Sitalakshmi Balakrishnan (1891–1931), was a woman of great intellectual attainment. He was the eldest of their four sons and the third of their ten children. Nobel Laureate Sir C.V. Raman was his uncle. Chandrasekhar and William Alfred Fowler (1911–1995) were awarded a Nobel Prize in Physics for their work on the theoretical structure and evolution of stars. Up to the age of 12, Chandrasekhar did not have any formal education and was taught by his parents and private tutors. His father was transferred to Madras (Chennai) in 1918. During 1922–1925, Chandrasekhar attended the Hindu High School, Triplicane, and then from 1925 to 1930 he studied at the Presidency College, where he obtained a B.S. (Hons.) degree in physics. He then received a Government of India scholarship for further studies in Cambridge. Here he was a research student of Ralph Howard Fowler (1889–1944), a noted astrophysicist. Chandrasekhar then spent a year at the Institut for Teoretisk Fysik in Copenhagen. He received his Ph.D. at Cambridge University in 1933. He held a Fellowship at Trinity College during 1933–1937. In 1936, Chandrasekhar married Lalitha Doraiswamy, whom he had met at Presidency College, Madras, where she was a student a year junior to him. In his Nobel autobiography, Chandrasekhar wrote: "Lalitha's patient understanding, support, and encouragement have been the central facts of my life." In 1937, Chandrasekhar joined the University of Chicago as an assistant professor. He remained at the university for his entire career, becoming Morton D. Hull Distinguished Service Professor of Theoretical Astrophysics in 1952 and attaining emeritus status in 1985, serving the university until the end of his life.

Unlike many renowned scientists, Chandrasekhar worked continuously in one specific area of physics for a number of years. His research career may be divided into the following seven periods. He focused on stellar structure, including the theory of white dwarfs during 1929–1939, and then concentrated on stellar dynamics, including the theory of Brownian motion, during 1938–1943. During 1943–1950, he studied the theory of radiative transfer, including the theory of stellar atmospheres, the quantum theory of the negative ion of hydrogen, and the theory of planetary atmospheres including the theory of the illumination and the polarization of the sunlit sky. He studied hydrodynamic and hydromagnetic stability, including the theory of the Rayleigh–Bénard (Henri Claude Bénard, 1874–1939) convection, during 1952–1961. He then addressed the equilibrium and the stability of ellipsoidal figures of equilibrium, partly in collaboration with Norman R. Lebovitz, during 1961–1968, the general theory of relativity and relativistic astrophysics during 1962–1971, and the mathematical theory of black holes during 1974–1983. Chandrasekhar's most notable work is the astrophysical Chandrasekhar limit, which he calculated in 1930 during his maiden voyage from India to Cambridge.

Some of the notable awards bestowed on Chandrasekhar were FRS (1944), AMS (American Mathematical Society) Gibbs Lecturer (1946), the Astronomical Society of the Pacific's Bruce Medal (first awarded in 1898 in honor of Catherine Wolfe Bruce, 1816–1900) in 1952, the Royal Society's Royal Medal in 1962, the Government of India's Padma Vibhushan in 1968, the Nobel Prize in Physics in 1983, and the Copley Medal in 1984. In 1999, NASA named the third of its four *Great Observatories* after Chandrasekhar. The Chandra X–ray Observatory was launched and deployed by Space Shuttle Columbia in 1999. The Chandrasekhar number, an important dimensionless number in magnetohydrodynamics, is named after him. The asteroid 1958 Chandra is also named after Chandrasekhar. The American astronomer Carl Edward Sagan (1934–1996), who studied mathematics under Chandrasekhar at the University of Chicago, praised him in his book *The Demon-Haunted World*: "I discovered what true mathematical elegance is from Subrahmanyan Chandrasekhar." Chandrasekhar passed away on August 21, 1995, in Chicago at the age of 84 due to heart failure and was survived by his wife, Lalitha Chandrasekhar. Roger John Tayler (1929–1997) wrote in the Biographical Memoirs of the Fellows of the Royal Society of London, "Chandrasekhar was a classical applied mathematician whose research was primarily applied in astronomy and whose like will probably never be seen again."

Alan Mathison Turing (1912–1954)

Alan Mathison Turing (1912–1954) was born in London, although he was conceived in Chatrapur, Orissa, India. His father, Julius Mathison Turing (1873–1947), was employed in the Indian Civil Service. At the age of six he enrolled at St Michael's, London, a day school. The headmistress recognized his genius right from the beginning, as did many of his later educators. As a boy, he was fascinated by mechanics and chemistry, and he performed a wide variety of experiments. At the age of 14, Turing attended Sherborne, an English boarding school. In 1928, Turing encountered Einstein's work; he not only grasped it, but also extrapolated Einstein's questioning of Newton's laws of motion from a text in which this was never made explicit. In 1931 he won a scholarship to King's College, Cambridge. He was elected a fellow of this college in 1935, after completing his dissertation, which included a rediscovery of the central limit theorem, a famous theorem in statistics. The computer room at King's College is now named after Turing. During this time, Turing became fascinated with the decision problem, a problem posed by Hilbert, which asked whether there is a general method that can be applied to any assertion to determine whether the assertion is true. Turing enjoyed running (later in life running as a serious amateur in competitions), and one day while resting after a run, he discovered the key ideas needed to solve the decision problem. His solution utilized what is now called the *Turing machine* as the most general model of a computing machine. Using these conceptual machines, he found a problem, involving what he called computable numbers, that could not be decided using a general method. From 1936 to 1938, Turing visited Princeton University where he worked with Alonzo Church, who had also solved Hilbert's decision problem. In 1938 he obtained his Ph.D. at Princeton; his dissertation introduced the notion of 'relative computing,' where Turing machines are augmented with the so-called oracles, allowing a study of problems that cannot be solved by a Turing machine. Perhaps the most remarkable feature of Turing's work on Turing machines was that with these machines he was describing a modern computer before the technology behind them had been invented. In 1939, Turing returned to King's College. However, at the outbreak of World War II, he joined the Foreign Office, performing cryptanalysis of German ciphers. He contributed to breaking the code of the Enigma, a mechanical German cipher machine, which played an important role in winning the war. After the war, Turing worked on the development of early computers. He was interested in the ability of machines to think, arguing that if a computer could not be distinguished from a person based on its written replies to questions, it should be considered to be "thinking." He was also interested in biology, and he wrote on morphogenesis, the development of form in organisms. He is also credited with popularizing the matrix formulation of the LU-decomposition in 1948. In 1952, Turing was convicted of "acts of gross indecency" after admitting to a sexual relationship with a man in Manchester. He was placed on probation and required to undergo hormone therapy (chemical castration). In 1954, Turing committed suicide by taking cyanide, without leaving a clear explanation. Legal troubles related to his homosexual relationship and the hormonal treatment that was mandated by the court to lessen his sex drive may have contributed to his decision to end his life. Turing's mother, Sara Turing (1881–1976), who survived him by many years, wrote a biography of her son in 1959. In addition to being a brilliant mathematician, Turing was also a world-class runner who competed successfully with Olympic-level athletes. Since 1966, the Turing Award, accompanied by a prize of $100,000, has been given annually by the Association for Computing Machinery to recognize lasting and major technical contributions to the computing community. It is widely considered to be the computing world's equivalent to the Nobel Prize.

Prabhu Lal Bhatnagar (1912–1976)

Prabhu Lal Bhatnagar (1912–1976) was born in a highly respectable family that had been advisors to the rulers of the princely state of Kota (India). His family was well known for its philanthropy, and Prabhu inherited high nationalistic ideals from his parents. He studied at Herbert College in Kota, and then at Maharaja's College in Jaipur, where he obtained his bachelor of science degree with the first rank in 1935, followed by an M.S. degree. Prabhu's research career started at the University of Allahabad, where he worked during 1937–1939 on Fourier and related series and on the solar system. In 1939 he was awarded a doctorate in mathematics for his thesis entitled *On the Origin of the Solar System*. In 1939 he joined St. Stephen's College, Delhi, on an invitation and remained there for 16 years. At St. Stephen's College he worked on the theory of white dwarfs, collaborating with Daulat Singh Kothari (1905–1993), an outstanding Indian physicist. In 1952, Prabhu was invited to Harvard University as a Fulbright scholar. At Harvard, he produced two very important works: a book *Stellar Interiors*, written with Donald Howard Menzel (1901–1976) and Hari Kesab Singh, and a research paper published in *Physical Review* in 1954 that contained the famous Bhatnagar–Gross–Krook (Eugene P. Gross, 1926–1991; Max Krook, died 1985) (BGK) collision model. This model was extensively developed for ionized gases with many applications, and was used to solve Boltzmann integro-differential equation of the kinetic theory of gases. During the last couple of decades, the BGK model has found an important new application, the derivation of numerical schemes, kinetic schemes to solve the hyperbolic conservation laws. These days, the BGK collision operator is essential for the recent development of lattice Boltzmann automata methods. Prabhu became a fellow of the Indian National Science Academy in 1950 and a fellow of the Indian Academy of Sciences in 1955. In 1956 he was invited to join the Indian Institute of Science (IISc), Bangalore, as the professor of the newly created department of mathematics. There he expanded his research into the field of non-Newtonian fluid mechanics, and he trained and supervised students selected from various parts of India for their doctoral degrees. He used to say that his department represented a mini-India. He was responsible for the establishment of the Computer Centre in IISc in 1970. Today (2013), IISc has grown to be one of the best hyper-computing facilities in India. Prabhu laid down the foundation of the Indian National Mathematics Olympiad. For his contributions to the nation, Prabhu was awarded the Padma Bhushan by the President of India on January 26, 1968. During the early 1960s, he developed complications in the lower spinal region and was operated on in the USA. In 1969, he moved to Rajasthan University in Jaipur as Vice Chancellor, and in 1971, he joined Himachal Pradesh University in Shimla as senior professor. In 1975, he accepted the post of Director of the newly created Mehta Research Institute in Allahabad. He drew his last breath at the age of 64 in Allahabad after a massive heart attack.

Andrzej Mostowski (1913–1975)

Andrzej Mostowski (1913–1975) was born in Lwów. He studied mathematics in Warsaw during 1931–1936. After receiving his master's degree he went to Vienna, where he studied under Gödel, and then to Zürich. In 1938, Mostowski defended his doctoral dissertation, written under the supervision of Tarski. During the Nazi occupation of Poland he worked as an accountant in a tile factory. During 1942–1944 he taught at the Underground University of Warsaw, where he was unofficially appointed a docent. After the Warsaw uprising of 1944 the Nazis tried to put him in a concentration camp; however, with the help of some Polish nurses he escaped from a hospital, choosing to take bread with him rather than his notebook containing his research. From 1946 until his death in Vancouver, Canada, he worked at the University of Warsaw. Mostowski spent the 1948–1949 academic year at the Institute for Advanced Study, Princeton, and in 1969–1970 he became a Fellow of All Souls College, Oxford. He took part in numerous congresses and conferences all over the world. Much of work during that time was on first order logic and model theory. His 1969 monograph *Constructible Sets with Applications* is highly appreciated all over the world. He is best remembered for the *Mostowski Collapse Lemma*. In 1956, Mostowski was elected an associate member and in 1963 a full member of the Polish Academy of Sciences. He received Polish State Prizes in 1953 and 1966 and the Jurzykowski Foundation Prize in 1972. In 1973, he was elected a member of the Finnish Academy of Sciences. He was on the editorial boards of several learning journals, including the *Fundamenta Mathematicae*, the *Journal of Symbolic Logic*; and the *Annals of Mathematical Logic*; he was also a coeditor of the Series for Mathematics, Physics, and Astronomy of the Bulletin of the Polish Academy of Sciences. His long association with the North Holland Publishing Company helped to raise the *Studies in Logic and the Foundations of Mathematics* series to its present eminence. His son Tadeusz is also a mathematician working on differential geometry. With Krzysztof Kurdyka and Adam Parusinski, Tadeusz Mostowski solved René Thom's (1923–2002) gradient conjecture in 2000.

Paul Erdös (1913–1996)

Paul Erdös (1913–1996) had no children, no wife, no house, no credit card, no job, no change of shoes, indeed nothing but a suitcase containing a few articles of clothing and some notebooks. To him, private property was a nuisance. He was not fussy about his food as long as he had coffee. A mathematician, he said, "is a machine for converting coffee into theorems." He was a constantly wandering Jew. For much of his life, Hungary, his homeland, was run by dictators. Many members of his family were murdered by the Nazis during World War II. Paul's father Lajos and his mother Anna had two daughters, aged three and five, who died of scarlet fever just days before Paul was born. This naturally had the effect of making Lajos and Anna extremely protective of Paul. He was introduced to mathematics by his parents, who were both teachers of mathematics. At the age of 3 he could multiply three-digit numbers in his head, and at four he had discovered negative numbers on his own. While he was in his late teens he made discoveries about prime numbers. Examples of prime numbers are 1913, the year in which Erdös was born, and 83, his age when he died of a heart attack at a mathematics conference in Warsaw. Erdös helped to recast a theory about prime numbers made by an earlier mathematician, using a more elementary approach. This, it was said, was analogous to creating the Panama Canal instead of shipping around South America.

Typically he would arrive in a city where he was to lecture, ring up a fellow mathematician, and announce, “My brain is in town.” He sounds like a guest from hell, but to his hosts this brain was a shared treasure and their collective responsibility. They would lodge him, feed him, and launder his clothing. In return for the comfort of friends who took care of his perfunctory needs, Erdös would cut gems of elegance from numbers, graphs, and logic. His problems often seemed simple when posed, yet they offered room for creativity and surprise. Suppose an infinite number of dots are painted on an infinite canvas in such a way that the distance in inches between any two dots is a whole number. What would be painting look like? Erdös brow would furrow as he showed that the result could only be a straight row of dots. But don’t ask for an explanation of his elegant proof unless you are interested in conic sections. He published over 1,500 research papers and had more than 500 coauthors, but he made no claim for their practicality. It was enough, he said, that a proof be “very nice.” Yet mathematics, however pure, has a way of turning up in useful places. For example, combinatorics, a branch of maths explored by Erdös, can be used to calculate the number of tiles needed to pave an irregular space. His work on graphs has been applied to the design of communication networks.

Erdös had his own special language, using terms such as epsilon (child), boss (woman), slave (man), captured (married), liberated (divorced), Supreme Fascist (God), Sam (United States), and Joe (Soviet Union). The extraordinary left side of his brain was put at the service of numerous young mathematicians at the start of their careers. To him a mathematician of promise was an epsilon, a small quantity. To an epsilon he was “Uncle Paul”. He gave them problems, paying them rewards of a few hundred dollars if they came up with solutions, giving away much of his modest income from lectures and prizes (among them the Wolf Prize, an Israel-based sort of Nobel Prize). A colleague likened Erdös to a honeybee: an industrious creature that buzzed about the world and pollinated the fields of mathematics. It is hard to exaggerate Erdös passion. For 19 h a day, 7 days a week, stimulated by coffee, and later by amphetamines, he worked on mathematics. He might start a game of chess, but would probably doze off until the conversation returned to maths. Often, Erdös has been compared to Euler, the most prolific mathematician who ever lived.

Samuel Eilenberg (1913–1998)

Samuel Eilenberg (1913–1998) was born in Warsaw, then part of the Russian Empire. Sammy, as Samuel was always called, studied at the University of Warsaw. He obtained his doctorate under the supervision of Borsuk from the University of Warsaw in 1936 at the age of 23, and he was awarded a master’s degree there in 1934. His dissertation, which was concerned with the topology of the plane, was published in *Fundamenta Mathematicae* in 1936. Its results were well received both in Poland and in the USA. In 1939, Sammy emigrated to the USA. Once there he went to Princeton. In 1940, Sammy became an instructor at the University of Michigan, where he was able to interact with renowned topologists. In 1941 he was promoted to assistant professor and then in 1945 to associate professor. He spent 1945–1946 as a visiting lecturer at Princeton before being appointed a full professor at the University of Indiana in 1946. In 1947, he moved to Columbia University, where he remained for the rest of his career. Sammy was a member of Bourbaki. In 1956, he wrote the book *Homological Algebra* with Henri Cartan, which became a classic. Sammy also wrote an important book on automata theory. The X-machine, a form of automaton, was introduced by him in 1974. In the later part of his life, Sammy worked mainly in pure category theory, being a founder of the field. He is widely known for the Eilenberg–Ganea (Tudor Ganea, 1922–1971) conjecture, the Eilenberg–Ganea theorem, Eilenberg–MacLane (Saunders MacLane, 1909–2005) space, the Eilenberg–Montgomery fixed point theorem, the Eilenberg–Moore (John Coleman Moore, born 1923) spectral sequence, the Eilenberg–Steenrod (Norman Earl Steenrod, 1910–1971) axioms, and the Eilenberg swindle. Sammy collected Asian art, specifically small sculptures and other artifacts from India, Indonesia, Cambodia, Nepal, Pakistan, Thailand, Sri Lanka, and Central Asia. In 1991–1992, the Metropolitan Museum of Art in New York staged an exhibition of more than 400 items that Sammy had donated to the museum, entitled "The Lotus Transcendent: Indian and Southeast Asian Art From the Samuel Eilenberg Collection." Sammy received several awards for his work. Among these was the Wolf Prize in 1986, which he shared with Atle Selberg (1917–2007), a Norwegian mathematician known for his work in analytic number theory and the theory of automorphic forms, especially for relating them to spectral theory. Sammy was elected to the National Academy of Sciences. Regarding Sammy's personality, Hyman Bass (born 1932) writes: "Though his mathematical ideas may seem to have a kind of crystalline austerity, Sammy was a warm, robust, and very animated human being. For him mathematics was a social activity, whence his many collaborations. He liked to do mathematics on his feet, often prancing while he explained his thoughts. When something connected, one could read it in his impish smile and the sparkle in his eyes." Sammy breathed his last in New York at the age of 84.

Israel Moiseevich Gelfand (1913–2009)

Israel Moiseevich Gelfand (1913–2009) was born into a Jewish family in the small town of Krasnye Okny in Ukraine. He was expelled from high school because his father had been a mill owner. At the age of 16, Israel went to Moscow, where he lived with his relatives and did some odd jobs, including door keeper of the Lenin Library. In 1932, Israel became a graduate student of Kolmogorov. He also started teaching at Moscow State University (MSU) as an Assistant Professor. In 1935, Israel defended his Ph.D. thesis, and in 1940, he became a Doctor of Science at MSU. During

1935–1940, he worked as an Associate Professor at MSU. Beginning in 1939, Gelfand also worked at the Mathematical Institute of the USSR Academy of Sciences. In 1941, he became a full professor at MSU. Gelfand held this position for the next 50 years, until he moved to the USA in 1990, where he worked as a Distinguished Professor at the Department of Mathematics at Rutgers University, in Piscataway. Gelfand made outstanding contributions to many branches of mathematics, including group theory, representation theory, and functional analysis. He is remembered for the Gelfand pair, the Gelfand triple, the Gelfand representation in Banach algebra theory, the Gelfand–Mazur theorem in Banach algebra theory, the Gelfand–Naimark (Mark Aronovich Naimark, 1909–1978) theorem, the Gelfand–Naimark–Segal (Irving Ezra Segal 1918–1998) construction, Gelfand–Shilov (Georgi Evgen'evich Shilov, 1917–1975) spaces, the Gelfand–Pettis integral (Billy James Pettis, 1913–1979), the Gelfand–Levitan (Boris Levitan, 1914–2004) theory, the Gelfand–Dikii equations, the Gelfand–Fuks (D.B. Fuks) cohomology of foliations, the Gelfand–Kirillov (Alexandre Aleksandrovich Kirillov, born 1936) dimension, the Gelfand–Tsetlin patterns, and the Gelfand–Tsetlin basis. Gelfand also published works on biology and medicine and worked extensively in mathematics education. He was the recipient of numerous awards and honors, including the Order of Lenin (three times), Foreign Membership of the Royal Society in 1977, the Wolf Prize in 1978, the Kyoto Prize (awarded annually since 1985) in 1989, and a MacArthur Foundation Fellowship in 1994. He was elected a foreign member of the US National Academy of Science, the American Academy of Arts and Sciences, the Royal Irish Academy, the American Mathematical Society, and the London Mathematical Society. In October of 2003, The New York Times described Gelfand as a scholar "among the greatest mathematicians of the twentieth century". Gelfand passed away at the age of 96 in the Robert Wood Johnson Hospital in New Brunswick, NJ. His legacy continues through his students Endre Szemerédi (born 1940), Kirillov, Joseph Bernstein (born 1945), and his son Sergei Gelfand.

Richard Wesley Hamming (1915–1998)

Richard Wesley Hamming (1915–1998) was born in Chicago. He obtained his bachelor of science degree from the University of Chicago in 1937, his master's degree from the University of Nebraska in 1939, and finally his doctorate from the University of Illinois at Urbana-Champaign in 1942 at the age of 27. He was appointed a professor at the University of Louisville during World War II, and in 1945, he moved to work on the Manhattan Project. Richard did extensive programming on one of the earliest (first generation) electronic digital computers to calculate the solutions of equations that were given to him by the project's physicists. The objective of the program was to discover if the detonation of an atomic bomb would ignite the atmosphere. The computation determined that this would not occur, and so the USA tested the bomb in New Mexico, USA, and then it was used twice against Japan on August 6, 1945, at Hiroshima and then on August 9, 1945 at Nagasaki. During 1946–1976 he worked at Bell Telephone Laboratories, where he collaborated with Shannon. During this period, Richard was an Adjunct Professor at the City College of New York's School of Engineering. During 1976–1997, he worked as an Adjunct Professor at the Naval Postgraduate School, Monterey, California, where he was made Professor Emeritus in 1997. He passed away there a year later, at the age of 82. He is widely known for the Hamming code, the Hamming window, Hamming numbers, sphere packing, and the Hamming distance. Among the many awards Richard received are the Turing Award of the Association for Computing Machinery (ACM), of which he was a founder and president, and the IEEE Emanuel R. Piore Award (established in 1976). According to Richard, "The purpose of computing is insight, not numbers."

John Wilder Tukey (1915–2000)

John Wilder Tukey (1915–2000) was born in New Bedford, Massachusetts. His parents, who were both teachers, decided that home schooling would best develop his potential. His formal education began at Brown University, where he studied mathematics and chemistry. He received bachelor's and master's degrees in chemistry from Brown in 1936 and 1937, and he continued his education at Princeton University, changing his field of study from chemistry to mathematics. He received his Ph.D. from Princeton in 1939 for work on topology under the supervision of Lefschetz, and he was then appointed an instructor of mathematics at Princeton. At the beginning of World War II, he joined the Fire Control Research Office, where he began working in statistics. Tukey found statistical research to his liking and impressed several leading statisticians with his skills. In 1945, at the end of the war, Tukey returned to the mathematics department at Princeton as a Professor of statistics, and he also took a position at AT&T Bell Laboratories. Tukey's first major contribution to statistics was his introduction of modern techniques for the estimation of spectra of time series. Tukey founded the Statistics Department at Princeton in 1966 and was its first chairman. Tukey made significant contributions to many areas of statistics, including the analysis of variance, the estimation of spectra of time series, inferences about the values of a set of parameters from a single experiment, and the philosophy of statistics. However, he is best known for his invention, with Cooley, of the fast Fourier transform. Tukey utilized his insight and expertise by serving on the President's Science Advisory Committee. He chaired several important committees dealing with the environment, education, and chemicals and health. He also served on committees that worked on nuclear disarmament. Tukey received many awards, including the Samuel Stanley Wilks (1906–1964) award from the American Statistical Association in 1965, the US National Medal of Science in 1973, and the Medal of Honor from the Institute of Electronic and Electrical Engineers in 1982.

Laurent Moïse Schwartz (1915–2002)

Laurent Moïse Schwartz (1915–2002) was born in Paris. While he attended Lycée Louis le Grand to enter the École Normale Supérieure, in 1934 he met Marie Héléne Lévy and fell in love with her. She was the daughter of the famous probabilist Paul Pierre Lévy (1886–1971), a Jewish French mathematician who introduced martingales and Lévy flights, Lévy processes, Lévy measures, Lévy's constant, the Lévy skew alpha stable distribution, the Lévy area, the Lévy arcsine law, and the fractal Lévy C curve. Marie Héléne herself was a mathematician, who contributed to the geometry of singular analytic spaces and taught at the University of Lille. During World War II they had to hide and change their identities. Laurent worked for the University of Strasbourg under the name Laurent Marie Sélimartin, while Marie Héléne used the name Lengé instead of Lévy. Laurent spent 1 year in Grenoble in 1944, and then he joined the University of Nancy, where he spent 7 years. He was an influential researcher and teacher. He joined the science faculty at the University of Paris in 1952, and in 1958, he became a teacher at the École Polytechnique. During 1961–1963, the École Polytechnique refused him the right to teach because he had signed the Manifesto of the 121, a document that criticized the Algerian war. He became a corresponding member of the French Academy of Sciences in 1973 and a full member in 1975.

Schwartz pioneered the theory of distributions. This theory clarified the mysteries of the Dirac Delta function and the Heaviside step function. It helped to extend the theory of Fourier transforms and is now of great importance to the theory of partial differential equations. Harald Bohr presented a Fields Medal (after John Charles Fields, 1863–1932) to Laurent, the first French mathematician to receive this honor, at the International Congress at Harvard on August 30, 1950, for his work on the theory of distributions. He described Laurent's 1948 paper as one: "...which certainly will stand as one of the classical mathematical papers of our times. ...I think every reader of his cited paper, like myself, will have left a considerable amount of pleasant excitement, on seeing the wonderful harmony of the whole structure of the calculus to which the theory leads and on understanding how essential an advance its application may mean to many parts of higher analysis, such as spectral theory, potential theory, and indeed the whole theory of linear partial differential equations...." Besides the Schwartz distribution, the Schwartz kernel theorem, Schwartz space, and the Schwartz–Bruhat (Francois Georges René Bruhat, 1929–2007) function are named after him. He also promoted public appreciation and understanding of science and mathematics. He said: "What are mathematics helpful for? Mathematics are helpful for physics. Physics helps us make fridges. Fridges are made to contain spiny lobsters, and spiny lobsters help mathematicians who eat them and have hence better abilities to do mathematics, which are helpful for physics, which helps us make fridges which..." Schwartz passed away at the age of 87 in Paris.

Claude Elwood Shannon (1916–2001)

Claude Elwood Shannon (1916–2001) was born in Petoskey, Michigan, and grew up in Gaylord, Michigan. His father was a businessman and a probate judge, and his mother was a language teacher and a high school principal. Shannon attended the University of Michigan, where he graduated in 1936. He continued his studies at M.I.T., where he was responsible for maintaining the differential analyzer, a mechanical computing device that consisting of shafts and gears that was built by his professor, Vannevar Bush. Shannon's master's thesis, written in 1936, studied the logical aspects of the differential analyzer. This thesis presents the first application of Boolean algebra to the design of switching circuits; it is perhaps the most famous master's thesis of the twentieth century. He received his Ph.D. from M.I.T. in 1940. Shannon joined Bell Laboratories in 1940, where he worked on transmitting data efficiently. He was one of the first people to use bits to represent information. At Bell Laboratories he worked on determining the amount of traffic that telephone lines can carry. Shannon made many fundamental contributions to information theory. In the early 1950s he was one of the founders of the study of artificial intelligence. He joined the M.I.T. faculty in 1956, where he continued his study of information theory. Shannon had an unconventional side. He is credited with inventing the rocket-powered Frisbee, and he is also famous for riding a unicycle down the hallway of Bell Laboratories while juggling four balls. Shannon retired when he was 50 years old, publishing papers sporadically over the following 10 years. In his later years he concentrated on some pet projects, such as building a motorized pogo stick. One interesting statement from Shannon, published in *Omni Magazine* in 1987, is "I visualize a time when we will be to robots what dogs are to humans. And I am rooting for the machines."

Edward Norton Lorenz (1917–2008)

Edward Norton Lorenz (1917–2008) was born in West Hartford, Connecticut. He studied mathematics at Dartmouth College in New Hampshire, and he began his career as a mathematics graduate student at Harvard University, but he turned his attention to meteorology during World War II. He received his Ph.D. from the Massachusetts Institute of Technology in 1948 and since then has been associated with this institution. In 1961, using a primitive computer, Lorenz attempted to build a simple model for weather prediction. His method stimulated real weather patterns quite well, but it also illustrated something much more important: When Lorenz changed the initial conditions slightly, the resulting weather patterns changed completely after a short time. Lorenz had discovered that simple differential equations can behave "chaotically." In his famous 1963 paper, Lorenz picturesquely explained that a butterfly flapping its wings in Beijing could affect the weather thousands of miles away some days later. This sensitivity is now called the "butterfly effect." He has received many medals and awards, which include the Carl Gustaf Rossby (1898–1957) Research Medal, membership of the American Meteorological Society in 1969, the Symons Memorial Gold Medal, membership of the Royal Meteorological Society in 1973, Fellowship of the National Academy of Sciences (USA) in 1975, membership of the Norwegian Academy of Science and Letters in 1981, the Crafoord Prize (established in 1980 in honor of Holger Crafoord, 1908–1982), membership of the Royal Swedish Academy of Sciences in 1983, honorary membership of the Royal Meteorological Society in 1984, the Kyoto Prize for his discovery of "deterministic chaos" in 1991, and the Buys Ballot medal in 2004. Lorenz died at his home in Cambridge at the age of 90, after a battle with cancer.

Julia Hall Bowman Robinson (1919–1985)

Julia Hall Bowman Robinson (1919–1985) was born in St. Louis, Missouri, but her family moved to Arizona and then to San Diego. She was slow to talk, and as a child she pronounced words so oddly that no one except her elder sister, Constance Bowman Reid (1918–2010), could understand her. Julia's mother, a kindergarten and first grade teacher, said that Julia was the most stubborn child she ever knew. Julia recalled: "I would say that my stubbornness has been to a great extent responsible for whatever success I have had in mathematics." When she was 9 years old, Julia had scarlet fever, which was followed by a bout with rheumatic fever, and she was forced to spend a year in bed at the home of a nurse. When she finally regained her health enough to return to school, she found herself 2 years behind. Working with a tutor three mornings a week, she completed the state syllabi for the fifth, sixth, seventh, and eighth grades in 1 year. In junior high school she was given an IQ test, and being a slow reader and unaccustomed to taking tests, she did poorly; her IQ was recorded as 98, two points below normal. Julia first became interested in mathematics when she learned that no matter how far the decimal expansion of the square root of two is carried out it will never become periodic. "I didn't see how anyone could know that—[I thought] all they could know is that the expansion had not become periodic in the part that was calculated. I tried to check it with my newly acquired skills at extracting square roots but finally gave up. After all, I couldn't go on forever either!" She studied at San Diego State University in 1936 and transferred as a senior to the University of California, Berkeley, in 1939. She received her bachelor's degree in 1940 and then continued to graduate studies. She obtained her doctoral degree in 1948 under Alfred Tarski, with a dissertation on "Definability and Decision Problems in Arithmetic." In 1975, Julia became a full professor at the University of California, Berkeley, teaching quarter time because she still did not feel strong enough for a full-time job. Julia's heart had been damaged by the rheumatic fever. She suffered from poor health and shortness of breath as an adult. She married a University of California, Berkeley, professor and American mathematician, Raphael Mitchel Robinson (1911–1995), in 1941. In 1961, Julia had an operation that removed the scar tissue from her mitral valve. The operation was a success and she became more active physically and started bicycling for exercise. In 1984, she was diagnosed with leukemia, she underwent treatment, and the cancer went into remission for several months. However, the disease recurred, and she passed away in Oakland, California at the age of 65. She is best known for her contributions to decision problems and Hilbert's tenth problem. She was the first female mathematician elected to the National Academy of Sciences and the first female president of the American Mathematical Society.

Raymond Merrill Smullyan (Born 1919)

Raymond Merrill Smullyan (born 1919) is from Far Rockaway, Long Island, New York. He dropped out of high school because he wanted to study what he was really interested in and not the standard high school material. After jumping from one university to the next, he earned an undergraduate degree in mathematics at the University of Chicago in 1955. He paid his college expenses by performing magic tricks at parties and clubs. He obtained a Ph.D. in logic in 1959 at Princeton, studying under Alonzo Church. After graduating from Princeton, he taught mathematics and logic at Dartmouth College, Princeton University, Yeshiva University, and the City University of New York. He joined the philosophy department at Indiana University in 1981, where he is now an Emeritus Professor. Smullyan has written many books on recreational logic and mathematics: *The Tao is Silent* (1977), *What is the Name of This Book*? (1978), *The Chess Mysteries of Sherlock Holmes* (1979), *This Book Needs No Title* (1980), *The Chess Mysteries of the Arabian Knights* (1981), *The Lady or the Tiger*? (1982), *Alice in Puzzle-land* (1982), *5000 BC* (1983), *To Mock a Mockingbird* (1985), *Forever Undecided* (1987), *Satan, Cantor, and Infinity* (1992), *The Riddle of Scheherazade: Amazing Logic Puzzles, Ancient and Modern* (1997), *Some Interesting Memories: A Paradoxical Life* (2002), and *Who Knows?: A Study of Religious Consciousness* (2003). Because his logic puzzles are challenging, entertaining, and thought provoking, he is considered to be a modern-day Lewis Carroll. Smullyan has also written three collections of philosophical essays and aphorisms, as well as several advanced books on mathematical logic and set theory: *Theory of Formal Systems* (1961), *First-Order Logic* (1968), *Gödel's Incompleteness Theorems* (1992), *Recursion Theory for Metamathematics* (1993), *Diagonalization and Self-Reference* (1994), and *Set Theory and the Continuum Problem* (1996). He is particularly interested in self-reference and has worked on extending some of Gödel's results, which show that it is impossible to write a computer program that can solve all mathematical problems. He is also interested in explaining ideas from mathematical logic to the public. Smullyan is a talented musician and often plays piano with his wife, who is a concert-level pianist. In fact, he has recently released a recording of his favorite classical piano pieces by composers such as Johann Sebastian Bach (1685–1750), Giuseppe Domenico Scarlatti (1685–1757), and Hermann Schubert (1848–1911). Making telescopes is one of his hobbies. He is also interested in optics and stereo photography. He has written an autobiography titled *Some Interesting Memories: A Paradoxical Life*. He stated, "I've never had a conflict between teaching and research as some people do, because when I'm teaching, I'm doing research." In 2001, documentary filmmaker Tao Ruspoli made a film about Smullyan called *This Film Needs No Title*.

Calyampudi Radhakrishna Rao (Born 1920)

Calyampudi Radhakrishna Rao (born 1920), a Fellow of Royal Society and one of the top Indian statisticians of the twentieth century, was born at Hadagali, Kingdom of Mysore, British India, on September 10, 1920. Prior to receiving his master's degree in statistics from Calcutta University, Kolkata, in 1943, he completed a master's degree in mathematics at Andhra University, Visakhapatnam. He worked at the Indian Statistical Institute, Kolkata, and the Anthropological Museum in Cambridge before receiving a doctorate degree at King's College in Cambridge University under the supervision of the renowned English statistician Ronald Fisher. He then obtained an Sc.D. (Doctor of Science, the highest degree) from Cambridge in 1965. He played an important role, under the direction of the doyen of Indian statistics, Mahalanobis, the founder of the Indian Statistical Institute, Kolkata, in the establishment of state statistical bureaus in different states of India and developing a network of statistical agencies at the district level for the collection of data. He is currently the Eberly Professor Emeritus of Statistics at the Department of Statistics at Eberly College of Science, Pennsylvania State University (PSU), USA, and the Director of the Center for Multivariate Analysis at PSU. Calyampudi has written around 400 articles and several books on statistics and related areas, such as *Matrix Algebra and Its Applications to Statistics and Econometrics*, *Statistics and Truth*, *Estimation of Variance Components and Applications*, and *Generalized Inverse of a Matrix and Its Applications* (with Sujit Kumar Mitra). He has received several medals, prizes, honors, and awards, such as the Shanti Swarup Bhatnagar (1894–1955) Award (1963) of the Council of Scientific and Industrial Research, the William Augustus Guy (1810–1885) Medal in Silver (1965) of the Royal Statistical Society, the Meghnad Saha Medal (1969) of the Indian National Science Academy, the Wilks Memorial Award (1989) of the American Statistical Association, the Padma Vibhushan (2001, the second highest civilian award given by the Government of India), the Srinivasa Ramanujan Medal (2003) of the Indian National Science Academy, an honorary D.Sc. from the University of Calcutta in 2003, the International Mahalanobis Prize (2003) of the International Statistical Institute, the India Science Award (2010, the highest award in a scientific field presented by government of India), and the Guy Medal in Gold (2011) of the Royal Statistical Society. President George Walker Bush (born 1946) bestowed on him the National Medal of Science, the US government's highest award for scientific achievements, on June 12, 2002, as a "prophet of new age" with the citation, "for his pioneering contributions to the foundations of statistical theory and multivariate statistical methodology and their applications, enriching the physical, biological, mathematical, economic and engineering sciences." In his honor, PSU has established the C.R. and Bhargavi Rao Prize in Statistics. A C.R. Rao Advanced Institute of Mathematics, Statistics and Computer Science (AIMSCS) was also established in Hyderabad, India. Calyampudi was President of the International Statistical Institute (The Netherlands), President of the Institute of Mathematical Statistics (USA), and President of the International Biometric Society (USA). He was inducted into the Hall of Fame of India's National Institution for Quality and Reliability (Chennai Branch) for his contributions to industrial statistics and his promotion of quality control programs in industry. He is remembered for the Rao Score test, the Cramér–Rao (Harald Cramér, 1893–1985) bound, the Rao–Blackwell (David Harold Blackwell, 1919–2010) theorem, the Rao distance, the Fisher–Rao theorem, Rao least squares, the Fisher–Rao metric, the Neyman–Rao (Jerzy Neyman, 1894–1981) test, the Rao covariance structure, the Rao U test, the Rao F test, the Rao Paradox, the Rao–Rubon theorem, the Kagan–Linnik–Rao theorem, the Lau–Rao–Shanbhag theorems, and the Rao–Yanoi inverse. The following sayings are due to Rao.

All knowledge is, in final analysis, history.
All sciences are, in the abstract, mathematics.
All decisions we make, have their rational in statistics.
All technologies, for efficient use, depend on computers.

Olga Alexandrovna Ladyzhenskaya (1922–2004)

Olga Alexandrovna Ladyzhenskaya (1922–2004) received many awards during her mathematical career, including First Prize from Leningrad State University in 1954 and again in 1961. From 1961 to 1991, she held the position of Head of the Laboratory of Mathematical Physics at the Steklov Mathematical Institute of the Academy of Sciences of the USSR. In 1969, she received the Chebyshev Prize of the USSR Academy of Sciences and the State Prize of the USSR. She was elected a corresponding member of the Academy of Sciences of the USSR (1981), a foreign member of the German Academy of Sciences Leopoldina (1985) and the Accademia dei Lincei (1989), a member of the Russian Academy of Sciences (1990), and a foreign member of the American Academy of Arts and Sciences (2001). She was awarded the Kovalevskaya Prize in 1992, an honorary doctorate from the University of Bonn on May 13, 2002, the Golden Lomonosov Medal, the Ioffe Medal, and the St. Petersburg University Medal in 2003. In 1998 she delivered the John von Neumann Lecture at the SIAM Annual Meeting in Toronto. She was a member of the St. Petersburg Mathematical Society after it was recreated in 1959, and she served as its Vice President from 1970 to 1990 and its President between 1990 and 1998, after which she was elected Honorary Member of the Society. In the Museum of Science (Boston, USA), Olga Ladyzhenskaya's name is among the names of other influential twentieth-century mathematicians, carved on a large marble desk in the mathematics Exhibition Hall.

Harish Chandra Mehrotra (1923–1983)

Harish Chandra Mehrotra (1923–1983), one of the renowned Indian mathematicians of the twentieth century, who is known for his contributions to Lie algebra, was born in Kanpur on October 11, 1923. He is known to the world as "Harish-Chandra". His father was a civil engineer with the railways, and he had to spend most of his time traveling to different parts of India inspecting canals. Harish-Chandra spent most of his childhood under the supervision of his maternal grandfather. Sometimes he spent time traveling with his father, despite his delicate health. His family gave his education at home the highest importance. He was admitted in Class Seven when he was just nine (younger than his schoolmates) and he completed Christ Church High School at fourteen. He remained in Kanpur and finished his intermediate college education at 16. Harish-Chandra then joined the University of Allahabad and studied theoretical physics, obtaining his bachelor's and master's degrees in 1941 and 1943. In 1943, he moved to Bangalore, and during 1943–1945, he worked as a research fellow on problems in theoretical physics under Homi Bhabha, a nuclear physicist, at the Indian Institute of Science. They published several joint papers, extending some of Dirac's results. On the recommendation of Homi Bhabha and Kariamanickam Krishnan, who was a teacher of Harish-Chandra at Allahabad University, Dirac accepted him as a Ph.D. student at Cambridge University in 1945. However, Harish-Chandra had very little interaction with Dirac. While attending one of Dirac's lectures, he observed that Dirac was essentially reading from one of his books. He wrote about Dirac, "... very gentle and kind and yet rather aloof and distant... [I decided] I should not bother him too much and went to see him once every term." At Cambridge, Harish-Chandra attended lectures by Wolfgand Ernst Pauli (1900–1958), the 1945 Physics Nobel Prize winner. During one of these lectures he pointed out a mistake in Pauli's work, and the two became lifelong friends. Harish-Chandra obtained his Ph.D. in 1947, and he moved to the USA in the same year. When Dirac visited Princeton during 1947–1948, Harish-Chandra worked as his assistant. He was greatly influenced by Weyl, Chevalley, and Artin, who were also in Princeton. During this period in Princeton, he realized that he was a mathematician and not a physicist. He later wrote, "Soon after coming to Princeton I became aware that my work on the Lorentz group was based on somewhat shaky arguments. I had naively manipulated unbounded operators without paying any attention to their domains of definition. I once complained to Dirac about the fact that my proofs were not rigorous and he replied, 'I am not interested in proofs but only in what nature does'. This remark confirmed my growing conviction that I did not have the mysterious sixth sense which one needs in order to succeed in physics and I soon decided to move over to mathematics." After Dirac left Princeton and returned to Cambridge, Harish-Chandra remained at Princeton for the second year (1948). He spent 1949–1950 at Harvard University, where he was influenced by Oscar Zariski. During 1950–1963, Harish-Chandra was on the faculty at Columbia University, where he carried out his best research on representations of semi-simple Lie groups. During this period he came into close contact with Andre

Weil. During these years, he also spent time at other institutions. In 1952–1953, he was at the Tata Institute of Fundamental Research (TIFR), Mumbai, India. In 1952 he married Lalitha Kale (whom he fondly called Lily), the daughter of a botanist from Allahabad. Lily, with her good spirits, generous affection, patience, and all-round competence, pampered him for 30 years. Harish-Chandra spent 1955–1956 at the Institute for Advanced Study at Princeton, then 1957–1958 as a Guggenheim Fellow (founded in 1925 by Olga and Simon Guggenheim in memory of their son) in Paris where he worked with Weil. Both Weil and Harish-Chandra often went walking together, exchanging their ideas and opinions. During 1961–1963, he was a Sloan Fellow (established in 1934 by Alfred Pritchard Sloan, Jr., 1875–1966) at the Institute for Advanced Study, and in 1968, he was appointed IBM von Neumann Professor at this institute.

Some of Harish-Chandra's major contributions deserve mention: the explicit determination of the Plancherel (Michel Plancherel, 1885–1967) measure for semi-simple groups, the determination of the discrete series representations, his results on Eisenstein series and in the theory of automorphic forms, and his "philosophy of cusp forms," as he called it, as a way of describing certain phenomena in the representation theory of reductive groups in a broad sense, including not only the real Lie groups, but also p-adic groups or groups over adele rings (invented by Chevalley). His scientific work, being a synthesis of analysis, algebra, and geometry, is of lasting influence. During his stay at Columbia University, he established as his special area the study of the discrete series representations of semi-simple Lie groups, which are the closest analogue of the Peter–Weyl theory in the non-compact case. He also worked with the Swiss mathematician Armand Borel (1923–2003), who was of the same age as Harish-Chandra, on the foundation of the theory of arithmetic groups. He enunciated a "philosophy of cusp forms" (a term he invented) that expressed his idea of reverse engineering of automorphic form theory, from a representation theory point of view, which was a precursor of Langlands' (Robert Phelan Langlands, born 1936) philosophy. This philosophy, suggested by Robert Phelan Langlands in 1967, consists of a web of far-reaching and influential conjectures that connect number theory and the representation theory of certain groups. It is a way of organizing number theoretic data in terms of analytic objects.

Harish-Chandra was awarded the Cole Prize of the American Mathematical Society in 1954. Some other notable awards bestowed on him are FRS (1973), the Srinivasa Ramanujan Medal (1974), Fellow of Indian Academy of Sciences, Fellow of the Indian National Science Academy (1975), and Fellow of the National Academy of Sciences of United States (1981). He passed away in 1983 at the age of 60 after a massive heart attack during a conference in Princeton in honor of Armand Borel's 60th birthday. He had suffered from three previous heart attacks, the first in 1969 and the third in 1982. On the last day of the conference, Sunday, October 16, 1983, he and Lily hosted many of the participants at their home. In the late afternoon, after the guests had departed, he went for his customary walk and never returned. A similar conference for his 60th birthday, which was scheduled for the very next year, became a memorial conference. Harish-Chandra is survived by his wife Lalitha and his two daughters, Premala and Devaki.

John Warner Backus (1924–2007)

John Warner Backus (1924–2007) was born in Philadelphia and grew up in Wilmington, Delaware. He attended the Hill School in Pottstown, Pennsylvania. He needed to attend summer school every year because he disliked studying and was not a serious student. He enjoyed spending his summers in New Hampshire, where he attended summer school and amused himself with summer activities like sailing. He obliged his father by enrolling at the University of Virginia to study chemistry. But he quickly decided chemistry was not for him, and in 1943, he entered the army, where he received medical training and worked in a neurosurgery ward at an army hospital. Ironically, Backus was soon diagnosed with a bone tumor in his skull, and he was fitted with a metal plate. His medical work in the army convinced him to try medical school, but he gave up after 9 months because he disliked the rote memorization required. After dropping out of medical school, he entered a school for radio technicians because he wanted to build his own high fidelity set. A teacher in this school recognized his potential and asked him to help with some mathematical calculations needed for an article in a magazine. Finally, Backus found what he was interested in: mathematics and its applications. He enrolled at Columbia University, from which he received both bachelor's and master's degrees in mathematics. Backus joined IBM as a programmer in 1950. He participated in the design and development of two of IBM's early computers. From 1954 to 1958, he led the IBM group that developed FORTRAN for the IBM 704 computer. Backus became a staff member at the IBM Watson Research Center in 1958. He was part of the committees that designed the programming language ALGOL, using what is now called the Backus–Naur form for the description of the syntax of this language. Later, Backus worked on the mathematics of families of sets and on a functional style of programming. The IEEE awarded Backus the W. Wallace McDowell Award (first in 1966) in 1967 for the development of FORTRAN. Backus became an IBM Fellow in 1963, he received the National Medal of Science in 1975, and he was given the prestigious Turing Award of the Association of Computing Machinery in 1977. He was also awarded a honoris causa from the University Henri Poincaré in Nancy (France) in 1989 and a Draper (Charles Stark Draper, 1901–1987) Prize in 1993.

Benoit Mandelbrot (1924–2010)

Benoit Mandelbrot (1924–2010) was born in Warsaw. His family emigrated to France in 1936. Mandelbrot was influenced by one of his uncles, who was professor of mathematics at the Collége de France, a great admirer of Hardy. He attended the Lycée Rolin at Paris until World War II began. His family then moved to Tulle in central France. In 1944, Mandelbrot began his studies at the École Polytechnique, where he was educated by Julia and Paul Lévi (among others). He then spent some time at the California Institute of Technology and the Institute of Advanced Study at Princeton. He obtained his doctorate in 1952 from the University of Paris. In 1958, Mandelbrot permanently moved to the USA and started working for IBM. In 1945, his uncle showed him a paper by Julia (published 1918); however, Mandelbrot showed no interest. It was not until 1970 that he started working on Julia's paper and displayed a Julia set graphically. For this, he developed new mathematical principles and wrote a graphical computer program. He published this work in the book *Les objets fractales: Forme, hasard et dimension* (1975), and more in-depth in *The fractal geometry of nature* (1982). In 1999 at the age of 75, Mandelbrot joined the Department of Mathematics at Yale and obtained his first tenured post. At the time of his retirement in 2005, he was Sterling Professor of Mathematical Sciences. His awards and honors include Knight in the French Legion of Honor (1990), the Wolf Prize for Physics (1993), the Lewis Fry Richardson (1881–1953) Prize of the European Geophysical Society (2000), the Japan Prize (2003), and the Einstein Lecturership of the American Mathematical Society (2006). The small asteroid 27500 Mandelbrot was named in his honor. An honorary degree from Johns Hopkins University was bestowed on Mandelbrot in the May 2010 commencement exercises. He died on October 14, 2010, in Cambridge, Massachusetts.

Maurice Karnaugh (Born 1924)

Maurice Karnaugh (born 1924) was born in New York City. During 1944–1948, he studied at City College of New York, and in 1949, he moved to Yale University to complete his B.S., M.S. (1950), and Ph.D. (1952). He was a member of the technical staff at Bell Laboratories from 1952 until 1966 and Manager of Research and Development at the Federal Systems Division of AT&T from 1966 to 1970. In 1970, he joined IBM as a member of its research staff. Karnaugh has made fundamental contributions to the application of digital techniques in both computing and telecommunications. He is famous for the Karnaugh map used in Boolean algebra. His current interests include knowledge-based systems in computers and heuristic (self-educating) search methods.

John Torrence Tate Jr. (Born 1925)

John Torrence Tate Jr. (born 1925) was born in Minneapolis, Minnesota. He received his bachelor's degree at Harvard University and a doctorate at Princeton University in 1950, under the direction of Artin. He was affiliated with Princeton University from 1950 to 1953 and with Columbia University from 1953 to 1954. He then taught at Harvard for 36 years before joining the University of Texas in 1990. He retired in 2009 and currently lives in Cambridge, Massachusetts, with his wife Carol. He has contributed several fundamental results in algebraic number theory and algebraic geometry, such as the Tate cohomology groups, the Poitou–Tate duality (Georges Poitou, 1926–1989), the Tate–Shafarevich (Igor Rostislavovich Shafarevich, born 1923) group, the Lubin–Tate (Jonathan Darby Lubin, born 1936) local theory of complex multiplication, his invention of rigid analytic spaces, the Hodge–Tate (William Vallance Douglas Hodge, 1903–1975) theory, the Tate curve, the Honda–Tate theorem, and Tate cycles. He is the recipient of several awards and honors, including the American Mathematical Society's Cole Prize (1956), the Wolf Prize in Mathematics (2002/03), and the Abel Prize (2010).

Peter David Lax (Born 1926)

Peter David Lax (born 1926), an American Jewish mathematician, was born to Klara Kornfield and Henry Lax (both medical doctors) in Budapest, Hungary. He emigrated to New York City in 1941 with his parents, where he attended Stuyvesant High School. In 1948, at the age of 22, he married Anneli Cahn (1922–1999), also a mathematician. Peter received his bachelor's degree in 1947 and his doctoral degree in 1949, both from New York University. He held a faculty position in the Department of Mathematics at the Courant Institute of Mathematical Sciences in New York. In 1970, the Control Data Corporation's supercomputer, CDC 6600, situated at the Courant Institute, was taken hostage by the Transcendental Students. Lax participated in the acquisition of the machine. Some of the students threatened to destroy the computer using bombs/incendiary devices, but Peter managed to defuse them and saved the computer. This incident resulted in the resignation of Jürgen Moser (1928–1999), director of the Courant Institute in 1967–1970. Lax contributed to integrable systems, fluid dynamics and shock waves, solitonic physics, hyperbolic conservation laws, and computational mathematics. He stated the "Lax conjecture" regarding matrix representations for third order hyperbolic polynomials in 1958, which was proved to be correct, after 45 years, in 2003. He is also known for the Lax–Wendroff (Burton Wendroff, born 1930) method, the Babuška–Lax–Milgram (Ivo M. Babuška, born 1926; Arthur Norton Milgram, 1912–1961) theorem, and the Lax equivalence theorem. He was awarded the National Medal of Science in 1986, the Wolf Prize in 1987, and, most notably, the Abel Prize in 2005.

Jean-Pierre Serre (Born 1926)

Jean–Pierre Serre (born 1926) was born in Bages, France. He began his education at the Lycée de Nimes, and then from 1945 to 1948 he attended the École Normale Supérieure in Paris. He obtained his doctorate at the Sorbonne in 1951 under the supervision of Henri Cartan. From 1948 to 1954, he held positions at the Centre National de la Recherche Scientifique in Paris. For 2 years, 1954–1956,

Serre worked at the University of Nancy. He was then elected professor at the Collége de France, where he remained until his retirement in 1994, when he became an honorary professor. He spent time at Harvard University and the Institute for Advanced Study at Princeton. Serre's most important work was on spectral sequences, which he applied in 1951 to his study of the relations between the homology groups of fibre, total space, and base space in a fibration. Based on this work and his work on complex variable theory in terms of sheaves, Serre was awarded a Fields Medal at the International Congress of Mathematicians in 1954. He has published many highly influential texts covering a wide range of mathematics. Among these texts, which show the variety of topics that Serre has worked on, are *Homologie singuliére des espaces fibrés* (1951), *Faisceaux algébriques cohérents* (1955), *Groupes d'algébriques et corps de classes* (1959), *Corps locaux* (1962), *Cohomologie Galoisienne* (1964), *Abelian l-adic representations and elliptic curves* (1968), *Cours d'arithmétique* (1970), *Représentations linéaires des groupes finis* (1971), *Arbres, amalgames, SL2* (1977), *Lectures on the Mordell–Weil theorem* (1989), and *Topics in Galöis theory* (1992). In the literature he is often cited for the Bass–Serre theory, Serre duality, Serre's multiplicity conjectures, Serre's property FA, the Serre spectral sequence, the Serre fibration, the Serre twist sheaf, and the thin set in the sense of Serre. In addition to the Fields Medal, Serre has received numerous other prizes and awards: the Prix Gaston Julia (1970), the Balzan (Eugenio Balzan, 1874–1953) Prize (1985), the Steele Prize (1995), the Wolf Prize (2000), and the Abel Prize (2003). He has been elected to the scientific academies of France, Sweden, the USA, the Netherlands, and the Royal Society of London. He has been awarded honorary degrees by the University of Cambridge (1978), the University of Stockholm (1980), the University of Glasgow (1983), the University of Harvard (1998), and the University of Oslo (2002).

John McCarthy (1927–2011)

John McCarthy (1927–2011) was born in Boston, Massachusetts. He grew up in Boston and in Los Angeles. He studied mathematics both as an undergraduate and a graduate student, receiving a B.S. in 1948 at the California Institute of Technology and his Ph.D. in 1951 at Princeton. After graduating from Princeton, McCarthy held positions at Princeton, Stanford, Dartmouth, and M.I.T. He held a position at Stanford from 1962 until his retirement at the end of 2000. At Stanford, he was the director of the Artificial Intelligence Laboratory, he held a named Chair in the School of Engineering, and he was a senior fellow of the Hoover Institute. McCarthy was a pioneer in the study of 'artificial intelligence,' a term that he coined in 1955. He worked on problems related to the reasoning and information needed for intelligent computer behavior. McCarthy was among the first computer scientists to design time-sharing computer systems. He developed LISP, a high-level programming language. He played an important role in using logic to verify the correctness of computer programs. McCarthy worked on the social implications of computer technology and on the problem of how people and computers make conjectures through assumptions. McCarthy was an advocate of the sustainability of human progress, and he was an optimist about the future of humanity. He also wrote science fiction stories. Some of his recent writings explore the possibility that the world is a computer program written by some higher force. Among the awards McCarthy had won were the Turing Award in 1971 from the Association for Computing Machinery, the Research Excellence Award of the International Conference on Artificial Intelligence, the Kyoto Prize, and the National Medal of Science.

Peter Naur (Born 1928)

Peter Naur (born 1928) was born in Frederiksberg, near Copenhagen. As a boy he became interested in astronomy. Not only did he observe heavenly bodies but he also computed the orbits of comets and asteroids. Naur attended Copenhagen University, where he received his degree in 1949. He spent 1950 and 1951 in Cambridge, where he used an early computer to calculate the motions of comets and planets. After returning to Denmark he continued working in astronomy but kept his ties to computing, for which he received his Ph.D. in 1957, but his encounter with computers eventually led to a change of profession. During 1959–1969 he worked at Regnecentralen, the Danish computing institute, and at the same time he gave lectures at the Bohr Institute and the Technical University of Denmark. At Regnecentralen, he worked on the development of compilers for ALGOL and COBOL. From 1969 to 1998 he was Professor of Computer Science at Copenhagen University, where he worked in the area of programming methodology. The BNF notation (Backus–Naur form) is now used in the description of the syntax for most programming languages. He pioneered the study of software engineering and software architecture. His book *Computing: A Human Activity* is a collection of his contributions to computer science. In this book he rejects the formalist school of programming, which views programming as a branch of mathematics. Naur won the 2005 ACM A.M. Turing Award for his work on the ALGOL 60 programming language. He is currently engaged in developing a theory of human thinking that he calls the Synapse-State Theory of Mental Life.

Sir Michael Francis Atiyah (Born 1929)

Sir Michael Francis Atiyah (born 1929), an outstanding British mathematician, was born to the Lebanese writer Edward Atiyah (1903–1964) and his Scottish wife, Jean Levens, in Hampstead, London. Michael grew up in Sudan and Egypt. He attended a primary school in Khartoum, Sudan during 1934–1941 and secondary schools in Cairo and Alexandria, Egypt during 1941–1945. Michael returned to England and studied at the Manchester Grammar School during 1945–1947. He then served with the Royal Electrical and Mechanical Engineers during 1947–1949. Atiyah studied at Trinity College, Cambridge, during 1949–1955. Michael received his Ph.D. in 1955 at the age of 26 for his dissertation on *Some Applications of Topological Methods in Algebraic Geometry*. He spent most of his academic career at the University of Oxford, Cambridge University, England, and the Institute for Advanced Study in Princeton, New Jersey, USA. Michael married Lily Brown on July 30, 1955, and they had three sons. He contributed to algebraic geometry, K theory, index theory, gauge theory, the Dedekind eta function from a topological point of view, and the connections between quantum field theory, knots, and Donaldson (Sir Simon Kirwan Donaldson, born 1957) theory. He introduced the concept of a topological quantum field theory. He also took part in the founding of the Isaac Newton Institute for Mathematical Sciences in Cambridge and was its first director during 1990–1996. Michael was president of the Royal Society of London during 1990–1995, Chancellor of the University of Leicester during 1995–2005, and president of the Royal Society of Edinburgh during 2005–2008. He is now retired and is an honorary professor at the University of Edinburgh.

Edsger Wybe Dijkstra (1930–2002)

Edsger Wybe Dijkstra (1930–2002) was born in the Netherlands. He began computer programming in the early 1950s while studying theoretical physics at the University of Leiden. In 1952, realizing that he was more interested in programming than physics, he quickly completed the requirements for his physics degree and began his career as a programmer, even though programming was not a recognized profession. In fact, in 1957, the authorities in Amsterdam refused to accept "programming" as his profession on his marriage license; however, they did accept "theoretical physicist." Dijkstra was a forceful proponent of programming as a scientific discipline. He made fundamental contributions to the areas of operating systems, including deadlock avoidance; programming languages, including structured programming; and algorithms. In 1972, Dijkstra received the Turing Award from the Association for Computing Machinery, one of the most prestigious awards in computer science. Dijkstra became a Burroughs Research Fellow in 1973, and in 1984, he was appointed to a Chair in Computer Science at the University of Texas, Austin.

Jacob Theodore Schwartz (1930–2009)

Jacob Theodore Schwartz (1930–2009) was born in the Bronx, New York. He received his bachelor's degree in 1948 from the City College of New York. He then received his master's and doctorate degrees at Yale University in 1949 and

1951, respectively. Jacob was known to his friends as Jack, and he occasionally wrote papers as Jack Schwartz. He designed the SETL programming language and the NYU Ultra computer. He founded the Department of Computer Science at the Courant Institute of Mathematical Sciences at New York University, and he served as its chair during 1969–1977. He also served as Chairman of the Computer Science Board of the National Research Council and was the Chairman of the National Science Foundation Advisory Committee for Information, Robotics and Intelligent Systems. During 1986–1989, he was the Director of the Information Science and Technology Office of the Defense Advanced Research Projects Agency (DARPA/ISTO) in Arlington, Virginia. He was elected to the National Academy of Science in 1976 and to the National Academy of Engineering in 2000. His core research interest was applied computational science. He worked on quantum field theory, the theory of linear operators, von Neumann algebras, time-sharing and parallel computing, programming language design and implementation, proof and program verification systems, robotics, set-theoretic approaches in computational logic, multimedia authoring tools, experimental studies of visual perception, and multimedia and other high-level software techniques for analysis and visualization of bioinformatic data. In 1999, Jacob took interest in molecular biology and began a multi-year collaboration with Michael Wigler (born 1947), a professor at Cold Spring Harbor Laboratory, New York. He authored at least 18 books and more than 100 research articles and technical reports. Jacob received the Steele Prize of the American Mathematical Society, the Wilbur Lucius Cross (1862–1948) Medal from Yale University, the Townsend Harris Medal of the City University of New York, and the Mayor's Medal for Contributions to Science and Technology from New York City. Jacob died in his sleep of liver cancer at his home in Manhattan, New York, at the age of 79.

Stephen Smale (Born 1930)

Stephen Smale (born 1930) is from Flint, Michigan, USA. He joined the University of Michigan in 1948. Smale was a marginal undergraduate student, getting mostly Bs, Cs, and even an F in nuclear physics in his sophomore and junior years. In spite of this performance, he was accepted as a graduate student of mathematics. Again, Smale performed poorly in his first years, earning a C average. At this point he received a letter from the department chair, Theophil Henry Hildebrandt (1888–1980), threatening to throw him out of the Michigan Ph.D. program if his grades did not improve. He finally received his Ph.D. in 1957 under the direction of Raoul Bott (1923–2005). In 1958, Smale astounded the mathematical world with a proof of a sphere eversion, that is, turning a sphere inside out without creating any crease. He made his reputation by proving the Poincaré conjecture for all dimensions greater than or equal to 5; he later generalized the Poincaré conjecture in a 107-page paper that established the h-cobordism theorem. He won the Fields Medal in 1966 for his work in low-dimensional topology. Smale then studied dynamic systems, discovering the Smale horseshoe, an important example of stability in a chaotic system. He also outlined a research program that was carried out by many others. His later work, applying Morse theory to mathematical economics, earned a Nobel Prize for his collaborator, Gerard Debreu (1921–2004). He then did deep and important work in theoretical computer science. In 1996 he was awarded the National Medal of Science. In 1998, he compiled a list of 18 mathematical problems to be solved in the twenty-first century, known as Smale's problems, in the spirit of the famous list of problems that Hilbert produced in 1900. It includes some of the original Hilbert problems, such as the Riemann hypothesis and the second half of Hilbert's sixteenth problem. Smale has always been politically active. He was involved in the Free Speech Movement with Mario Savio (1942–1996) and participated in the protest against the Vietnam War. He even gave a speech on the steps of the Kremlin in 1966 in protest of US policies in Vietnam. At one time he was subpoenaed by the House Un-American Activities Committee.

Smale retired from his long-held position at the University of California at Berkeley in 1995, and he took up a post as professor at the City University of Hong Kong. He is currently a professor at the Toyota Technological Institute at Chicago, a research institute closely affiliated with the University of Chicago. In 2007, Smale was awarded the Wolf Prize in mathematics. Smale is arguably one of the most remarkable living mathematicians. He has amassed over the years one of the finest private mineral collections in existence.

Chen Jing-Run (1933–1996)

Chen Jing–Run (1933–1996) was a native of Fuzhou. He graduated from Xiamen University in 1953 and later joined the Institute of Mathematics of the Chinese Academy of Sciences. While studying Goldbach's Conjecture, he proved that every sufficiently large even number has the form $p + q$, where p is prime and q is either prime or the product of two primes. During the so-called 'cultural revolution' of the 1960s, this kind of mathematics was frowned upon in China for being far removed from any conceivable application to industry or agriculture. Because he stubbornly stuck to his esoteric research at the risk of neglecting his teaching, Chen was discriminated against during the reign of the so-called 'gang of four' and may have lost his academic position. After the overthrow of the gang of four, he was rehabilitated and even declared a 'hero of the revolution.' He published almost 70 research papers and was invited to deliver 45-min lectures in 1978 and 1982 to the International Congress of Mathematicians.

Paul Joseph Cohen (1934–2007)

Paul Joseph Cohen (1934–2007), a brilliant Jewish Polish-American mathematician, was born to Abraham and Minnie Cohen in Long Branch, New Jersey, USA. He studied at Stuyvesant High School, New York City, and graduated there in 1950. During 1950–1953, he attended Brooklyn College in New York City, but he left the college without completing his bachelor's degree. He then attended the University of Chicago and obtained his master's and doctorate degrees in 1954 and 1958, respectively. His doctoral dissertation was entitled *Topics in the Theory of Uniqueness of Trigonometric Series*. He proved the independence of the continuum hypothesis in 1963. He is also widely known for the axiom of choice from Zermelo–Frankel set theory. Paul was appointed as an assistant professor at Stanford University in 1961, was promoted to the post of associate professor in 1962, and became full professor there in 1964. He was awarded the Bôcher Prize (1964), the Fields Medal (1966), and the National Medal of Science (1967). Paul was deeply influenced by Gödel and explained how he came to the idea of forcing (a mathematical technique) while reading Gödel's *The Consistency of the Continuum Hypothesis*, a book that consists of notes of a course given at the Institute for Advanced Study at Princeton in 1938–1939. Paul became Marjorie Mhoon Fair Professor in Quantitative Science at Stanford in 1972. Although he officially retired in 2004, he continued teaching at Stanford until shortly before his death due to a rare lung disease at Stanford Hospital in Palo Alto.

Edayathumangalam Venkatarama Krishnamurthy (1934–2012)

Edayathumangalam Venkatarama Krishnamurthy (1934–2012) was a physicist and computer scientist who has done multifarious, innovative work in computer and computational science. He received his bachelor's and master's degrees, both in physics, at Madras University. He then completed a doctorate around 1960 under Gopalasamudram Narayana Iyer Ramachandran (1922–2001), an FRS. His doctorate work involved the design of the computer Leelavati. He (along with his coresearcher N.P. Murarka) developed the logical design of India's first second generation computer, ISIJU, in 1962. ISIJU stands for the Indian Statistical Institute (ISI) and Jadavpur University (JU). Both institutions are situated in Kolkata, ISI in the north and JU in the south. The computer ISIJU operated for about 10 years (1965 to 1975). Krishnamurthy spent about 8 years at ISI during the 1960s, and then he spent about 16 years (1971–mid-1980s) at the Indian Institute of Science (IISc), Bangalore. During this period, he also spent a significant amount of time at several universities, including the University of Illinois at Urbana Champaign, the University of Maryland in College Park, Maryland (USA), the Technion-Israel Institute of Technology, and the University of Lagos, Nigeria. After leaving IISc, he spent some time at the University of Waikato in New Zealand, and then the Australian National University, in Canberra, as a professor.

Of his numerous contributions to computational and computer science, his research in quantum computing is most significant. Quantum computing, a revolutionary computing technique, is based on theoretical quantum bits, called qubits, that unlike traditional bits can take quantum values. Many books and research articles have been published about quantum computing, both by physicists and by computer/computational scientists, but it has not become a reality even after 32 years of research. There is no realization of a quantum computer so far for any meaningful laboratory or commercial use. Krishnamurthy clearly pointed out the negative wealth generated over the years in quantum computing research.

Charles Antony Richard Hoare (Born 1934)

Charles Antony Richard Hoare (born 1934) was born in Colombo (Sri Lanka) to British parents. He received his bachelor's degree in the classics at the University of Oxford in 1956. He then studied computer translation of human languages at Moscow State University in the school of Kolmogorov. In 1960, he left the Soviet Union and joined Elliott Brothers, Ltd, where he implemented ALGOL 60 and began developing major algorithms. He became a Professor of Computing Science at the Queen's University of Belfast in 1968, and in 1977, he was made Professor of Computing at Oxford University Computing Laboratory. He is currently an Emeritus Professor there, and he is also a principal researcher at Microsoft Research in Cambridge, England. Hoare has made many important contributions to the theory of programming languages and to programming methodology. He was the first person to define a programming language based on how its programs could be proved to be correct. He is also the creator of quicksort (in 1960, at age 26), one of the most commonly used and studied sorting algorithms. Hoare is a noted writer in the technical and social aspects of computer science. He is the recipient of several awards, including the ACM Turing Award (1980), the Harry H. Goode (1909–1960) Memorial Award (1981), FRS (1982), and an Honorary Doctorate of Science from Queen's University Belfast (1987). Hoare was knighted for his services to education and computer science in 2000, received the Kyoto Prize for Information Science (2000), and was made a Fellow of the Royal Academy of Engineering (2005).

Donald Ervin Knuth (Born 1938)

Donald Ervin Knuth (born 1938) grew up in Milwaukee, where his father taught book keeping at a Lutheran high school and owned a small printing business. He was an excellent student and earned several academic achievement awards. He applied his intelligence in unconventional ways; Knuth won a contest when he was in eighth grade by finding over 4,500 words that can be formed from the letters in the phrase "Ziegler's Giant Bar". This won a television set for his school and a candy bar for everyone in his class. Knuth had a difficult time choosing physics over music as his major at the Case Institute of Technology. He then switched from physics to mathematics, and in 1960, he received his bachelor of science degree, simultaneously receiving a master of science degree by a special award of the faculty, who considered his work outstanding. At Case, he managed the basketball team and applied his talents by considering his mule for the value of each player. This novel approach was covered by *Newsweek* and by Walter Cronkite on the CBS television network. Knuth began graduate work at the California Institute of Technology in 1960 and received his Ph.D. there in 1963. During this time he worked as a consultant, writing compilers for different computers. Knuth joined the staff of the California Institute of Technology in 1963, where he remained until 1968, when he took a job as a full professor at Stanford University. He retired as Professor Emeritus in 1992 to concentrate on writing. He is especially interested in updating and completing new volumes of his series *The Art of Computer Programming*, a work that has had a profound influence on the development of computer science, which he began writing as a graduate student in 1962, focusing on compilers. In common jargon, "Knuth," referring to *The Art of Computer Programming*, has come to mean the reference that answers all questions about such topics as data structures and algorithms. Knuth is the founder of the modern study of computational complexity, and he has made fundamental contributions to the subject of compilers. His dissatisfaction with mathematics typography inspired him to invent the now widely used TeX and Metafont systems. TeX has become a standard language for computer typography. Two of the many awards Knuth has received are the 1974 Turing Award and the 1979 National Medal of Technology, awarded to him by President Jimmy Carter (born 1924). Knuth has written for a wide range of professional journals in computer science and mathematics. However, his first publication, when he was a college freshman in 1957, was a parody of the metric system called "The Potrzebie Systems of Weights and Measure," which appeared in *MAD Magazine* and has been in reprint several times. He is a church organist, as his father was. He is also a composer of music for the organ. Knuth believes that writing computer programs can be an aesthetic experience, much like writing poetry or composing music. He pays $2.56 for the first person to find each error in his books and $0.32 for significant suggestions.

Neil James Alexander Sloane (Born 1939)

Neil James Alexander Sloane (born 1939) studied mathematics and electrical engineering at the University of Melbourne on a scholarship from the Australian state telephone company. He mastered many telephone-related jobs, such as erecting telephone poles, in his summer work. After graduating, he designed minimal cost telephone networks in Australia. In 1962, he came to the USA and studied electrical engineering at Cornell University. His Ph.D. thesis, in 1967 under Frederick Jelinek (1932–2010) and Wolfgang Heinrich Johannes Fuchs (1915–1997), concerned what are now called neural networks. He took a job at Bell Labs in 1969, working in many areas, including network design, coding theory, and sphere packing. After AT&T split up in 1996, he moved from Bell Labs to AT&T Labs, where he now works. One of his favorite problems is the *kissing problem* (a name he coined), which asks how many spheres can be arranged in n dimensions so that they all touch a central sphere of the same size. In two dimensions the answer is 6, since 6 pennies can be placed so that they touch a central penny. In three dimensions, 12 billiard balls can be placed so that they touch a central billiard ball. Two billiard balls that just touch are said to "kiss" giving rise to the terminology "kissing problem" and "kissing number." Sloane, together with Andrew Odlyzko (born 1949), showed that in 8 and 24 dimensions the optimal kissing numbers are, respectively, 240 and 196,560. The kissing number is known in dimensions $1, 2, 3, 8$, and 24, but not in any other dimensions. Sloane's books include *Sphere Packings, Lattices and Groups*, 3rd ed. with John Conway (born 1937); *The Theory of Error-Correcting Codes* with Jessie MacWilliams (1917–1990); *The Encyclopedia of Integer Sequences* with Simon Plouffe (born 1956); and *The Rock-Climbing Guide to New Jersey Crags* with Paul Nick. The last book demonstrates his interest in rock climbing. Sloane is an IEEE Fellow and a member of US National Academy of Engineering. He is former editor of the IEEE Transactions on Information Theory and a recipient of the IEEE Centennial Medal, the IEEE Information Theory Society Prize Paper Award, and the Chauvenet Prize of the Mathematical Association of America.

Nicolas Bourbaki (Born 1939)

Nicolas Bourbaki (born 1939) was born in France; however, his name is Greek. He is ranked as one of the most influential mathematicians. His works, a comprehensive set of volumes in advanced mathematics that reflect the trends of twentieth-century mathematics, are much read and quoted. He had enthusiastic supporters and seething critics, and, most curious of all, he did not exist. Nicolas Bourbaki, who has an office at the École Normale Supérieure in Paris, is a collective pseudonym employed by an informal group of mathematicians. The founding members included Henri Cartan, Chevalley, Jean Coulomb (1904–1999), Jean Deslartes (1903–1968), Jean Dieudonné (1905–1992, the chief scribe of Bourbaki's works), Charles Ehresmann (1905–1979), René de Possel (1905–1974), Szolem Mandelbrojt (1899–1983), and André Weil. The group's membership has varied over the years, sometimes involving as many as 20 mathematicians, but the members are required to resign by age 50. The Bourbaki perspective sees modern mathematics as a ball of many tangled strands of yarn, where those strands in the center of the ball react to one another in a nearly unpredictable manner. In this tangled ball, the ends of the strands that issue outward in various directions have no intimate connection with anything within. The Bourbaki method is to snip off these free strands and to concentrate only on the tight core of the ball from which all the rest unravels. Thus much of mathematics is purposefully avoided by the Bourbaki group.

Enrico Bombieri (Born 1940)

Enrico Bombieri (born 1940) was born in Milan, Italy. He received his Ph.D. from the University of Milan in 1963. He held positions at the University of Milan, the University of Cagliari, the University of Pisa, and the Scuola Normale Superiore in Pisa. Currently, Bombieri is IBM Von Neumann Professor at the Institute for Advanced Study, Princeton. He has made pioneering contributions to various branches of mathematics, such as number theory, algebraic geometry, complex analysis, and minimal surfaces. He is known for Bombieri's mean value theorem and the Bombieri–Vinogradov theorem, which deals with the large sieve method and its applications to the distribution of prime numbers in arithmetic progressions. In 1974, he was awarded the Fields Medal for his major contributions to the study of prime numbers, univalent functions and the local Bieberback conjecture, the theory of functions of several complex variables, and the theory of partial differential equations and minimal surfaces, in particular, the solution of Bernstein's problem in higher dimensions. He is also a recipient of several awards, including the Feltrinelli Prize in 1976, the Balzan Prize in 1980, the Cavaliere di Gran Croce al Merito della Repubblica, Italy in 2002, AMS Doob Prize for his book *Heights in Diophantine Geometry* in 2008, and the King Faisal Prize (established in 1976) in 2010. He is a member of the Accademia Nazionale, Rome; the Accademia Nazionale dei Lincei, Italy; and the Academia Europaea, and he is a foreign member of the Royal Swedish Academy and the French Académie des Sciences. He is also a member of the US National Academy of Sciences and a fellow of the American Academy of Arts and Sciences.

Sathamangalam Ranga Iyengar Srinivasa Varadhan (Born 1940)

Sathamangalam Ranga Iyengar Srinivasa Varadhan (born 1940), an FRS, and an Indian American mathematician known for his contributions to probability theory, specifically his unified theory of large deviations, was born in Madras. He obtained his undergraduate degree in 1959 from Presidency College (Madras) and his doctorate in 1963 from the Indian Statistical Institute (Calcutta), under the supervision of C.R. Rao. Since 1963, he has worked at the Courant Institute of Mathematical Sciences at New York University, where he began as a postdoctoral fellow (during 1963–1966), strongly recommended by Monroe David Donsker

(1925–1991). Srinivasa is currently a professor at the Courant Institute. Here he worked on diffusion processes with Daniel W. Stroock (born 1940) and on large deviations with Donsker. Srinivasa married Vasundra, an academic in media studies at the Gallatin School of Individualized Study. They had two sons; one died during the September 11, 2001, terrorist attack, and the other is a trader in New York City. Srinivasa has received many awards and honors, including the Birkhoff Prize (1994), the Margaret and Herman Sokol Award of the Faculty of Arts and Sciences, New York University (1995), the Leroy P. Steele Prize for Seminal Contribution to Research (1996) from the American Mathematical Society, awarded for his work with Stroock on diffusion processes, the Abel Prize (2007) for his work on large deviations with Donsker, Padma Bhushan (2008), the third highest civilian honor bestowed by the Government of India, and the National Medal of Science (2010), the highest honor bestowed by the United States government on scientists, engineers and inventors, from President Barack Obama (born 1961). Srinivasa is a member of the US National Academy of Sciences (1995) and the Norwegian Academy of Science and Letters (2009). He was elected a Fellow of the American Academy of Arts and Sciences (1988), the Institute of Mathematical Statistics (1991), the Indian Academy of Sciences (2004), and the Society for Industrial and Applied Mathematics (2009).

Mikhail Leonidovich Gromov (Born 1943)

Mikhail Leonidovich Gromov (born 1943) was born in Boksitogorsk, a town about 125 miles east of St. Petersburg (Leningrad), and has been a French citizen since 1992. Mikhail studied at Leningrad University, where he received his master's degree in mathematics in 1965. While continuing his research, he married Margarita Gromov in 1967, was appointed an Assistant Professor at Leningrad University in the same year, and obtained a doctorate in 1969 from Leningrad University. He continued to serve the university until 1974, when he left Russia and went to the USA, where he was appointed Professor of Mathematics at the State University of New York at Stony Brook. In 1979, he gave lectures on metric structures for Riemannian and non-Riemannian spaces, which have been remarkably influential. He is best known for his contributions to geometric group theory in which he characterized groups of polynomial growth and created, along with James Cannon (born 1943), the notion of a hyperbolic group. His contributions to symplectic topology (he introduced pseudo-holomorphic curves) and Riemannian geometry are also notable. Gromov carried out extensive research in analysis and algebra, and he would often formulate a problem in "geometric" terms. For instance, his homotopy principle (h-principle) on differential relations is the basis for a geometric theory of partial differential equations. Mikhail is also interested in mathematical biology. Gromov's theorem on groups of polynomial growth, Gromov's compactness theorem, Gromov's theorem on almost flat manifolds, Gromov's inequality for complex projective space, Gromov's systolic inequality for essential manifolds, the

Lévy–Gromov inequality, and the Gromov norm are some of the contributions associated with his name. His work is relevant to problems in applied mathematics in a way that reflects his extraordinary creativity and analytic power. He is the recipient of many awards, such as the Prize of the Mathematical Society of Moscow (1971), the Oswald Veblen Prize in Geometry (AMS) (1981), the Wolf Prize in Mathematics (1993), the Janos Bolyai International Mathematical Prize (2005), and the Abel Prize (2009) for his revolutionary contributions to geometry.

Mitchell Jay Feigenbaum (Born 1944)

Mitchell Jay Feigenbaum (born 1944) was born in Philadelphia. In 1964, he received a bachelor's degree in electrical engineering from the City College of New York. In 1970, Feigenbaum earned his doctorate in elementary particle physics at the Massachusetts Institute of Technology, and then he took positions at Cornell, Virginia Polytechnic Institute, and later at the National Laboratory in Los Alamos, New Mexico. Feigenbaum returned to Cornell University in 1982, and in 1986, he joined Rockefeller University as professor of mathematics and physics. He has also spent some time at the Institute for Advanced Study in Princeton and the French IHES. He gave a great impetus to the mathematical study of fractals, and his pioneering studies in chaos theory (apparently disordered behavior of nearby points) led to the discovery of the Feigenbaum constants. He is the recipient of a MacArthur foundation award in 1984 and a Wolf Foundation Prize in physics in 1986.

Leonard Max Adleman (Born 1945)

Leonard Max Adleman (born 1945) was born in San Francisco, California. He received a B.S. in mathematics in 1968, and his Ph.D. in computer science in 1976 from the University of California, Berkeley. Adleman was a member of the mathematics faculty at M.I.T. from 1976 until 1980, where in 1977 he was a coinventor of the RSA (the last initials of three partners, Rivest, Shamir, and Adleman) cryptosystem. This system is now in widespread use in security applications, including digital signatures and our everyday e-mails. They were awarded the ACM Turing Award for this work in 2002. In 1980, he took a position in the computer science department at the University of Southern California (USC). He was appointed to a chaired position at USC in 1985. Adleman has worked on computer security, computational complexity, immunology, and molecular biology. He invented the term "computer virus." In 1994, his paper *Molecular Computation of Solutions to Combinatorial Problems* described the experimental use of DNA as a computational system. In this paper he solved a seven-node instance of the Hamiltonian graph problem, an NP-complete problem similar to the *traveling salesman problem* (a salesman wants to pass through every city, but each city only once, and return to his city of origin). This paper was the first instance of the successful use of DNA in computation. In 2002, he and his research group solved a 'nontrivial' problem using DNA computation. Specifically, they solved a 20-variable 3-SAT problem having more than one million potential solutions. This technique has now sparked great interest. Adleman was a technical advisor for the movie *Sneakers*, in which computer security played an important role.

Ronald Linn Rivest (Born 1947)

Ronald Linn Rivest (born 1947) received his bachelor's degree from Yale in 1969 and his Ph.D. in computer science from Stanford in 1974. Rivest is a computer science professor at MIT, and was a confounder of RSA Data Security, which held the patent on the RSA cryptosystem that he invented along with Adi Shamir and Leonard Adleman. Areas that Rivest has worked in besides crptography include machine learning, VLSI design, and computer algorithms. He is a coauthor of a popular text on algorithms [CoLeRiSt01]. He is a member of several academies and organizations. Together with Adi Shamir and Len Adleman, Rivest has been awarded the 2000 IEEE Koji Kobayashi Computers and Communications Award (established in 1986) and the Secure Computing Lifetime Achievement Award. He also shared with them the Turing Award.

Shing-Tung Yau (Born 1949)

Shing–Tung Yau (born 1949) was born in Shantou, Guangdong Province, China, to a family of eight children. When he was only a few months old, his family emigrated to Hong Kong. He earned his undergraduate degree in 1969 at the Chinese University of Hong Kong, and his Ph.D. in 1971 at the University of California, Berkeley, under the direction of Shiing-Shen Chern, the legendary geometer. After spending a few years at the Institute for Advanced Study, Princeton, the State University of New York at Stony Brook, and Stanford University, Yau joined Harvard University in 1987. Yau's first major contribution to differential geometry was his proof of the Calabi (Eugenio Calabi, born 1923) conjecture, which concerns how volume and distance can be measured not in four, but in five or more dimensions. In 1979, Yau and Richard Schoen (born 1950) proved Einstein's positive mass conjecture. Their proof was based on their work with minimal surfaces, which had an impact on topology, algebraic geometry, general relativity, and provided a tool for understanding how black holes form. He also settled the Smith (Paul Althaus Smith, 1900–1980) conjecture (with William H. Meeks), the Frankel conjecture (with Yum-Tong Siu, born 1943), and the Mirror conjecture (with Bong Lian and Kefeng Liu, born 1965). To help develop Chinese mathematics education, Yau started by educating students from China, then established mathematics research institutes and centers, organized conferences at all levels, initiated outreach programs, and raised private funds for these purposes. Yau has been the recipient of several awards and prizes, including the Fields Medal in 1982, the Crafoord Prize, and the Wolf Prize in Mathematics in 2010. He is an honorary professor of over ten universities, has been awarded doctorates by numerous universities, and is a member/fellow of several organizations.

Adi Shamir (Born 1952)

Adi Shamir (born 1952) was born in Tel Aviv, Israel. He received an undergraduate degree in mathematics at Tel Aviv University in 1972 and then his master's and doctorate degrees in computer science at the Weizmann Institute of Science in 1975 and 1977, respectively. His thesis was titled, "Fixed Points of Recursive Programs." After a year of post doctorate at Warwick University, Shamir did research at MIT from 1977–1980 before returning to be a member of the faculty of mathematics and computer science at the Weizmann Institute. He is currently a Professor in the Applied Mathematics Department at the Weizmann Institute, and he leads a group studying computer security. Since 2006, he has been an invited professor at the École Normale Supérieure in Paris. He is one of the inventors of the RSA algorithm (along with Ron Rivest and Len Adleman), one of the inventors of the Feige–Fiat–Shamir Identification Scheme (along with Uriel Feige and Amos Fiat, born 1957), and one of the inventors of differential cryptanalysis (along with Eli Biham). The RSA algorithm enables secure communication between computers using public-key cryptography. Today, the RSA system is used in email systems, Web browsers, secure shells, virtual private networks, mobile phones, and in many other applications requiring the secure exchange of information. The Feige–Fiat–Shamir Identification Scheme is a type of parallel zero-knowledge proof that allows one party to prove to a second party that the first party possesses some secret information, without revealing to the second party what that secret information is. Differential cryptanalysis is a general method for attacking block ciphers. It recently become known that differential cryptanalysis was already known and was kept a secret, by both IBM and the NSA. Shamir also made numerous contributions to the fields of cryptography and computer science, such as the Shamir secret sharing scheme, the breaking of the Merkle–Hellman cryptosystem, visual cryptography, the TWIRL and TWINKLE factoring devices, and the proof of the equivalence of the complexity classes PSPACE and IP. In recognition of his contributions to cryptography, Shamir (along with Rivest and Adleman) was awarded the 2002 ACM Turing Award. He has also received ACM's Kanellakis Theory and Practice Award (established in 1996 in memory of Paris Christos Kanellakis, 1953–1995), the Erdös Prize of the Israel Mathematical Society, the IEEE's W.R.G. Baker Prize (created in 1956 from a donation of Walter R.G. Baker, 1892–1960), the UAP Scientific Prize, the Vatican's PIUS XI Gold Medal, and the IEEE Koji Kobayashi Computers and Communications Award.

Narendra Karmarkar (Born 1957)

Narendra Karmarkar (born 1957) was born in Gwalior, India, in 1957. He obtained his bachelor's degree in electrical engineering at the Indian Institute of Technology, Bombay, in 1978. He completed his master's degree at the California Institute of Technology and his Ph.D. in computer science at the University of California, Berkeley. Karmarkar is known for developing the first polynomial-time $O(n^{3.5})$ projective transformation algorithm for the solution of a linear program (LP): *Maximize* $c^t x$ *subject to* $Ax = b$, $x \geq 0$ (*null column vector*). He published his famous algorithm, an interior-point method, in 1984 while working for Bell Laboratories in New Jersey. Karmarkar received several prizes and awards for this algorithm. They include (i) the Lanchester Prize of the Operations Research Society of America for Best Published Contributions to Operations Research in 1984, (ii) the Marconi International Young Scientist Award (established in 1974 by Gioia Marconi Braga to commemorate the centennial of the birth of her father Guglielmo Marconi, 1874–1937) in 1985, (iii) the Texas Instruments Founders' Prize in 1986, (iv) Fellowship of Bell Laboratories in 1987, (v) the Fulkerson Prize in Discrete mathematics given jointly by Mathematical Programming Society and American Mathematical Society in 1988, (vi) the Ramanujan Prize for Computing given by Asian Institute Informatics in 1989, (vii) the Distinguished Alumnus Award given by the department of Computer Science and Engineering at the University of California, Berkeley, in 1993, and (viii) the Kanellakis Award of the ACM (Association for Computing Machinery) in 2000. After developing the projective transformation algorithm, Karmarkar worked on a new architecture for supercomputing, based on concepts from projective geometry over finite fields. Currently, he is applying these concepts to "sculpturing free space," which may be viewed as a nonlinear analogue of folding the perfect corner. This model was presented at IVNC, Poland, on July 16, 2008, and at the MIT center for bits and atoms on July 25, 2008.

Manindra Agrawal (Born 1966)

Manindra Agrawal (born 1966) was born in Allahabad, India. He obtained his bachelor's and doctorate degrees in computer science from the Indian Institute of Technology, Kanpur (IIT-K) in 1986 and 1991, respectively. He currently holds the N. Rama Rao Chair Professorship (established in 1998) and the Headship of the Department of Computer Science and Engineering of IIT-K. His main research interests are complexity theory and computational number theory. His peripheral interests include cryptography, complex analysis, and combinatorics. He is best known for his polynomial-time primality test algorithm, known as AKS (Agrawal–Kayal–Saxena), which he (with his students Kayal and Saxena) developed in 2004.

The problem of testing the primality of a number in polynomial time has been a very long-standing issue. The time required to run his algorithm is at most the number of digits to the power of twelve times a polynomial evaluated at the log of the number of digits. Prime numbers are used in error-free computations using multiple-modulus residue arithmetic as well as p-adic arithmetic. Besides, these (usually large prime numbers) are also employed in cryptography and computer security. Consequently, his algorithm has significant practical importance in the current security scenario. Agrawal has been the recipient of several awards and prizes, which include the Young Engineer Award in 1998, the Clay Research Award in 2002, the Shanti Swarup Bhatnagar Award in 2003, the Dr. Meghnad Saha Award in 2003, the ICTP Prize in 2003, the Fulkerson Prize (established by friends of the late Delbert Ray Fulkerson 1924–1976) in 2006, the Gödel Prize in 2006, the Infosys Prize for mathematics in 2008, the G.D. Birla Award in 2009, the P.C. Mahalanobis Birth Centenary Award in 2009, and the Padma Shri (given by Government of India) Award in 2013.

Terence Chi-Shen Tao (Born 1975)

Terence Chi–Shen Tao (born 1975) was born in Adelaide, Australia. Tao was a child prodigy, at the age of two he attempted to teach a 5-year-old child mathematics and English; he took university-level mathematics courses at the age of nine, after scoring a 760 on the mathematics section of the SAT at the age of eight; he competed in the International Mathematical Olympiad in 1986, 1987, and 1988, winning bronze, silver, and gold, respectively. At 13, he was the youngest gold medalist in the history of the competition, and he holds this record to this day. Tao received his bachelor's and master's degrees at the age of 17 from Flinders University. In 1992, he won a Fulbright Scholarship for postgraduate study in the USA. He completed his Ph.D. at the age of 21 at Princeton University under Elias Menachem Stein (born 1931). He was appointed full professor at UCLA at 24, the youngest person to hold that rank at the university. Tao's areas of research include harmonic analysis, partial differential equations, combinatorics, number theory, and signal processing. During 1999–2001 and 1999–2006, he held Sloan Foundation and Packard fellowships. He is the recipient of the Salem Prize (Raphaël Salem, 1898–1963) in 2000 for his work in L^p harmonic analysis; the Bôcher Prize in 2002 for his breakthrough on the problem of critical regularity in Sobolev (Sergei Sobolev) spaces of the wave maps equations; the Clay Research Award in 2003; the American Mathematical Society's Levi Leonard Conant (1857–1916) Prize in 2005; the Fields Medal in 2006 for his contributions to partial differential equations, combinatorics, harmonic analysis, and additive number theory; the SASTRA Ramanujan Prize in the same year; the MacArthur Fellowship and the Ostrowski Prize in 2007; he was elected an FRS in 2007; the Alan Tower Waterman (1892–1967) Award in 2008; and the King Faisal Prize in 2010. He is a corresponding member of the Australian Academy of Sciences, and a Foreign Member of the National Academy of Sciences and the American Academy of Arts and Sciences. His most famous result is the Green–Tao theorem (with Ben Joseph Green, born 1977), which states that there exist arbitrarily long arithmetic progressions of prime numbers. Tao currently lives with his wife and son in Los Angeles, California.

Bibliography

A. Aaboe, *Episodes from the Early History of Mathematics* (Random House, New York, 1964)
A.D. Aczel, *Fermat's Last Theorem: Unlocking the Secret of an Ancient Mathematical Problem* (Four Walls Eight Windows, New York, 1996)
D. Adamson, *Blaise Pascal: Mathematician, Physicist, and Thinker About God* (St. Martin's Press, New York, 1995)
R.P. Agarwal, H. Agarwal, S.K. Sen, Birth, Growth and Computation of Pi to ten trillion digits. Adv. Differ. Equat. **2013**, 100 (2013)
A.A. Al-Daffa', *The Muslim Contribution to Mathematics* (Humanities Press, Atlantic Highlands, 1977)
A.A. Al-Daffa', J.J. Stroyls, *Studies in the Exact Sciences in Medieval Islam* (Wiley, New York, 1984)
E.J. Aiton, *Leibniz: A Biography* (A. Hilger, Bristol, Boston, 1984)
R.E. Allen, *Greek Philosophy: Thales to Aristotle* (The Free Press, New York, 1966)
G.J. Allman, *Greek Geometry from Thales to Euclid* (Arno Press, New York, 1976)
E.N. da C. Andrade, *Sir Issac Newton, His Life and Work* (Doubleday & Co., New York, 1954)
W.S. Anglin, *Mathematics: A Concise History and Philosophy* (Springer, New York, 1994)
W.S. Anglin, *The Queen of Mathematics* (Kluwer, Dordrecht, 1995)
H.D. Anthony, *Sir Isaac Newton* (Abelard-Schuman, New York, 1960)
H.G. Apostle, *Aristotle's Philosophy of Mathematics* (The University of Chicago Press, Chicago, 1952)
R.C. Archibald, Outline of the history of mathematics.Am. Math. Monthly **56** (1949)
B. Artmann, *Euclid: The Creation of Mathematics* (Springer, New York, 1999)
C.N. Srinivasa Ayyangar, *The History of Ancient Indian Mathematics* (World Press Private Ltd., Calcutta, 1967)
A.K. Bag, *Mathematics in Ancient and Medieval India* (Chaukhambha Orientalia, Varanasi, 1979)
W.W.R. Ball, *A Short Account of the History of Mathematics* (Dover, New York, 1960)
S.F. Barker, *Philosophy of Mathematics* (Prentice-Hall, Englewood Cliffs, 1964)
M.E. Baron, *The Origin of Infinitesimal Calculus* (Dover, New York, 1993)
J.D. Barrow, *Pi in the Sky: Counting, Thinking, and Being* (The Clarendon Press, New York, 1992)
C. Baumgardt, *Johannes Kepler: Life and Letters* (Victor Gollancz, London, 1952)
P. Beckman, *A History of* π (St. Martin's Griffin, New York, 1976)
A. Beiler, *Recreations in the Theory of Numbers* (Dover, New York, 1964)
B. Belhoste, *Augustin–Louis Cauchy: A Biography*, Translated by F. Ragland (Springer, New York, 1990)
E.T. Bell, *Men of Mathematics* (Touchstone Books, New York, 1986)

R.P. Agarwal, S.K. Sen, *Creators of Mathematical and Computational Sciences*,
DOI 10.1007/978-3-319-10870-4

E.T. Bell, *The Development of Mathematics*, 2nd edn. (Dover, New York, 1992)

D.C. Benson, *The Moment of Proof: Mathematical Epiphanies* (Oxford University Press, New York, 1999)

J.L. Berggren, *Episodes in the Mathematics of Medieval Islam* (Springer, New York, 1986)

D. Berlinski, *A Tour of Calculus* (Vintage, New York, 1995)

B.E. Berndt, *Notebooks of S. Ramanujan* (Springer, New York, 1985)

B.C. Berndt, R.A. Rankin, *Ramanujan Letters and Commentary* (Affiliated East West Press, New Delhi, 1997)

E.W. Beth, *The Foundations of Mathematics* (North-Holland, Amsterdam, 1959)

E.W. Beth, *Mathematical Thought: An Introduction to the Philosophy of Mathematics* (D. Reidel, Dordrecht, 1965)

T.G. Bhaskar, D. Kovach, V. Lakshmikantham, The hybrid set theory. Nonlinear Anal. Hybrid Syst. **1**, 414–416 (2007)

M.G. Bishop, *Pascal, the Life of Genius* (Reynal & Hitchcock, New York, 1936)

M. Black, *The Nature of Mathematics: A Critical Survey* (Routledge & Kegan Paul, London, 1965)

D. Blatner, *The Joy of Pi* (Walker Publishing Co., New York, 1997)

D.M. Bose, S.N. Sen, B.V. Subbarayappa, *A Concise History of Science in India* (Indian National Science Academy, New Delhi, 1971)

C. Boyer, *The History of the Calculus and Its Conceptual Development* (Dover, New York, 1959)

C. Boyer, *A History of Mathematics*, 2nd edn. (Wiley, New York, 1991)

J.W. Brewer, M. Smith, *Emmy Noether, a Tribute to Her Life and Work* (Marcel Dekker Inc., New York, 1981)

D.E. Brody, A.R. Brody, *The Science Class You Wish You Had: The Seven Greatest Scientific Discoveries in History and the People Who Made Them* (Berkeley Publishing Group, New York, 1997)

R.S. Brumbaugh, *Plato's Mathematical Imagination* (Indiana University Press, Bloomington, 1954)

L.N.H. Bunt, P.S. Jones, J.D. Bedient, *The Historical Roots of Elementary Mathematics* (Prentice-Hall, Englewood Cliffs, 1976)

D.M. Burton, *The History of Mathematics: An Introduction*, 4th edn. (McGraw-Hill, New York, 1999)

F. Cajori, The history of the notations of calculus. Ann. Math. **25**, 1–46 (1923)

F. Cajori, Leibniz, the master builder of notations. Isis **7**, 412–429 (1925)

F. Cajori, *A History of Mathematics*, 4th edn. (Chelsea Publishing, New York, 1985)

R. Calinger, *Gottfried Wilhelm Leibniz* (Rensselaer Polytechnic Institute, Troy, 1976)

R. Calinger (ed.), *Classics of Mathematics* (Prentice-Hall, Englewood Cliffs, 1995)

R. Calinger, *A Contextual History of Mathematics: To Euler* (Prentice Hall, Upper Saddle River, 1999)

G. Cantor, *Contributions to the Founding of the Theory of Transfinite Numbers*, Translated by P.E.B. Jourdain (Dover, New York, 1952)

M. Cantor, *Vorlesungen über Geschichte der Mathematik* (Teubner, Leipzig, 1907)

G. Cardano, *The Book of My Life*, Translated by J. Stoner (Dover, New York, 1962)

G. Cardano, *The Great Art*, Translated by T.R. Witmer (MIT Press, Cambridge, 1968)

S. Chandrasekhar, *Newton's Principia for the Common Reader* (Oxford University Press, Oxford, 1995)

M. Ranga Charya (ed.), Mahavira (author), *Ganita Sara Sangraha* (Madras Government Publications, Madras, 1912)

D. Chattopadhyaya, *History of Science and Technology in Ancient India: The Beginnings* (Firma KLM PVT, Calcutta, 1986)

M. Clagett, *Greek Science in Antiquity* (Collier, New York, 1963)

W.E. Clark (ed.), *The Aryabhatia of Aryabhata* (The University of Chicago Press, Chicago, 1930)

C.C. Clawson, *Mathematical Mysteries: The Beauty and Magic of Numbers* (Plenum Press, New York, 2000)

M.P. Closs (ed.), *Native American Mathematics* (University of Texas Press, Austin, 1986)

I.B. Cohen, *Sir Issac Newton's Papers and Letters on Natural Philosophy* (Harvard University Press, Cambridge, 1958)
J.R. Cole, *Pascal: The Man and His Two Loves* (New York University Press, New York, 1995)
L. Conant, *The Number Concept. Its Origin and Development* (Macmillan, New York, 1923)
R. Cooke, *The Mathematics of Sonya Kovalevskaya* (Springer, New York, 1984)
J.L. Coolidge, *The Mathematics of Great Amateurs* (Dover, New York, 1963)
N. Copernicus, *On the Revolutions of Heavenly Spheres*. Great Mind Series (Prometheus Books, New York, 1995)
R. Courant, H. Robbin, I. Stewart, *What Is Mathematics?* (Oxford University Press, New York, 1996)
J.G. Crowther, *Six Great Scientists* (Barnes & Noble, USA, 1995)
C. Cullen, *Astronomy and Mathematics in Ancient China: The Zhou Bi Suanjing* (Cambridge University Press, Cambridge, 1996)
P. Damerow, *Mathematics in Ancient Iraq: A Social History* (Princeton University Press, Princeton, 2008)
B. Datta, The Bakhshali manuscript. Bull. Calcutta Math. Soc. **21**, 1–60 (1929)
B. Datta, *The Science of the Sulbas: A Study in Early Hindu Geometry* (Calcutta University Press, Calcutta, 1932)
B. Datta, A.N. Singh, *History of Hindu Mathematics*, vol. 2 (Asia Publishing House, Bombay, 1962)
B. Datta, *Ancient Hindu Geometry: The Science of the Sulba* (South Asia Books, Bombay, 1993)
J.W. Dauben, *Georg Cantor: His Mathematics and Philosophy of the Infinite* (Harvard University Press, Cambridge, 1979)
J.W. Dauben (ed.), *The History of Mathematics from Antiquity to the Present. A Selective Bibliography* (Garland, New York, London, 1985)
M. Davis, *The Universal Computer: The Road from Leibniz to Turing* (W.W. Norton & Co., New York, 2000)
P.J. Davis, R. Hersh, *The Mathematical Experience* (Birkhäuser, Boston, 1981)
P. Dedron, J. Itard, *Mathematics and Mathematicians*, vol. 2 (The Open University Press, Milton Keynes, England, 1973)
R. Descartes, *The Geometry*, Translated by D.E. Smith and M.L. Latham (Dover, New York, 1954)
R. Descartes, *Discourse on Method and the Meditations on First Philosophy*, 4th edn. (Hackett Publishing Company, Indianapolis, 1998)
A. Dick, *Emmy Noether 1882–1935* (Birkhäuser, Basel, 1970)
L.E. Dickson, *History of the Theory of Numbers* (Chelsea, New York, 1971)
E.T. Drake, *Restless Genius: Robert Hooke and His Earthly Thoughts* (Oxford University Press, Oxford, 1996)
J.L.E. Dreyer, *A History of Astronomy from Thales to Kepler* (Dover, New York, 1953)
J.M. Dubbey, *The Mathematical Work of Charles Babbage* (Cambridge University Press, Cambridge, 1978)
W. Dunham, *Journey Through Genius: The Great Theorems of Mathematics* (Penguin, New York, 1991)
W. Dunham, *The Mathematical Universe: An Alphabetical Journey Through the Great Proofs, Problems, and Personalities* (Wiley, New York, 1994)
W. Dunham, *Euler: The Master of Us All* (MAA, Washington, DC, 1999)
G.W. Dunnington, *Carl Friedrich Gauss, Titan of Science: A Study of His Life and Work* (Hafner, New York, 1955)
M. Dzielska, *Hypatia of Alexandria, Revealing Antiquity* (Harvard University Press, Cambridge, 1996)
S. Edeen, J. Edeen, *Women Mathematicians* (Dale Seymour Publications, Parsippany, 1990)
C.H. Edwards Jr., *The Historical Development of the Calculus* (Springer, New York, 1979)
H.M. Edwards, *Fermat's Last Theorem. A Genetic Introduction to Algebraic Number Theory* (Springer, New York, 1977)
A. Einstein, *Ideas and Opinions* (Bonanza Books, New York, 1986)

A. Einstein, *World As I See It* (Citadel Press, New York, 1993)
H.W. Eves, *An Introduction to the History of Mathematics* (Holt, Rinehart and Winston, New York, 1969)
H.W. Eves, *In Mathematical Circles: A Selection of Mathematical Stories and Anecdotes* (Weber & Schmidt, Boston, 1969)
H.W. Eves, *In Mathematical Circles: A Second Collection of Mathematical Stories and Anecdotes* (Weber & Schmidt, Boston, 1971)
H.W. Eves, *Great Moments in Mathematics: Before 1650* (The Mathematical Association of America, Washington, DC, 1981)
H.H. Eves, *Great Moments in Mathematics: After 1650* (The Mathematical Association of America, Washington, DC, 1982)
H.W. Eves, *Return to Mathematical Circles: A Fifth Collection of Mathematical Stories and Anecdotes* (PWS-Kent Publishing, Boston, 1988)
J.J. Fahie, *Galileo, His Life and Work* (John Murray, London, 1903)
J. Fauvel, J.J. Gray, *The History of Mathematics: A Reader* (The Open University, London, 1987)
L.P. Fibonacci, *The Book of Squares*, annotated and Translated by L.E. Sigler (Academic, Boston, 1987)
J.V. Field, J.J. Gray, *The Geometrical Work of Girard Desargues* (Springer, New York, 1987)
M. Fierz, *Girolamo Cardano, 1501–1576* (Birkhäuser, Boston, 1983)
D.H. Fowler, *The Mathematics of Plato's Academy: A New Reconstruction* (Clarendon Press, Oxford, 1987)
H.A. Freebury, *A History of Mathematics* (Macmillan, New York, 1961)
W. Frucht, *Imaginary Numbers* (Wiley, New York, 2000)
G. Galilei, *Dialogue on the Great World Systems* (The University of Chicago Press, Chicago, 1953)
G. Galilei, *Dialogues Concerning Two New Sciences*. Great Mind Series (Prometheus Books, New York, 1995)
M.J. Gazale, *Number: From Ahmes to Cantor* (Princeton University Press, Princeton, 2000)
C. Gillespie, *Pierre–Simon Laplace 1749–1827: A Life in Exact Science* (Princeton University Press, Princeton, 1997)
R.J. Gillings, *Mathematics in the Time of the Pharaohs* (Dover, New York, 1982)
H.H. Goldstine, *A History of Numerical Analysis from the 16th Through the 19th Century* (Springer, New York, 1977)
H.H. Goldstine, *A History of the Calculus of Variations from the 17th Through the 19th Century* (Springer, New York, 1977)
J. Gow, *A Short History of Greek Mathematics* (Hafner, New York, 1923)
R.P. Graves, *Life of Sir William Rowan Hamilton*, vol. 3 (Hodges, Figgis, Dublin, 1882)
J.V. Grabiner, *The Origins of Cauchy's Rigorous Calculus* (MIT Press, Cambridge, 1981)
I. Grattan-Guinness, *The Development of the Foundations of Mathematical Analysis from Euler to Riemann* (MIT Press, Cambridge, 1971)
I. Grattan-Guinness (ed.), *From the Calculus to Set Theory, 1630–1910. An Introductory History* (Duckworth, London, 1980)
I. Grattan-Guinness, *The Fontana History of the Mathematical Sciences* (Fontana Press, Glasgow, 1997)
R. Grimsley, *Jean d'Alembert [1717–1783]* (The Clarendon Press, Oxford, 1963)
L.S. Gristien, P.J. Campbell, *Women of Mathematics: A Biobibliographic Sourcebook* (Greenwood Press, New York, 1987)
M. Guillen, *Five Equations that Changed the World* (MJF Books, New York, 2000)
J. Gullberg, P. Hilton, *Mathematics: From the Birth of Numbers* (W.W. Norton & Co., New York, 1997)
R.C. Gupta, Madhava's and other medieval Indian value of pi. Math. Educ. **9**, B45–B48 (1975)
J. Hadamard, *An Essay on the Psychology of Invention in the Mathematical Field* (Princeton University Press, Princeton, 1945)
E. Hairer, G. Wanner, *Analysis by Its History* (Springer, New York, 1996)
A.R. Hall, *From Galileo to Newton 1630–1720* (Harper & Row, New York, 1963)

A.R. Hall, *Philosophers at War: The Quarrel Between Newton and Leibniz* (Cambridge University Press, Cambridge, 1980)
T. Hall, *Gauss, a Biography* (MIT Press, Cambridge, 1970)
T.L. Hankins, *Sir William Rowan Hamilton* (Johns Hopkins University Press, Baltimore, 1980)
G.H. Hardy, Mathematical proof. Mind **38**, 1–25 (1928)
G.H. Hardy, *Ramanujan* (Chelsea, New York, 1959)
G.H. Hardy, *A Mathematician's Apology* (Cambridge University Press, Cambridge, 1967)
R. Harre (ed.), *Early Seventeenth Century Scientists* (Pergamon, Oxford, 1965)
T.L. Heath, *The Thirteen Books of Euclid's Elements*, vol. 3 (Dover, New York, 1956)
T.L. Heath, *The Works of Archimedes* (Dover, New York, 1953)
T.L. Heath, *A Manual of Greek Mathematics* (Dover, New York, 1963)
T.L. Heath, *A History of Greek Mathematics*, vol. 2 (Dover, New York, 1981)
J.L. Heilbron, *Geometry Civilized: History, Culture and Technique* (Clarendon Press, Oxford, 1998)
R. Hersh, *What Is Mathematics, Really?* (Oxford University Press, New York, 1997)
D. Hilbert, *Foundations of Geometry*, Translated by E.J. Townsend (Open Court, Chicago, 1902)
P. Hoffman, *Archimedes' Revenge: The Joys and Perils of Mathematics* (W.W. Norton, New York, London, 1988)
A. Hooper, *Makers of Mathematics* (Random House, New York, 1948)
D. Hutchings (ed.), *Late Seventeenth Century Scientists* (Pergamon, Oxford, 1969)
G. Ifrah, *From One to Zero: A Universal History of Numbers* (Viking, New York, 1985)
G. Ifrah, *The Universal History of Numbers: From Prehistory to the Invention of the Computer* (Wiley, New York, 1999)
P. Jha, *Aryabhata I and His Contributions to Mathematics* (Bihar Research Society, Patna, 1988)
G.G. Joseph, *The Crest of the Peacock* (Princeton University Press, Princeton, 2000)
V. Kagan, *N. Lobachevski and His Contributions to Science* (Foreign Languages Publishing House, Moscow, 1957)
R. Kaplan, *The Nothing that Is: A Natural History of Zero* (Oxford Press, New York, 1999)
L.C. Karpinski, *The History of Arithmetic* (Rand McNally, Chicago, 1925)
D.S. Kasir (ed.), *The Algebra of Omar Khayyam* (AMS Press, New York, 1972)
V.J. Katz, *A History of Mathematics: An Introduction* (Harper Collins, New York, 1993)
V.J. Katz, Ideas of calculus in Islam and India. Math. Mag. **68**, 163–174 (1995)
G.R. Kaye, *Indian Mathematics* (Amd Simla, Calcutta, 1915)
A. Keller, *Expounding the Mathematical Seed.* Vol. 2 (The Translation & the Supplements): A Translation of Bhaskara 1 on the Mathematical Chapter of the Aryabhatiya (Birkhäuser, Basel, 2006)
B. Kemp, *Ancient Egypt: Anatomy of a Civilization* (Routledge, London, 1991)
H. Kennedy, *Peano: Life and Work of Giuseppe Peano* (D. Reidel, Dordrecht, Holland, 1980)
J. Kepler, *Epitome of Copernican Astronomy & Harmonies of the World.* Great Mind Series (Prometheus Books, New York, 1995)
F. Klein, *Development of Mathematics in the Nineteenth Century*, Translated by M. Ackerman (Math Science Press, Brookline, 1979)
M. Kline, *Mathematics in Western Culture* (Oxford University Press, New York, 1953)
M. Kline, *Mathematics: A Cultural Approach* (Addison-Wesley, Reading, 1962)
M. Kline, *Mathematics, the Loss of Certainty* (Oxford University Press, New York, 1980)
M. Kline, *Mathematics and the Physical World* (Dover, New York, 1981)
M. Kline, *Mathematical Thought from Ancient to Modern Times*, 2nd edn. (The Clarendon Press, New York, 1990)
I.L. Kushyar, *Principles of Hindu Reckoning* (University of Wisconsin Press, Madison, 1966)
D.F. Lach, *Asia in the Making of Europe*, vol. 2 (Chicago University Press, Chicago, 1965)
V. Lakshmikantham, *The Origin of Human Past (Children of Immortal Bliss)* (Bharateeya Vidya Bhavan, Bombay, 1999)
V. Lakshmikantham, S. Leela, J. Vasundhara Devi, *The Origin and History of Mathematics* (Cambridge Scientific Publishers, Cambridge, 2005)

V. Lakshmikantham, J. Vasundhara Devi, *What India Should Know* (Bharateeya Vidya Bhavan, Bombay, 2006)
L.Y. Lam, *A Critical Study of the Yang Hui Suan Fa, a 13th Century Mathematical Treatise* (Singapore University Press, Singapore, 1977)
L.Y. Lam, T.S. Ang, *Fleeting Footsteps: Tracing the Conception of Arithmetic and Algebra in Ancient China* (World Scientific, Singapore, 1992)
J.P. LaSalle, L. Solomon (eds.), *Recent Soviet Contributions to Mathematics* (Macmillan, New York, 1962)
Y. Li, S.R. Du, *Chinese Mathematics: A Concise History*, Translated from the Chinese and with a preface by J.N. Crossley and A.W.C. Lun (The Clarendon Press, New York, 1987)
U. Libbrecht, *Chinese Mathematics in the Thirteenth Century* (MIT Press, Cambridge, 1973)
A. Macfarlane, *Lectures on Ten British Mathematicians of the Nineteenth Century* (Wiley, New York, 1916)
D. MacHale, *George Boole. His Life and Work* (Boole Press, Dublin, 1985)
M.S. Mahoney, *The Mathematical Career of Pierre de Fermat, 1601–1665* (Princeton University Press, Princeton, 1973)
E. Maor, *E: The Story of a Number* (Princeton University Press, Princeton, 1998)
J.C. Martzloff, *The History of Chinese Mathematics*, Translated by S. Wilson (Springer, New York, 1994)
K.O. May, *Bibliography and Research Manual of the History of Mathematics* (University of Toronto Press, Toronto, 1973)
E.A. Maziarz, T. Greenwood, *Greek Mathematical Philosophy* (Frederick Ungar, New York, 1968)
J. McLeish, *Number: The History of Numbers and How They Shape Our Lives* (Fawcett Columbine, New York, 1991)
K. Menninger, *Number Words and Number Symbols: A Cultural History of Numbers* (Dover, New York, 1992)
J. Merz, *Leibniz* (Hacker Press, New York, 1948)
H. Meschkowski, *Ways of Thought of Great Mathematicians* (Holden-Day, San Francisco, 1964)
H.O. Midonick, *The Treasury of Mathematics* (Philosophical Library, New York, 1965)
Y. Mikami, *The Development of Mathematics in China and Japan* (Chelsea, New York, 1913/1974)
R.E. Moritz, *On Mathematics and Mathematicians* (Dover, New York, 1958)
J. Muir, *Of Men and Numbers: The Story of Great Mathematics* (Dover, New York, 1996)
T.S.B. Murthy, *A Modern Introduction to Ancient Indian Mathematics* (Wiley Eastern, New Delhi, 1992)
P.J. Nahin, *An Imaginary Tale* (Princeton University Press, Princeton, 1998)
S.H. Nasr, *Science and Civilization in Islam* (Harvard University Press, Cambridge, 1968)
J. Needham, *Science and Civilization in China* (Cambridge University Press, Cambridge, 1959)
R. Netz, *The Shaping of Deduction in Greek Mathematics: A Study in Cognitive History*. Ideas in Context, vol. 51 (Cambridge University Press, Cambridge, 1999)
O. Neugebauer, *The Exact Sciences in Antiquity*, 2nd edn. (Dover, New York, 1969)
O. Neugebauer, *A History of Ancient Mathematical Astronomy*, vol. 3 (Springer, New York, 1975)
J.R. Newman (ed.), *The World of Mathematics* (Simon and Schuster, New York, 1956)
O. Ore, *Niels Henrik Abel* (University of Minnesota Press, Minneapolis, 1957)
O. Ore, *Number Theory and Its History* (Dover, New York, 1988)
L.M. Osen, *Women in Mathematics* (MIT Press, Boston, 1974)
M.D. Pandit, *Mathematics as Known to the Vedic Samhitas* (Sri Satguru Publications, New Delhi, 1993)
T. Pappas, *The Joy of Mathematics* (Wide World Publishing/Tetra, San Carlos, 1989)
T. Pappas, *Mathematical Scandals* (Wide World Publishing/Tetra, San Carlos, 1997)
S. Paramesvaran, *The Golden Age of Indian Mathematics* (Swadeshi Science Movement, Kerala, 1998)
D. Pedoe, *Geometry and the Visual Arts* (Dover, New York, 1976)
T. Perl, *Women and Numbers* (Wide World Publishing/Tetra, San Carlos, 1993)
J.-P. Pier, *Developments in Mathematics 1900–1950* (Birkhäuser, Boston, 1994)

D. Pingree, *Census of the Exact Sciences in Sanskrit*, vol. 4 (American Philosophical Society, Philadelphia, 1970–1981)
K. Plofker, *Mathematics in India* (Princeton University Press, Princeton, 2008)
G. Polya, *How to Solve It: A New Aspect of Mathematical Method* (Princeton University Press, Princeton, 1945)
G. Polya, *Mathematical Discovery*, vol. 2 (Wiley, New York, 1962/1965)
G. Prasad, *Mathematical Research in the Last Twenty Years* (Walter de Gruyter, Berlin, 1923)
G. Prasad, *Some Great Mathematicians of the Nineteenth Century, Their Lives and Their Works*, vol. 2 (Benares Mathematical Society, Benares, 1933–1934)
C.T. Rajagopal, M.S. Rangachari, On medieval keralese mathematics. Arch. Hist. Exact Sci. **35**, 91–99 (1986)
S. Ranganathan, *Ramanujan: The Man and the Mathematician* (Asia Publishing House, Bombay, 1967)
S.B. Rao, *Indian Mathematics and Astronomy: Some Landmarks* (Jnana Deep Publications, Bangalore, 1994)
C. Reid, *Hilbert* (Springer, New York, 1970)
C. Reid, *Courant in Göttingen and New York: The Story of an Improbable Mathematician* (Springer, New York, 1976)
E. Rosen, *Three Imperial Mathematicians* (Abanis, New York, 1986)
R. Rucker, *Infinity and the Mind* (Bantam Books, New York, 1983)
B. Russell, *The Autobiography of Bertrand Russell*, vol. 2 (Little, Brown & Co., Boston, 1951)
S. Saliba, *A History of Arabic Astronomy* (New York University Press, New York, 1994)
T.A. Sarasvati, *Geometry in Ancient and Medieval India* (Motilal Banarsidass, Delhi, 1979)
K.V. Sarma, *A History of the Kerala School of Hindu Astronomy* (Viveshvarand Institute Publications, Hoshiarpur, 1972)
G. Sarton, *The Study of the History of Mathematics and the Study of the History of Science* (Dover, New York, 1957)
W.L. Schaaf, *A Bibliography of Recreational Mathematics. A Guide to the Literature*, 3rd edn. (National Council of Teachers of Mathematics, Washington, DC, 1970)
J.F. Scott, *The Scientific Work of René Descartes* (Taylor & Francis, London, 1952)
J.F. Scott, *A History of Mathematics from Antiquity to the Beginning of the Nineteenth Century* (Taylor & Francis, London, New York, 1969)
J.F. Scott, *The Mathematical Works of John Wallis, D.D., F.R.S. (1616–1703)* (Chelsea, New York, 1981)
A. Seidenberg, The origin of mathematics. Arch. Hist. Exact Sci. **18**, 301–342 (1978)
C. Seife, *Zero: The Biography of a Dangerous Idea* (Viking, New York, 2000)
G.E. Sen, *The Pageant of Indian History* (David McKay Co., New York, 1948)
S.K. Sen, R.P. Agarwal, *π, e, ϕ with MATLAB: Random and Rational Sequences with Scope in Supercomputing Era* (Cambridge Scientific Publishers, Cambridge, 2011)
S.N. Sen, A.K. Bag, *The Sulbasutras* (Indian National Science Academy, New Delhi, 1983)
K. Shen, J.N. Crossley, A.W.C. Lun, *The Nine Chapters on the Mathematical Art* (Oxford University Press and Science Press, Beijing, 1999)
K.S. Shukla, *Aryabhatiya of Aryabhata with the Commentary of Bhaskara I and Somesvara* (Indian National Science Academy, New Delhi, 1976)
G.F. Simmons, *Calculus Gems: Brief Lives and Memorable Mathematics* (McGraw-Hill, New York, 1992)
J. Singh, *Great Ideas of Modern Mathematics: Their Nature and Use* (Dover, New York, 1959)
S. Singh, J. Lynch, *Fermat's Enigma: The Epic Quest to Solve the World's Greatest Mathematical Problem* (Bantam Books, New York, 1998)
D.E. Smith, L.C. Karpinski, *The Hindu–Arabic Numerals* (Ginn, Boston, 1911)
D.E. Smith, Y. Mikami, *A History of Japanese Mathematics* (Open Court, Chicago, 1914)
D.E. Smith, *History of Mathematics*, vol. 2 (Dover, New York, 1958)
D.E. Smith, *A Source Book in Mathematics* (Dover, New York, 1959)

D.E. Smith, J. Ginsburg, *A History of Mathematics in America Before 1900* (Arno, New York, 1980)
D. Solow, *How to Read and Do Proofs* (Wiley, New York, 1990)
D.A. Somayaji, *A Critical Study of the Ancient Hindu Astronomy* (Karnataka University, Dharwar, 1971)
N. Spielberg, B. Anderson, *Seven Ideas that Shook the Universe* (Wiley, New York, 1987)
L.A. Steen, *On the Shoulders of Giants: New Approaches to Numerary* (National Academy Press, Washington, DC, 1990)
L.A. Steen (ed.), *Mathematics Today* (Springer, New York, 1978)
S. Stein, *Archimedes: What Did He Do Besides Cry Eureka*? (MAA, Washington, DC, 1999)
S. Stein, *Mathematics: The Man Made Universe* (Dover, New York, 2000)
J. Stillwell, *Mathematics and Its History* (Springer, New York, 1989)
D.J. Struik, *A Sourcebook in Mathematics, 1200–1800* (Princeton University Press, Princeton, 1986)
D.J. Struik, *A Concise History of Mathematics*, 4th edn. (Dover, New York, 1987)
B.V. Subbarayappa, K.V. Sarma (eds.), *Indian Astronomy, a Source Book* (Nehru Center, Bombay, 1985)
J.W.N. Sullivan, *The History of Mathematics in Europe, from the Fall of Greek Science to the Rise of the Conception of Mathematical Rigour* (Oxford University Press, New York, 1925)
J. Suzuki, *A History of Mathematics* (Prentice Hall, Upper Saddle River, 2002)
F.J. Swetz, T.S. Ang, A brief chronological guide to the history of Chinese mathematics. Hist. Math. **11**, 39–56 (1984)
F.J. Swetz (ed.), *From Five Fingers to Infinity* (Open Court, Chicago, 1994)
I. Todhunter, *A History of the Mathematical Theory of Probability from the Time of Pascal to that of Laplace* (Chelsea, New York, 1949)
H.W. Turnbull, *The Great Mathematicians* (New York University Press, New York, 1969)
B.L. van der Waerden, *Science Awakening*, Translated by A. Dresden (Kluwer, Dordrecht, 1975)
B.L. van der Waerden, *Geometry and Algebra in Ancient Civilizations* (Springer, Berlin, 1983)
B.L. van der Waerden, *A History of Algebra* (Springer, Berlin, 1985)
B. van Nooten, Binary numbers in Indian antiquity. J. Indian Philos. **21**, 31–50 (1993)
V.S. Varadarajan, *Algebra in Ancient and Modern Times* (American Mathematical Society, Providence, 1998)
K. Venkatachalam, *Bharateeya Sakamula* (Arya Vijnanadeepa Publications, Vijayawada, 1950)
K. Venkatachalam, *The Plot in Indian Chronology* (Arya Vijnanadeepa Publications, Vijayawada, 1950)
F. Viéte, *The Analytic Art*, Translated by T.R. Witmer (Kent State University Press, Kent,, 1983)
A. Weil, *Number Theory: An Approach Through History: From Hammurabi to Legendre* (Birkhäuser, Boston, 1983)
D.G. Wells, *You Are a Mathematician: A Wise and Witty Introduction to the Joy of Numbers* (Wiley, New York, 1997)
R.S. Westfall, *Never at Rest: A Biography of Isaac Newton* (Cambridge University Press, Cambridge, 1980)
H. Weyl, *Philosophy of Mathematics and Natural Science* (Princeton University Press, Princeton, 1949)
E. Whittaker, *From Euclid to Eddington* (Dover, New York, 1958)
N. Wiener, *I Am a Mathematician: The Later Life of a Prodigy* (Doubleday, New York, 1956)
R.L. Wilder, *The Foundations of Mathematics* (Wiley, New York, 1965)
R.L. Wilder, *The Evolution of Mathematical Concepts: A Historical Approach* (Wiley, New York, 1968)
M. Williams, *A History of Computing Technology* (Prentice-Hall, Englewood Cliffs, 1995)
E. Zebrowski, *A History of the Circle: Mathematical Reasoning and the Physical Universe* (Rutgers University Press, Piscataway, 1999)
M.C. Zeller Sr., *The Development of Trigonometry from Regiomontanus to Pitiscus* (Edwards Brothers, Ann Arbor, 1946)

Name Index

R.P. Agarwal, S.K. Sen, *Creators of Mathematical and Computational Sciences*,
DOI 10.1007/978-3-319-10870-4

C

D

E

H

I

J

N

Subject Index

R.P. Agarwal, S.K. Sen, *Creators of Mathematical and Computational Sciences*,
DOI 10.1007/978-3-319-10870-4

O

P

U

V

W

MIX
Papier aus verantwortungsvollen Quellen
Paper from responsible sources
FSC® C105338

If you have any concerns about our products,
you can contact us on
ProductSafety@springernature.com

In case Publisher is established outside the EU,
the EU authorized representative is:
Springer Nature Customer Service Center GmbH
Europaplatz 3, 69115 Heidelberg, Germany

Printed by Libri Plureos GmbH
in Hamburg, Germany